ENGINEERING AND ENVIRONMENTAL IMPACTS OF SINKHOLES AND KARST

PROCEEDINGS OF THE THIRD MULTIDISCIPLINARY CONFERENCE ON SINKHOLES
AND THE ENGINEERING AND ENVIRONMENTAL IMPACTS OF KARST
ST.PETERSBURG BEACH / FLORIDA / 2-4 OCTOBER 1989

ENGINEERING AND ENVIRONMENTAL IMPACTS OF SINKHOLES AND KARST

Edited by
BARRY F.BECK
with the invaluable assistance of
ADRIANNE HAGEN, SCOTT CAVIN, BRIAN BARFUS & VIRGINIA MERKLE
Florida Sinkhole Research Institute, University of Central Florida, Orlando

Sponsored by the
Florida Sinkhole Research Institute, Division of Sponsored Research,
University of Central Florida

A.A.BALKEMA / ROTTERDAM / BROOKFIELD / 1989

The texts of the various papers in this volume were set individually by typists under the supervision of each of the authors concerned.

Authorization to photocopy items for internal or personal use, or the internal or personal use of specific clients, is granted by A.A.Balkema, Rotterdam, provided that the base fee of US$1.00 per copy, plus US$0.10 per page is paid directly to Copyright Clearance Center, 27 Congress Street, Salem, MA 01970. For those organizations that have been granted a photocopy license by CCC, a separate system of payment has been arranged. The fee code for users of the Transactional Reporting Service is: 90 6191 987 8/89 US$1.00 + US$0.10.

Published by
A.A.Balkema, P.O.Box 1675, 3000 BR Rotterdam, Netherlands
A.A.Balkema Publishers, Old Post Road, Brookfield, VT 05036, USA

ISBN 90 6191 987 8

Table of contents

3. Ground water quality and pollution in karst terranes

4. Engineering geology of karst terranes

5. Geophysical investigations of karst terrane

6. *Case studies of engineering in karst terrane*

7. *Planning, governmental and legal implications of karst terrane*

EDITOR'S NOTE

Because of the lengthy time required for postal communication with China, the last two articles in this volume were received from Song Lin Hua of the Yunnan Institute of Geography, Kunming, People's Republic of China, after the deadline for publication. Because of this, there was no opportunity to edit, correct, or polish the manuscripts as we would normally do.

However, the information content is not diminished by this and it is obvious that Mr. Song has devoted considerable effort to contributing these communications regarding karst in China and some of its practical impacts. In order to include these articles at the last moment, we are printing them just as they arrived so that the results of Mr. Song's experience will be available to our participants and readers.

Keynote address:
Karst, caves and engineering – The British experience

A.C.WALTHAM   *Civil Engineering Department, Nottingham Polytechnic, Nottingham, UK*

ABSTRACT

Cavernous limestone creates engineering difficulties only in isolated cases in Britain, as little of its outcrop is in areas of extensive development. Less cavernous rocks have caused more widespread hazards : failures on the chalk are common but are generally small except for some dramatic collapses associated with old mines. Major subsidences in the Cheshire saltfield have largely ceased due to the reduction in brining, while sandstone caves in Nottingham have been conserved beneath buildings which bridge them. Research on the Yorkshire limestone karst includes mapping of caves and sinkholes which suggest that studies of their distribution and correlation are of limited value to engineering predictions of failure sites. More applicable are preliminary results from model testing of cave roof stability; the importance of bedding and jointing is demonstrated, but many caves only have a major influence on ground bearing capacity where their width approaches or exceeds their cover thickness.

Introduction

Cavernous limestone karst creates an environment which the geologist may find interesting, the explorer may find challenging, but the civil engineer finds a wretched nuisance. It is the terrain type which least lends itself to the mathematical analysis which can yield the firm design criteria beloved by the engineer. It is the terrain type whose geomorphological understanding becomes an art, matured only through experience, and then applied sensitively and individually at each new site assessment.

The British experience of karst engineering has been constrained by the fact that our green and pleasant land has only a limited extent of dynamically active karst in areas of population and engineering development. Britain has seldom experienced the sinkhole hazards that have confronted Florida, China or South Africa, but a few problems have been tackled and a few lessons have been learned. This paper outlines Britain's karst environments, reviews some of the current research, and refers briefly to some specific projects.

Subsidence Hazards in the Karst of Britain

Britain has karst developed on strong cavernous limestone, on soft chalk limestone and on salt and gypsum, but except for the chalk none occupies any large area. The main karst zones which have created any environmental impact in England and Wales are shown in figure 1; Scotland has very little karst. Current understanding of the various forms of subsidence hazard has been assessed by the writer (Waltham, 1989) with some emphasis on the situation in Britain and the remedial measures appropriate in civil engineering works. The distribution of underground cavities, both natural and artificial, has been reviewed by Culshaw and Waltham (1987). What Britain lacks in natural cavities, it more than compensates for with artificial ones; extensive mining subsidence is related to both current extraction and the failure of old workings (Whittaker and Reddish, 1989). Following in the wake of a nationwide survey of old mine cavities, the UK Government Department of the Environment has commissioned a survey of natural cavities; data collection is still in progress, but, though many cavities exist, the engineering difficulties caused by them appear to be slight except in the urban areas on chalk.

Strong Paleozoic limestones, mostly of Carboniferous age, support the only significant cavernous karst in Britain. Their erosional resistance leaves their main outcrops as upland, largely free of development and therefore creating minimal hazard impact. The largest outcrops, in and around the Yorkshire Dales, contain hundreds of caves and thousands of sinkholes; the subsidence hazard has trivial economic impact on the open hillsides, and is outweighed by the problems of the less intelligent sheep falling down open shafts. Most of the Dales sinkholes are subsidence dolines formed in the mantle of boulder clay; many are still slumping and expanding, but new sinkholes are rare, due partly to the undisturbed state of the fell land.

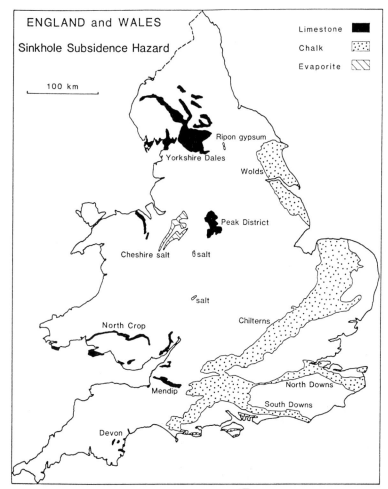

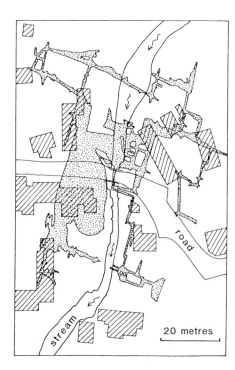

Figure 2: The houses of West Scrafton and the caves beneath, in the limestone karst of the Yorkshire Dales.

Figure 1

Various roads, farms and villages are underlain by caves, both in the Dales and elsewhere on Britain's limestones. In some cases the rock cover is only a few metres, but most British caves are small, and the ratio of roof thickness/cave width is rarely less than 1; no collapses of cave roofs have been recorded, and no potential failures are recognised. However, the North Crop limestones of South Wales dip beneath grit plateaus which are pocked by hundreds of large collapse depressions; many of these are 100 metres above the cavernous limestone and developed by a massive scale of cavity migration Thomas, 1974). Some explored caves in the same area show roof collapse and upward migration of 10-20 metres, in limestone with beds 30-50 cm thick. Even this scale of collapse could threaten a village like West Scrafton (figure 2) except that its underlying cave is roofed by massive limestone quite capable of bridging the void.

Previously unknown caves have been unroofed at a number of construction sites on limestone, notably on road projects crossing the small outcrops of Devonian limestone in South Devon. In all cases, the caves have been only a few metres across, and have been filled with concrete to allow construction to continue. Buried sinkholes in an eroded rockhead were revealed on a building site in South Wales, but were shallow enough to allow column bases to be extended down to sound rock at minimal extra cost (Statham and Baker, 1986). Pinnacled rockhead is not a feature of the British karst.

The English chalk is a soft porous limestone with limited cavern development. Sinkholes on it are mainly subsidence dolines only a few metres across, though larger fossil depressions occur in some zones; they are most common in areas with a thin cover of permeable soils (Edmonds, 1988). Buried sinkholes, filled with clay or sand and known as chalk pipes, can pose a subsidence hazard, and are locally common enough to warrant the precautionary use of raft foundations. The largest individual ground failures on chalk are the collapses over existing cavities, notably old chalk or flint mines; these include the spectacular collapses at Bury St. Edmunds and Norwich (Waltham, 1989). They have sometimes been referred to as crown holes, formed simply by the progressive failure of the mine roofs, but some have occurred in unmined areas where the only voids are natural. The rapid failure and large scale of these collapses is a function of drainage disturbance and ultimate liquefaction of the deep layer of frost-shattered putty chalk which is prevalent on the outcrop in Southern England just outside the ice cover achieved during the last Pleistocene glaciation. The disastrous combination of soakaway drains and existing cavities into which the chalk can run is now a widely recognised engineering hazard; where old mines are even suspected to exist, efficient storm drainage is essential.

There are other outcrops of limestone in Britain, omitted from figure 1. These have neither the cavernous nature of the older limestones, nor the structural weakness of the chalk, and have therefore created engineering hazards in only a few isolated cases.

The salt of the Cheshire Plain has created its own distinctive karst with the meres (lakes) occupying shallow subsidence hollows in the boulder clay terrain. Around the turn of this century, destructive subsidences and sinkhole collapses achieved a spectacular notoriety, especially in the town of Northwich, though nearly all were caused directly by mining or brining operations. The subsidence descriptions by Calvert (1915) are classics (reviewed by Waltham, 1978, 1989; Whittaker and Reddish, 1989). In later years the reduction of uncontrolled brining has virtually eliminated the subsidence problem in all the English salt regions. Remedial and precautionary engineering works had become almost routine. Steel or timber building frames and their repeated jacking, reballasting of railways with adjustable overhead gantries, jackable bridge decks, and the periodic raising of roads, are now largely of the past; only concrete rafts continue in use for houses in some small areas of continuing subsidence.

Gypsum only occurs at a few discontinuous horizons in England, and only in the Ripon area have sinkholes developed on any significant scale (Cooper, 1988). These have formed by collapse of soils and rocks into fissures and small caves rapidly dissolved from the gypsum by local concentrations of groundwater flow. Some buildings have been damaged, and pipes of collapse breccia reaching the surface from gypsum at depths of 100 metres indicate the scale of subsidence which can develop.

Environnmental Hazards in the Karst of Britain

Exploitation of karst aquifers is currently not a major challenge in Britain. With declining demand by heavy industry, flooding by rising water tables is more common than any difficulty due to over-abstraction. The chalk is Britain's single most important groundwater resource, but it behaves largely as a diffuse aquifer with only modification by conduit flow. There is little abstraction from cavernous limestones in Britain, except for many farm and village supplies which have long utilised small springs. British experience with such aquifers has been abroad, including successful exploration for rural water supplies from cavernous limestone in Java (Waltham et al., 1985) and assessment of the problems of dams inside caves in China (Smart and Waltham, 1987).

Problems of aquifer pollution are more immediate. Rapid pollutant transmission through fissures has long been recognised as a threat in the chalk aquifer (Atkinson and Smith, 1974), and current work is concerned with drainage off the new motorways around London. Pollution can be even more serious from old limestone quarries used as landfill sites, especially in the porous limestone of the English Cotswolds (Smart, 1985). However, quarries in the more massive Carboniferous limestone commonly have relatively impermeable floors in rock below the natural subcutaneous zone of high transmissivity; unless blast fractures create an artificial subcutaneous zone, these sites may offer less pollution hazard (Smart and Friederich, 1986), and research into this field continues. Sadly Britain has had its share of pollution through sinkhole dumping (Stanton, 1982), where a wider understanding of the dangers is the best remedy to the hazard. The same may apply to currently fashionable concern over the radon hazard, where monitoring and research is in progress (Gunn et al., 1989).

Building over the Sandstone Caves of Nottingham

Nottingham is well known for the many caves which underlie its city centre, occupying the site of the medieval town. The caves are in Triassic Sherwood Sandstone and are not natural, but were all cut artificially during the last 800 years; they were used as storehouses, workshops etc. The poorly cemented sandstone has an U.C.S. of 5-30 MN/m² (725-4350 lbf/in²), but is very little fractured and forms steep cliffs overlooking the Trent Valley. Caves were easily worked by hand, and spans of 5 metres were commonly left unsupported; most cave roofs are only at depths of a few metres. As the caves fell disused and records were lost, they became a hazard to modern rebuilding.

Current site investigation practice in Nottingham is to drill the site of each column base or load bearing structure to prove at least 3 metres of sound rock. Splayed bores and water pressure tests are optional extras, and exposed caves are normally filled with concrete. A column base directly over a span of no more than 5 metres would cause failure

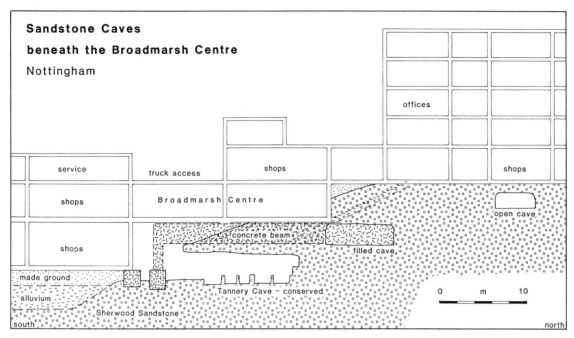

Sandstone Caves

beneath the Broadmarsh Centre

Nottingham

offices

service · truck access · shops · shops

shops · Broadmarsh Centre

shops · concrete beam · open cave · filled cave

made ground · Tannery Cave - conserved

alluvium

Sherwood Sandstone

south · north

0   m   10

Figure 3

as a plug, flared over the bottom metre. In this situation, 3 metres of sound rock will generally give a factor of safety of 6 over the Safe Bearing Pressure of 2 MN/m² (290 lbf/in²) normally used for the Sherwood Sandstone. The 3 metre guideline appears to be appropriate; there has been no failure of a modern building over a cave in Nottingham.

The Broadmarsh Centre is a split-level shopping complex built on a sandstone slope containing a series of caves which were deemed worthy of conservation for their historical

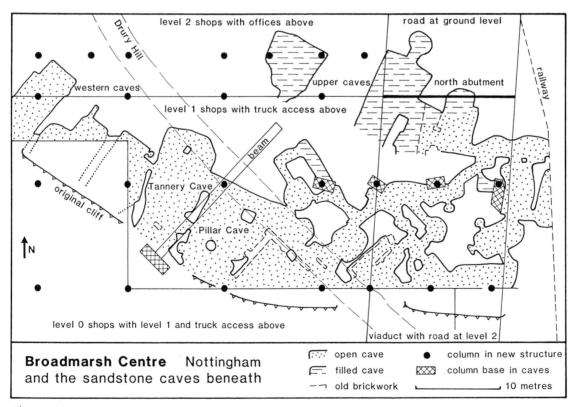

Drury Hill · level 2 shops with offices above · road at ground level

western caves · upper caves · north abutment · railway

level 1 shops with truck access above

beam

original cliff · Tannery Cave

Pillar Cave

N

level 0 shops with level 1 and truck access above

viaduct with road at level 2

Broadmarsh Centre Nottingham
and the sandstone caves beneath

open cave · column in new structure
filled cave · column base in caves
old brickwork · 10 metres

Figure 4

4

values (figure 3). The structure was therefore designed to span the caves, with minimal intrusion on them. The columns of the southern line in figure 4 are founded beyond the caves. The northern two lines and the bridge abutment were founded over some unimportant caves which were filled; the column over the western cave stands on a 5 metre thick roof. The central columns were founded on concrete bases formed in the sandstone cut to below the cave floors, and now visible in the caves. The exception is the column just east of the Tannery Cave where a base pad would have been unacceptable in the finest of the caves; with a load of over 6 MN (600 tons) it could not be founded on the rock above. It therefore stands on a deep, rigid, reinforced concrete beam which spans the entire cavernous zone (figure 3). The Broadmarsh caves are still open beneath the shopping complex, and they survive as an example of environmentally aware engineering in cavernous ground. Only the eastern end of the caves were destroyed by less aware railway engineers a century ago.

Sinkhole and Cave Distribution
    In the karst upland of the Yorkshire Dales sheep grazing is more important than civil engineering. But the high level of cave exploration creates an unusual opportunity to compare cave distribution, bedrock geology as exposed in the caves, and sinkhole distribution in areas of drift mantle.

    Ingleborough has 35 km. of karst, containing over 3500 subsidence sinkholes (locally known as shakeholes) and nearly 55 km of mapped cave. As part of a site assessment all features are now logged in a database working on hectare units (with sides of 100 metres); figures 5 and 6 are partial printouts. The caves within 20 metres of the surface are concentrated along the shale margin which provides allogenic drainage, and less so along the basement spring line. Sinkholes show some concentration close to the shale margin, but they primarily follow the distribution of the boulder clay in which they are formed, with the higher densities in the thinner overburden. There is no real correlation between the sinkholes and the caves. Nearly half the hectare blocks with caves have no sinkholes, and only 20% of the sinkhole sites have known caves beneath (though more unknown caves must exist). The few active sinkhole failures reveal no pattern, but are associated with locally increased or modified drainage input.

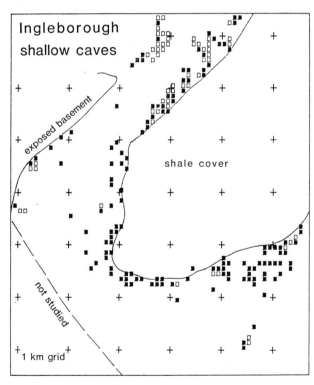

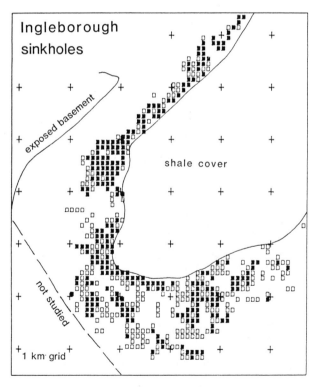

Figure 5: Caves within 20 meters of the surface on Ingleborough, plotted in hectare site units. Solid squares indicate caves with openings to the surface; open squares indicate cave passages with no surface connection.

Figure 6: Subsidence sinkholes on Ingleborough. Open squares indicate 1-4 sinkholes in the hectare unit; solid squares indicate more than 4 sinkholes.

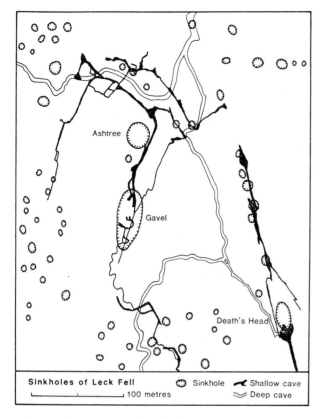

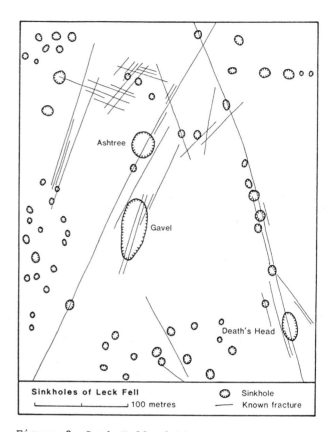

Figure 7: Leck Fell sinkholes and caves

Figure 8: Leck Fell sinkholes and fractures

The nearby karst plateau of Leck Fell has many shakeholes in a continuous boulder clay mantle, part of which (remote from any shale margin) is shown in figures 7 and 8. Bedrock fractures are mapped from cave exposures and bedding is nearly horizontal. A line of sinkholes occurs along the Death's Head fault, and the Gavel and Ashtree sinkholes have formed by collapse and run-in into a large old cave passage. Otherwise the sinkholes show little correlation with either the fractures or the caves.

Neither of these sites offers any real scope for predictions or zoning of the hazards due to potential sinkhole failure or from the existence of shallow caves. An honest assessment of the hazard offered by underground cavities in limestone may be the old saying "caves is where you find them" (said in a thick Yorkshire accent with local unconventional grammar).

Failure of Loaded Cave Roofs
With the risk of unknown caves underlying a construction site, the ultimate questions are how thick the cave roof needs to be to remain stable, and how deep exploration boreholes should be taken into solid rock. Neither beam nor plate theory is easily applied to these problems and the simple calculations concerning punching failure are inappropriate for well bedded limestone. Destructive testing of scale models offers some scope for evaluating the problems, the principles having already been applied to ground movement in deep mines (Hobbs, 1966).

Models with a length scale factor of 1:50 have been made with plaster to represent the cavernous Carboniferous limestone (U.C.S. 90 MN/m², 13,000 lbf/in²). Stress and strength are at a scale factor of 1:90 (Hobbs, 1966). Testing involves various cave widths and shapes, roof thicknesses, bed thicknesses and fracture densities, under loads applied to pads of various sizes at different positions relative to the cave. Preliminary results have been obtained for loading of a small square pad centrally over a cave passage 80 mm wide at varying depth beneath beds of different thickness and fracturing (figure 9). The failure stress for similar non-cavernous material is shown for comparison.

Normal Safe Bearing Pressure for this limestone is 4 MN/m² (580 lbf/in²). By direct analogy with figure 9, roughly half that figure would maintain comparable factors of safety in most cases if caves were no wider than 4 metres and sound rock could be proven for

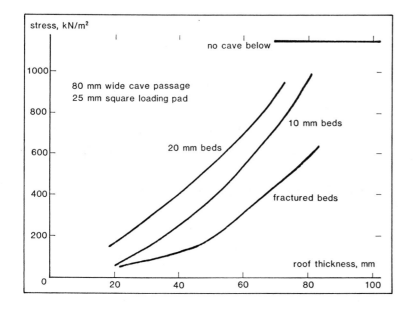

Figure 9: Failure of roof spans over scale model caves, showing the effects of roof thickness and structural detail on the bearing capacity.

3 metres with boreholes above them. Bedding and jointing are critical to this evaluation and 4 metres of cover would need to be proven in well jointed rock (scaled to 80 mm roof thickness in fractured beds). Similarly caves of this size at depths greater than 6 metres would appear irrelevant to structural loading. Further testing should refine evaluation of the parameters involved in cave roof stability, and a more thorough analysis will be published in due course.

References

Atkinson, T C and Smith, D I, 1974. Rapid groundwater flow in fissures in the chalk : an example from South Hampshire. Quart. Journ. Eng. Geol., 7, 197-205.

Calvert, A F, 1915. Salt in Cheshire. Spon, London, 1206pp.

Cooper, A H, 1988. Subsidence resulting from the dissolution of Permain gypsum in the Ripon area : its relevance to mining and water abstraction. Geol. Soc. Eng. Geol. Spec. Pub., 5, 387-390.

Culshaw, M G and Waltham, A C, 1987. Natural and artificial cavities as ground engineering hazards. Quart. Journ. Eng. Geol., 20, 139-150.

Edmonds, C N, 1988. Induced subsurface movements associated with the presence of natural and artificial underground openings in areas underlain by Cretaceous chalk. Geol. Soc. Eng. Geol. Spec. Pub., 5, 205-214.

Gunn, J, Fletcher, S, Prime, D and Middleton, T, 1989. Radon daughter concentrations in British caves : implications for cavers and tourist cave operators. Proc. 10th Int. Speleo. Cong., Hungary, in press.

Hobbs, D W, 1966. Scale model studies of strata movement around mine roadways. Int. Journ. Rock Mech. Min. Sci., 3, 101-127.

Smart, P L, 1985. Applications of fluorescent dye tracers in the planning and hydrological appraisal of sanitary landfills. Quart. Journ. Eng. Geol., 18, 275-286.

Smart, P L and Friederich, H, 1986. Water movement and storage in the unsaturated zone of a maturely karstified carbonate aquifer, Mendip Hills, England. Proc. Conf. Environmental Problems in Karst Terranes and Their Solutions, Bowling Green, Kentucky, 59-87.

Smart, P L and Waltham, A C, 1987. Cave dams of the Guanyan System, Guangxi, China. Quart. Journ. Eng. Geol., 20, 239-243.

Stanton, W I, 1982. Mendip - pressures on its caves and karst. Trans. Brit. Cave Res. Assoc., 9, 176-183.

Statham, I and Baker, M, 1986. Foundation problems on limestone : a case history from the Carboniferous limestone at Chepstow, Gwent. Quart. Journ. Eng. Geol., 19, 191-201.

Thomas, T M, 1974. The South Wales interstratal karst. Trans. Brit. Cave Res. Assoc., 1, 131-152.

Waltham, A C, 1989. Ground subsidence. Blackie, London; and Chapman and Hall, New York, 202pp.

Waltham, A C, Smart, P L, Friederich, H and Atkinson, T C, 1985. Exploration of caves for rural water supplies in the Gunung Sewu karst, Java. Ann. Soc. Geol. Belg., 108, 27-31.

Waltham, T (A C), 1978. Catastrophe, the violent earth. Macmillan, London; and Castle, New York, 170pp.

Whittaker, B N and Reddish, D J, 1989. Subsidence, occurrence, prediction and control. Elsevier, Oxford and New York, 528pp.

1. The geologic and hydrologic assessment of karst areas

Surficial karst patterns: Recognition and interpretation

ERNST H.KASTNING   *Department of Geology, Radford University, Radford, Va., USA*

ABSTRACT

Surficial karstic landforms commonly exhibit easily recognizable patterns, such as alignments of dolines (sinkholes) and solution valleys. Recognition of patterns and interpretation of these features in context with the regional geologic setting often gives considerable insight into the nature of subsurficial configurations including groundwater flowpaths and networks. Influences that control surficial karstic landforms include lithologic and stratigraphic character of the bedrock, geologic structure, hydrodynamic variables, and topographic evolution of the land surface. Analysis of surficial features is a valuable tool in assessing hidden underground conditions and potential environmental problems in karst regions.

Introduction

Karst landforms of all scales are guided in their development and morphology by various geologic factors including lithologic and stratigraphic properties of the soluble bedrock; geologic structures such as folds, faults, and joints; patterns and dynamics of groundwater flow; evolution of the surficial topography and geochemistry of karst waters. In karst science, as in all geologic investigations, careful and accurate analysis of existing forms and structures in context with the regional environment will disclose the processes and sequential mode of origin of landforms, both underground and on the surface. Surficial karstic phenomena are typically products of subsurficial mechanisms. Therefore, recognition and mapping of surficial landforms provides substantial insight toward understanding the underground components of a karst region. Readily discernible karstic features on the surface are thereby a means of interpreting conditions below the ground that are normally hidden from view, including geologic structure and patterns of groundwater flow (e.g. Kastning, 1983a,b, 1984; Kastning and Kastning, 1981).

Patterns of Surficial Karst

Maximum dissolution of carbonate rock occurs along zones of enhanced solubility or along avenues of secondary porosity and permeability. In many cases, karstic landforms, such as dolines (sinkholes) and solutionally widened fractures, form easily recognizable patterns including lineaments and dendritic networks. These occur at various scales ranging from those confined to less than a square kilometer to some extending tens of kilometers. Linear patterns of karst features are expected on the basis that many planar features occur in a geologic setting with sedimentary rocks, including beds and bedding planes, fractures and fracture systems, and potentiometric surfaces. Many of these planar structures are non-horizontal and typically intersect one another and the land surface. Because intersections of planes are straight lines, it follows that linear features abound in terrane characterized by folded and fractured sedimentary rocks.

Like fluvial drainage networks on the surface, many patterns of groundwater flow in carbonate terrane are dendritic, especially where a well integrated flow network has evolved, concentrating tributary flow from points of discrete recharge into master conduits that convey water to single discharge points (Palmer, 1975). Under some circumstances, well established groundwater patterns are expressed on the land surface.

The usefulness of assessing subsurficial conditions through interpretation of the surficial karstic landscape can easily be demonstrated by examples. The following cases have been selected from the author's research in Texas, Kentucky, and Virginia to illustrate these concepts for differing geologic settings.

Lithostratigraphic Influences

Carbonate rocks are much more soluble than other sedimentary rocks. However, there is considerable variation among carbonate units with respect to how easily solutional enlargement occurs. At any one locality, particular beds may be highly prone to cavern

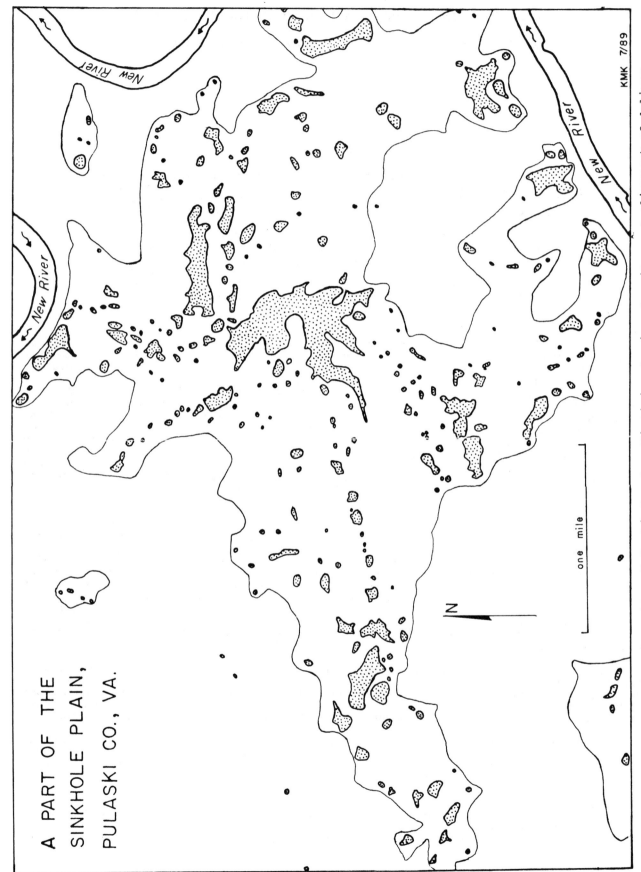

A PART OF THE
SINKHOLE PLAIN,
PULASKI CO., VA.

New River

New River

New River

one mile

N

KMK 7/89

Figure 1: Sinkhole plain of eastern Pulaski County, Virginia showing density and alignment of dolines.

12

development, whereas adjacent units are not. Factors that contribute to enhanced solubility are high primary porosity, high calcite content, and low content of less soluble material including dolomite, clay, chert, and the like. Bedding thickness is also an important influence on the rates of dissolutional enlargement of fracture porosity. Commonly, thinly bedded limestone will promote large solutional conduits owing to a higher frequency of horizontal fractures (bedding planes) and to mechanical breakdown of the roofs of conduits as these openings grow in cross section. These factors are well documented for several large karst regions (Rauch and White, 1970; Dreiss, 1984; Palmer, 1962; Kastning, 1975, 1983a).

Dolines are generally confined to zones where highly soluble carbonate rocks are exposed at the surface. (The principal exception would be where less soluble material has collapsed into intrastratal cavities within underlying cavernous rocks.) Sinkhole plains commonly form in wide expanses of exposed limestone, especially where beds are relatively flat-lying such as in the low rolling karst of central Kentucky (Quinlan, 1970; White and others, 1970), western Kentucky (Kastning and Kastning, 1981), and Indiana (Palmer and Palmer, 1975) or in wide valley bottoms of the Appalachian Valley and Ridge Province (Hubbard, 1984, 1988; see also Figure 1).

Alternatively, where a relatively thin, highly soluble, but steeply inclined carbonate unit is exposed at the surface, the outcrop pattern may be very narrow and dolines may be confined solely to the band of exposed rock (Figure 2). Dolines may also be confined to a narrow strip along the floor of a steep-walled valley that has intersected a highly soluble horizontal bed of limestone.

Correctly assessing the underlying reasons for zonation of dolines along narrow bands of exposed carbonate rock necessitates geologic mapping. This requires determination of the attitude of the beds in relation to the topographic surface. Note that in the example of Longhorn Cavern, Texas illustrated in Figure 2, dolines are constrained by the structural rather than topographic setting.

Structural Influences

Geologic structures in carbonate rock terrane are often observable as traces or lines, commonly referred to as "lineaments." Mapping and interpretation of lineaments in karst regions is a relatively recent approach, but it has seen wide geographic application (e.g. Lattman and Parizek, 1964; Warren and Wielchowsky, 1973; Ogden, 1974; Wilson, 1977; Ogden and Reger, 1977; Wermund and others, 1978; Barlow and Ogden, 1980; Kastning and Kastning, 1981; Kastning, 1983a,c, 1984).

Lineaments may simply represent traces of fractures along the surface. In semi-arid regions, such as the Edwards Plateau of central Texas, lineaments are pronounced in aerial photography because much of the sparse vegetation takes root in fractures and what is viewed in the photographs are lines of shrubs and small trees (Kastning, 1983a, 1984). Elsewhere, lineaments in karst may represent alignments of dolines. This is very pronounced in the relatively flat lying limestone of western Kentucky for example, where doline alignments are consistent with mapped fractures including both faults and joints (Kastning and Kastning, 1981). Correlation of doline alignments with mapped faults has been shown in one area and this has led to being able to map inferred faults in adjacent areas where doline alignments are strong, yet faults are heavily obscured by overburden (Figure 3). Similar doline alignments occur in the New River drainage basin of Virginia (Kastning, 1988, 1989) where orientations reflect strike-oriented drainage and fractures consistent with regional tectonism (Figure 1).

Drainage Influences

In many settings dolines occur as strings of depressions. As a preliminary method of interpretation, such sequences may provide information on the extents of underground drainage basins or on likely patterns of groundwater flow. Dolines initially developed along stream valleys may take on a dendritic configuration after the surface drainage has been pirated underground. There are suggestions of this process in Pulaski County, Virginia (Figure 1) and in western Kentucky (Figure 3). In cases where alignments of dolines are linear and parallel to the strike of the limestone beds, it may not be possible at first to ascertain if dolines have formed along a fracture related to local tectonism or if dolines have formed in a bed favorable to dissolutional activity. Careful mapping of the topography in relation to exposed beds, determination of lithologic properties of the beds containing the dolines, and mapping of joints and other fractures in caves or doline walls should resolve the problem.

Environmental and Engineering Applications

Once mapping and interpretation of surficial karst features has been accomplished in an area of study, it is possible to incorporate the findings into an assessment of the site for the purposes of land-use management, construction, groundwater determinations, waste disposal, and the like. It must be emphasized that recognition of patterns from aerial photographs and topographic maps in itself is usually not sufficient for reaching conclusions. The data must be corroborated where possible with on-site fieldwork, including verification of bedrock lithology, geologic structure, tracing of underground water with dyes, and other techniques. The main usefulness of mapping surficial karst landforms is as a quick means of estimating the nature of the local karst prior to further and more expensive studies. In a few cases, information gleaned from the surficial determinations may be the only practical way of obtaining data under the constraints of time and expense often imposed on the researcher.

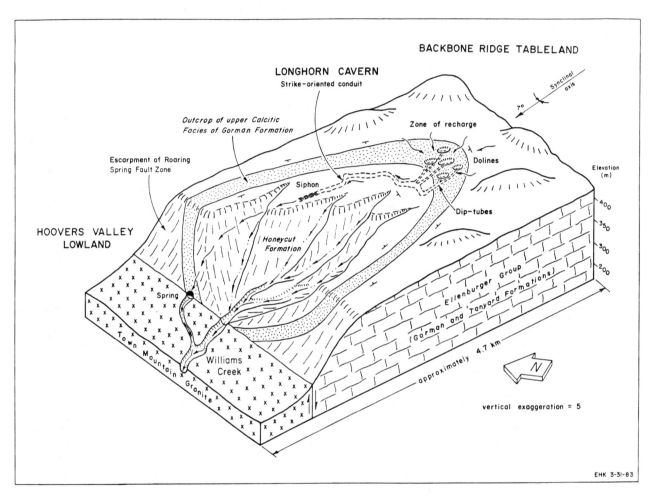

Figure 2: Block diagram illustrating relationship of Longhorn Cavern and associated dolines, Burnet County, Texas, to bed of highly soluble limestone within a plunging syncline. From Kastning (1983a).

References Cited

Barlow, C.A. and Ogden, A.E., 1982, A statistical comparison of joint, straight cave segment, and photo-lineament orientations: National Speleological Society Bulletin, v. 44, p. 107-110.

Dreiss, S.J., 1984, Effects of lithology on solution development in carbonate aquifers: Journal of Hydrology, v. 70, p. 295-308.

Hubbard, D.A., Jr., 1983, Selected karst features of the northern Valley and Ridge province, Virginia: Virginia Division of Mineral Resources Publication, no. 44, one sheet (scale 1:250,000).

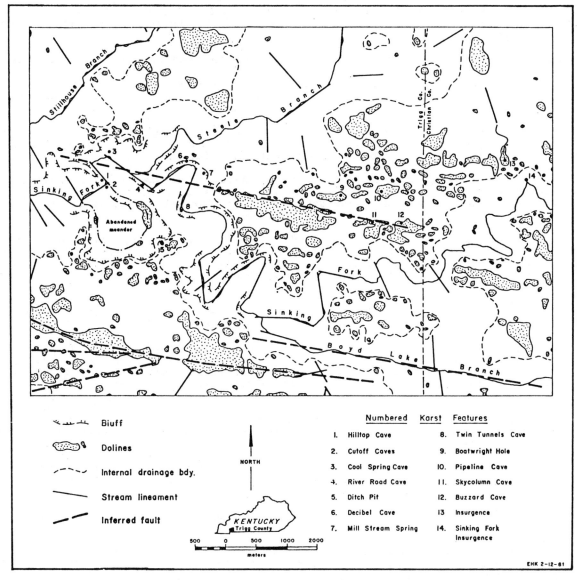

Figure 3: A part of eastern Trigg County, Kentucky, showing dolines, locations of major caves, internally drained areas, stream and doline lineaments, inferred faults, and bluffs. From Kastning and Kastning (1981).

Hubbard, D.A., Jr. 1988, Selected karst features of the central Valley and Ridge province, Virginia: Virginia Division of Mineral Resources Publication, no. 83, one sheet (scale 1:250,000).

Kastning, E.H., 1975, Cavern development in the Helderberg Plateau, east-central New York: New York Cave Survey Bulletin 1, 194 p. plus 8 plates.

Kastning, E.H., 1983a, Geomorphology and hydrogeology of the Edwards Plateau karst, central Texas: Ph.D. dissertation (unpublished), The University of Texas at Austin, 656 p. plus 6 plates.

Kastning, E.H., 1983b, Karstic landforms as a means to interpreting geologic structure and tectonism in carbonate terranes (abstract): Geological Society of America Abstracts with Programs, v. 16, p. 608.

Kastning, E.H., 1983c, Relict caves as evidence of landscape and aquifer evolution in a deeply dissected carbonate terrain: Southwest Edwards Plateau, Texas, U.S.A., in Back, W. and LaMoreaux, P.E. (guest editors), V.T. Stringfield Symposium--Processes in Karst Hydrology: Journal of Hydrology, v. 61, p. 89-112.

Kastning, E.H., 1984, Hydrogeomorphic evolution of karsted plateaus in response to regional tectonism, in LaFleur, R.G. (editor), Groundwater as a Geomorphic Agent: Proceedings of the Thirteenth Annual ("Binghamton") Geomorphology Symposium, Troy, New York, 1982: International Series no. 13: George Allen and Unwin, London, p. 351-382.

Kastning, E.H., 1988, Karst of the New River drainage basin, in Kardos, A.R. (editor), Proceedings, Seventh New River Symposium, Oak Hill, West Virginia, April 7-9, 1988: New River Gorge National River, Oak Hill, West Virginia, p. 39-49.

Kastning, E.H., 1989, Environmental sensitivity of karst in the New River drainage basin, in Kardos, A.R. (editor), Proceedings, Eighth New River Symposium, Radford, Virginia, April 21-22, 1989: New River Gorge National River, Oak Hill, West Virginia, p. 103-112.

Kastning, K.M. and Kastning, E.H., 1981, Fracture control of dolines, caves, and surface drainage: Mississippian Plateau, western Kentucky, U.S.A., in Beck, B.F. (editor), Proceedings of the Eighth International Congress of Speleology, Bowling Green, Kentucky, July 18 to 24, 1981: National Speleological Society, Huntsville, Alabama, v. 2, p. 696-698.

Lattman, L.H. and Parizek, R.R., 1964, Relationship between fracture traces and the occurrence of ground water in carbonate rocks: Journal of Hydrology, v. 2, p. 73-91.

Ogden, A.E., 1974, The relationship of cave passages to lineaments and stratigraphic strike in central Monroe County, West Virginia, in Rauch, H.W. and Werner, E. (editors), Proceedings of the Fourth Conference on Karst Geology and Hydrology: West Virginia Geological and Economic Survey, Morgantown, West Virginia, p. 29-32.

Ogden, A.E. and Reger, J.P., 1977, Morphometric analysis of dolines for predicting ground subsidence, Monroe County, West Virginia, in Dilamarter, R.R. and Csallany, S.C. (editors), Hydrologic Problems in Karst Regions: Western Kentucky University, Bowling Green, Kentucky, p. 130-139.

Palmer, A.N., 1962, Geology of the Knox Cave System, Albany County, New York: B.Sc. thesis (unpublished), Williams College, Williamstown, Massachusetts, 95 p. plus 2 plates.

Palmer, A.N., 1975, The origin of maze caves: National Speleological Society Bulletin, v. 37, p. 57-76.

Palmer, M.V. and Palmer, A.N., 1975, Landform development in the Mitchell Plain of southern Indiana: Origin of a partially karsted plain: Zeitschrift für Geomorphologie, v. 19, p. 1-39.

Quinlan, J.F., 1970, Central Kentucky karst: Actes de la Réunion Internationale Karstologie, Languedoc-Provence, 8-12 juillet 1968: Études et Traveaux de Méditeranée, no. 7, p. 235-253.

Rauch, H.W. and White, W.B., 1970, Lithologic controls on the development of solution porosity in carbonate aquifers: Water Resources Research, v. 6, p. 1175-1192.

Warren, W.M. and Wielchowsky, C.C., 1973, Aerial remote sensing of carbonate terranes in Shelby County, Alabama: Ground Water, v. 11, p. 14-26. (reprinted as Geological Survey of Alabama Reprint Series, no. 26, 13 p.)

Wermund, E.G.; Cepeda, J.C.; and Luttrell, P.E., 1978, Regional distribution of fractures in the southern Edwards Plateau and their relationship to tectonics and caves: University of Texas at Austin, Bureau of Economic Geology, Circular 78-2, 14 p.

White, W.B.; Watson, R.A.; Pohl, E.R.; and Brucker, R., 1970, The central Kentucky karst: The Geographical Review, v. 60, p. 88-115.

Wilson, J.R., 1977, Lineaments and the origin of caves in the Cumberland Plateau of Alabama: National Speleological Society Bulletin, v. 39, p. 9-12.

Environmental aspects of the development of Figeh Spring, Damascus, Syria

P.E. LAMOREAUX   *P.E. LaMoreaux & Assoc., Inc., Tuscaloosa, Ala., USA*

ABSTRACT

Hydrogeological studies at Figeh Spring were directed to determine ground-water flow paths, recharge, storage and discharge units, and the maximum reliable yield. The project was designed to provide information upon which to base pumpage to augment low-season flows from the spring which is the major water supply for the city of Damascus, Syria and will provide guidelines for environmental protection.

Introduction

A hydrogeological assessment and stress pumping test at Figeh Spring were performed to determine ground-water flow paths, recharge, storage and discharge characteristics and the maximum reliable yield. The project was designed to augment low-season flows from the spring and to supplement the water supply for the City of Damascus, Syria. The evaluation was directed toward installation of permanent pumping facilities at the spring.

Before additional modification to the spring could be made, it was necessary to understand, in detail, the recharge, storage and discharge, as well as preferential ground-water flow patterns in the karst limestone system. Evaluation included studies of satellite imagery, sequential aerial photography, geomorphology, stratigraphy, and geologic structure (folding, faulting and jointing). The studies were performed to determine the relationships among geologic control, karstification, preferential ground-water flow patterns, and storage characteristics of the aquifer system.

Interpretation of remote-sensed data, including satellite imagery and high and low level air photography, was verified by field studies. Geologic and geomorphic parameters controlling discharge and preferential flow were described. A well and spring inventory, stream flow measurements and sequential sampling and analyses of surface water and ground water provided water quality parameters, which were correlated with natural geologic phenomena and stress pumping from wells and springs. The final phase of work included a series of synchronized pumping tests at Figeh Spring, Side Spring, Ain Harouch, and PELA test wells.

Geomorphology

Associated with the landscape in the recharge area for Figeh Spring there are many karst features. Although karren, solution pits, small dolines, and short cave systems exist, there are areas in the recharge zone that do not contain well-developed karst systems. Karst features in the uplands are most common in the outcrop areas of thicker Cenomanian limestones and dolomites. The karstification ends at the base of the massively-bedded carbonates because the underlying, thin-bedded Cenomanian limestones impede ground-water movement and development of significant solution cavities with depth.

Paleokarst features had their origin along bedding planes in the Cenomanian and Turonian limestone during Paleocene and Eocene, at a time of major structural movement. During Pliocene and Pleistocene major changes in ground-water base level occurred and karstification was accentuated.

Paleokarst features have been preserved since Pleistocene and are abundant. However, of major significance to the Figeh karst system is the presence of extensive solution cavity systems and caves in the Turonian limestone and dolomite, as well as collapse features in Senonian strata which have been buried by Pleistocene alluvial deposits of the Barada River. The most significant karst feature of the area is the cave which serves as the conduit for discharge of ground water at Ain Figeh. The cave was formed by ground-water discharge that rises subparallel to the Khadra anticlinal axis in Turonian dolomite. The cave floor is underlain by as much as 70 meters of breccia which indicates that the cave has migrated upward through the dolomite by a process of roof collapse. Continuation of cave evolution, through roof collapse, will provide changes in position of the discharge point of Ain Figeh with time. For this reason any construction or development of the spring must be carefully planned and carried out with extreme caution.

Geology

Stratigraphic Sequence

Essentially all exposed rock units in the area of recharge for Figeh Spring belong to the Cretaceous System and include important aquifers and aquicludes. The most complete and continuous section of these Cretaceous rocks, more than 1800 meters thick, is exposed between Bloudan and Wadi Hubeidi.

The Cretaceous system was subdivided and described by Dubertret (1949). Later work under Ponikarov (ed., 1968) provided extensive information about the Cretaceous system, but resulted only in minor modification to the stratigraphy. The present study included interpretation of aerial photographs and field mapping to verify the distribution of lithostratigraphic units of the Cretaceous System.

Dubertret distinguished the Lower and Upper Cretaceous series and divided them into stages as follows:

```
Upper Cretaceous Series
      Senonian Subseries
            Maastrichtian Stage
            Campanian Stage
            Santonian Stage
            Coniacian Stage
      Turonian Stage
      Cenomanian Stage
Lower Cretaceous Series
      Albian Stage
      Upper Aptian Stage
      Pre-upper Aptian
```

Geologic Structure

The recharge area for Figeh Spring includes a northeast trending part of the Anti-Lebanon Range that is bounded on the northwest by the Bekka Valley rift-system and by the Barada River to the south; the southeastern boundary is marked by the limits of large Tertiary basins. The area has a complex history of compressional folding associated with its location near the apex of curvature of the Dead Sea-Bekka rift systems and has been affected by multi-phase vertical, left-lateral movement along the rifts.

The northwest boundary of the recharge basin occurs along a branch of the major rift system which forms the Bekka Valley. The branch fault system is well exposed near Serghaya and will be referred to as the Serghaya fault zone. The zone itself is 1 to 3 kilometers wide, however, intense fracturing of adjacent Jurassic and Cretaceous rocks has occurred. The Serghaya fault zone behaves as a hinge fault with normal, left-lateral movement. Stratigraphic throw increases to the northeast. Maximum throw in the Zebdani quadrangle is over 1,000 meters.

The southeastern border of the recharge area is coincident with a major lineament (Jarajir lineament) which is sub-parallel to the Serghaya and Bekka rift faults. The Jarajir lineament extends more than 100 kilometers northeast from Halboun. The Jarajir lineament is presumed to be the surface expression of a major fault system at depth.

The Jarajir lineament serves as the boundary between the Cretaceous rocks, exposed in the recharge area for Figeh Spring, and the large Tertiary-Quaternary basin (13 by 50 kilometers) in which Assal el Ward is located. Northeast of the Assal el Ward basin the Jarajir lineament appears to bifurcate. One branch follows the central axis of the Tertiary-Quaternary basin in the vicinity of Jarajir and Quara. The other branch follows the western margin of the basin.

Dominant folds of the area include the Huraire Syncline, the Hassiya Anticline, the Khadra Anticline, and the Dome of Figeh. Within the recharge area, faults and joints are numerous, and no exposure of rock is devoid of fractures. The major faults in the southern part of the recharge area strike northwest, almost perpendicular to fold axes, and consist of both normal and reverse faults, some of which have a right-lateral component of movement. North, northeast and east-west trending normal faults may have significant displacement, but are not as numerous as the northwest-trending faults.

The Huraire Syncline and Hassiya Anticline are northeast-trending, subparallel folds, which have been intermittently active since late Cretaceous (Cenomanian). The Huraire Syncline and Hassiya Anticline are dominant structural features for a distance of about 20 kilometers north of the Barada River. The syncline is nearly symmetrical with a dip of 45 to 60 degrees on each limb. However, the anticline is asymmetrical with a dip of 45 to 60 degrees on the northwest limb. The southeast limb has a dip of 15 to 20 degrees. Both folds plunge gently toward the southwest, except near their southern limits where plunge steepens and cross-folding causes the axes to trend almost due south.

The Huraire Syncline is terminated by faulting at its southern end near the Village of Huraire. Cenomanian, Turonian, and Senonian strata, exposed in the syncline, terminate abruptly at the fault. South of the fault, approximately 300 meters of Pliocene conglomerate are exposed in the downthrown block. The Pliocene conglomerate extends approximately 5 kilometers to the Barada River.

The Khadra Anticline has a total exposed length of 12 kilometers. The axis of the fold plunges 11 degrees in a direction of north 78 degrees west. The anticline is asymmetrical with a near vertical southern limb, the fold being well exposed north of Ain Khadra. The Khadra Anticline is one of the most recent folds (Pliocene-Pleistocene) in the recharge area and is transverse to the dominant folds of the Anti-Lebanon Range.

Intersection of the northwestern trending Khadra Anticline with the northeastern trending structures of the Anti-Lebanon Range has created the Dome of Figeh; has refolded the Hassiya Anticlines and the Huraire Syncline; and has created southeastern plunging minor folds exposed west of the Dome.

The path of the Barada River is controlled by faults parallel to the Khadra Anticlinal axis. At the southeast end of the fold axis faults controlled deposition in the Neogene basin, marginal to the Anti-Lebanon Range.

Straight segments of wadis have characteristically developed along faults or joints. Fracture patterns can be accurately interpreted from aerial photographs or from drainage maps. Fracture-controlled surface morphology is characteristic of the northern two-thirds of the recharge area. Ground-water infiltration and ground-water flow in the central part of the recharge basin is controlled by the south to southeastern stratigraphic dip, southward-dipping, normal faults in valleys, and north-south trending, minor fracture systems. The Cenomanian and Turonian strata are intensely fractured, thus causing these beds to be porous and permeable. The significant directional component of flow is toward Figeh Spring.

Hydrogeology of the Figeh Area

Geologic Setting Controls Karst Development
For karstification to develop and for solution action to progress to form caverns, it is necessary:

1. That water rich in carbon dioxide be available to recharge the system;

2. That sufficient permeability (in the form of fractures or bedding plane or both) be available for water to move in the rocks, and

3. That water be able to discharge from the system.

A fourth (4) criterion which expedites the process, is a steady source of recharge water such as from snow melt, or an overlying blanket of sediments. Under water-table conditions, the zone of higher permeability tends to develop in the zone of greatest circulation and solution, which is commonly at or just below the water table. Topography and position of carbonates below ground are important; carbonates that allow at least moderate circulation and are entrenched by perennial streams tend to develop solution openings and to increase circulation. Some circulation of water and to a certain extent solution cavities may locally occur at a depth of several hundred meters below the level of the major stream of a region provided that a good discharge system and well-established gradient are available, and if the water is chemically aggressive. This is true in the case of the Barada River Valley which flowed at a lower elevation during the late Pleistocene.

The relation of the recharge to the discharge area in a karst region determines, to a large extent, the patterns of the lateral solution channels or openings. The size and frequence of these channels will depend on many factors involving the conditions in the recharge area, the volume of water that enters the recharge area, the solubility of the karst rocks, and the rate at which the base level is lowered as perennial streams entrench.

Where the discharge area is along a more or less straight line, such as the Barada River Valley, the lateral solution channels tend to be more or less parallel to each other and at right angles to the line of discharge. Discharge is therefore in a line of springs like Figeh, Ain Harouch, and Ain Kadra into the Barada. The direction of movement of the water between the recharge areas and discharge areas is affected by faulting, folding, jointing and also other geologic structures. Cave passages may occur at more than one level.

Water-level behavior in space and time is a primary consideration for interpreting karst hydrology. The position of the water table is important because:

1. The water table defines generally the zones of greatest circulation and solution.

2. The configuration of the water table aids in identifying the general direction of flow, the hydraulic gradient, and areas of recharge and discharge.

3. Information about the water table provides general information about permeability of the aquifer system.

4. The position of the water table locally indicates the extent to which caverns are filled with air or water.

Large seasonal variation in water level beneath the carbonate uplands may result in the movement of ground water to one basin in dry weather and to another basin in wet weather. A controlling factor in the range of seasonal ground-water levels is the great infiltration capacity of karst terranes. Large volumes of water from heavy storms infiltrate into the air-filled caverns of karst lands, reaching the zone of saturation quickly, and causing the water table to rise rapidly. From a single storm or snowmelt event, the rise in the water table in relatively impermeable parts of the saturated carbonate rocks can be as much as 10 or more meters, whereas the rise of the water table in permeable zones, of the same carbonate rocks, can be less than a meter.

Big springs, such as Ain Figeh, one of the world's major springs, rather than small springs and diffuse seepage, are the general rule in karst regions, and big springs often emerge from underground streams or caves. Most ground-water flow, however, occurs in large solution openings near the top of the saturated zone, which carry much of the ground water to the springs. Zones of solution openings may tend to branch upgradient, along fractures, which in arterial fashion represent the more permeable, upper parts of the saturated zone. The water table is depressed along the arterial system so that ground-water discharge meets the surface stream almost at grade.

In karst areas the distribution of permeability beneath streams causes them to lose or gain water, depending on the position of the water table with reference to stream level. A surface stream may lose water where bedrock in the losing stretch is very permeable and the water table is low, and it will gain water where the water table is above stream level. For example, the Barada River gains water where it crosses an area immediately south of the Wadi Huraire Syncline and loses water in the reach west of Deir Moukarren.

An understanding of the geologic history and paleokarstification is critical in the Figeh karst system in the vicinity of Ain Figeh where the fractured limestone and dolomites dip toward the Barada River Valley. Karstification of these rocks began during the Pleistocene along fractures and bedding of the more pure limestone. These beds were subsequently folded and fractured during tectonic uplift in the area. During the Pleistocene the Barada River near Ain Khadra became impounded, probably by a landslide, and the base level of the river was raised. Erosion subsequently has taken place, dropping the base level to the present level of the Barada River. At Figeh, alluvial deposits extend as much as 45 meters below present land surface. These beds of sand, gravel, and boulders that fill a buried Paleokarst feature at Ain Figeh comprise an important segment of the Karst aquifer storage system.

Excellent plan views and cross sections of three (3) major solution features in the immediate vicinity of Figeh Spring (Gallery Cave, Cheikh Cave, and the Karst Cavity) have been mapped and surveyed in detail (Sogreah). Each cave is subparallel to the bedding of Turonian rocks, however, each contains small openings associated with fractures. The lower elevation of each cave is 835 meters above mean sea level and corresponds to elevations of late Pleistocene to recent river terraces. The presently active karst system of Figeh, Gallery Cave and Cheikh Cave were originally paleokarst features of late Pleistocene to Recent age.

Gallery Cave and Cheikh Cave differ from the Figeh Karst Cavity, as they occur at higher elevations than the Figeh Karst Cavity, and because the floors of these two (2) caves are underlain by solid bedrock, whereas the lower part of the Figeh Karst Cavity is filled with breccia.

The developmental history of the Figeh Karst Cavity is more complex than for Gallery and Cheikh Caves as two (2) different, but interrelated, karst features are associated with Figeh Karst Cavity. The primary feature is the Karst Cavity, however, above the Karst Cavity is a depression in the top of the bedrock. The depression is filled with breccia overlain by soil, and/or alluvial deposits, as demonstrated by the logs for test wells in the area.

The Karst Cavity is overlain by Turonian strata. Although the overlying beds are fractured, they have not been significantly displaced from their stratigraphic position except where overlying depressions occur. However, drill holes that have penetrated the bottom of Karst Cavity tapped an underlying karst breccia. For example, 27 meters of karst breccia exists beneath the cavity.

A theory of origin for the Figeh Karst Cavity incorporates the upward movement of ground water along fractures connecting Cenomanian and Turonian strata. Ground water in Turonian strata is under artesian pressure, and has eroded, by dissolution and mechanical action, a cave system subparallel to bedding of Turonian strata. A progressive evolution of the cave system in which the physical cavity migrates upward through the Turonian rocks is due to intermittent collapse of the cave roof and back filling the cave floor with karst breccia. The stratigraphic interval at which the cave system originated is unknown as drilling has not penetrated the base of the breccia in the karst.

Environmental Constraints to Future Use of Figeh System
1. The Cenomanian and Turonian strata comprise a karst aquifer system through which water is conveyed from recharge to storage and ultimately to discharge at the springs at Ain Figeh, and Ain Harouch. There are many additional springs along the Barada River that are points of discharge. There is likewise a substantial upward flow from the Turonian rocks through fractures and solution cavities into the Barada which maintains its base flow.

2. The overlying Senonian marls act as aquitards that restrict discharge to land surface where they occur.

3. Detailed photogeologic and ground-truth studies have established that virtually the entire recharge area for the Figeh Spring system is underlain by rocks of Cenomanian Age, that are fractured and karstified in this area and that a major source of recharge is from the snow fields and snow accumulation at altitudes above 1,500 meters (see map) upgradient from the springs.

4. Geologic mapping and hydrogeologic studies during the summer of 1981 and 1982 verified that the Cenomanian strata also contains a complex karstified aquifer system that supplies large quantities of water to the Figeh system. Ground-water movement is from the outcrop of these rocks into interconnected openings along bedding, fractures, and locally in solution cavities to the underlying storage system. The specific movement of water is controlled by the dip of bedding, the plunge of folds, the location of faults, and the system and intensity of fracturing.

5. Major structural features influence the direction of ground-water movement from recharge to storage to discharge. Intersection of the anticlinal axis of the Hassiya and Khadra folds create an interference structure (the dome of Figeh). These three structures control the outcrop pattern of the Turonian rocks in the area and thus along with the faults and fractures, control the present location of Figeh and the other springs that discharge into the Barada River.

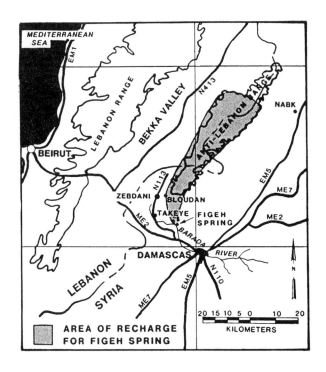

Figure 1: Location and recharge area of Figeh Spring

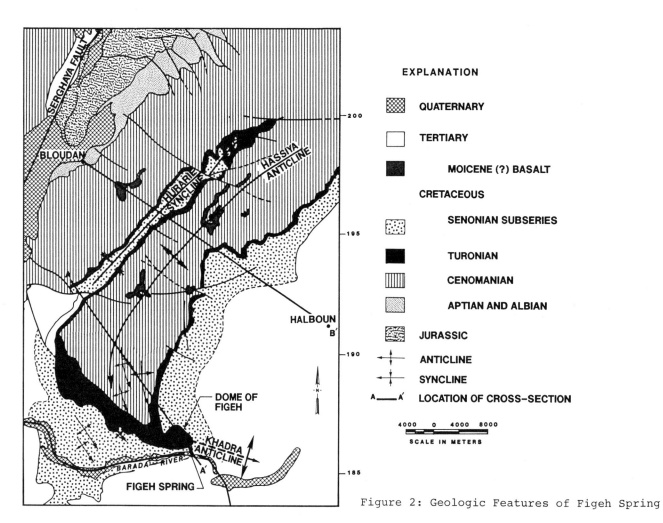

EXPLANATION

QUATERNARY

TERTIARY

MOICENE (?) BASALT

CRETACEOUS

SENONIAN SUBSERIES

TURONIAN

CENOMANIAN

APTIAN AND ALBIAN

JURASSIC

ANTICLINE

SYNCLINE

A——A' LOCATION OF CROSS-SECTION

Figure 2: Geologic Features of Figeh Spring

6. Interpretation of the structural and stratigraphic setting indicated that ground water in the area should be divided into two different flow paths. Both regimes flow southward but part of the water is diverted around the western flank of Figeh dome. Some of this water rises at Ain Harouch and some is believed to be lost as underflow beneath the Barada River. The second major flow regime consists of the water that moves around the eastern side of Figeh dome, and rises at Ain Figeh. A portion of this water moves southeastward along the Khadra fold axis, and some may underflow the Barada River. The Side Spring at Figeh receives water from both flow regimes, plus flow through the alluvial beds of an ancient Barada River channel that now parallels and is adjacent to the Barada River.

7. Flow directions of ground water are shown in the map. Points of discharge are shown by the location of the springs. Figeh Spring discharges near a Pleistocene meander of the Barada River. Figeh Spring, during Roman occupation (63 B.C. to 633 A.D.) discharged at a lower elevation than it does today. Figeh Spring at present discharges through an opening in the Turonian rocks. At this point, a reservoir system has been developed diverting the water into the channels that supply the city of Figeh. There is also withdrawal of ground water from sump pumps that have been installed at the Spring. This water is under semi-artesian pressure.

8. The importance of understanding the reservoir system from which the water flows, the flow direction, and the rate of water withdrawn allows determination of the area around the Spring in which environmental restraints should be imposed on development.

   A. In the closest semi-circular area around Ain Harouch, Side Spring and Figeh, two karst flow systems are involved. Maximum security must be imposed to limit development that could cause physical (discharge) or chemical change, or cause pollution to access near surface solution cavities and rock fractures that are connected to the water in the Spring. This area has been titled, "Maximum Security Area." No development other than those activities associated with the production of water from the spring should be undertaken.

   B. A second semi-circle delineates an area upgradient area in the spring-flow system in which development such as housing, agriculture, grazing and other activities should be restrained. There should be no placement of landfills, hazardous, toxic, or radioactive waste, and all construction or other development should be carefully reviewed and approved by representatives of Figeh Spring.

   C. The third outlying semi-circular area is an area that would be safe and proper for development of light agriculture and/or light grazing with minimum use of pesticides, insecticides and fertilizers. Minimum housing, urban development, or construction should be allowed all by approval of Figeh. This area should be restricted from the assignment or placement of hazardous, toxic and/or radioactive waste.

REFERENCE

Dubertret, L., 1949 Carte Geologique au 50,000 du Liban. Feuille de Zebdani et Notice Explicative, Beyrouth.

Ponikarov, V.P., ed., A.V. Vinogradskey (translator), 1968, The geological map of Syria, scale 1:50,000, with explanatory notes: Minestry of Industry, Syriani Arab Republic, 120 p.

NOTE: A more extensive version of this paper will be published in Env. Geol. and Water Sci., V. 13, no. 2, p. 73-127.

Sinkhole formation and its effect on Peace River hydrology

THOMAS H.PATTON   *Patton & Associates Inc., Gainesville, Fla., USA*
JEAN-GEORGES KLEIN   *Santa Fe Community College, Gainesville, Fla., USA*

ABSTRACT

Prior to extensive pumping and consumptive use of groundwater in the upper Peace River region of Polk County, Florida, potentiometric levels in the Floridan Aquifer were about three to six meters above the land surface along the river from Bartow to Fort Meade. This condition resulted in an area of artesian flow along the Peace River from north of Bartow to as far south as the Polk-Hardee county line.

Such flow contributed significantly to the annual flow of the Peace River; for example, in 1933 Kissengen Springs alone had a flow of 28.2 million gallons per day (43.6 cfs), translating to a 1.24 m^3/s contribution to the discharge of the river.

Due to heavy pumpage, the potentiometric surface declined, and in 1950 Kissengen Springs ceased to flow when this surface dropped below the level of the spring outlet. By 1965 the potentiometric level in the Floridan stood at six to eight meters below the land surface along the Peace River between Bartow and Fort Meade. By the early 1970's the decline exceeded fifteen meters. This dramatic decline in the potentiometric surface and reversal of artesian flow into the river has had a far-reaching impact on the hydrology of the area, radically altering the groundwater flow and influencing the regimen of the upper Peace River.

Coincident with, and as a result of, regional hydrological changes in the Peace River, there has been an associated development of numerous sinkholes. All available data indicate that increased water table drawdown and declining potentiometric levels in the aquifers of the upper Peace River basin have led to intensified solutional activity and increased collapse.

An analysis by the authors of sinkhole formation in the flood-plain of the upper Peace River basin indicates that sinkholes and other solutional conduits act as influent channels into the Floridan Aquifer and that the number of these karst features and the water quantities involved are sufficient to affect the flow of the Peace River in both high- and low-water stages. Because most of the sink-holes are in the bed of the high-water channel of the river, the greatest loss of flow from the Peace River occurs during its high-water stage, resulting in a reduction of both the amplitude and the duration of the high-water flow. This in turn has narrowed the lateral extent of the high-water channel.

Introduction

In 1969 Kaufman described the consequences of ground water withdrawal in the upper reaches of the Peace River in some detail. In that paper he considered the effect of this withdrawal on lake levels, aquifer levels, salt-water encroachment, sinkhole development and long-range adequacy of water supplies. He also noted the reversal of hydrological conditions along the Peace River in Polk County and went on the say: "This decline in the water level has caused a reversal of the hydrologic conditions, in that a potential now (1965) exists for water to move from the Peace River and the shallow water table to the Floridan Aquifer, as opposed to the flow of water from the Floridan into the river in 1934." Although several studies have been done on the Peace and the Alafia rivers, none have specifically addressed the effects of this flow reversal on the regimen of either of these rivers.

This paper is an outgrowth of a geologic study carried out by the authors between 1979 and 1981 in the upper reaches of the Peace River between Bartow and Fort Meade, in Polk County, Florida, at the behest of the Board of Trustees and the Department of Natural Resources of the State of Florida.

The Peace River begins its course within the Polk Uplands, a broad, generally poorly drained, relatively undissected expanse created by earlier eustatic fluctuations of sea-level. Its southward course is restricted between the Lakeland Ridge to the west and the Winter Haven and Lake Henry Ridges to the east (White, 1970). In this upper section, the Peace River is a relatively shallow, low-gradient, island-braided stream that flows through a rather narrow valley incised into the surficial undifferentiated sands and the underlying phosphate-bearing Bone Valley Formation. The lower clay-rich member of this formation is a relatively impermeable unit that aids somewhat in keeping regional water perched above the present lowered potentiometric surface of the Floridan Aquifer. In places, the scoured low-water channel of the river has worn through these basal clays of the Bone Valley Formation and appears to have been arrested by the top of the more resistant dolomites and limestones of the underlying Hawthorn Formation.

The Peace River has long been a seasonally dimorphic stream system. At low-discharge stages, the river occupies a mostly single, narrow, meandering channel. At high-discharge stages, the river occupies a wide, anastomosing, island-braided channel (Patton and Klein, ms.). These two distinct channel morphologies are responses to the large differences in discharge between high-water and low-water flows. Since the floodplain, including its channel systems, is a feature of a river valley associated with a particular climate or hydrologic regimen of the drainage basin, an alteration of the conditions of equilibrium by changes in the hydrologic regimen will result in altering the level and/or extent of the floodplain. Because of the low slopes lateral to the low- and high-water channels, any process that reduces the elevation of the high-water stage also reduces the lateral extent of the reach of the high water.

Qualitative Assessment of Loss of Artesian Flow on River Hydrology

In 1930 the Peace River received significant amounts of water from the regional aquifers, especially the Floridan, by way of seeps and springs, the largest of which was Kissengen Springs. Conservatively estimating the flow of Kissengen Springs at 0.88 cubic meters per second (31 cfs or 20 mgd) as reported by Heath (1961) and estimating the average stream flow of the Peace River to be at 8.72 m^3/s (308 cfs), reported by Heath and Wimberly (1971), shows that Kissengen alone contributed at least ten percent and perhaps as much as fifteen percent of the water in the upper Peace River. Certainly other springs and karst conduits also contributed to the total flow of the river.

During the dry season, when rainfall runoff is significantly reduced, this artesian effluent flow no doubt contributed significantly to the river's base flow. During the rainy season, there is not only greater overland runoff into the stream, but also greater recharge to the aquifers. Under these higher recharge conditions, the potentiometric surface of the aquifers rises, leading to still greater artesian flow during these seasonally wet periods. This additional flow also would have augmented the high stages of the river, particularly contributing to the latter stages of high-water flow because of the lag time induced by slower underground travel from areas of recharge. Thus, the artesian flow along the valley not only influenced the mean annual flow of the river but the high-water stages as well.

When hydrologic conditions were reversed because of industrial and agricultural drawdown, the Peace River no longer received effluent water from the aquifers, and because such artesian flow provided a substantial amount of the total volume of water in the Peace River, after 1950 the mean annual flow was significantly reduced. The high-water stage of the river was also reduced; however, because artesian flow was a more constant contributor to base flow than runoff, it represented a smaller percentage of the high-water flow and, therefore, its loss altered the low-water stages more than the high-water stages. In sum, the cessation of artesian flow caused by the lowering of the potentiometric surface has led to a decrease in flow of the Peace River, but has affected low-water discharge relatively more than high-water discharge. However, the overall effect would be to lower the flow at all river stages.

At those times when artesian conditions prevailed, new sinkhole formation proceeded at a limited pace; however, in the 1940s and early 1950s, when effluent conditions changed to influent conditions, numerous sinkholes began to open up in the Peace River valley. The occurrence and areal distribution of sinkholes in Polk County during the period of 1953 to 1960 are discussed by Stewart (1966). Kaufman's (1969) re-analysis of Stewart's data shows that two-thirds (12 of 18) of the reported sinkholes occurred between 1954 and 1956, at a time when artesian water levels were declining. Moreover, nearly ninety percent (16 of 18) occurred within the area of greatest ground water circulation. In the spring of 1965, Polk County experienced very low rainfall. Consequently, consumptive use of ground water increased and the aquifers reached record-low levels. In this same period, numerous collapse episodes occurred in the northern Peace River basin. Kaufman ultimately concluded that a "causal relationship may exist between lowered artesian water levels and the occurrence of sinkholes." Later reports have been less cautious about the relationship between the lowered potentiometric surface and sinkhole formation. In a report prepared for the Southwest Florida Water Management District, Holzschuh (1972) estimated that 150 to 200 major sinkholes have formed in the upper Peace River basin since 1949. Of this figure, he estimates that perhaps "half would not have occurred had the potentiometric surface remained at its 1949 levels." A flurry of collapse episodes during 1967 and 1968 caused widespread concern in Polk County. Sowers (1974), using some of these 1967 failures as classical examples of ravelling, attributed collapse episodes to lowered water tables followed by increased infiltration. The Civil Defense Office in Polk County has on file numerous episodes of additional collapse.

Sinkhole Survey

Because of the potential role of sinkholes as recharge avenues to the underlying aquifers and their concomitant role in lowering the flow of the Peace River, a sinkhole survey was carried out to establish the extent of collapse activity and ultimately allow evaluation of its effect on the regimen of the Peace River.

From March 21 to April 7, 1980 a field reconnaissance took place along the upper reaches of the Peace River valley for the purpose of locating, describing, and marking for later surveying all sinks and pipes, influent sloughs, and other karst-related features. These were then located by survey and individually contoured. The area covered by the investigation included the entire 22.5-kilometer stretch of the river from Bartow to Fort Meade and from the edge of the low-water channel to the lateral edge of the high-water channel.

The survey located approximately ninety actual and/or probable solution features. Of these ninety features, fifty-one (57 percent) are larger than three meters in diameter and nine (10 percent) are over twelve meters in diameter. The largest feature has a diameter of over twenty-three meters and exposes the Bone Valley and Hawthorn Formations down to the Floridan Aquifer.

Sinkholes are not evenly distributed along the upper Peace River valley. Forty-one of the fifty-one larger features noted above occur between Bartow and Homeland, in the upper half of the 22.5 kilometer stretch of the river. Sixteen lie along a 0.6 kilometer stretch approximately 2.6 to 3.2 kilometers downstream from Bartow. Another twelve, including the largest feature noted, are clustered between 4 and 5.6 kilometers below Bartow. An additional seven occur in the vicinity of Kissengen Springs. Only ten occur in scattered locations between Homeland and Fort Meade. Most of these sinkholes lie within the high-water channel, although we also noted several complexes within the low-water channel. Such a distribution is to be expected because the areal extent of the meandering low-water channel represents but a small fraction of the entire valley floor.

However, the several sinkholes located within the low-water channel represent a greater percentage of conduits per unit area than simple probability calculation would dictate. This is likely due to the influence of local geologic structure. The partly rectangular nature of the course of the Peace River channel and the linear nature of several of the sinkhole complexes clearly indicate that local drainage and solution are influenced by underlying structure. The coincidence of channel location and structural lineation (joints and fractures) would account for the greater-than-expected sink occurrence in the low-water channel.

No attempt was made during this initial reconnaissance to measure the volume of flow into those sinks still taking water. Because the river was then in the process of rapidly subsiding from a higher stage, many of the higher-elevation sinks receiving water from the river only a couple of days before were beginning to infill with sediments and other debris, precluding any estimate of the water flowing into them. However, imbrication and orientation of leaves, woody debris, and sedimentary particles occurring in the sloughs connecting the river and the sinks indicate that the direction of flow in every instance where it could be ascertained was from the river into the sinks. In lower-elevation sites, river water was observed clearly flowing from the river into the sinks through well defined channels and sloughs. In no case was there any indication of flow from the sinks into the river.

Subsequent visits during times of medium- and high-water stages again revealed the dramatic loss of river water into those sinks occurring within the high-water channel. Virtually all of the larger sinkholes showed evidence of inflow, indicating that during high-flow stages they act as access conduits to the underlying Floridan Aquifer.

In order to ascertain the subsequent fate of the water seen disappearing into the sinkholes, tracer dye studies were carried out at the instigation of the senior author by another state contractor. On January 8, 1981, Rhodamine WT was injected at three separate locations: into the northernmost cluster, and two more into the second cluster, with one injection to the east of the river and one to the west into the largest feature. Between the injection sites and Fort Meade, six river sites and seven monitoring wells were subsequently sampled three times per day for fifteen days.

Fluorometric analysis of the river and well samples showed the following: 1) no dye was found in any of the river samples, and 2) only one well located nearly due south of the injection sites detected any dye. This well monitors the Floridan Aquifer. Calculations of the rates of dye dispersal based on injection and detection times indicate ground water flow velocities of 7.6 cm/s or approximately 6.6 km/day (McQuivey et al., 1981). This information leads to the following conclusions:

1. Once the water leaves the river, the water does not return to the surface but enters the Floridan Aquifer.

2. The water flows down-gradient (southward) in the Floridan as can be expected from the potentiometric contours in the Floridan and,

3. The rapid flow of water in this aquifer suggests that the surface sinkholes that act as influent avenues are connected to other voids within the Floridan.

Given the lack of general knowledge of this area, precise dating of the time of formation of many of the sinkholes is unlikely. Geological evidence, however, indicates that they most likely formed within the past thirty-five years, and on the basis of geomorphic evidence, some within the past decade.

One additional striking geomorphic development is the formation of accessory channels and sloughs that drain the higher-stage water into these newly-formed sinks. Originating from the edge of the sinkholes, some of these sloughs have eroded back toward the main low-water channel to the point where they have breached its banks and effectively "captured" part of its flow. Unlike the high-water channel anabranches, which become floored with silts and clays as the river subsides, the sinkhole channels leading from the river remain swept clean of finer sedimentary particles and are paved by sand and pebbles.

Whether it be directly or via channels and sloughs, these sinkholes drain substantial amounts of water and act as influent channels to the underlying aquifers. Depending on their location, they affect the flow of the river in different ways. Water loss through sinkholes in the bed of the low-water channel would be expected to have its greatest influence on low-water flow, although it would also affect high-water flow. On the other hand, sinkholes located in the high-water channel would act as conduits to the underlying aquifers only at higher water stages, and thus would influence only high-water discharge.

Thus, these sinkholes are not merely the random surface expressions of immanent geochemical processes, but are the results of specific and detectable hydrologic and geomorphic events. The geological literature and field observations clearly point out that the hydrology of the river and the hydrology of ground water in the upper Peace River valley are interdependent. It is also apparent that sinkhole and other solution features are the primary hydrologic connections between surface and ground water. It is evident, therefore, that the large number of sinkholes located within the valley floor of the Peace River has a direct bearing on the regimen of the Peace River and on the lateral extent, depth, and persistence of the high-water channel. As such, they contribute significantly to the reduction of the river's peak discharge and have played a significant role in the lowering of the river's high-water channel line since 1950.

Quantitative Assessment of Loss of Artesian Flow on River Hydrology

It is apparent from the size of the sinkholes described above and the evidence of rapid inflow associated with the larger ones that the amount of river water lost during high-water stage represents a significant fraction of the total volume of water flowing down the Peace River. In order to obtain a more quantitative assessment of river water loss through sinkholes, we estimated the potential flow into the sinks in two different ways. First, we estimated the amount of water which could be diverted from the river by calculating how much water the various sinkhole-feeding sloughs would carry. Second, we compared flows in different segments of the river.

We arrived at the amounts of water withdrawn from the river at moderate high water by taking rough cross-sectional measures of the influent sloughs and by using a conservative high-water velocity of 0.31 meters per second. The estimates included below are intended to represent only a rough approximation; differences in head, velocity, permeability, etc., at higher flows obviously would increase the actual inflow. Also not taken into account is the point where the transmissivity of the sink-aquifer connection is exceeded by the volume of the water supplied to it, thus causing the conduit to reject the excess. However, it is unlikely that the underlying aquifer is saturated at high flow. Also, we must point out that the 25-odd sinks chosen are merely the more obvious ones and do not include the 65-plus smaller but cumulatively important sinks also mapped during this study. Further, this list does not include any sinks occurring in the bed of the low-water channel. Additional conservative methods used in obtaining these results included the following: 1) if a sink was fed by multiple sloughs, we measured only the primary one; 2) we assumed the water in the influent sloughs to be no greater than bankfull, that is, we did not consider the flow during the times of normal high water when the hydraulic head would have been higher than the depth of the slough.

Nor does this take into account water lost into the various sinkholes filled with sediments. Pumping tests carried out at one such sinkhole located near Kissengen Springs indicated a rate of loss in excess of 11.35 liters/second (180 gal/min). During the four hours that water was pumped into this sink, no water was detected in a monitoring well located three meters (10 ft) away from the edge of the sink, indicating that all the water being lost was entering the underlying aquifer. These figures indicate that while sinkholes filled with sediments do not transmit as much water as do sinkholes with cleared passageways, they nevertheless act as significant conduits for recharge of the aquifer. Given the solution-riddled nature of the top of the Hawthorn, there are no doubt many sinkholes which have been filled in with sediments and hence were not identified during the survey. Even though sinkholes cease to be indentifiable topographic features, they continue to accept water and, even when buried, may have a significant impact on the local hydrology. Using the above parameters, the total slough-to-sink inflow for the twenty-five largest solution features was estimated at 15.21 m^3/s (537 cfs). While these estimates are undeniably crude, they provide a reasonable, conservative approximation of the loss that takes place during high-water stages.

On 9 August 1979, measured discharge at Bartow was 11.19 m^3/s (393 cfs). Downstream discharge dropped to 7.93 m^3/s (280 cfs) at Homeland, while from Homeland to Fort Meade, discharge remained at 7.93 m^3/s (280 cfs) (McQuivey et al., 1981). Normally, a stream picks up discharge in the downstream direction, but in this instance there is a loss of nearly thirty percent of the river's flow. More significantly, however, this loss occurs in that stretch of the river where the sinkholes are clustered; south of Homeland there are few sinkholes and the discharge measurements remain constant.

A second, more detailed series of discharge measurements were undertaken on January 9, 1981. At the Highway 60 gaging station in Bartow discharge was 0.79 m^3/s (28 cfs). Just below the Bartow sewage treatment plant, some 2 kilometers downstream from the gaging station, discharge had increased to 0.85 m^3/s (30 cfs). This increase was due to the water being released by the treatment plant. Below the first sinkhole cluster and just above the second cluster, discharge was 0.51 m^3/s (18 cfs). Within a stretch of 300 meters, 0.09 m^3/s (3 cfs) disappeared into a sinkhole within the river and another 0.26 m^3/s (9 cfs) were taken up by a single sinkhole in the high-water channel. Some 900 meters below the second cluster of sinkholes, discharge had decreased to 0.09 m^3/s (3 cfs). Between that point and the US 98 bridge in Fort Meade, the river discharge once again increased by almost 0.26 m^3/s (9 cfs) for a total of nearly 0.34 m^3/s (12 cfs) (McQuivey et al., 1981).

It is instructive to compare these measurements to the published mean daily flow values.

Day	Flow @ Bartow (cfs)	Flow @ Ft. Meade (cfs)
January 6	27	14
7	27	14
8	27	18
9	28	17
10	29	14
11	28	13
12	27	13

Table 1: Published daily mean flow values, January 6-12, 1981.
Source: U.S.G.S. Water resources data for Florida, water year 1980, volume 3A: Southwest Florida.

While these figures clearly demonstrate that water loss takes place, they mask the magnitude of the loss. On the basis of these daily discharge measurements by the U.S.G.S., the average loss appears to be 0.37 m^3/s (13 cfs), when in fact the measured loss by McQuivey (1981) is twice that.

During a span of ten water years (1977-78 to 1986-87) there was a total of 1,518 days (4.16 yrs.) when the average daily flow at Fort Meade was less than that at Bartow. Although these totals do not take into account the lag induced by travel time, they nevertheless indicate that for nearly five months out of the year water volume gains from runoff contribution in the basin between Homeland and Fort Meade do not compensate for the discharge lost to the sinkholes upstream between Bartow and Homeland.

Yearly average daily flows at both stations reveal the same pattern of water loss:

Water Year	Avg. Discharge @ Bartow (cfs)	Avg. Discharge @ Fort Meade (cfs)	%Bartow Flow @Fort Meade	% Lost
78-79	254	160	63.0	37.0
79-80	252	169	67.1	32.1
80-81	247	150	60.7	39.3
81-82	247	160	64.8	35.2
82-83	250	187	74.8	25.2
83-84	249	187	75.1	24.9
84-85	244	174	71.3	28.7
85-86	240	167	69.6	30.4
86-87	238	164	68.9	31.1

Table 2: Comparison of average daily discharge at Bartow and Fort Meade.
Source: U.S.G.S. Water resources data for Florida, water years 1978 through 1986, volume 3A: Southwest Florida.

McQuivey et al. (1980) originally calculated a series of best fit exponential functions on plots of drainage area vs. discharge and concluded that, on the basis of these curves, discharge at Fort Meade should be approximately 130 percent of the Bartow discharge. Simple drainage-area ratios (1204 km² or 465 sq. mi. vs. 1010 km² or 390 sq. mi.) would lead one to expect at least a nineteen percent increment in discharge, whereas in fact there is up to thirty-nine percent loss. Whatever the method, it is evident that discharge at Fort Meade should exceed that at Bartow. Even though the intensive mining activities in the area have effectively limited the actual size of the drainage basin and its discharge, discharge should at the very least remain constant, especially since the Bartow sewage treatment plant and mining activities actually return some water to the river. Thus, even discounting whatever gain in discharge might be expected in this stretch of the river, there has been a minimum of twenty-five percent and a maximum of thirty-nine percent loss in average discharge directly attributable to inflow into the underlying Floridan Aquifer via sinkholes.

From late September to mid-October of 1982 the upper Peace River experienced a moderate high-water event. On September 29, 1982, daily discharge in Bartow peaked at 45.88 m^3/s (1620 cfs). Discharge at Fort Meade on that and the following day remained at 33.98 m^3/s (1200 cfs). This 11.89 m^3/s (420 cfs), or twenty-six percent, measured loss compares favorably with the estimated loss of 15.21 m^3/s (537 cfs) calculated above, as well as with the percentage loss values given in Table 2.

Because of this loss of water via sinkholes through the high-water channel, the amplitude of the high-water stage is decreased. It would also be reasonable to expect that the duration of the high-water stage has decreased as well, for the length of time during which the high-water channel is wetted decreases and thereby the duration of the high-water stage is lessened.

Conclusions

Consumptive use of ground water in the region of the upper Peace River has led to a lowering of the potentiometric surface by over fifteen meters. This lowering had the following effects:

A. It has led to a cessation of artesian flow in the Peace River valley, decreasing the total flow of the Peace River, and to a reversal of conditions from effluent to influent.

B. It has led to the formation of numerous new sinkholes which act as conduits of influent flow into the underlying aquifers.

C. These sinkholes withdraw water from the Peace River. Those located within the low-water channel decrease water flow at all times.

D. However, because most of the sinkholes exist in the bed of the high-water channel, the greatest water loss from sinkholes occurs during high-water stage.

E.   The combined effect of the cessation of artesian flow and water loss into sink-
      holes has changed significantly the regimen of the Peace River. Average annual
      flow and low-water discharge have decreased and the high-water stage has been re-
      duced both in amplitude and duration.

F.   The reduction in amplitude and duration of the high-water stage has narrowed the
      lateral extent of the high-water channel and has caused the abandonment of its
      natural scarp. The river has eroded a new terrace scarp which reflects the edge
      of the new and now restricted high-water channel.

References

Heath, R.C., 1961. Surface water resources of Polk County, Florida: Fla. Geo. Surv. Inf.
    Circ. No. 25, 123 p.

Heath, R.C., and Wimberly, E.T., 1971. Selected flow characteristics of Florida streams
    and canals: Fla. Bur. of Geo. Inf. Circ. 69.

Holzschuh, J.C., 1972. State of water resources within the Peace River basin with emphasis
    on ground water: Tech. Rept. to Southwest Florida Water Management District.

Kaufman, M.I., 1967. Hydrologic effects of ground-water pumpage in the Peace and Alafia
    river basins, Florida, 1934-1965: Fla. Div. of Geo., Rept. of Inv. No. 49, 32 p.

McQuivey, R.S., et al., 1980. Hydrologic considerations of the Ordinary High Water line,
    Peace River, Bartow to Fort Meade, Florida: Unpublished Rept., prepared for the State
    of Fla., The Sutron Corporation, Fairfax, Va., 22030.

McQuivey, R.S., et al., 1981. Dye and discharge study, Peace River, Bartow to Fort Meade,
    Fla.: Unpublished Rept. prepared for the State of Fla., The Sutron Corporation,
    Fairfax, Va., 22030.

Patton, T.H., and Klein, J.G. The geomorphology of the upper Peace River valley: Unpub-
    lished manuscript, Patton and Associates Inc., Gainesville, Fla., 32606.

Sowers, G.F., 1974. Foundation subsidence in soft limestones in tropical and subtropical
    environments: Law Engineering Special Tech. Publ. G-7, 18 p.

Stewart, H.G., Jr., 1966. Ground water resources of Polk County, Fla.: Fla. Geo. Surv.
    Rept. Inc. No. 44, 170 p.

White, W.A., 1970. The geomorphology of the Florida peninsula: Fla. Bur. of Geo. Bull.
    No. 51, 164 p.

Two-dimensional numerical simulation of the relationship between sinkhole lakes and ground water along the Central Florida ridge

KEITH W.RYAN   *Storch Engineers, Florham Park, N.J., USA*

ABSTRACT

An observation well network has been installed around a small lake in central Florida to investigate whether "plugged" sinkholes provide an avenue for contamination to reach potable water supplies in the Floridan Aquifer. Numerical modeling has been employed to evaluate the field data and a number of regional numerical simulations have been conducted to examine leakance from sinkhole lakes under a wide range of geologic settings typical of the central Florida ridge.

Regional and site-specific numerical simulations indicate that leakance from lakes to the Floridan Aquifer occurs under the majority of geologic settings typical of the study area, regardless of the presence or absence of fractures and other high permeability conduits associated with solution or subsidence features. In particular, leakance through the lake bottom can be expected to occur where: the anisotropy ratio $(Kh/Kv) \leq 100$, the hydraulic gradient ≤ 0.02, or a water table mound is absent on the regional downgradient side of the lake.

INTRODUCTION

The sinkhole lakes in central Florida are often thought of as "plugged"; that is, holding water and no longer hydraulically connected to the underlying limestone aquifer. However, the degree to which these sinkhole lakes are connected to the underlying aquifer is variable. Hydraulic isolation of surface and ground waters is not necessary to maintain a consistent lake stage; flow into the lake need only exceed outflow.

To investigate the potential for contamination of potable ground water supplies in the Floridan Aquifer by leakance from "plugged" sinkholes (sinkhole lakes), it is important to understand the hydrogeologic relationship between these lakes and the underlying ground water flow systems. Toward this end, a number of regional numerical simulations have been conducted to examine leakance from sinkhole lakes under a wide range of geologic settings typical of the central Florida ridge. In addition, an observation well network has been installed around Lake Confusion, a small sinkhole lake in central Florida, to evaluate the effectiveness of numerical modeling in the study of individual lakes.

DESCRIPTION OF THE STUDY AREA

The study area, the central Florida ridge, is located in the mid-peninsular physiographic province of Florida (White, 1970). Topographically, this province comprises a discontinuous highland of sub-parallel north-south trending ridges separated by narrow valleys and broad flat uplands. Lake Confusion is a 14 acre lake with a circular shoreline which lies along the boundary between the Polk Upland and the Lake Wales Ridge at the southern terminus of the Green Swamp (Fig. 1).

Structurally, the study area is underlain by two major elements, the Peninsular Arch and the Ocala Uplift, anticlinal features which strike NW-SE from southeast Georgia to Lake Okeechobee. The Peninsular Arch deforms Precambrian - Mesozoic age igneous, metamorphic, and sedimentary rocks. The Ocala Uplift deforms the Cenozoic sediments which overlie the western flank of the Peninsular Arch, specifically the Eocene carbonates that comprise the Floridan Aquifer System.

The stratigraphic sequence within central Florida which is of interest to this study consists of siliclastics and carbonates of Holocene through Eocene age. Hydrologically, the stratigraphic section can be separated into three units: Surficial, Intermediate, and Floridan Aquifer Systems (Gilboy, 1985). Along the central Florida ridge, as the stratigraphic sequence thickens to the south, the relationship between these hydrostratigraphic units becomes more complex. In the vicinity of Lake Confusion, the

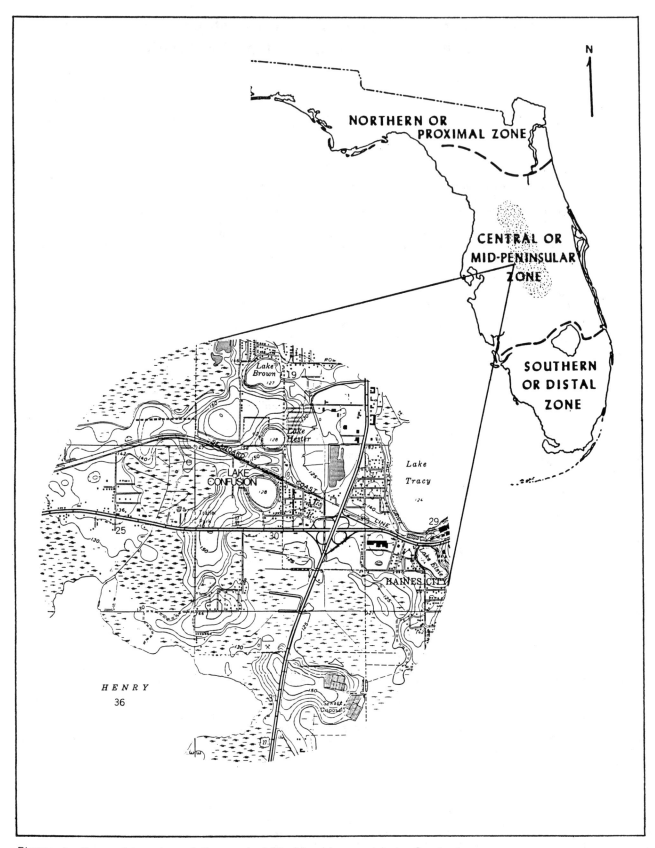

Figure 1   General location of the central Florida ridge and Lake Confusion

Surficial Aquifer System consists of Holocene through Pleistocene sediments; Pliocene units are absent for the most part. The Hawthorn Formation, which constitutes the Intermediate Aquifer System at the southern end of the ridge, acts as a confining layer for the Floridan north of central Polk County. Throughout the majority of the Florida peninsula, the primary ground water supply is the Floridan Aquifer System, a sequence of Eocene carbonates.

NUMBERICAL MODELING

Two scales of investigation were employed to conduct the numerical modeling: regional and local (site-specific). A series of simulations were conducted based upon data available in the literature, each simulation representing a hydrogeologic setting considered to be typical of conditions common to central Florida. Simulations conducted in this manner are referred to as regional. Data from the geotechnical and geophysical investigations of Lake Confusion were used to conduct a site-specific numerical simulation.

Solution Technique
The numerical modeling technique used in this study combines Darcy's Law and the equation of continuity to describe steady-state flow of ground water through a saturated porous medium:

$$\frac{d}{dx}\left(Kx\frac{dh}{dx}\right) + \frac{d}{dy}\left(Ky\frac{dh}{dy}\right) = \frac{-R(x,y)}{K(x,y)b} \quad [1]$$

where: Kx and Ky = principal components of the hydraulic conductivity tensor;

$\frac{dh}{dx}$ and $\frac{dh}{dy}$ = potentiometric gradient along the respective dimensional axis;

$R(x,y)$ = volume of recharge/unit time/unit area;

b = thickness of the aquifer.

Equation [1] is often referred to as Poisson's equation, describing steady-state flow with an external source or sink. The assumptions made in both the regional and local investigations include: 1) laminar flow; 2) incompressible fluid of constant density; 3) rigid porous medium; 4) dimensional axes are coincident with the principal components of the hydraulic conductivity tensor; 5) water table, lake level, and the potentiometric surface of the Floridan Aquifer System are known and invariant with respect to time; 6) ground water flow can be adequately described in terms of two dimensions, x and y.

Equation [1] is solved as a two-dimensional finite-difference approximation in which the head at a specific node is a function of the head at the four surrounding nodes. Under homogeneous isotropic conditions, the nodal head is the simple arithmetic mean of the four surrounding heads. Where heterogeneity and anisotropy are present, the equation describing the nodal head becomes more complex, incorporating the harmonic mean of the hydraulic conductivity. The set of nodal equations comprising is solved iteratively until the difference between the head at each node for subsequent iteration steps falls within a pre-determined error tolerance. A total model error (the difference between the sum of all of the nodal heads for two subsequent iterations) of 0.0003 meters (0.001 ft) of head has been chosen for both the regional and site-specific simulations.

General Model Construction
The designs of the regional and site-specific models are basically the same, patterned after the models developed by Winter (1976). Each model is configured as a vertical profile through the stratigraphic section with the water table and lake surface forming the upper boundary and the potentiometric surface of the Floridan forming the base of the model. Lateral boundaries are vertical and located so that the profile extends a sufficient distance along the x-axis to encompass the surface/ground water flow system under study. A general schematic of the model format is presented in Figure 2.

Each model comprises a geologic profile with a finite-difference grid overlay. The continuous surface of the profile is approximated by discrete cells, each cell bounded by four grid nodes (Fig. 3). Such a format is referred to as block-centered. The size of the finite-difference grid varies between individual models but generally employs rectangular cells with an x:y dimensional ratio of 40:1. Those models incorporating fracture networks employ variable grid cells with x:y dimensional ratios ranging from 10:1 to 40:1. The maximum number of cells comprising a model is 4800 in a configuration of 15 rows and 320 columns.

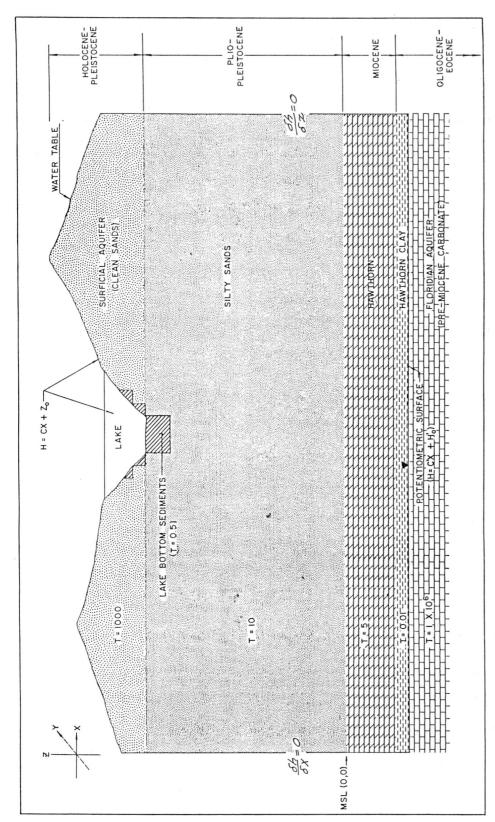

Figure 2. Hydrostratigraphy of the regional model

36

There are a large number of programs available for the solution of the finite-difference approximation. In this study, equation 1 is solved by employing a commercially available spreadsheet program. The spreadsheet method of finite-difference model construction is described in detail by Olsthoorn (1985).

One of the primary difficulties in applying general models to the analysis of specific geologic settings is dimensional disparity; the vertical and/or horizontal dimensions of the model do not fit those of the specific setting under consideration. When dimensional units are employed, a regional model becomes site-specific and its usefulness as a general indicator is limited. To avoid this problem, a scaling technique presented in Winter (1976) is employed, in which model dimensions are expressed as functions of the stratigraphic thickness (T) of the model (Fig. 2).

Hydraulic conductivity and transmissivity are treated as synonymous terms; in a vertical profile, the aquifer thickness becomes a unit thickness. In the regional models, hydraulic conductivity is expressed as a dimensionless ratio where clean surficial sands are assigned an arbitrary hydraulic conductivity value of 1000. Hydraulic conductivities of all remaining strata are presented in Figure 2. The ratio method eliminates the necessity of determining actual hydraulic conductivities and allows for a typical or regional solution. Hydraulic conductivity ratios also more accurately reflect the degree of confidence that should be placed in regional estimates of hydraulic parameters. The hydraulic conductivities used in the site-specific model are derived from actual field measurements and, in the case of the Floridan, the results of pumping tests reported in the literature (Callahan, 1964; Robertson, 1973; Stewart, 1966).

Boundary conditions are generally the same for all of the regional simulations as well as the site-specific case. The upper boundary of each model is defined by the water table within the Surficial Aquifer System. The lower boundary of each model is formed by the potentiometric surface of the Floridan. Both horizontal model boundaries are defined as constant head (Dirichlet) conditions. The vertical boundaries which establish the horizontal extent of each model have been defined in two different forms; as no-flow boundaries (Neumann condition) and as a mixed (Roberts) condition in which the flux through each node is a function of nodal cross-sectional area, hydraulic conductivity, and the local hydraulic gradient. Vertical flow between the Surficial and Floridan Aquifer Systems occurs as flow through a confining layer to a constant head boundary. The implications of modeling the potentiometric surface in the Floridan as a constant head rather than a head-dependent boundary are discussed in Ryan (1989).

Within each model, geologic units such as a confining layer are represented by variations in the nodal hydraulic conductivity. Ground water flux between the interior nodes of each model is a function of the Laplace equation, a subset of equation [1] in which the sink/source term is equal to zero. All nodes within the regional and site-specific models exhibit anisotropy; the degree of anisotropy varies between models, but remains uniform within the same model.

REGIONAL SIMULATIONS

Model Configuration
    A total of 60 regional simulations were conducted to determine the effect of variations in individual and multiple model parameters. The general hydrostratigraphy of the regional models comprises a three component system: surficial aquifer system, confining unit, carbonate aquifer system (Fig. 4). If the variable "T" represents the total stratigraphic thickness of the general flow system model, the thickness of the surficial aquifer (unconsolidated overburden) is 0.8T and the confining unit, which comprises the remainder of the model, is 0.2T in thickness. The carbonate aquifer system is represented by the potentiometric surface of the Floridan, a single row of cells with fixed head values.

The upper two components of the general model have been further subdivided into separate hydrostratigraphic units with hydraulic conductivities representative of sediments commonly encountered in central Florida. Lacustrine sedimentation is modeled as a stratum of low permeability material underlying the profundal zone of the lake. The thickness of this unit grades from 0.04T at the shoreline to 0.08T beneath the center of the lake. Low permeability organic sediments were omitted from the littoral portion of the model, reflecting conditions encountered at the study site and reported in the literature (McBride and Pfannkuch, 1975).

Simulations have been conducted with anisotropy ratios (Kh/Kv) ranging from 10:1 to 1000:1. Modeling and field investigations indicate that anisotropy ratios for alluvial and coastal plain sediments may vary from 4:1 (Mansur and Dietrich, 1965) to as high as 1000:1 (Bennett and Giusti, 1971). Lacking any definitive data concerning anisotropy ratios for the sediments of the study area, a default Kh/Kv of 10:1 was chosen as an appropriate value for

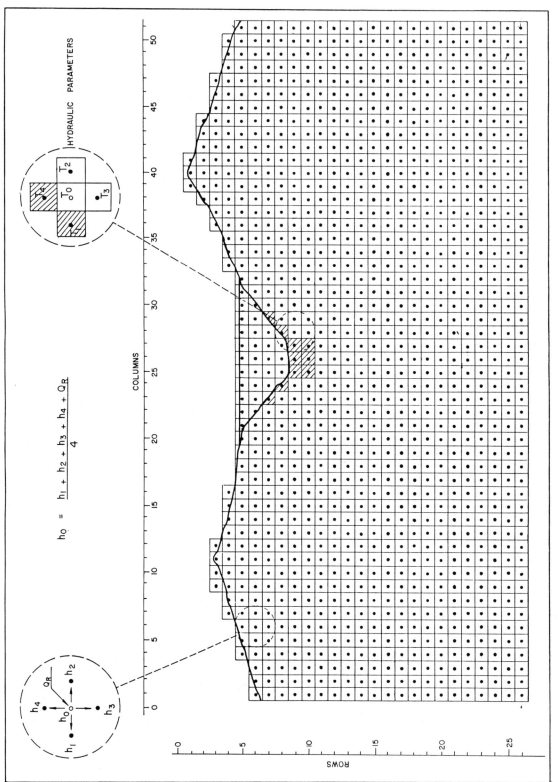

Figure 3. General discretization format for the regional model

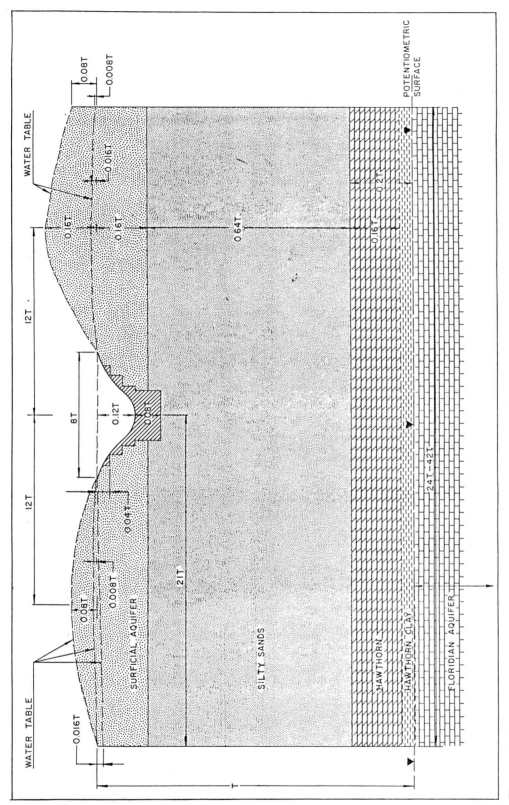

Figure 4. Dimensional relationships within the regional model

39

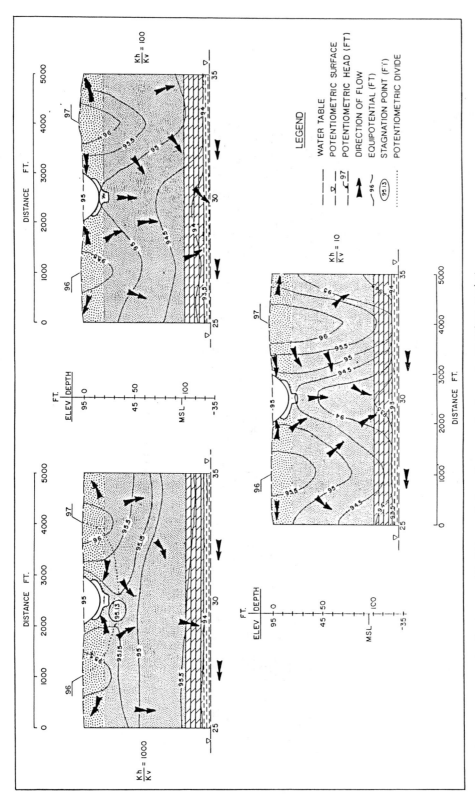

Figure 5. Relationship between Kh/Kv and development of a stagnation point

siliclastic and carbonate sediments deposited in shallow marine and marine/fluvial environments.

Results of the Regional Simulations

In his benchmark paper, Winter (1976) found that the factor controlling the interaction between lakes and ground water is the continuity of the local flow system boundary beneath the lake. Specifically, the point of minimum hydraulic potential along this boundary (stagnation point) is the key to ground water flow into and out of a lake. This conclusion is reaffirmed by the results of the regional and site-specific simulations conducted as a part of this study.

It should be noted that the regional simulations conducted as a part of this study comprise only a small segment of the population of "typical" geologic settings of central Florida. Although the hydrostratigraphic setting of a particular site may differ from those presented, the general observations and conclusions should be applicable. The following observations summarize the outcome of the regional computer simulations:

1. Virtually all of the simulations resulted in leakance through the lake bottom. No stagnation point could be maintained in the absence of a downgradient water table mound (flow-through conditions). A stagnation point develops in simulations where:

   $Kh/Kv \geq 100$; $i \geq 0.02$; downgradient water table mound exists;

2. The majority of ground water within the overburden will ultimately discharge through the semi-permeable confining layer into the Floridan Aquifer System; leakance from the Surficial Aquifer to the Floridan will occur even though the sinkhole lake is hydraulically closed;

3. Strength of the stagnation point is proportional to Kh/Kv (Fig. 5);

4. The ratio of vertical to horizontal flow within the model is proportional to Kh/Kv;

5. Discharge from ground water to the lake is confined to the uppermost littoral zone when upgradient and downgradient water table mounds are present;

6. Where the difference in head between the water table and the potentiometric surface exceeds $0.08T$, a hydraulic divide develops beneath both internally drained and flow-through lakes preventing the development of an intermediate or regional flow system at depth within the surficial sediments;

7. Fractures located at a distance greater than $4T$ from a lake on either the upgradient or downgradient side have little or no effect on leakance from the lake into the Floridan;

8. Given a fracture set located beneath a lake, the magnitude of leakance is greatest when fractures are centrally located and least when they are located at the downgradient edge. Increased leakance induced by upgradient fractures appears to occur as a result of the interception of local ground water flow which recharges the lake (Fig. 6);

9. The head differential between the water table and the potentiometric surface of the Floridan is an important factor controlling the amount of leakance from the lake. Under the simulation in this study, a 12-fold increase in the distance between the water table and the potentiometric surface resulted in a 40-fold increase in the gradient across the bottom of the lake;

10. Thickness of the confining unit separating the Surficial and Floridan Aquifer Systems appears to exert somewhat less of an impact on leakance from a lake than does head differential. A 10-fold increase in the thickness of the confining layer resulted in an 8-fold decrease in the gradient across the lake bottom.

SITE-SPECIFIC SIMULATION

Water budget analysis and numerical modeling have both been employed to examine the relationship between Lake Confusion, a small sinkhole lake, and the surrounding ground waters. The water budget technique provided inconclusive results, primarily because of significant errors in the measurement of time-dependent climatic variables. The source and magnitude of these errors are discussed at length by Winter (1981). The numerical model, on the other hand, indicates a downward leakance of water from Lake Confusion to the Floridan coupled with a flux of ground water through the lake in a shallow intermediate-regional

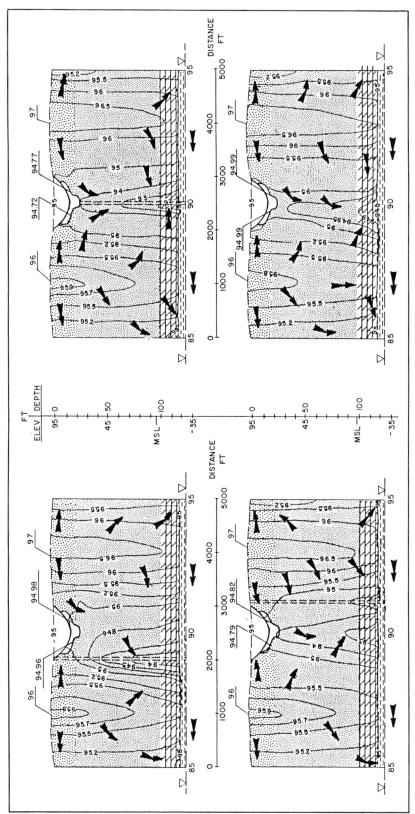

Figure 6. Effect of fractures on leakance from a lake

42

unconfined flow system. A detailed comparison of the water budget and numerical modeling techniques used to study Lake Confusion is presented in Ryan (1989).

Model Configuration

The site-specific numerical model constructed for Lake Confusion is based upon a two-dimensional profile through the lake from west to east (Fig. 7). The model consists of seven different hydrostratigraphic units defined by roughly 950 block-centered nodes arranged in 23 rows and 41 columns. Column width was varied between 15, 23, and 30 meters (50, 75, and 100 ft) while row height was maintained at 1.5 meters (5 ft) throughout the model. The vertical no-flow boundaries which delimit the model have been placed a distance of roughly 9T upgradient and downgradient of the edge of the topographic basin to minimize boundary effects within the body of the model. The dimensions of the model, distribution of geologic units, and distribution of hydraulic potential are presented in Figure 7.

The hydrostratigraphic units comprising the model are based upon the results of the geotechnical investigation and falling head tests conducted in the field. The different lithologic units were grouped into seven categories, each represented by a transmissivity. The transmissivity assigned to the Floridan Aquifer System is typical of that reported in the literature (Stewart, 1966) for the general area of the study site. All units within the model have been assigned an anisotropy ratio (Kh/Kv) of 10:1.

Four fractures are included in the model of Lake Confusion. The number and position of fractures is based upon a fracture-trace analysis. The fractures were assigned a transmissivity equal to that of the silty sand stratum. Each fracture is simulated as a column, 15 meters (50 ft) in width, extending downward from the silty sand stratum through the underlying Hawthorn Formation to the top of the Floridan. The transmissivity and width of the fractures were assigned an assumed value for convenience of modeling; however, there is no field data to support the assumption.

Results of the Site-Specific Simulation

The behavior of the site-specific model of Lake Confusion is very similar to the regional simulations which incorporate a fracture or group of fractures (Fig. 6). Some of the difference in the distribution of head between regional and site-specific simulation results from the distribution of stratigraphic units. The following observations summarize the outcome of the finite-difference simulation of Lake Confusion:

1. Ground water inflow to Lake Confusion occurs along the upgradient 1/3 of the lake bottom. Leakance from the lake to the underlying strata occurs over the remaining 2/3 of the lake bottom;

2. The percentage of leakance from the lake which is discharged to the Floridan has not been quantified. However, the distribution of hydraulic potential within the model suggests that intermediate and regional flow within the overburden are limited to the immediate vicinity of the water table leaving the majority of ground water in the overburden to discharge to the Floridan;

3. The fractures beneath and adjacent to Lake Confusion form the centers of individual local flow systems (no-flow boundaries) which preclude the development of intermediate or regional flow systems at depth;

4. Distribution of head within the model suggests that precipitation infiltrating to the water table beyond a distance of 3-4T upgradient of the lake will be more likely to discharge to the Floridan than into Lake Confusion. One of the implications of this is the potential for a contaminant spilled near the upgradient topographic divide to find its way into the Floridan without being detected in a downgradient lake.

CONCLUSIONS

An examination of the interaction between surface and ground waters along the central Florida ridge indicates that leakance from lakes to the Floridan Aquifer System will occur under the majority of geologic settings typical of the area. Leakance of surface waters to the Floridan does not require fractures or other hydraulic interconnections normally associated with solution or subsidence features and therefore is not confined to what are commonly referred to as sinkhole lakes. As noted by Winter (1976), the interaction between lake and ground waters is a function of the existence and strength of the stagnation point, the point of minimum head along the potentiometric divide separating the lake from underlying

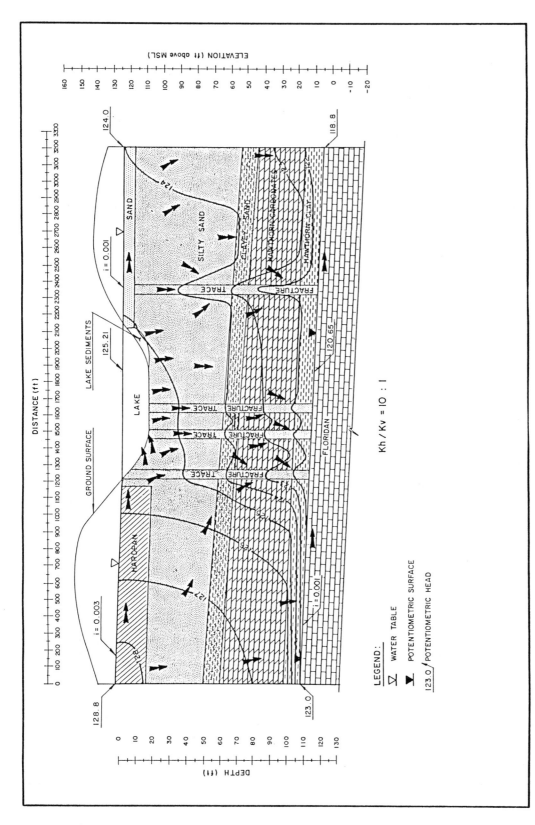

Figure 7. Distribution of hydraulic potential and ground water flow directions predicted by the site-specific simulation of Lake Confusion

44

ground water. Numerical modeling indicates that a stagnation point will develop if the following conditions are met:

1. anisotropy ratio (Kh/Kv) $\geq$ 100;

2. hydraulic gradient $\geq$ 0.02;

3. a water table mound is present on the regional downgradient side of the lake (internally-drained condition).

Violation of any one of these conditions will generally result in leakance from the lake bottom to the Floridan.

REFERENCES

Bennett, G.D. and E.V. Giusti, 1971. Coastal ground-water flow near Ponce, Puerto Rico. U.S.G.S. Professional Paper 750-D, pp. 206-211.

Callahan, J.T., 1964. The yield of sedimentary aquifers of the Coastal Plain, southeast river basins. U.S. Geological Survey Contributions to the Hydrology of the United States, Water-Supply Paper 1660W, 56 pp.

Gilboy, A.E., 1985. Hydrogeology of the Southwest Florida Water Management District, Southwest Florida Water Management District Regional Analysis Section, Technical Report 85-01, 18 pp.

Johnston, R.H., P.W. Bush, and J.A. Miller. 1986. IN: Regional Aquifer Systems of the United States - Aquifers of the Atlantic Coastal Plain, J. Vecchioli and A. I. Johnson, eds. AWRA Monograph Series No. 9. pp. 153-167.

Mansur, C.I., and R.J. Dietrich, 1965. Pumping test to determine permeability ratio. Journal of the Soil Mechanics and Foundation Division, Proceedings of the American Society of Civil Engineers, vol. 91, no. SM4, pp. 151-183.

Olsthoorn, T.N., 1985. The power of the electronic worksheet: modeling without special programs. Ground Water, vol. 23, no. 3, pp. 381-390.

Prickett, T.A. and C.G. Lonnquist, 1971. Selected digital computer techniques for groundwater resource evaluation. Illinois State Water Survey, Urbana, Bulletin 55, 62 pp.

Robertson, A.F., 1973. Hydrologic conditions in the Lakeland ridge area of Polk County, Florida. Florida Bureau of Geology, Report of Investigation, no.64, 54 pp.

Ryan, K.W., 1989. Ground Water Pollution Associated with Sinkholes: Phase II - Evaluation of Ground Water Contamination From Sinkhole Lakes. Florida Sinkhole Research Inst., 94 pp.

Stewart, H.G., 1966. Ground-water resources of Polk County. Florida Bureau of Geology, Report of Investigation, no. 44, 170 pp.

Trescott, P.C. and G.F. Pinder, 1975. Finite difference model for aquifer simulation in two dimensions and results of numerical experiments. U.S.G.S. Administrative Report.

White, W.A., 1970. The geomorphology of the Florida peninsula. Florida Bureau of Geology, Geological Bulletin no. 51, 164 pp.

Winter, T.C., 1976. Numerical simulation analysis of the interaction of lakes and ground water. U.S.G.S. Professional Paper 1001, 45 pp.

Winter, T.C., 1981. Uncertainties in estimating the water balance of lakes. Water Resources Bulletin, vol. 17, no. 1, pp. 82-115.

Sinkholes (dolines) in Bhander Limestones around Rewa, central India, and their environmental significance

YAMUNA SINGH   *Atomic Minerals Division, Begumpet, Hyderabad, India*

ABSTRACT

The Bhander Limestones of the Vindhyan Supergroup (Precambrian), exposed around Rewa in northeastern Madhya Pradesh, Central India, show excellent development of sinkholes (dolines) of various shapes and sizes. They are commonly funnel-, bowl-, trough-, oval-, and ir-regular-shaped. The diameter of the dolines varies between 0.40 m and 10 m with a depth up to 2.5 m, and the surface area between 0.13 m^2 to 80 m^2. Density of the dolines varies from 20d/km^2 to 1550 d/km^2.

Sinkholes in the area are mainly solutional and collapse types. Solutional sinkholes have formed primarily due to pronounced karsti-fication of limestones around some favourable points, generally at the intersections of joints, while collapse sinkholes are the prod-uct of cavities lying near the surface.

Sinkholes either appear singly or form doline fields and long-drawn-out rows on joints. Normally, sinkholes of the high topogra-phic areas are in line with those of low topographic areas, direction of which commonly matches with the prevalent joint direction.

Groundwater in karst aquifers is potable and suitable for various uses. However, shows of pollution in groundwater are noticed in the vicinity of streams and sinkholes that are interconnected and structur-ally aligned. It is observed that the pollutants are carried to ground-water through an intricate network of interconnected sinkholes. These sinkholes receive pollutants from those streams and associated depres-sions which are used for disposal of industrial and municipal effluents.

Those areas where sinkholes are well interconnected have been identified, and it is suggested that non-disposal of the industrial and municipal effluents in such areas would prevent groundwater pol-lution, ensure fresh water supply and maintain aquatic environmental balances in the area.

Introduction

The Bhander Limestones of the Vindhyan Supergroup (Precambrian) are well exposed in Rewa area of the northeastern Madhya Pradesh, Central India (Fig. 1). They are generally fine to medium grained, flaggy to massive and thinly to thickly bedded. Chertification of limestones, in the form of bedded and nodular chert, is common. Presence of algal stromatolites of various forms characterise certain variety of the Bhander Limestones. Various types of limestones observed in the study area, from bottom to top, are given below (Table I).

Table I. Various types of the Bhander Limestones in Rewa area, northeastern Madhya Pra-desh, Central India.

Greenish grey and purple stromatolitic limestone.
Dark grey, well bedded and cross bedded limestone, and dark grey stromatolitic limestone.
A band of flat pebble conglomerate.
Light and dark grey stromatolitic limestone with algal bedding.
A band of flat pebble conglomerate.
Pink and light yellow laminated limestone.
Dirty white and dirty yellow argillaceous limestone.

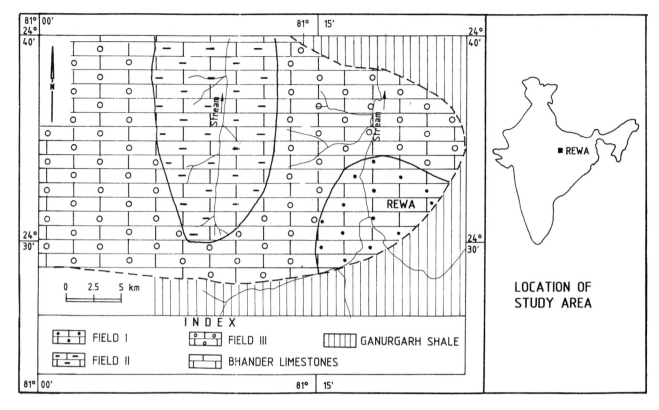

Fig.1 Geological map of the area around Rewa, showing sinkhole fields.

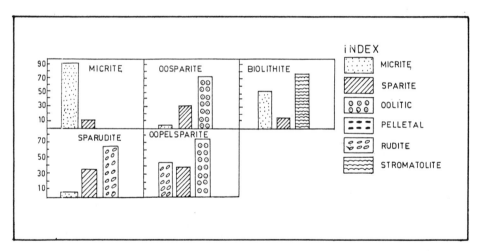

Fig.2 Petrographic composition of microfacies of the Bhander Limestones around Rewa.

Under the microscope, the stromatolitic limestone shows mainly alternating laminae of micrite and solomicrite with sparsely distributed quartz grains throughout the micrite. Non-stromatolitic limestone consists chiefly of silty and micritic calcite, with a few detrital grains of quartz, biotite and muscovite. Petrographic composition of microfacies of the Bhander Limestones is shown in Figure 2.

Hydrogeology and various karst features of the Bhander Limestones have been described in a preliminary way by Singh (1985, 1989). This paper deals with the various types of sinkholes in Bhander Limestones and their origin, hydrogeology and environmental significance.

48

Sinholes (dolines) in Bhander Limestones

Sinkholes are small depressions of low to moderate dimensions whose diameter is greater than their depth. They are well developed in Bhander Limestone terranes, both in the plain areas and in the hills. They are small, but generally well developed, geomorphological features to give rise to structures referable to a set of terminology. Various structures observed are abime (slit), aven (grike leading to a cavern), swallowhole (irregular sinkhole), doline (lensoidal elongated depression), gouffre (naturally carved out well), grotte (giant cavernous structure), katavothre (irregular grike), resurgence (gaining spring), perte (losing spring) and embut (oozing spring without any ground relief) (Fig. 3-6). The sinkhkoles often lead to caves (Fig. 4) and caverns of various shapes and dimensions, and there are often benches of karstification with different sets of carving (Fig. 6). Steep walls or scarp faces in the hills and the river chasms are also beautifuly furrowed to give rise to karren (Fig. 6).

Morphology of sinkholes

As stated earlier, the sinkholes show large variations in their morphology. Some of the common types of sinkholes are funnel-shaped, bowl-shaped, trough-shaped, oval-shaped and irregular-shaped. The funnel-and bowl-shaped sinkholes are best developed in dark grey limestone, while trough-, oval-and irregular-shaped sinkholes are excellently observed in argillaceous limestone. In general most of the sinkholes are asymmetrical. The sinkholes are either exposed or covered under soil. Those sinkholes which are bare are, at times, filled up partly with terra-rossa.

Sinkhole density and fields

The area shows wide variations from place to place in sinkhole density. Based on sinkhole density, the investigated area has been divided into three fields (Fig. 1). They are (i) Field I (less than 500 d/km^2), (ii) Field II (500-1000 d/km^2) and (iii) Field III (more than 1000 d/km^2). Details of each sinkhole field are given in Table II.

Table II. Sinkhole field, sinkhole density, sinkhole diameter, sinkhole depth, sinkhole surface area and sinkhole type in Bhander Limestones around Rewa, northeastern Madhya Pradesh, Central India.

Sl. No.	Sinkhole Field	Sinkhole density (d/km^2)	Sinkhole diameter (m)	Sinkhole depth (m)	Sinkhole surface area (m^2)	Sinkhole type
1.	Field I	20-150	0.4- 3.8	0.1-1.0	0.13-11.5	Mainly solutional
2.	Field II	560-955	0.5- 5.5	0.5-2.0	0.2 -24.0	Mainly solutional; rare collapse.
3.	Field III	1000-1550	0.4-10.0	0.8-2.5	0.13-80.0	Both solutional and collapse;rare subsidence.

Perusal of data given in Table II reveals that sinkhole density, sinkhole diameter, sinkhole depth and sinkhole surface area gradually increase from sinkhole field I to III. Similarly, in sinkhole field I, mainly solutional sinkholes are found while in field III both solutional and collapse sinkholes are found, with rare subsidence sinkholes. It clearly suggests that field III has undergone pronounced karstification followed by field II and I. Indeed, density of the sinkholes is directly proportional to density of joints and fractures in limestones. Wherever density of joints and fractures increases, density of sinkholes increases and vice-versa.

In sinkhole field I sinkholes generally appear singly, while in field II and III, they commonly form doline fields and long-drawn-out rows on joints (Fig.5). Normally, sinkholes of the high topographic areas are in line with those of low topographic areas, direction of which commonly matches with the prevalent joint direction.

Types of sinkholes and their origin

The sinkholes in the study area are of three types (i) solutional, (ii) collapse, and (iii) subsidence. Among these, solutional and collapse dolines are predominant. The solutional sinkholes have formed primarily due to pronounced karstification of limestones either under a soil cover or a vegetation cover, around some favourable points, generally at the intersections of the joints, fractures, interstices and bedding planes; and also by a karren phenomenon. The collapse sinkholes have resulted by cavities which lie near the surface. The formation of subsidence sinkholes, which are rare, was possible by a slow downward movement of limestone mass.

Hydrogeology of sinkholes

Hydrologically pervious sinkholes act as a feeder of rain water to the groundwater reservoir. Hydrologically impervious sinkholes can be made hydrologically pervious by

Fig. 3 Intricate network of interconnected irregular shaped sinkholes (swallowholes) in Bhander Limestones. Note a few bowl-and through-shaped sinkholes and abime (slit).

Fig.4 Grikes and clints in bedded Bhander Limestones, with U-shaped sinkholes. Grike forming an aven, leading to a cavern with the perennial river generally dry at this location. Note a few abime (slit).

Fig.5 A broad view of sinkhole field and long-drawn-out rows of sinkholes on karstified joints. Note sinkholes of different shapes and sizes.

Fig.6 Three benches of karstification in Bhander Limestones in the river bed, with river flowing after monsoon over the last bench. Sinkholes are exposed on the second bench. Vertically karstified surface is beautifully furrowed to give rise to karren.

cleaning clayey and silty soils (terra-rossa) deposited in them, and used to enrich the adjoining alluvium and karstic limestone aquifer bodies. It would constitute an extensive groundwater reservoir and can be planned for future use. By taking advantage of shallow impermeable beds, a large number of check walls can be suitably planned to enrich storage and thereby groundwater is prevented from returning to the streams. Karst springs, which are common in field II and III (zone of high secondary permeability) and are seen in the form of submergence and resurgence (Singh, 1981), can be developed for augmented ground-water supply by cleaning silty and clayey materials filled in their conduits, through sodium carbonate, sodium hexametaphosphate and acid treatments followed by shooting by mild explosives.

Quality of groundwater in karst aquifers

Data on chemical analyses of representative groundwater samples from karst aquifers are given in Table III.

Table III. Data on chemical analyses of groundwater samples from karst aquifers in Rewa area, northeastern Madhya Pradesh, Central India.

Sample No.	pH	Sp.Elec. Cond. (micro mohs at 25°C)	All values are in ppm									
			T.D.S.	T.H. as CaCO3	Si	Na	K	Ca	Mg	Cl	SO4	HCO3
1.	7.5	796.0	585.1	256.7	8.0	68.5	1.5	52.5	30.5	38.6	194.0	200.0
2.	7.3	2078.4	1454.9	925.6	19.5	42.5	3.2	190.0	109.6	171.6	117.0	800.0
3.	7.4	2103.7	1430.5	923.1	19.0	40.0	3.1	190.0	109.0	114.4	234.0	720.0
4.	7.4	634.6	437.9	277.4	10.0	20.5	2.7	57.5	32.5	56.2	117.0	140.0
5.	7.7	2323.1	1602.9	850.0	18.0	105.0	5.9	187.5	92.5	171.6	220.0	800.0
6.	7.3	751.1	510.7	314.5	8.0	21.5	3.5	65.0	37.0	562.0	78.0	240.0
7.	7.5	3356.6	2282.5	1463.8	31.0	90.0	29.2	300.0	173.6	557.7	440.5	668.0
8.	7.2	1154.5	858.3	561.0	14.0	13.5	3.5	115.0	66.5	85.8	78.0	480.0
9.	7.4	2320.3	1601.0	838.4	14.0	110.5	6.2	172.5	99.0	85.8	312.0	800.2
10.	7.6	1624.3	1172.0	375.0	18.0	72.5	1.5	135.0	77.4	114.4	312.0	400.0
11.	7.5	595.2	417.6	275.0	9.5	20.5	3.5	57.5	32.0	28.6	105.0	160.0
12.	7.3	1384.2	941.2	578.2	18.0	15.7	4.5	120.0	69.1	42.9	200.5	480.7
13.	7.4	2893.7	1996.7	1229.9	24.5	55.0	3.7	252.5	145.6	114.4	400.5	1000.0
14.	7.3	637.5	465.4	289.6	8.5	20.5	2.7	16.0	34.0	57.2	142.0	140.0

From chemical analyses data (Table III) it is apparent that water is in general weakly alkaline and very hard (hardness measured as CaCO3; 0-60 ppm, soft; 61-120 ppm, moderately hard; 121-180 ppm, hard; and more than 180 ppm, very hard; Newcomb, 1972). Total dissolved salts-wise, unpolluted groundwater is fresh, as total dissolved salts are less than 1000 ppm (T.D.S. up to 1000 ppm, fresh water; 1000-3000 ppm, slightly saline water; Winslow and Kister, 1956). The alkaline earths content exceeds both alkalies as well as bicarbonate ions. The water is potable and suitable for various usages as concentrations of calcium, magnesium, sodium, potassium, sulphate, bicarbonate and chloride ions are generally well within the permissible limits.

Groundwater pollution and its impact on aquatic environment

From chemical data (Table III) it is evident that 50% of water samples (sample nos. 2, 3,5,7,9,10 and 13) are slightly saline (T.D.S. 1000-3000 ppm) and contain anomalously high ions of calcium, magnesium, sodium, potassium, sulphate, bicarbonate and chloride, delimiting thereby the scope of such waters for various usages. Such increase in local salinity in groundwater has been ascribed to local surface pollution (Singh, 1987), caused by disposal of untreated industrial and municipal wastes in surface streams and associated depressions in limestone terrain. Sinkholes, which are present along the stream courses and in their vicinity, being interconnected with each other through an intricate network of karstified joints and fractures, etc., gradually carry these pollutants to groundwater in karst aquifers. The pollutants do not get properly diffused throughout the groundwater reservoir because of varying order of interconnectivity of seconday pore systems, generated by uneven karstification of limestones. This is the prime reason for more concentration of pollutants locally in the vicinity of those streams and sinkholes which are used for disposal of industrial and municipal effluents.

Since density and interconnectivity of the sinkholes are not uniform in all the fields, those fields where sinkhole density is low and their interconnectivity is poor can be used for disposal of industrial and municipal effluents. For example, in field I sinkhole density is low and sinkholes appear singly and show poor interconnectivity and therefore, field I (Fig. 1) can be suitably planned for disposal of industrial and municipal effluents, after giving effective treatments to them. If due attention is not paid to site selection for disposal of effluents and their effective treatments before they are disposed, continued influx of pollutants to groundwater through interconnected sinkholes would cause serious groundwater pollution and make the groundwater in karst aquifers unsuitable for various usages. Moreover, industrial effluents are known to pollute the water to the extent that it becomes septic and cannot sustain life (Singh, et al., 1987). Therefore, disposal of industrial and municipal effluents in a planned way is suggested to prevent groundwater pollution, ensure fresh water supply and maintain aquatic environmental balances in the area.

References

Newcomb, R.C. (1972) Quality of the groundwater in basalt of the Columbia River Group, Washington Oregon and Idaho, U.S.G.S. Water Supply Paper No.1999, 71 p.

Singh, Y.(1981) Karstic springs as a source of groundwater around Rewa district, M.P. National Seminar, Rewa University, Rewa, M.P. (abstract), p.22.

Singh, Y.(1985) Hydrogeology of the karstic area around Rewa, M.P., India, Karst Water Resources (Proceedings of the Ankara-Antalya Symposium, July, 1985), IAHS Publ.No. 161, pp. 407-416.

Singh, Y.(1987) Groundwater pollution in karst aquifers near Rewa Town, Madhya Pradesh. Role of Earth Sciences in Environment (Proceedings of the I.I.T. Bombay Symposium, December, 1987), pp. 293-297.

Singh, Y.(1989) Karst landforms around Rewa district, Madhya Pradesh, Proceedings I.S.C.A., Madurai (abstract), p.37.

Singhal, B.B.S., R.P.Mathur, D.C.Singhal, B.S.Mathur, V.K.Minocha, A.K.Seth and S.K.Sinha (1987) Effects of industrial effluents on the water quality of river Hindon, district Saharanpur, U.P. Jour. Geol. Soc. India, v.30, no.5, pp. 356-368.

Winslow, A.G. and L.R.Kister (1956) Saline water resources of Texas, U.S.G.S. Water Supply Paper No. 1365, 105 p.

Two methods for representing sinkholes in a three-dimensional ground water flow model

MARK D.TAYLOR   *Camp Dresser & McKee Inc., Atlanta, Ga., USA*

PETER J.RIORDAN & BRENDAN M.HARLEY   *Camp Dresser & McKee Inc., Cambridge, Mass., USA*

ABSTRACT

Sinkholes can have a significant impact on the ground water flow characteristics of a confined/unconfined aquifer system. This is particularly true in the Floridan Aquifer system in Central Florida. Most of the sinkholes in this area breach the confining unit between the surficial and Floridan Aquifers and thus provide a mechanism for large quantities of water to flow between the two aquifers. From a regional perspective, the sinkholes provide for a major source of ground water recharge to the Floridan aquifer which can increase water levels miles away. Locally, sinkholes cause a depression in the water levels of the surficial aquifer. Therefore, when and how to represent sinkholes is an important consideration when modeling ground water flow in sinkhole terrain.

Two methods of representing the Floridan sinkholes in Central Florida were investigated with a three-dimensional finite element model recently developed by Camp Dresser & McKee Inc. Both methods proved successful in fulfilling their objectives. The first method was through the use of a one-dimensional element linking the confined Floridan Aquifer with the unconfined surficial aquifer at the center of each sinkhole. This method was used in a regional model where only the water levels in the Floridan Aquifer were of concern. The one-dimensional elements superpose the normal three-dimensional element structure of the finite element grid system thus providing conduits for ground water flow. Flow through each one-dimensional element is governed by the conductance and storage properties assigned to the element and the head difference between the two nodes defining the element. This method allowed for easy representation of over 70 sinkholes in the regional model study area.

The second method of sinkhole representation was through the normal definition of three-dimensional elements outlining each sinkhole, except that a "pond" scheme was added to those elements to take into account the surface water aspects of the sinkhole lakes. This method was used in a site specific model where surficial water levels and lake elevations were of concern. The pond computations are analogous to standard reservoir routing procedures where a mass balance is performed on the pond at the end of each time step. On the basis of the change in volume and the stage-area-volume relationship for the pond, a new pond stage is computed. This stage is then assigned as a specified head boundary condition to all nodes connected to the pond. This method of sinkhole representation allowed for a detailed analysis of surficial water level fluctuations around the sinkhole lakes.

Introduction

Sinkholes can have a significant impact on the ground water flow characteristics of a confined/unconfined aquifer system. This is particularly true in Central Florida where large sinkhole lakes are almost as common as orange groves. Most of the sinkholes in Central Florida provide a mechanism for large quantities of water to flow between the confined and unconfined aquifers. When and how to represent these sinkholes is, therefore, an important consideration when attempting to model the ground water flow system in this area. The purpose of this paper is to discuss the two methods used in a recent investigation conducted by Camp Dresser & McKee Inc.

Hydrogeology

Ground water flow in Central Florida occurs primarily in two aquifers: the surficial (water table) aquifer and the Floridan (artesian) Aquifer. Separating the two aquifers is a relatively impermeable sandy clay to clay confining unit known as the Hawthorn Formation.

The surficial aquifer consists mainly of quartz sand with varying amounts of clay and shell. The base of the surficial aquifer is considered to be approximately 15 meters (50 feet) below land surface, on the average, throughout most of Central Florida. Ground water flow in the surficial aquifer is generally toward local points of discharge such as lakes, streams, and sinkholes. Flow is also vertical, as water leaks through the underlying confining unit. Most natural recharge to the surficial aquifer comes from rain. The mean annual precipitation in Central Florida is approximately 127 centimeters (50 inches) per year.

In general, the Floridan aquifer consists of alternating layers of limestone and dolomite or dolomitic limestone, containing solution-enlarged fractures and bedding planes that commonly yield large quantities of water to wells. The actual thickness of the Floridan aquifer is not known, but is estimated to be approximately 600 meters (2,000 feet). Most of the recharge to the Floridan Aquifer in Central Florida is from surficial aquifer leakage through the confining unit, and through local sections where sinkhole collapse has ruptured the confining unit.

Sinkholes

Sinkholes are a natural and common geologic feature in areas such as Central Florida which are underlain by carbonate rocks (i.e., limestone and dolomite). Rainfall combines with carbon dioxide from the atmosphere and from decaying vegetation to form weak carbonic acid. As this water percolates into and through the carbonate rocks, dissolution takes place and cavities are gradually formed. When dissolution weakens the roof of a cavern to the extent that it can no longer support the overburden, the overlying layers fall into the cavity and a sinkhole forms at the surface. This type of sinkhole, where the confining layer separating the water table and artesian aquifers is breached suddenly, is commonly referred to as a collapse sinkhole. The characteristics of a collapse sinkhole and its effect on ground water movement are shown schematically in **Figure 1**. Most of Central Florida's natural lakes, ponds, and closed depressions were formed in this manner.

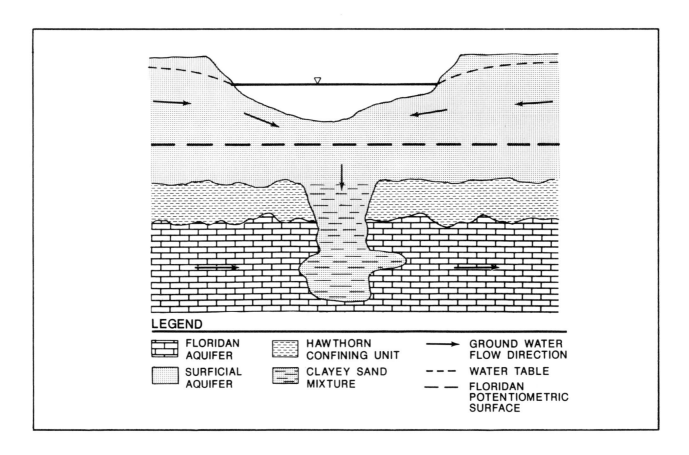

Figure 1 - Schematic Cross-Sectional Diagram of a Collapse Sinkhole

In the long term, geologic processes, such as the natural dissolution of limestone and dolomite, set the stage for the occurrence of sinkhole collapse. These geologic processes, however, usually have little to do with the final cause of collapse. In the short term, sinkhole collapse may be triggered through other natural processes, it may be induced, or it may be caused by a combination thereof. Induced sinkholes are those related to man's activities, whereas natural sinkholes are not. Generally, induced sinkholes develop in a shorter time span than natural sinkholes. The development and character of both, however, are dependent on the extent of carbonate rock dissolution.

All natural causes of sinkhole collapse can be traced back to extremes in precipitation. A lack of rainfall, for instance, results in lowered ground water levels causing a loss of buoyant support of the unconsolidated materials overlying the cavities in the limestone. This leads to general soil stress which can cause the materials that bridge the cavity to fail and sinkholes to appear. An abundance of rainfall, on the other hand, can accelerate vertical seepage, thus increasing piping activity through the carbonate rocks as well as the unconsolidated materials. This piping activity can also cause the materials bridging the limestone cavity to fail and sinkholes to appear.

Induced sinkholes may be divided into two general types: those resulting from a decline in water levels due to ground water withdrawals and those resulting from increased loading due to construction. Large withdrawals of water for water supply, irrigation, or frost protection can lower ground water levels enough to provide a triggering mechanism for sinkhole collapse in the same manner as a lack of rainfall. Increased loading due to construction may overstress the unconsolidated materials overlying the limestone cavities enough to also cause the materials bridging the cavities to fail and sinkholes to form. The most common construction activity found to trigger sinkhole collapse is the ponding of water in reservoirs and artificial lakes. Not only does this impoundment of water increase the soil loading, it provides a continuous source of percolating water which increases piping activity that can trigger sinkhole collapse.

As shown in Figure 1, the natural ground water flow regime is altered significantly by a collapse sinkhole. The two-dimensional flow in both the surficial and Floridan aquifers becomes three-dimensional as the two aquifers become interconnected. The magnitude of ground water flow between the two aquifers depends on both the size of the sinkhole, particularly at its base, and the hydraulic conductivity of the clayey sand mixture remaining after sinkhole collapse.

Locally, collapse sinkholes have the greatest impact on the surficial aquifer. Water levels in the surficial aquifer are generally significantly depressed around collapse sinkholes. The local effect on Floridan Aquifer water levels, however, is minor due to the high transmissivity of this aquifer. The regional effect of several collapse sinkholes on the Floridan aquifer, however, is significant as the cumulative recharge to the Floridan Aquifer through several sinkholes can increase water levels miles away.

Two methods of representing the sinkholes in Central Florida were investigated with a three-dimensional finite element model recently developed by Camp Dresser & McKee Inc. The first method was through the use of a one-dimensional element linking the confined Floridan Aquifer with the unconfined surficial aquifer at the center of each sinkhole. This method was used in a regional model where only the water levels in the Floridan Aquifer were of concern. The second method was through the normal definition of three-dimensional elements outlining each sinkhole, except that a "pond" scheme was added to those elements to take into account the surface water aspects of the sinkhole lakes which formed as a result of sinkhole collapse. This method was used in a site specific model where surficial water levels and lake elevations were of concern. Both these methods proved successful in fulfilling their objectives.

Regional Models
    The goal of most regional models is to simulate ground water flow over a large area without considering the local variations of flow due to nonuniformities in the geologic matrix. Only general flow directions and quantities are of concern. This is particularly true of regional ground water models for Central Florida which concentrate on simulating the flow in the Floridan Aquifer. In order to develop a flow model which provides a reasonable representation of the Floridan Aquifer system, however, all major sources of recharge must be incorporated. Therefore, although the sinkholes in Central Florida may be considered nonuniformities, they must be incorporated in the model since they provide a mechanism for major aquifer recharge.

A major problem in CDM's investigation was how to incorporate into the regional model over 70 sinkholes which are relatively small and are spread out over a large area. The regional model is a three-dimensional finite element model and therefore can handle the three-dimensional flow created by the sinkholes. The problem was that the elements were too large to represent the smaller sinkholes adequately with the normal definition of three-dimensional element properties, and to decrease the size of the elements would create an undesirable computational and data management burden. The solution was the use of one-dimensional elements linking the surficial and Floridan aquifers at the center of each sinkhole.

The one-dimensional elements superpose the normal three-dimensional element structure of the finite element grid system thus providing conduits for ground water flow. This superposition is shown schematically in **Figure 2**. Essentially, from a regional perspective, flow from the surficial aquifer to the Floridan Aquifer through sinkholes is one-dimensional in the vertical direction, and thus physically, the one-dimensional elements provide accurate representation of the sinkholes. From a regional perspective, the sinkholes are actually conduits of flow.

Flow through each one-dimensional element is governed by the conductance and storage properties assigned to the element and the head gradient between the two nodes defining the element. The one-dimensional element properties are slightly different from the normal three-dimensional element properties. Conductance is defined as the hydraulic conductivity times the area of flow. The storage properties include compressibility for confined flow and element yield for unconfined flow. Compressibility is defined as specific storativity times the area of flow, while element yield is defined as specific yield times the area of flow.

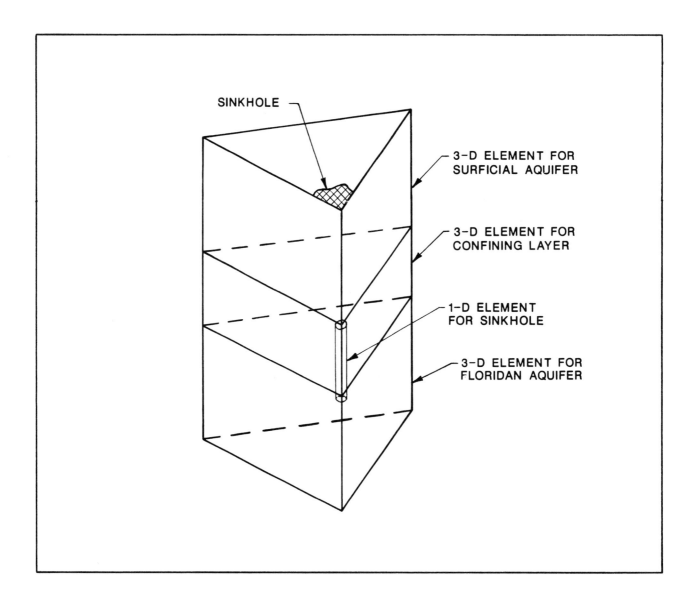

Figure 2 - One-Dimensional Element Superposition for Sinkhole Representation

Site Specific Models

The goal of most site specific models is to simulate ground water flow within a local area as accurately as possible. All nonuniformities are accounted for to the best extent possible and all local variations in flow are considered. For most site specific ground water models for areas in Central Florida, the goal is to represent flow both in and between the surficial and Floridan aquifers as accurately as possible. Sinkholes must therefore be represented in more detail than with the one-dimensional elements described previously. The sinkholes must be represented through the normal definition of three-dimensional elements outlining each sinkhole. Because most of the sinkholes in Central Florida have lakes above them, some routine is also needed to represent the interaction between surface water and ground water. In CDM's investigation, a "pond" scheme was added to those elements outlining each sinkhole to take into account this interaction.

The pond element computations are analagous to standard reservoir routing procedures. A mass balance is performed on the pond at the end of each time step in the simulation as follows:

$$V = [Q_{in} + Q_{out} + Q_{rain} + Q_{evap} + Q_{seep} + Q_{ret}] \ t$$

where

V	=	Change in Volume
Q_{in}	=	Specified inflow (average for time step) (positive)
Q_{out}	=	Specified outflow (average for time step) (negative)
Q_{rain}	=	Free surface rainfall accretion and watershed runoff over time step (positive)
Q_{evap}	=	Free surface evaporation over time step (negative)
Q_{seep}	=	Seepage loss (or gain) from nodes in ground water system connected to pond, computed at end of time step
Q_{ret}	=	Flows returning to pond from any boundary node in the system, computed at end of time step
t	=	Time step

On the basis of the change in volume and the stage-area-volume relationship determined for the pond, a new pond stage is computed. This stage is then assigned as a specified head boundary condition to all nodes connected to the pond. The model then continues on to the next time step and calculates new ground water levels in the aquifer system with this boundary condition imposed.

The addition of the "pond" scheme to the three-dimensional finite element model provides a method for performing a detailed and accurate analysis of surficial aquifer water level fluctuations around the sinkhole lakes, which could not be performed with the regional model. In addition, the fluctuation of water in the lakes themselves can be investigated with this "pond" scheme. These detailed analyses are usually key to the success of a site specific model investigation, particularly in Central Florida.

Summary

Due to the significant impact sinkholes can have on ground water flow in a confined/unconfined aquifer system, when and how to represent sinkholes in ground water models is an important consideration. Two methods of representing sinkholes in Central Florida were investigated with a three-dimensional finite element model. The first method was through the use of a one-dimensional element linking the confined aquifer with the unconfined aquifer. This method allowed for easy and adequate representation of over 70 sinkholes in a regional model developed to evaluate ground water flow in the confined aquifer. The second method was through the normal definition of three-dimensional elements outlining each sinkhole, except that a "pond" scheme was added to those elements to take into account the surface water aspects of the sinkhole lakes. This method allowed for a detailed analysis of surficial aquifer and lake water level fluctuations around the sinkholes in a site specific model.

References

Camp Dresser & McKee Inc., 1984, "DYNFLOW - A Three-Dimensional Finite Element Ground Water Flow Model, Description and User's Manual," unpublished.

Sinclair, W.C., Stewart J.W., Knutilla, R.L., Gilboy, A.E., Miller, R.L., 1985. Types, Features, and Occurrences of Sinkholes in the Karst of West-Central Florida. USGS Water Resources Investigation, 85-4126.

Sinkholes: Their Geology, Engineering, and Environmental Impact. 1984. Proceedings of the First Multidisciplinary Conference on Sinkholes, Orlando, Florida: B.F. Beck (Ed.)

Geological and geophysical investigations for delineating karstic structures in southwestern portion of Cuddapah Basin, Andhra Pradesh, India

B.VENKATANARAYANA & T.VENKATESWARA RAO   *NGRI, Hyderabad, India*

ABSTRACT

A large number of karstic structures and cavernous zones occurring in calcareous formations of lower Cuddapahs and Kurnools of Proterozic period are reported from the southwestern portion of Cuddapah basin, Andhra Pradesh, India. These structures are abundantly found around Tadpatri (Anantapur district) and Betemcherla (Kurnool district) towns and are aligned in NNE-SSW direction. This region is infested by a large number of ENE-WSW, NNE-SSW and NW-SE lineaments as well as intersected by NE-SW and NW-SE trending major faults. Geological, structural and geomorphological set-up suggest the presence of many more such cavernous zones in this region.

Geological (structural, geomorphological) and geophysical (geoelectric resistivity sounding, profiling, total magnetic intensity and shallow seismic refraction) studies were carried out around Chillavaripalle, Kona (Anantapur district), and Kuntalapadu (Kurnool district) to identify and delineate a few more new karstic structures in this region.

Geological studies revealed the presence of a few shear and fault zones in this area. Geomorphological observations have been useful to identify a few horizontal cavernous zones occurring at different levels above the present river beds indicating the paleo-drainage system. Some new karstic zones lying at shallow depths, viz. around 25 m, have been identified on the basis of the detailed geophysical investigations.

The most important finding of these studies is the close correlation of the levels of cavernous zones with planation surfaces. Similar observations were made in Japan, UK, USA, Canada and Africa. Such correlation appears to be a global phenomenon. It appears that the eustatic movements which occurred during the Quaternary period might have played a significant role in causing karstification of carbonate rocks of this area.

Introduction

Karstic structures, developed through leaching and dissolution of limestones, dolomite and other soluble rocks along lineaments, shear zones, etc., play an important role in the field of geohydrology. Intense karstification creates secondary porosity and enhances the permeability which control both movement and storage of groundwater.

A large number of karstic structures, sinkholes and cavernous zones are extensively developed in calcareous deposits of India in general and more particularly in Kurnool district, Andhra Pradesh. The longest and deepest caves, so far known in India, namely Belum caves (Lat. 15 06'N; Long. 78 07'E) with a total extent of 3,500 m and Borra caves (Lat. 18 10'N; Long. 83 00'E) extending up to a depth of 83 m, are occurring in Andhra Pradesh. These karstic structures are abundantly found around Bethemcherla, Banaganapalle and Tadpatri towns aligned in NNE-SSW directions (Fig. 1). This region is infested by a large number of ENE-WSW, NNE-SSW and NW-SE lineaments and intersected by NE-SW and NW-SE faults. The morphological, tectonic, and geological set up suggests the existence of many more cavernous zones in this region and there is a need to identify and explore them in

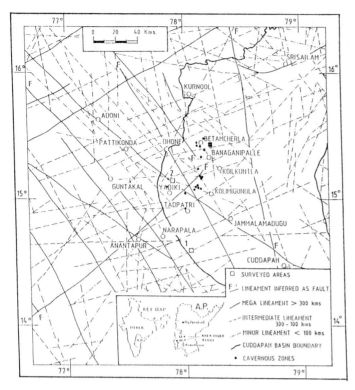

Fig. 1: Lineament map of surveyed areas and neighbouring areas

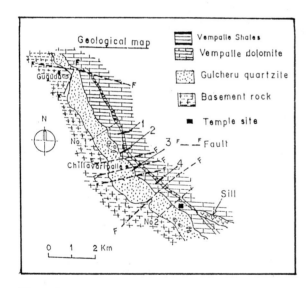

Fig. 2: Geological map of Chillavaripalle

this drought prone region of India to mitigate problems of water scarcity. Two regions, namely Chillarvaripalle (Lat. 14 38'N; Long. 77 50'E) and Kona (Lat. 15 10'N; Long. 77 15'E) (shown as 1 and 2 in Fig. 1, respectively) exhibiting favourable lithology and tectonic set up and occurring in the southwestern portion of the Cuddapah basin of Proterozoic period are selected for detailed geological and geophysical investigations. These investigations involving geomorphological, lithological, structural, hydrogeological, electrical resistivity soundings and profilings, total magnetic and shallow seismic refraction studies, led to the identification of new karstic zones having good groundwater resources. Bore wells drilled at the recommended sites are found to be yielding a stupendous amount of water.

Geology of the Areas Under Investigations

Chillavaripalle and Kona lie in Anantapur district, Andhra Pradesh. The former is situated about 40 km SE of Anantapur town, while the latter is about 15 km NNE of Yadiki. Tectonic features of the area (Fig.1) reveal that the area is infested by a large number of lineaments and faults. Majority of the mega lineaments are broadly trending in NW-SE direction. One of these is found to be passing through Adoni-Pattikonda-Kona-Jammalamadugu-Cuddapah towns and beyond up to the eastern coast of India. Another NW lineament, passes through Anantapur and Chillavaripalle and intersects the earlier lineament around Jammalamadugu. These mega lineaments are interesected by a set of faults trending in SW-NE, and NW-SE and also intersect each other around Banaganapalle. In addition, a large number of intermediate and minor lineaments are criss-crossing the area. It is evident from this, that the surveyed areas and region around Banaganapalle are tectonically affected and became viable for karstification.

In Chillavaripalle, rock types belonging to the Papagni series of the lower Cuddapah group are encountered (Fig. 2). The Gulcheru quartzites, the oldest formation of the Cuddapah group, unconformably overlie the archaean crystalline basement. It consists of conglomerates, quartzitic sandstone and quartzites both ferruginous and siliceous varieties, with intercalations of thin shales. Vempalle formations, conformably overlying

60

the Gulcherus consists of dolomitic limestones and is well developed all along the western margin from Vemula near Pulivendla in the south to Dhone in the north. These formations, near Chillavaripalle, strike in the NW direction and gently dip in the NE direction. However, the formations are steeply dipping in the valley near Katamiah temple, 3 km south of Chillavaripalle. A major dip fault in which the NW block shows a downward displacement of about 50m is observed. A few minor dip faults are also encountered in this area (Shibasaki et al., 1985).

In the Kona region, the Kurnool group (Banaganapalle sandstone, Narji limestone, Auk shales and Paniam quartzites) unconformably overlie the Tadpatri series of Cuddapah group forming large flat topped hills. Narji series is represented by both reddish and grey limestone of considerable thickness. Auk shale is exposed along the scarps of flat topped hills. Paniam quartzites occur as a capping on the higher hills. Along valley portions secondary deposition of quaternary carbonate material is observed. A fault trending in N 60 W is suspected along the river course with 60 m vertical displacement.

Karst Topography

Topography of the Chillavaripalle area is characterized by the presence of ridges and cuestas. The ridges, strike in NW-SE and the cuestas slopes gently towards NE. The valley portion, where geophysical investigations were carried out, is formed by gentle sloping (Gulcheru quartzites) in SW and steep cliffs (Vempalle dolomites limestone) in NE. Three topographic flat surfaces which appear to be remanants of river erosional terraces are observed in the quartzitic slopes. On the other side a cavernous system occurs in the dolomitic cliffs. The topographic terraces A, B, and C (Fig.3) are formed in three levels viz. 75-80m, 41-44m, and 17-18m respectively above the present river bed. Correspondingly three cavernous levels A, B and C occur in the dolomitic cliff. Among them, level B is quite pronounced, while A is collapsed and C is relatively less prominent. The cavernous feature at level B is controlled by the three main vertical joint systems and cuts the dolomites obliquely.

The Kona valley exhibits a steep slope up to 90 to 100 m above the present river bed extending in NW-SE. Three upland erosional surfaces are situated at 92m, 85m and 80m above the present river bed (Fig.4) Correspondingly, cavernous zones are formed in the steep slopes of level III. These caverns A, B, C and D are distributed located at 92-75m, 40m, 15-10m, and 7m above the present river level respectively. Cavern A consists of three sub levels at heights of 92, 87 and 76m. Cavern B is located in the Central Portion of the limestone cliff. 5 m thick sediments bearing plant fossils are found in cavern B. Level C, around temple, covers a small area and D shows two sub levels at 11 and 7 m height. The latter corresponds with the present river terraces.

Topographic analyses of these two surveyed areas suggest the occurrence of 3 to 4 peneplain levels in the quartzites with corresponding cavernous zones in the carbonate terrain. It is believed that this could be a regional phenomenon and this can be found elsewhere in other parts of the Cuddapah basin.

Geophysical Investigations

Four geoelectrical traverses (shown in Fig.2) of 1 km apart with a sounding station interval of 50m were carried out for selecting a suitable location for conducting detailed geophysical surveys. Even though resistivity lows are observed in all the traverses, they are more prominently developed in 3 and 4 traverses and are found to be extending up to a depth of 30m. Geohydrological features like occurence of springs in the neighbouring uplands and subsurface water flows further suggest feasible conditions for karstification in this part of the carbonate terrain. Detailed geophysical surveys were carried out to test the viability of the above inferences.

Fifty-two geoelectrical profiling traverses (confined within the traverses of 3 and 4) were laid with a traverse interval of 20 m and a station interval of 10 m. The Wenner configuration with an electrode separation of 10 to 50 m in steps of 5 m was deployed in the present studies. Apparent resistivity contour map for each separation was prepared and one of such maps for a=20 m is shown in Fig. 5. A 100 ohm-m resistivity contour line starting at station No. 8 is found dividing the surveyed area into two halves, viz., a high resistivity zone (I) in the NE part and a low resistivity zone (II) in the SW portion. Zone I is the reflection of a barrier like feature which might have been formed due to dip faulting. Such barrier like features are also inferred by geohydrological observations near the Katamiah temple. This high resistivity zone is further segmented into 'A' ranging from 100 to 700 ohm-m and 'B' from 300 - 2,000 ohm-m, separated by a relatively low resistivity zone (200 ohm-m) along the traverse 12. This demarcation of A and B suggests a localized lithological variation in the dolomitic limestone. A dip fault observed along traverse 12 is understood to be responsible for causing the lithological variations. Zone II, extending up to 20,000 sq m, might have been formed due to intense weathering or intense karstification and filled up with some low resistivity material. The nearest quartzite

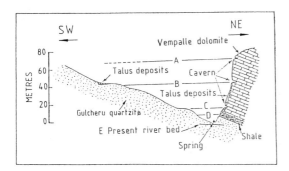

Fig. 3: Schematic topographic section at Achillavaripalle area

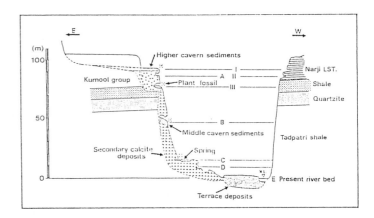

Fig. 4: Schematic topographic section at Kona area

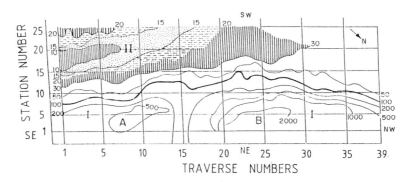

Fig. 5: Apparent resitivity map of Chillavaripalle (a=20m)

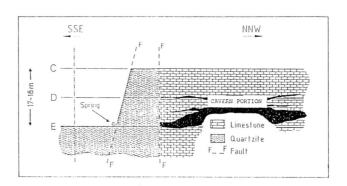

Fig. 6: Geological section along the valley portion showing two karstic zones at a depth of 17-18m

outcrop is about 400 m away and the thickness of the limestone formation at this site is calculated to be about 50m. The dip fault coinciding with traverse 12 might have acted as a good aquifer and the rainwater infiltered along the bedding planes, joints and other structural discontinuities might have intensified the karstification process.

Magnetic Survey

Total magnetic intensity data was collected in the grid pattern, along the same traverses laid for electrical resistivity profiling, with a station interval of 5m in an area of 250x1,020 m. The total magnetic intensity map exhibits two distinguishable magnetic anomalous zones viz., low and higher magnetic intensity zones. The magnetic high elongated zone extending linearly in NE-SW more or less coincides with that of the electrical low resistivity zone. This feature may either be attributed to a long and narrow trough-like structure filled with ferruginous material or a subsurface basic intrusive body. Occurrence of the latter is of remote possibility as geological evidences were not found. Computer modeling of the magnetic intensity values suggest the presence of a linear trough filled with magnetic materials at a depth of 20 m dipping 70 NE. The presence of a low magenetic zone particularly around traverse 16, 24, and 32 in the remaining portion encircling a magnetic high may be assumed as a magnetically depleted zone. These linear bands suggest a possible fracture zone. The magnetic material in the fractures might have been leached, leading to the transformation of massive carbonates into a good aquifer. The magnetic zone at traverse 24 between F2 and F3 is found to be a good groundwater potential zone which was evident from the performance of existing bore wells.

Seismic Investigations

Four seismic profiles were laid, deploying a 12 channel portable digital seismic unit ES 1210, keeping the geophone interval at 5m. The lengths of the profiles laid were up to 280m, 220m and 330m along traverses 24, 2 and perpendicular to them respectively. The fourth profile for a length of 220m was laid at about 220m away from the 2nd profile towards SE. Time-distance plots for different profiles were prepared. Velocities and thicknesses for different layers were calculated using intercept time method and also by critical distance method. The results indicate a three layered structure in which the average seismic velocities range from 0.3 to 1.10 km/sec for highly weathered or soil; 1.7 to 2.2 km/sec for fractured rock and 4.0 km/sec or above for less fractured rock. Sometimes, the third layer is not discernible on the time distance graphs due to attenuation of the elastic energy. The average thickness of the fractured or karstified zone is found to be about 26 m. The seismic traverse at S-2 depicts a highly weathered zone and the absence of the hard rock corroborates the presence of a highly distributed weathered zone. Similar observations can also be made for traverse 3. The low velocities may be attributed to the water saturation in the second layer.

A schematic geological depth profile of the valley portion based on geohydrological and detailed geophysical observations, was drawn along the NNW-SSE direction and is shown in Fig.6. It depicts two karstic zones at a depth of 18m coinciding with the cavernous levels of C and D. Groundwater from these two karstic zones appear to be flowing through the semipermeable zone in the Gulcheru quartzites and emerges out as a spring near the temple.

Discussion and Conclusions

In the Chillavaripalle and Kona regions, the river terraces and planation surfaces can be correlated with the levels of cavernous zones. These are found to be interrelated. Hall, in Sierra Leone, has identified five different river terraces ranging from 0-5, 5-10, 10-40, 50-80 and 80-100 ft. The last level is more or less very widely observed in many parts of the world and this is believed to have been formed during the pleistocene period. It is now generally agreed that there were eustatic movements during the geological past. During late Tertiary period the sea level was about 400 ft (135 m) higher than the present level. These large scale sea level variations may be responsible for karstification.

Even in the close vicinity of the surveyed areas, there were evidences that the Penna river which was flowing due north from its origin as a tributary of Hageri river has taken a sudden perpendicular turn towards the east due to headward erosion (Vaidyanadhan, 1962) of Chitravati river. This deviation and additional water of the Penna raised the water table around Belum Betemcherla and surveyed areas and became a cause of karstification. Decreasing of the water table due to active erosion of the Penna river bed might have caused different levels of karstification, particularly around Belum.

In conformity with the global picture, the cavernous zones of the Cuddapah basin could be classified into five levels. Level A is the highest cavern formation at 80 to 100 m above the present river bed. Level B is the middle cavern formation at 40 to 50m. Levels C, D and E are the lower cavern formations occurring at heights between 0 to 20m above the present river bed. Permanent springs are noticed at the same height of the lower cavern formation. The heights of the cavern entrances of Taishaku karst plateau, West-Japan could also be divided into various levels similar to those of Cuddapah karst terrains (Kitabingao plateau Research Group, 1969). The horizontal arrangements of caverns which have different heights were found along Tojo river. Similar observations were also made in England (Sweeting, 1950) and in North America (Palmer, 1984). This vertical arrangement of cavernous formations especially their relative heights above the present river bed is of vital significance as outlined below.

The large number of caverns which are being formed at present are expected to be near the water table or close to the natural discharging points along the river. Conversely speaking, the cavernous formations distributed at the higher levels above the present water table indicates the position of older water tables of geologic past. The cavernous formation could be regarded as a kind of "fossil" or paleosystem of hydrological flow regime (Shibasaki, 1985). These facts imply that the vertical distribution of the cavernous formation has resulted from regular geologic processes during the Quaternary period.

The reasons for the coincidence of the phenomenon of coplanar arrangement of caverns and river terraces in different parts of the world are not yet clear. From the field evidences, it may be inferred that these caverns were formed essentially due to eustatic movements since Middle Pleistocene. A detailed geochronological study could perhaps help our understanding of these complex processes.

Acknowledgement
    The authors express their deep sense of gratitude to Prof. V.K. Gaur, Director Dr. C.P. Gupta, Deputy Director, NGRI, Hyderabad for providing all facilities and encouragement. The help received from Mr. Prabhakara Prasad, Mr. S. Sankaran and members of Indo-Japanese team is gratefully acknowledged. Sincere help received from Mr. G.R. Babu and Mr. P.T. Varghese and Mr. M. Jayarama Rao is acknowledged.

References

Hall, P.K. The diamond fields of Sierra Leone, Published by GS of Sierra Leone, pp. 1-133.

Kitabingo Plateau Research Group, 1969. On the age of the formation of limestone cave: An example, Jour. Geol. Soc., Japan, 75(5), pp 281-287 (in Japanese with English abstract)

Palmer, A.N., 1984. Geomorphic interpretation of Karst features, 173-209, In Lafleur, R.G., ed., Groundwater as a Geomorphic Allen and Unwin, Boston.

Shibasaki, T., Balakrishna, S., Yoshimura, T., Venkatanarayana, B., Furukawa, H., Venkateswara Rao, T., Chuman, N., Ramamohan Rao, Y. and Venkatewarlu, K., 1985. Karstification process of carbonate rocks in the Cuddapah sedimentary basin, Indian Peninsular Shield, J. Fac. Mar. Sci. Tech. Tokai Univ., no.21, p. 31.45.

Sweeting, M.M., 1950. Erosion Cycle and limestone caverns in Ingleborough district, Geogr. Jour., p. 115, 63-78.

Vaidyanadhan, R. 1962. Effect of uplift and structdure on drainage in the southern parts of the Cuddapah basin. Jou. Geol. Soc. India, V.3, p.70-85

Partial reclassification of first-magnitude springs in Florida

WILLIAM L. WILSON & WESLEY C. SKILES   *Karst Environmental Services, Inc., High Springs, Fla., USA*

ABSTRACT

The most widely cited list of first-magnitude springs in Florida contains errors, omissions and inconsistencies that warrant extensive revision. At this time, only a partial revision is possible because of incomplete knowledge of spring discharge, especially for submarine and subriver springs that have no individual surface channel. In order to account for the complexity of springs, four separate lists are used for various types of features. The lists are for 1) single vent springs and groups of hydrologically-related springs that represent the total outflow of clear water from a ground-water drainage basin, 2) river rises, which usually represent the resurgences of nearby sinking streams (the rises have brown water that is more characteristic of surface streams), 3) a combined list of first magnitude springs regardless of the water source, and 4) karst windows, which are not springs, but constitute portals to significant underground streams. The proposed classification system eliminates double-counting of water resources which occurred frequently on the old list, and allows a more accurate compilation of total spring flow.

Four karst windows were removed from the list of first-magnitude springs. Two river rises were transferred from the list of first-magnitude springs and placed, along with five others, on a new list of first-magnitude river rises. Four hydrologically invalid spring groups were removed from the list of first-magnitude springs, whereas two newly identified springs were added. One single-vent spring was recommended for deletion from the list of first-magnitude springs.

As a result of these changes, the largest known spring, or spring group, in Florida, is Silver Springs (23.24 cms or 820 cfs), in Ocala, although the main spring at Spring Creek, south of Tallahassee, may be larger. Deletion of the karst windows and river rises, in order to avoid double-counting water resources, has resulted in a net reduction of the total spring flow in Florida by approximately 38.35 cms (1,371 cfs), or nearly 11 percent. At this time, 19 first-magnitude, clear-water springs, or spring groups, are known in Florida; 4 other springs may or may not have first-magnitude discharges; 7 river rises are first-magnitude, and 4 karst windows have first-magnitude flows. The karst windows should not be counted as ground-water resources. The river rises can be counted as ground-water resources only if their associated sinking streams are not counted as surface-water resources.

Introduction

The State of Florida is world famous for its numerous, beautiful, large karst springs. The most commonly cited list of first-magnitude springs in Florida is the one compiled by Rosenau and others (1977). They listed 27 springs, and groups of springs, each having an average discharge in excess of 2.83 cubic meters per second (cms) or 100 cubic feet per second (cfs). The list needs extensive revision because it contains errors, omissions and inconsistencies. Also, the nature of cave springs and other cavernous drainage features is sufficiently complex so that one list does not adequately cover the range of geomorphic features that are present in the karst of Florida. The purpose of this article is to revise the list of first-magnitude springs in Florida, as much as existing data allow, and to point out the drainage basins where knowledge of the spring discharge is, as yet, insufficient for proper classification.

Complications

The definition of individual springs and hydrologically-related spring groups as the mouth of ground-water drainage basins is complicated by the fact that 100 percent of the drainage may not discharge through the spring(s). Underwater cave exploration and dye tracing (Fisk and Exley, 1977 ; Ruppert and Wilson, 1989) have repeatedly revealed that many springs are merely tap-off passages from larger conduits. At Devil's Eye Spring Group, a newly recognized first-magnitude spring group, the authors found that 27 percent of the water in the phreatic cave does not emerge at the known spring outlets, but passes downstream through siphon tunnels in the cavernous aquifer. Thorough study of the cavernous drainage system, including direct exploration and quantitative dye tracing, is usually necessary to fully delineate the cavernous ground-water drainage basins.

Another problem with compiling a list of single vent springs is that discharges reported in the literature for some, supposedly individual, springs, actually represent spring groups, especially where two or more vents are located in the same pool and do not have separate surface stream channels. Where one or more small spring vents are associated with a very much larger, main spring vent, then it is simpler to accept the reported discharge as the approximate value for the main spring. In many cases this assumption has no effect on the ranking of the main spring. The authors have attempted to distinguish between multiple spring vents in a single pool , as much as possible.

The lack of discharge measurements for large vents in the same pool, or for submarine springs in the same embayment, poses a major difficulty for consistent spring discharge ranking. Greater effort needs to be made at measuring individual spring flow by subaqueous measurement techniques, rather than relying entirely on surface stream gauging techniques, which cannot provide individual spring discharges in many cases. Pipe-full discharge measurement techniques, utilizing scuba diving, need to be implemented on a routine basis in order to provide complete and reliable spring discharge data.

River Rises

In contrast to the famous, clear-water, karst springs of Florida, a number of springs are classified as "river rises" because they discharge brown water with a tannin content that compares closely with that of nearby surface streams. The river rises are the resurgences of sinking streams or, more commonly, major surface streams that sink partially. They are herein defined as springs which discharge water composed at least 50 percent, by volume, of water derived from sinking streams. In Florida, the underground course between the sinking stream, and its associated rise, is usually rather short; less than three miles in the cases examined to date, so the quality of the water is little affected by passage through the underground course.

River rises are identified separately in order to avoid double-counting surface-stream flow as a ground-water resource. However, the discharge of the river rises is augmented, to some extent, by ground water. Usually, the proportion of ground water in the discharge is less than 25 percent. Very little data is available on the input and output of the sinking stream/river rise systems, so the amount of ground water discharged through these springs is poorly known, but may be substantial. It is imperative that input/output studies be conducted at river rises and their associated sinking streams in order to fully assess Florida's ground-water resource.

Rosenau and others (1977) included some river rises in their list of first-magnitude springs, but ignored other, very large rises. The inconsistent treatment of river rises has led to mis-statements about the total amount of ground-water resources available from the Floridan Aquifer. River rises will be treated as a separate class of spring in this paper, and, for the most part, will not be counted as part of the total ground-water resource.

Karst Windows

Karst windows are a type of sinkhole in which an underground stream is temporarily exposed to the surface because of cave-roof collapse (von Osinski, 1935). Although, the cave stream is at the surface for a short distance, the fact that it sinks within the same topographic depression in which it resurges, indicates that the stream has not returned to flow in surface channels in a meaningful way. An underground stream should not be called a spring just because it can be seen through a hole in the cave roof. To do so leads to the illogical situation of counting the same stream of water more than once when compiling a list of springs. However, some karst windows are significant features, in that they expose cave streams of first-magnitude rank, so they will be entered on a separate list of first-magnitude karst windows, rather than combining them with springs.

Deletion of Karst Windows from the List of First-Magnitude Springs
Four of the "springs" presented on the list of first-magnitude springs by Rosenau and others (1977) are karst windows, rather than true springs. Kini "Spring," River Sink, Natural Bridge "Spring" and Falmouth "Spring" are transferred from the list of first-magnitude springs to the list of first-magnitude karst windows. The discharges are deleted from the ground-water resources of Florida, because the flow is accounted for at other springs or streams.

Kini "Spring" (4.99 cms or 176 cfs) and River Sink (4.65 cms or 164 cfs) are almost certainly upstream segments of flow that resurges at Wakulla Spring, although no dye trace has yet been performed to confirm this hypothesis. Natural Bridge "Spring" (3.00 cms or 106 cfs) is a brown-water flow system that very probably contributes flow to St. mark's River Rise. Falmouth "Spring" (4.48 cms or 158 cfs) is an entrance to the Falmouth-Cathedral Cave System, which extends northwest-southeast as a single major trunk passage for over 3,000 m (10,000 feet) (Exley, 1989). Water in the cave system flows northwest toward the Suwannee River located approximately 6.4 km (4 miles) from Falmouth "Spring". A major portion of the flow may resurge along the south bank of the Suwannee River at a little-known spring called Line-Eater Spring, which may be first-magnitude. Line-Eater Spring contains 2.4 km (1.5 miles) of surveyed passage extending southeast toward Falmouth "Spring." The flow through the cave is so strong that even underwater scooters cannot advance in some places.

Deletion of Invalid Spring Groups
Four of the spring groups, on the list of first-magnitude springs prepared by Rosenau and others (1977), discharge water of various sources and quality. The lack of valid hydrologic relations is reason to delete Ichetucknee Springs, Crystal River Springs, Spring Creek Springs and Wacissa Springs from the list of first-magnitude springs. The discharge from these spring groups remains as part of the ground-water resources of Florida, but the ranking of spring discharge should be changed.

Ichetucknee Spring Group (25.0 cms or 361 cfs) consists of at least nine individual springs that occur along a 3.7 km (2.3 mile) long segment of the Ichetucknee River, on the border between Suwannee and Columbia counties. None of the springs are known to be connected by caves and they probably convey flow from different sides of the river. Jug Hole (also known as Blue Hole) may be a marginal first-magnitude spring, by itself, but individual discharge measurements are not available.

Crystal River Spring Group (25.0 cms or 916 cfs) consists of at least 23 individual springs located in a two mile square area of Crystal Bay along the central-west coast of Florida, in Citrus County. All of the springs are submarine and no direct measurements of individual spring discharges have been made. Discharge for the springs comes from a wide area north, east and south of the bay, indicating that many of the springs are hydrologically unrelated. Tarpon Spring, in the central-east part of the bay, is the largest individual spring in the group, but it probably is not a first-magnitude spring.

Spring Creek Group (56.76 cms or 2,003 cfs) consists of at least 14 submarine springs in two separate branches of a tidal creek, at Oyster Bay, along the coast south of Tallahassee. The springs are discharging water with chloride contents ranging from less than 250 mg/L to 1,200 mg/L (Rosenau and others, 1977). Because of the chemical differences, and the lack of known cave connections, these springs cannot, at this time, be considered as a valid, hydrologically-related spring group. The main spring in Spring Creek (also known as Spring No. 1, or Spring Creek Rise) is definitely a first-magnitude spring. The vent at Spring Creek Spring is approximately 10 m (30 feet) in diameter and discharges water with such a high velocity that it is difficult for divers to hold onto the walls. Divers have descended through the vent to a depth of 55 m (180 feet), with no decrease in flow (Exley, 1989). Measurements performed by the USGS on May 30, 1974, indicate that the discharge of the main spring is less than 21.7 cms (764 cfs), but direct observations by divers indicate that the average discharge may exceed 28.3 cms (1,000 cfs), which would make it the largest spring in the United States. Additional study of this large spring group is greatly needed in order to understand the karst ground-water resources of the Woodville Karst Plain.

Wacissa Springs (11.0 cms or 389 cfs), in Jefferson County, northern Florida, consists of 12 springs, but the discharge reported by Rosenau and others (1977) is for a total of at least 18 springs. Many of the spring vents in the group have different quality water indicating that they represent different ground-water drainage basins. Big Spring commonly discharges green water; Cassidy, Little Blue and Minnow springs usually discharge blue water; whereas Log, Thomas and Spring No. 1 commonly discharge brown water. Big Spring may be a first-magnitude spring by itself, but no individual discharge measurements are available.

Newly Identified First-Magnitude Springs and Questionable Rankings

Devil's Eye Spring Group is a newly identified first-magnitude spring group (Wilson and Skiles, 1988) located along the Santa Fe River in section 34, T. 7 N., R.16 E. Three springs, Devil's Eye, Devil's Ear, and July Spring, are connected by cave passages and serve as distributary outlets for a single ground-water drainage basin that extends north and northeast from the springs. More than 10.9 km (6.8 miles) of underwater cave passages have been surveyed. July Spring occurs on the north bank of the Santa Fe River, in Columbia County. Devil's Ear Spring occurs in the floor of the Santa Fe River, on the south side of the river, and Devil's Eye Spring occurs in a spring run on the south side of the river in Gilchrist County. All of the springs discharge clear water into the river, which is dark brown because of tannin. The combined discharge of the spring group averages 8.0 cms (284 cfs). Devil's Ear, the largest single vent in the group, discharges an average of 3.5 cms (124 cfs) and is a first-magnitude spring by itself. Rosenau and others (1977) reported these springs, estimated that they had second-magnitude discharges, and knew that they were connected by caves, but did not recognize them as a first-magnitude spring group.

Croaker Hole is a spring in the floor of the St. Johns River, near the southwest corner of Little Lake George, approximately 0.3 km (0.2 miles) north of Norwalk Point, in Putnam County. At a depth of 12 m (40 feet) is a horizontal cave entrance, 3.7 m (12 feet) high and 7.6 m (25 feet) wide. A strong current of clear water issues from the cave. The authors estimate the discharge is approximately 10.2 cms (360 cfs). The passage inside the entrance is approximately 12 feet high and 15 feet wide. It extends approximately 100 feet to a collapse where boulders prevent further penetration.

Rosenau and others (1977) reported Hornsby Spring (4.6 cms or 163 cfs), along the Santa Fe River, in Alachua County, as a first-magnitude spring based on the average of two discharge measurements. The discharges were 2.2 and 7.08 cms (76 and 250 cfs). The authors have no additional measurements to present, but based on years of observation, are of the opinion that Hornsby Spring has an average discharge closer to 2.2 cms, and that the 7.08 cms discharge represents an unusually high flood flow. Hornsby Spring should be studied further to determine if it is a first-magnitude spring. The authors recommend deleting it from the list until additional supporting data become available.

River Rises

Springs that discharge brown water are the resurgences of sinking streams that usually have passed short distances underground, as described above. In order to avoid double-counting surface water as ground water, it seems advisable to place first-magnitude river rises on a separate list. Rosenau and others (1977) reported two first-magnitude springs, Alapaha Rise (17.2 cms or 608 cfs) and St. Mark's Spring (14.7 cms 519 cfs) that are, in fact, river rises. These two springs are transferred to the list of first-magnitude river rises. The discharge of these two rises is deleted from the ground-water resources of Florida. The proportion of ground water that augments the flow of the rises is completely unknown except at Santa Fe Rise where ground-water contributes 25 to 43 percent of the discharge based on two measurements (Skirvin, 1962).

Five known river rises of first-magnitude were not included on the list of first-magnitude springs by Rosenau and others (1977). Santa Fe River Rise, Suck Hole Rise, Siphon Creek Rise, Steinhatchee River Rise, and Chipola River Rise are added to the list of first-magnitude river rises. The first three rises occur along the Santa Fe River, in northcentral Florida. The Santa Fe River repeatedly sinks, or sinks partially, along its course and has significant surface and subsurface components of flow.

Santa Fe River Rise is located along the border of Columbia and Alachua County in northcentral Florida. The average discharge of the rise is at least 16.6 cms (586 cfs) (Hunn and Slack, 1983). The Santa Fe River sinks complete in O'Leno State Park, at a point 4.8 km (3 miles) northeast of the rise. Based on average flow, the Santa Fe River is probably the largest sinking stream in the United States. According to Skirvin (1962), when the volume of water entering the sink is between 5.67 and 16.2 cms (200 and 570 cfs), then the amount of groundwater augmentation at the rise is fairly constant: 4.25 to 5.38 cms (150 to 190 cfs), respectively. This implies that as the stage of the river rises, the potentiometric surface of the ground water rises in nearly direct proportion.

Suck Hole Rise is located several hundred yards upstream from the US Highway 27 Bridge across the Santa Fe River, northwest of the town of High Springs, in Alachua County. Approximately 4.5 cms (160 cfs) discharge from an underwater cave opening 1.2 m (4 feet high) and 6.1 m (20 feet) wide. Even during low water, the current is so strong that it is difficult to enter the cave. The water discharging from Suck Hole

Rise probably sinks at Suck Hole, located on the north side of the river approximately 1,040 m (3,400 feet) northeast from the rise.

Siphon Creek Rise is located along the south side of the Santa Fe River in north-central Gilchrist County. Approximately one-third of the flow in the Santa Fe River sinks at a group of three swallow holes called Big Awesome Suck, Little Awesome Suck and Track 1 Swallow Hole. The sinks are located on the north side of the river approximately 1 mile upstream from Siphon Creek Rise. The swallow holes and rise are connected by a humanly passable connection that is 2.4 km (7,800 feet) long. The cave system contains 6.4 km (21,000 feet) of surveyed passage. Most of the cave system is 27 to 37 m (90 to 120 feet) underwater. Several side passages contribute clear streams of ground water to the brown, sinking-stream water, but the proportion is unknown. Siphon Creek Rise is an underwater cave entrance 3 m (10 feet) high and 9 m (30 feet) wide discharging approximately 17 cms (600 cfs).

The Steinhatchee River sinks completely along the border between Dixie and Taylor counties, northeast of the small town of Tenille, approximately 16 km (10 miles) inland from the northwest coast of peninsular Florida. A USGS gauging station, 0.8 km)0.5 mile) upstream from the sink, has recorded an average flow of 9.3 cms (328 cfs). The discharge of the rise, located 1.2 km (0.75 mile) southwest of the sink, may be slightly higher, but no gauging records are available.

Chipola River sinks in Florida Caverns State Park, approximately 3.2 km (2 miles) north of Marianna in Jackson County, in the central part of the Florida Panhandle. The underground course of the stream is only about 0.4 km (0.25 mile) long, and the stream flows from north to south. The average discharge of the Chipola River Rise is not known, but very likely exceeds 2.83 cms (100 cfs), based on the authors' observations.

Summary and Conclusions

Based on the changes described above, the clear-water, first-magnitude springs of Florida are ranked by discharge in Table 1. River rises are defined and treated separately in order to avoid double-counting water resources. The first-magnitude river rises are shown in Table 2. The clear-water springs and the brown-water river rises are, in fact, all true springs, so a combined list is shown in Table 3, to illustrate the ranking of Florida's largest springs, regardless of the water source. Karst windows were formerly, erroneously, counted as springs, but this resulted in double-counting of water resources. First-magnitude karst windows are ranked, by the discharge of the exposed cave stream, in Table 4.

The most frequently quoted list of first-magnitude springs in Florida was prepared by Rosenau and others (1977). Six of the features on the list had to be deleted in order to avoid double-counting water resources. Four of the features are karst windows and the other two are river rises. Deletion of these features, from the list of springs, reduces the total discharge of springs in Florida by 49.05 cms (1,731 cfs) or 13.7 percent less than the total amount estimated by Rosenau and others (1977).

Addition of Croaker Hole, a previously unreported first-magnitude spring, increases the total spring flow by approximately 10.2 cms (360 cfs), so the net change in the total spring flow is 38.85 cms or 1,371 cfs (10.9 percent) less than that estimated by Rosenau and others (1977). The Devil's Eye Spring Group was known to Rosenau and others (1977), but was not recognized as a first-magnitude spring by them, so its discharge is not considered to be a net change in the total spring flow.

Individual spring discharge measurements are not available for many major springs in Florida, especially submarine springs such as those at Spring Creek and Crystal river. An expanded program of periodic spring measurement is needed to accurately and completely characterize the renewable ground-water resources in Florida. Underwater gauging techniques need to be developed and standardized for use at springs that discharge into bodies of water and have no individual surface stream channel.

The inflow at major swallow holes and sinking streams needs to be measured and compared to the discharge at river rises, in order to determine the amount of ground-water augmentation. Significant amounts of ground water join the pirated stream water, but the proportions are almost completely unknown. The river rises are an important type of first-magnitude spring in Florida.

TABLE 1

FIRST-MAGNITUDE, CLEAR-WATER, SINGLE VENT AND
HYDROLOGICALLY-RELATED SPRING GROUPS, IN FLORIDA

Spring Rank and Name	County	Average Discharge (cms)	
1 Silver Springs	Marion	23.2	(820 cfs)
2 Rainbow Springs	Marion	21.6	(763 cfs)
3 Spring Creek Main Spring	Wakulla	<21.7	(764? cfs)
4 Wakulla Spring	Wakulla	11.1	(390 cfs)
5 Croaker Hole	Putnam	10.2	(360 cfs)
6 Holton Spring	Hamilton	8.16	(288 cfs)
7 Devil's Eye Springs Group	Gilchrist and Columbia	8.05	(284 cfs)
8 Blue Springs	Jackson	5.38	(190 cfs)
9 Manatee Spring	Levy	5.13	(181 cfs)
10 Weeki Wachee Springs	Hernando	4.99	(176 cfs)
11 Homosassa Springs	Citrus	4.96	(175 cfs)
12 Troy Spring	Lafayette	4.70	(166 cfs)
13 Blue Springs	Volusia	4.59	(162 cfs)
14 Gainer Springs	Bay	4.51	(159 cfs)
15 Chassahowitzka Springs	Citrus	3.94	(139 cfs)
16 Alexander Springs	Lake	3.40	(120 cfs)
17 Blue Spring	Madison	3.26	(115 cfs)
18 Silver Glen Springs	Marion	3.17	(112 cfs)
19 Fannin Springs	Levy	2.92	(103 cfs)

Other clear-water springs that might be first-magnitude:

20 Jug Hole Spring	Columbia
21 Big Spring	Jefferson
22 Tarpon Spring	Citrus
23 Line-Eater Spring	Suwannee

Note: The list above specifically excludes river rises, which discharge brown water
derived mostly from nearby sinking streams.

TABLE 2

FIRST-MAGNITUDE RIVER RISES IN FLORIDA

River Rise Rank and Name	County	Average Discharge (cfs)	
1 Alapaha Rise	Hamilton	17.2	(608 cfs)
2 Siphon Creek Rise	Gilchrist	≈17.0	(600 cfs)
3 Santa Fe Rise	Columbia/Alachua	>16.5	(584 cfs)
4 St. Marks Spring	Leon	14.7	(519 cfs)
5 Steinhatchee Rise	Dixie	>9.30	(328 cfs)
6 Suck Hole Rise	Alachua	≈4.53	(160 cfs)
7 Chipola	Jackson	>2.83	(100 cfs)

Note: River rises are cave springs that discharge brown water having a tannin content
comparable to nearby surface streams. The discharge is composed dominantly of
water from nearby sinking streams, and should not be counted as a ground-water
resource unless the flow in the sinking stream is not counted. Ground water
augments the discharge at the rises, but the proportion of ground water in the
flow is unknown except at Santa Fe Rise where it ranges from 25 to 43 percent,
based on two measurements (Skirvin, 1962).

TABLE 3

FIRST-MAGNITUDE SPRINGS IN FLORIDA, REGARDLESS OF
WATER SOURCE

Spring Rank and Name	County	Average Discharge (cms)
1 Silver Springs	Marion	23.2 (820 cfs)
2 Rainbow Springs	Marion	21.6 (763 cfs)
3 Spring Creek Main Spring	Wakulla	<21.7 (764?cfs)
4 Alapaha Rise	Hamilton	17.2 (608 cfs)
5 Siphon Creek Rise	Gilchrist	≈17.0 (600 cfs)
6 Santa Fe Rise	Columbia/Alachua	>16.5 (584 cfs)
7 St. Marks Spring	Leon	14.7 (519 cfs)
8 Wakulla Spring	Wakulla	11.1 (390 cfs)
9 Croaker Hole	Putnam	10.2 (360 cfs)
10 Steinhatchee Rise	Dixie	>9.30 (328 cfs)
11 Holton Spring	Hamilton	8.16 (288 cfs)
12 Devil's Eye Springs Group	Columbia and Gilchrist	8.05 (284 cfs)
13 Blue Springs	Jackson	5.38 (190 cfs)
14 Manatee Spring	Levy	5.13 (181 cfs)
15 Weeki Wachee Springs	Hernando	4.99 (176 cfs)
16 Homosassa Springs	Citrus	4.96 (175 cfs)
17 Troy Spring	Lafayette	4.70 (166 cfs)
18 Blue Springs	Volusia	4.59 (162 cfs)
19 Suck Hole Rise	Alachua	≈4.53 (160 cfs)
20 Gainer Springs	Bay	4.51 (159 cfs)
21 Chassahowitzka Springs	Citrus	3.94 (139 cfs)
22 Alexander Springs	Lake	3.40 (120 cfs)
23 Blue Spring	Madison	3.26 (115 cfs)
24 Silver Glen Springs	Marion	3.17 (112 cfs)
25 Fannin Springs	Levy	2.92 (103 cfs)
26 Chipola Rise	Jackson	>2.83 (100 cfs)

Other springs that might be first-magnitude:

27 Jug Hole Spring	Columbia	
28 Big Spring	Jefferson	
29 Tarpon Spring	Citrus	
30 Line-Eater Spring	Suwannee	

Note: The springs and spring groups listed above include both clear-water cave
springs, the discharge from which can be counted as a ground-water resource, and
brown-water springs, herein termed "river rises", that represent the resurgence
of sinking streams. In order to avoid double counting surface water as ground
water, the river rises should not be counted as ground-water resources.

TABLE 4

FIRST-MAGNITUDE KARST WINDOWS IN FLORIDA

River Rise Rank and Name	County	Average Discharge (cms)
1 Kini "Spring"	Wakulla	4.99 (176 cfs)
2 River Sink	Wakulla	4.65 (164 cfs)
3 Falmouth "Spring"	Suwannee	4.48 (158 cfs)
4 Natural Bridge "Spring"	Leon	3.00 (106 cfs)

Note: Karst windows are collapse sinkholes which expose a segment of cave stream.
karst windows are not springs. Their discharge should not be counted as a
ground-water resource, because doing so will result in double counting the water
if it is also counted at the true spring. A good example of this is Kini
"Spring" and River Sink. Water flows from Kini "Spring" to River Sink and
probably resurges at Wakulla Spring.

References
Exley, I.S., 1989, personal communication.

Fisk, D.W., and Exley, I.S., 1977, Exploration and environmental investigation of the Peacock Springs Cave System in Dilamarter, R.R., and Csallany, S.C. (eds.), Hydrologic Problems in Karst Regions: Western Kentucky University, Bowling Green, Kentucky, p. 297-302.

Hunn, J.D., and Slack, L.J., 1983, Water Resources of the Santa Fe River Basin, Florida: US Geological Survey, Water-Resources Investigations Report 83-4075, 105 p.

Jennings, J.N., 1971, Karst: Australian National University Press, Canberra, 252 p.

Parker, C.G., Fergeson, G.E., Love, S.K., and others, 1955, Water resources of southeastern Florida: US Geological Survey, Water Supply Paper 1255, 965 p.

Rosenau, J.C., Faulkner, G.L., Hendry, C.W., Jr., and Hull, R.W., 1977, Springs of Florida: Florida Bureau of Geology, Bulletin No. 31 (Revised), 461 p.

Ruppert, F., and Wilson, W.L., 1989, The geology and hydrology of Wakulla Spring in Stone, W.C. (ed.), The Wakulla Springs Project: Raines Graphics, Austin, Texas, p. 163-174.

Skirvin, R.T., 1962, The underground course of the Santa Fe River near High Springs, Florida: unpublished Master of Science Thesis, University of Florida, Gainesville.

von Osinski, W., 1935, Karst windows: Indiana Academy of Science Proceedings, v. 44, p. 161-165.

Wilson, W.L., and Skiles, W.C., 1988, Aquifer characterization by quantitative dye tracing at Ginnie Spring, northern Florida in Proceedings of the Second Conference on Environmental Problems in Karst Terranes and Their Solutions: National Water Well Association, Dublin, Ohio, p. 121-141.

2. Sinkhole development and occurrence

Sinkhole density of the Forest City Quadrangle

AIDA BAHTIJAREVIC   *Geography Department, Faculty of Science and Mathematics, University of Sarajevo, Yugoslavia*
Currently: Florida Sinkhole Research Institute, University of Central Florida, Orlando, USA

ABSTRACT

Sinkhole density maps were constructed for the Forest City Quadrangle, Central Florida, and used for terrane evaluation.

Earlier investigations showed that the study area has the highest frequency of new sinkhole collapses in the Orlando area. New sinkholes and old sinkholes were treated as subpopulations of the total sinkhole population in the area. The area contains 385 sinkholes with the area ranging from 0.0002-0.57 km^2. The area affected by sinkholes, based on the average size, varies from 10,000 to 150,000 m^2/km^2 or 1 to 15% area/km^2.

According to the results of the terrane evaluation, 0.8% of total study area has a very high sinkhole density (13-15 sinkholes/km^2), 5.9% a high sinkhole density (10-12 sinkholes/km^2), 14.2% a medium sinkhole density (7-9 sinkholes/km^2), 32.5% a low sinkhole density (4-6 sinkholes/km^2), 22.4% a very low sinkhole density (1-3 sinkholes/km^2) and 24.2% of the area has no sinkholes.

The terrane evaluation is part of the geomorphological quantitative analysis of karst relief. A lack of correlation of old and new sinkhole densities eliminate it's application to sinkhole hazard predictions in thickly covered karst. This method could expand the data base for sinkhole-risk models and may be applicable to sinkhole risk assessement in thinly to moderatly covered karst where old and new sinkhole densities do correlate, as well as in other land use and population distribution terrain evaluations.

REZIME

Za podrucje topografske karte Forest City Quadrangle u Orlando regiji, Centralna Florida, konstruisane su karte gustine vrtaca i primijenjene u evaluaciji terena.

Ranija istrazivanja su pokazala da izabrano podrucje ima najvecu frekvenciju recentnog pojavljivanja vrtaca u Orlando regiji. Podrucje sadrzi 385 vrtaca cija povrsina varira od 0.0002 do 0.57 km2 i prosjecno iznosi 0.01 km2. Prosjecna pokrivenost vrtacama iznosi 10.000 - 150.000 m2/km2 ili 1-15% povrsine/km2.

Prema rezultatima evaluacije terena , 0.8% tretiranog podrucja ima vrlo veliku gustinu vrtaca (13-15 vrtaca/km2), 5.9% veliku gustinu (10-12), 14.2% srednju gustinu (7-9), 32.5% malu gustinu (4-6), 22.4% vrlo malu gustinu (1-3), dok je 24.2% podrucja bez vrtaca.

Izvrsena evaluacija terena je dio geomorfoloske kvantitativne analize kraskog reljefa. Njena primjena u procjeni rizika recentnog pojavljivanja vrtaca u debelo pokrivenom krasu je ogranicena zbog nesaglasnosti gustina neaktivnih i recentnih vrtaca. U kraskim terenima sa tankim do srednje debelim pokrivacem u kojima navedene gustine koreliraju, primjenjeni metod moze dati bazu podataka za modele rizika i primjenjiv je u procjeni rizika od pojavljivanja kolapsa. Takojer je moguca primjena u evaluacijama za potrebe prostornog planiranja (land-use, distribucija stanovnistva i sl.).

Introduction

In regions where sinkholes are developing, such as Florida's covered but active karst area, sinkhole investigations should be one of the very important factors in any land use or land planning evaluations. Recent problems in the Orlando area, including damage caused by new sinkhole development and reactivation, show that special attention should be given to the evaluation and prediction of sinkhole appearance for the purpose of proper land planning, which would assist in avoiding future problems and subsequent damages.

The purpose of this work was to evaluate sinkhole density in a sample area and to investigate the possibility of using "old" sinkhole density maps to predict "new" sinkhole development.

The terms "ancient sinkhole" or "old sinkhole" and "new sinkhole" are used herein to distinguish between "inactive" sinkholes - those that formed in premodern time, but have not suffered collapse in recent time - and "active" sinkholes, which are known to have developed during the last 30 years. From the geomorphological aspect these terms are completely arbitrary, because collapse is but one brief, and usually repeated, stage in the development of sinkholes in a mantled karst terrane. However, because of their practical meaning, these terms are important in planning studies. To inspect the relationship between old and new sinkholes and to provide comparability of these results with results of other sinkhole investigations, sinkhole density maps and analyses were done separately for old and new sinkholes.

The Forest City Quadrangle was chosen as the sample area because of its high frequency of recent sinkhole collapses (Wilson and others, 1987). The area contains 385 sinkholes which range in area from 0.0002 to 0.57 km^2 with an average of 0.01 km^2.

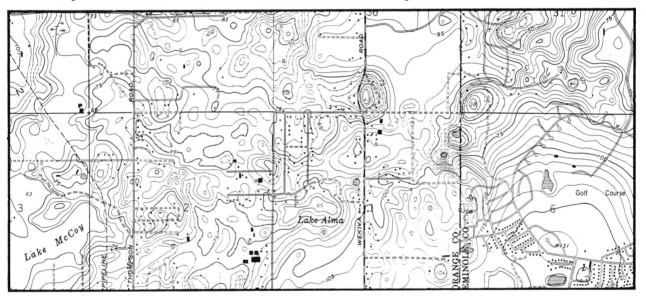

Figure 1. Typical topography of sinkhole affected area in the Forest City Quadrangle.

Methodology

The data for the old sinkhole density maps is principally derived from the Forest City 7.5 minute quadrangle map (1:24,000 scale) by cartomorphometric interpretation. As this map was made by photogrammetric techniques in 1959, it does not contain information about new sinkholes. The source of data for new sinkholes is the Florida Sinkhole Research Institute's computerized data base which contains information on new sinkholes reported in last 30 years. These records were used for locating new sinkholes on the topographic map by their longitude and latitude.

Sinkhole densities were mapped by the cartomorphometric method suggested by Miskovic & Musa (1985). This method consists of taking cartomorphometric values, in this case the number of sinkholes per square kilometer. Measurement units are organized as a hexagonal netweork. Horizontal and vertical overlapping provided double measuring of the treated area and multiplied the number of data points by three, thus increasing the map precision. Each map is based on the 1,377 data points. The interval between data points is 0.5 km.

The sinkhole density isolines are constructed mathematically by interpolation between data points. The hexagonal network provides 6 equal interpolation axes so the constructed

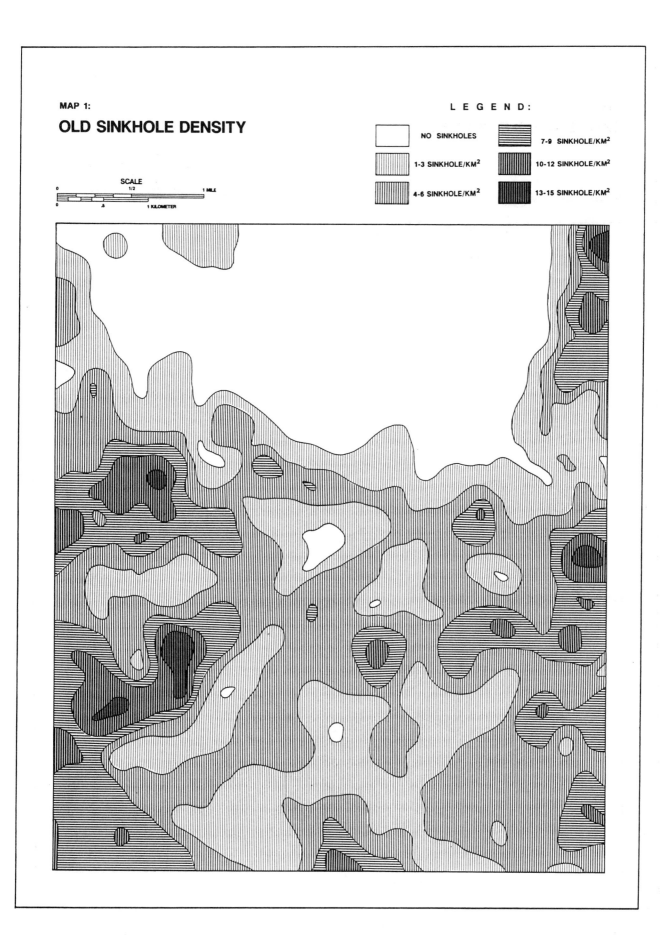

MAP 1:

OLD SINKHOLE DENSITY

LEGEND:

NO SINKHOLES

1-3 SINKHOLE/KM²

4-6 SINKHOLE/KM²

7-9 SINKHOLE/KM²

10-12 SINKHOLE/KM²

13-15 SINKHOLE/KM²

SCALE

0   1/2   1 MILE

0   .5   1 KILOMETER

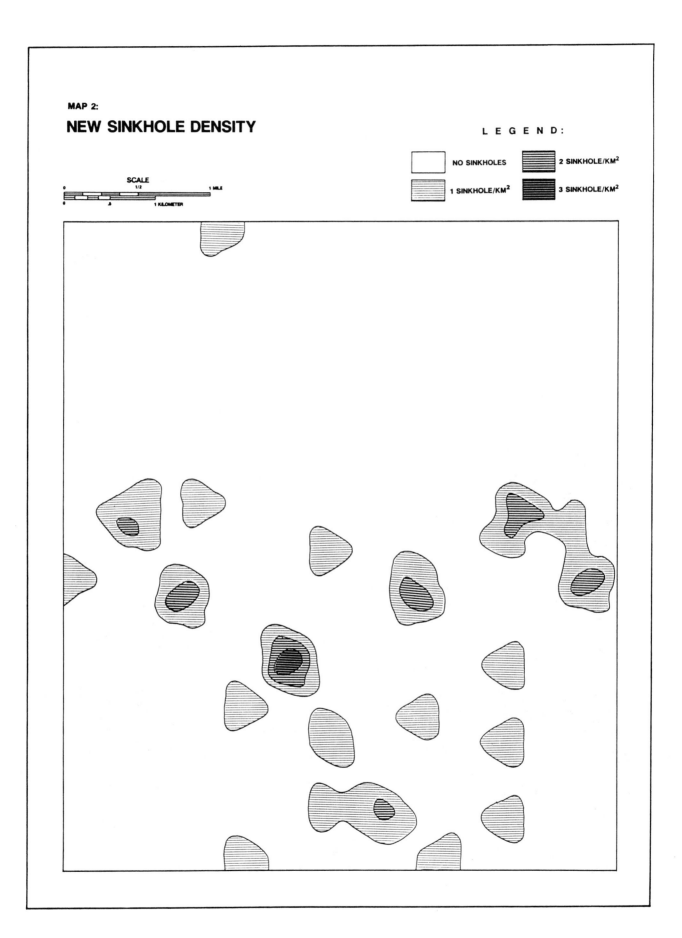

MAP 2:

NEW SINKHOLE DENSITY

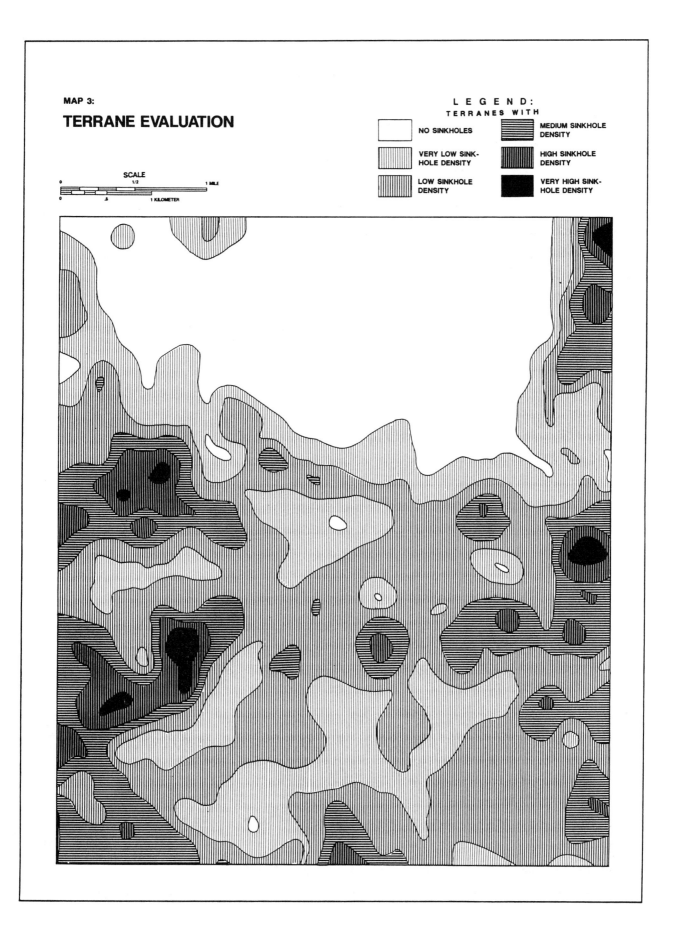

MAP 3:

TERRANE EVALUATION

SCALE

0   1/2   1 MILE

0   .5   1 KILOMETER

L E G E N D:
TERRANES WITH

NO SINKHOLES

VERY LOW SINK-
HOLE DENSITY

LOW SINKHOLE
DENSITY

MEDIUM SINKHOLE
DENSITY

HIGH SINKHOLE
DENSITY

VERY HIGH SINK-
HOLE DENSITY

isolines are not deformed and maps are comparable to the topographic map from which the data base is derived.

The final result of the described procedure is a basic sinkhole density map at a 1;24,000 scale. Isoline values are 0.5, 1.5, 2.5, ... 13.5, and 14.5 sinkholes/km^2.

The old sinkhole density map (Map 1) was made by generalization of the basic map. For this purpose sinkhole density values were classified as follows:

Class 0   No sinkholes
Class 1   1-3 sinkholes/km^2
Class 2   4-6 sinkholes/km^2
Class 3   7-9 sinkholes/km^2
Class 4   10-12 sinkholes/km^2
Class 5   13-15 sinkholes/km^2

The associated isoline values are 0.5, 3.5, 6.5, 9.5 and 12.5 sinkholes/km^2.

The new sinkhole density map (Map 2) is not generalized and it contains the basic sinkhole density values.

The terrane evaluation map (Map 3) is based on both old and new sinkhole density data compiled by the same method.

For better utility, the maps are completed with different patterns for the various classes so that the maps automatically show sinkhole density either as black-white or color version.

Results and Conclusions

The highest old sinkhole density in the Forest City Quadrangle is 15 and the average density is 2.2 sinkholes/km^2. The area where sinkholes appear is 116 km^2 or 72.5% of the total. The density is irregularly spread over the area, varing from 1.2 km^2 or 0.7% of the total area for the class 5 (12-15 sinkholes/km^2) to 46.3 km^2 or 28.9% of the area for class 2 (4-6 sinkholes/km^2). The results are tabulated below.

Table 1. Density of old sinkholes in the Forest City Quadrangle*

Class number	Number of sinkholes per km^2	Area km^2	Percentage of the total area
0	0	44.0	27.5
1	1-3	36.3	22.7
2	4-6	46.2	28.9
3	7-9	24.3	15.2
4	10-12	8.0	5.0
5.	13-15	1.2	0.7
Average	2.2	-	-
Total	-	160	100

* All measurements done by Resource Measurement System (Bausch & Lomb) with precision of 3%.

The maximum new sinkhole density in the study area is 3 sinkhole/km^2 and the average density is 0.2 sinkhole/km^2. The area where new sinkholes appear is 22.7 km^2 or 11.7% of the total area. Density classes vary in area from 0.3 km^2 or 0.2% for class 3 (3 sinkhole/km^2) to 17.4 km^2 or 10.2% of the area for class 1 (1 sinkhole/km^2). The results are tabulated in Table 2.

The comparative analysis of the old and new sinkhole densities in the Forest City Quadrangle showed that there is no correlation. That is, new sinkholes are appearing most commonly in the areas that do not correspond with areas of high old sinkhole density.

New sinkholes appear in the areas with the lowest old sinkhole density or without any old sinkholes, as well as in the areas with medium old sinkhole density, but they do not appear at all in the areas with highest old sinkhole density.

As the Forest City Quadrangle has an average of 30-60 m of insoluble cover (Wilson et

Table 2. Density of new sinkholes in the Forest City Quadrangle*

Class number	Number of sinkholes per km²	Area km²	Percentage of the total area
0	0	140	88.3
1	1	17.4	10.2
2	2	2.3	1.3
3	3	0.3	0.2
Average	0.2	-	-
Total	-	160	100

*The new sinkhole density map and analysis is done on the basis of data for 30 new sinkholes recorded by the Florida Sinkhole Research Institute.

Table 3. Terrane evaluation based on the combined old and new sinkhole densities

Type of terrane	Percentage of total area
Terranes with no sinkholes	24.2
Terranes with very low sinkhole density	22.4
Terranes with low sinkhole density	32.5
Terranes with midium sinkhole density	14.2
Terranes with high sinkhole density	5.9
Terranes with very high sinkhole density	0.8

al., 1987) and is typical of thickly covered karst, these conclusions support the earlier conclusions of Upchurch & Littlefield (1987) that in thickly covered karst areas the location of sinkholes can not be predicted by the locations of old sinkholes.

Table 3 shows that the terranes with low sinkhole density appear with the highest percentage of the total area (32.5). Other terranes areas vary from 0.8 to 22.4% of the total area with areas having a very high sinkhole density occupying the smallest percentage (0.8) of the terrane.

The average area affected by sinkholes varies from 10,000 to 150,000 m²/km² or 1-15% area/km².

This terrane evaluation represents only one aspect of the geomorphological quantitative analysis and evaluation of karst relief. In conjunction with the analysis of the other morphometric parameters it can give significant geomorphological conclusions. It is a preliminary step in terrain evaluation, but it is not applicable to new sinkhole hazard prediction in thickly covered karst areas where the old and new sinkhole densities do not correlate. In the areas of thin to moderately covered karst where the old and new sinkhole densities may correlate (Upchurch & Littlefield, 1987), this evaluation could be applicable for sinkhole-risk models and risk assesment. Application in land use and population distribution planing requires assessment of other morphometric parameters.

Aknowledgements
This work was supported by Florida Sinkhole Research Institute, University of Central Florida, Orlando. I want to thank Barry F. Beck for providing financial support as well as William L. Wilson for useful suggestions and language improvement and T. Scott Cavin for the help in drafting the illustrations.

References

Miskovic, M. & Musa. S., 1985. Kartografske metode u fizickoj geografiji, XII kongres geografa Jugoslavije - Zbornik, Novi Sad, p.p. 273-278. (Cartogrphic methods in Physical Geography, In Proceedings of XII Yugoslav Geography Congres, Novi Sad, p.p. 273-278).

Upchurch, S. B. & Littlefield, J. R. , 1987. Evaluation of data for sinkhole development
    risk models, In B.F. Beck & W.L. Wilson (ed.) Karst Hydrogeology: Engineering and
    Environmental Applications, A.A. Balkema, Rotterdam, p.p. 359-364.

Wilson, W. L. et al., 1987. Hydrogeologic factors associated with recent sinkhole
    development in the Orlando Area, Florida, Florida Sinkhole Research Institute Report
    No. 87-88-5, University of Central Florida, Orlando, 104 p.

Collapse sinkhole at the inlet tunnel of a powerhouse, Pont-Rouge, Quebec

MICHEL BEAUPRÉ   Hydro-Québec, Montréal, Canada
JACQUES SCHROEDER   UQAM, Montréal, Canada

ABSTRACT

In 1988, a road was closed near Pont-Rouge due to a nearby collapse of the roof of an inlet tunnel leading to the Mc Dougall hydroelectric powerhouse owned and operated by Domtar Inc. The almost uneconomical plant had to be closed down because of a massive influx of sediments, the presence of trees in the tunnel and debris railings. This new sinkhole has given us insights in what geologic, geomorphologic and hydraulic factors were favorable to the initiation and development of such a phenomenon over the original karstic conduit (meander cut-off), partially excavted and protected during the 1927's powerhouse construction works. Topographic surverys of 1927, 1952 and 1988 have given us information on how and how fast the underlying dome has formed before the final collapse in May, 1989. In this part of the gallery, the top and roof of the wall are made up of very compact till, ice-pushed and folded limestone slabs, and blocks on a length of 40 meters. Because of the till's high compaction, no remedial work was suggested; rather a surveillance program was proposed and a geotechnical investigation was put forward to decide how the closed road should be moved.

RESUME

Une route a été fermée durant l'été 1988 près de Pont-Rouge à la suite de l'effondrement du toit d'une conduite forcée. Celle-ci alimentait la centrale électrique de Mc Dougall utilisée par la compagnie Domtar Inc. Il a fallu fermer la centrale à cause des sédiments et de la végétation qui bouchaient la conduite. Cependant grâce à cet effondrement, nous avons pu identifier les facteurs géologiques, géomorphologiques et hydrauliques responsables du sinistre. Ce dernier dérive du calibrage d'une galerie karstique recoupant un méandre, aménagement exécuté en 1927 lors de la construction de la centrale électrique. Grâce aux relevés topographiques de 1927, 1952 et 1988, on a pu évaluer comment et à quelle vitesse s'est établi le dome remontant à l'origine de l'effondrement. Dans cette partie de la conduite, sur 40 mètres de long le toit consistait en un till très compact et en copeaux de calcaires glaciotectonisés. Vu la compacité élevée du till, aucun travail pour remédier à la situation n'est suggéré après le second effondrement en mai 1989. Par contre un programme de surveillance et des investigations géotechniques sont nécessaires pour décider du nouvel emplacement de la route.

On May 20th, 1988, a 5 x 10 m collapse sinkhole was reported on tne upstream part of the inlet tunnel of the hydroelectric powerplant operated by the Domtar Company, a few meters from the Dery road, which crossed the underground gallery at nearly a right angle. The proximity of the 12 m deep vertical shaft compelled the Quebec Department of Transport to temporarily close the part of Dery Road closest to the collapse. A long debate started as to whom was responsible for the collapse and who would pay for future costs of road closure and relocation. A sudded enlargement of the sinkhole on May 5, 1989, led to the definitive shut-down of the powerplant and closure of the Dery Road until construction work could be initiated.

Location

The small Mac Dougall powerplant was built in 1926-27 in the neck of great meander on the Jacques Cartier River, downstream and in the southern part of the Pont-Rouge township, some 30 km west of Quebec city.

This plant is composed of an upstream cofferdam, a water intake, a straight and gently sloping inlet gallery, a distributor with upstream debris railings and a powerhouse. Water goes back in the river through a small outlet canal blasted in the riverbank limestones.

Upstream reservoir is at ± 75 m a.s.l. and total head around 20 m. Ground elevations nearby range between 85 and 100 m a.s.l. (fig. 1).

Geological and geomorphological contexts

The Jacques Cartier River takes its source in the Canadian Shield and flows southwest toward the St Lawrence River. In the powerplant area, it has embanked itself some 10 to 30 m in unconsolidated sediments and underlying middle Ordovician limestones. A small canyon with vertical or overhanging walls and great horizontal slabs was cut in the lowest part of the valley (fig. 1).

Underlying limestones belong to the Deschambault Formation (Trenton Group). These are massive, soluble, slightly fractured and competent. On top is the Neuville Formation (Trenton Group) in which limestones are much thinner, have finer grain size, are more brittle, less soluble and show more shaly partings. Dips are less than 3° to 6° toward south or

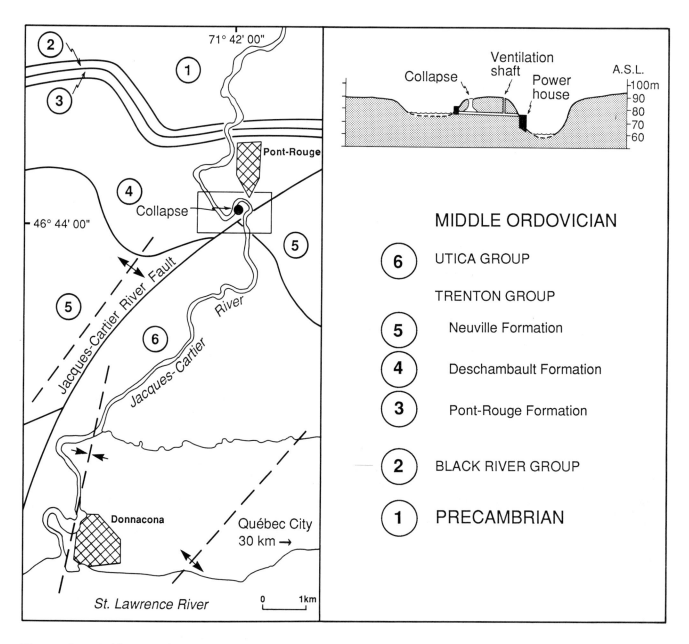

Figure 1: Collapse sinkhole location, topography and bedrock geology.

southwest. Some great open folds plunge southwestwardly. Just southeast from the power-plant is a northeast trending gravity fault (N 055° - 65° S.E.) with a downthrown block on the southern side. Deschambault limestone outcrops on the northern side while Neuville Formation and overlying shales can be seen on the south side. Vertical joints have a metric spacing, can be followed over tenths of meters and belong mainly to two sets, N 129° E/N 142° E and N 032° E/N 045° E.

Unconsolidated sediments can be divided into two facies. On top is a 1-2 m thick layer of stratified and clean sands and gravels of low to medium compacity. Underneath is a typical grey and clayey lodgment till less than 10 m thick (fig. 2). It is a calcareous, homogeneous, very compact, almost impervious and fissile till with a low water content. In the lower half, it has a clayey and silty matrix with limestone fragments (10-15 cm), while the upper part shows a more sandy matrix with some 25 to 30% of Precambrian boulders.

Right on top of the nearly flat-lying Deschambault limestones and in the basal part of this very dense till, one can see fractured, sheared and folded limestone blocks

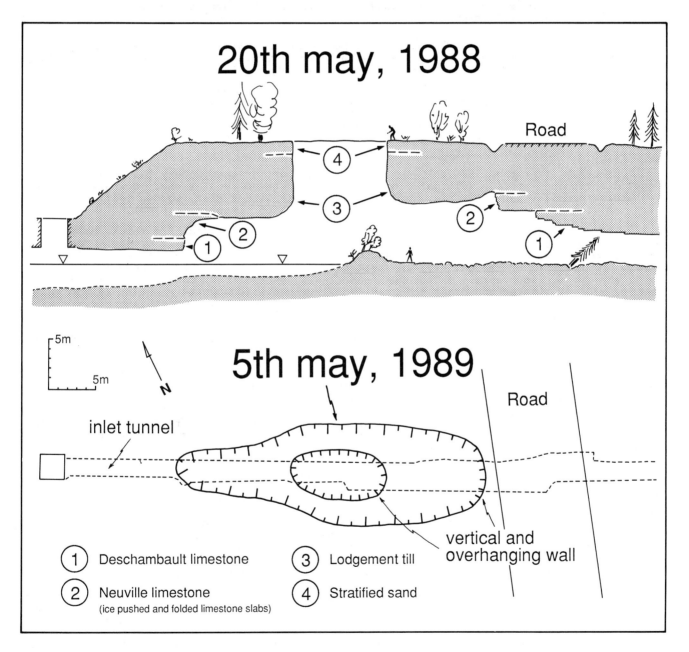

Figure 2: The 1988 collapse in longitudinal section and in plan; limits of the 1989 collapse.

($\leqslant$1 m) from the Neuville Formation. Analysis of slickensides between Deschambault and Neuville limestones and of other structural features (joints, faults, folds, dips) show that we have glacially thrusted and folded Neuville limestone on top of the massive underlying flat-lying Deschambault Formation (Schroeder et al., in press).

The inlet tunnel

This tunnel is perfectly parallel and follows vertical open joints of the first set. Study of 1926-27 construction drawings seems to show that the original cavity was of a partly karstic, partly glacio-karstic origin. It is a typical meander cut off. Prior to construction, flood waters from the Jacques Cartier River were probably flowing through this cave that had vertical elliptical sections in its upstream part and rectangular ones downstream. The cave outlet was hanging some 15 m over the river bed.

The N.W. - S.E. trending, 170 m long, gallery had a very gently southward sloping (1%) streambed at + 68 m a.s.l. with a horizontal ceiling located some 8 to 10 m underneath ground level. Prior to the 1926-27 construction work, the cave ceiling was flat, straight and almost parallel to the floor. A natural pit was located at the ventilation shaft site. There does not seem to have been any excavation work in that part of the gallery, except for the ventilation shaft area.

Construction work included concreting parts of the riverbed or lateral wall cavities; construction of a steel ventilation shaft with a wood framing; construction of wood ceilings with 10" x 10" and 2" x 8" wooden pieces; and concreting of the water inlet and outlet.

Minor excavation work was done in the upstream part of the gallery; we do not know exactly what was done. These were probably aimed at getting a minimum hydraulic section along the tunnel. During our 1988 survey, we found some wooden roof sections in the upstream part of the tunnel but we do not know when, where or why these were put there. No - where are these shown on the 1926-27 and 1952 drawings.

The collapse

Shortly after the 1988 collapse, the upstream part of the gallery had an irregular, upper longitudinal profile (elongated dome) and a downstream regular one. Downstream, the transverse profiles were sill rectangular (2.5 x 4.0 m) with a flat, wooden ceiling and straight, flat vertical walls, showing only minor solution features ($\leq$10 cm).

Overlying the wooden roof were flat lying limestones less than 7 m thick, except around the ventilation shaft where we had a very dense till with a small erosion dome (1 m). Comparison of 1926-27, 1952 and 1988 drawings has indicated that there has been very little change in this part of the gallery, and that remedial or protective works have fulfilled their aim for over 60 years. In the upstream area, upper parts of walls and ceilings were made of till or sheared and fractured Neuville limestone. Transverse profiles (4 x 6 m) were of the keyhole type with a sudden submetric widening where the underlying massive Deschambault Formation and the overlying till or glacially deformed limestones met. The thickness of the till was some 5-6 m south of the collapse and 7-8 m north of it. The width of the gallery was around 3.0 to 4.0 m. Water seepage through till was minimum. Neuville limestones were mainly seen downstream and around the sinkhole and were badly fractured, sheared and folded. Walls and ceiling in there were irregular and loosely broken. Some continuous vertical open joints filled (<10 cm) with clayey silt (till matrix) were running in the ceiling and had a small angle (25° - 30°) with the gallery axis, thus creating significant weakness zones in the already shaken ceiling.

The first collapse was thus formed in and at the upstream limit of the glacially deformed limestones, in a place where the tunnel was widest and where ceilings were thinnest, weakest and most erodible by incoming waters.

The 1988 mapping has indicated that till surfaces were fresh and unaltered. We also noted some 2 m of fresh till fallen on a wooden roof section resting on the Deschambault limestones and some 4 m away from the existing till ceiling. What was found was an elongated dome less than 40 m long, 4 m wide and 4 m high, whose formation seems to have started shortly after the 1926-27 excavation work and water circulation. Since then, it has slowly progressed for more than 60 years in these mostly unprotected till and sheared limestones. The 1952 drawing shows a rising dome some 5 to 6 m under ground level in the downstream limit of the 1988 elongated dome. Erosion thus progressed upstream and upward in this till, between the north and south flat-lying limestone ceilings.

The 1988 observations also confirmed the existence of a small dome right under the sinkhole location. From 1926 to 1952, the 1952 dome was slowly excavated by the flowing water completely filling the tunnel. Between 1952 and 1988, it has migrated upstream and upward at "mean" rates of $\pm$ 1 m/yr and 0.1 m/yr to form the 1988 elongated dome. A central upper cylindrical dome was also formed during this 36 year period (fig. 2). We do not know what was the role of the roots of the trees now stuck midway in the tunnel. Removed volume is around 700 m³, of which 300 m³ was part of the collapse. In 1989, the vertical shaft has progressed upstream and downstream, to the flat-lying limestone ceilings.

Conclusion

The 1988 collapse was formed over a water-filled gallery, where mechanical, geological and geometrical factors were most favorable, in spite of the protective work that seemed to have been done in this area or close by. Combined action of a variable upward water head, water absorption by till, stress release in walls, and erosion of soaked sediments has slowly created a rising dome between 1926 and 1952, which has "rapidly" migrated upstream (+ 1 m/yr) and upward (1.1 m/s), thus forming an elongated (40 x 4 x 4 m) dome in which a secondary, upper and cylindrical dome was formed exactly at the sinkhole place.

The 1989 enlargement of this sinkhole occurred in springtime just over the elongated dome (frosting/thawing, increase of water content and pressure in the rock, ceiling weakening, stress redistribution around the vertical shart). In less than one year, this small sinkhole had enlarged many times, so that one could say that the rate of formation of such a phenomenon has followed a more or less accelerating trend between 1926-27 and 1989.

The reactivation of such a semi-fossil cave, and excavation work done in it, have produced stresses and created new processes that were surely responsible for the initiation or at least the acceleration of such a collapse phenomenon. This case history is a good example of the kind of adverse impact uncontrolled interventions in karstic terrain can have. Curiously enough, this terrain seems to have taken, at least temporarily, its revenge against our land managers.

References

Schroeder, J., Beaupré, M., Cloutier, M. (in press). Substratum glaciotectonisé et till syngénétique à Pont-Rouge, Québec. Géographie physique et quaternaire.

Qualitative modelling of the cover-collapse process

JIAN CHEN & BARRY F.BECK   *Florida Sinkhole Research Institute, University of Central Florida, Orlando, Fla., USA*

ABSTRACT

Cover-collapse sinkholes develop rapidly and unexpectedly. Most of the processes culminating in the final surface collapse occur below ground. In order to observe and study the mechanisms by which dissolved voids in limestone propagate upward to cause sinkholes, a parallel plate model was utilized. Natural sediments were layered in the tank and holes were opened in the bottom. Homogeneous and stratified settings in both dry and saturated conditions were examined in 23 different modelling trials. Sandy sediments collapse rapidly in either a dry or saturated condition, but are more stable when only partially saturated. Clayey sediments may span voids when dry, but when saturated will erode. The collapse process occurs upward in a cylindrical zone above the "drain" until it reaches the surface where lateral erosion will occur until a stable slope angle is reached. Upward collapse in sand will be stopped by cohesive strata, such as clay, and lateral erosion forming a soil void will continue until the stratum collapses or the erosion ceases. Such collapse of cohesive strata may plug the erosion zone and stop the subsidence process. Under some conditions soil voids may be stable. Slab loading of the surface does not increase the collapse process. While only qualitative and preliminary, the modelling studies appeared to yield conclusions which may be useful in sinkhole studies.

Introduction

Cover-collapse sinkholes, which occur in mantled karst terranes, are a serious geologic hazard which have begun to receive international attention in recent years. This catastrophic land surface subsidence is endemic to karst areas where the cavernous carbonate bedrock is overlain by unconsolidated sediments, as is shown in Figure 1. While sinkholes do occasionally develop due to the collapse of the bedrock roof over a large karstic cavity, sinkholes which develop due to the downward erosion of unconsolidated overburden into such cavities are much more common and much more damaging. This paper will discuss a simple model of the physical processes of cover-collapse.

The conditions under which sinkholes form, their study and investigation, the damages they cause, and the prediction or prevention of catastrophic land surface collapse, or post-collapse repair were the focus of two recent conferences (Beck, 1984; Beck and Wilson, 1987) and have been a major component of several other recent meetings concentrating on karst hydrogeology and engineering (NWWA, 1986: NWWA, 1988; Sitar, 1988). At these meetings the processes and mechanisms of cover-collapse have been described by several authors, but in almost all cases the processes have been inferred from the field investigation of conditions after the sinkhole had collapsed. Because most of the formative process occurs beneath the ground surface, it can only be witnessed after the surface opening has already collapsed. In most instances, when a sinkhole is reported in someone's backyard, the geologist arrives only to find an immobile surface pit. (Note: the authors will assume that the reader is familiar with the hypothesized mechanisms of sinkhole collapse as postulated by Jennings, 1971; Sowers, 1975; and Newton, 1987; and summarized by Beck, 1988.)

In this respect the geologic process of sinkhole formation differs from other common geologic hazards which can be seen occurring at the earth's surface such as landslides or floods. Sinkhole collapse is more akin to the sudden disruption of an earthquake which originates underground. However, the sinkhole erosion process begins underground, is of short duration and local effect, and typically occurs unexpectedly. Detailed study of earth-quake occurrence has led to the discovery of many factors which may become predictors of future disasters, but present research does not seem to promise any reliable prognosticator of sinkhole collapse.

The goal of this sinkhole modelling project is to provide data not obtainable from the study of already collapsed sinkholes by observing and documenting the mechanical process of

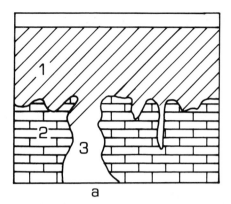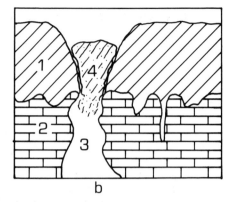

Figure 1: Generalized geologic sketch of a cover-collapse sinkhole. a = before; b = after. 1--unconsolidated overburden sediment; 2--carbonate rocks; 3--karstic cavity, usually solution pipe or karren; 4--collapsed sediment filling the cavity.

cover collapse as it occurs. Modelling can be used to evaluate how different geologic factors, such as overburden thickness, state of water saturation, and the nature of the sediment, will determine the final form of the land surface subsidence. Modelling can also be useful in examining the actual mechanism of upward propagation and collapse. The current modelling effort is a very simple, preliminary experiment but it may potentially provide a theoretical framework for more rigorous quantitative physical or mathematical modelling, eventually leading to the desired predictions.

Previous Work
    The only previous sinkhole modelling effort, to the author's knowledge, was that of Holm, 1986. Holm was studying the geomorphic development of the Southern Lake Wales Ridge in Florida and began modelling the sinkhole collapse process in an attempt to understand that terrane. Holm's work was conducted as an M.S. thesis in geology at the University of Florida under Dr. Robert Lindquist. Some of our present work duplicates and substantiates Holm's findings. However, Holm's thesis is unpublished and, due to a career change, will probably remain unpublished.

The Modelling Equipment and Procedure
    A narrow, parallel-plate type tank was constructed with wooden bottom and ends and plexiglass sides. The clear sides were 52" long by 36" high and the width was adjustable to 2.5", 4.0", 5.5", or 7.0". The tank was open on the top to permit the arrangement of the experimental materials. The interior conditions could clearly be seen in cross-section through the sides.

    Holes 0.25", 0.5", 0.75", 1.0" 1.25" and 2.0" in diameter were drilled in the bottom of the tank and plugged. The holes were to imitate the openings in the limestone at its interface with the overburden. The model considered only the opening and not the geometry of the underlying conduits. This limitation was based on both practical and theoretical considerations: It would be very difficult to model the sinuous, irregular pattern of the limestone conduits, and, the opening itself is the critical factor. It is well-known that even when karstification of limestone is very intense and there are large, subterranean river systems developed, these will not result in the development of collapse sinkholes unless there are openings to the surface through which the overburden may be eroded (Fig. 2).

    Overburden sediment was simulated using various natural sediments: medium-fine sand, silty sand, and sandy clay. The medium-fine sand was between 0.59 mm and 0.23 mm grain size diameter. The consistency of the sediment was varied by adding water. In order to observe the collapse process, marker beds were created with sand dyed pink.

    Semi-saturated and saturated conditions were created by sprinkling water from a common garden watering can onto the sediment surface prior to the modelling episode. During the trial no additional water was added so that the water draining downward was limited to the original contents of the tank; this is referred to as a limited recharge condition. In some trials, static loads were placed on top of the sediments to determine the effect of external loading on the sinkhole development.

    The modelling trials began by selecting the geologic conditions--thickness, arrangement, and nature of overburden strata, and the size of the limestone opening--which

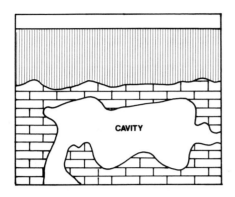

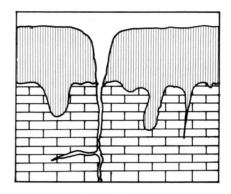

Figure 2a: A large karstic cavity not connected to the surface will not cause the development of a cover-collapse sinkhole.

Figure 2b: An opening in contact with the unconsolidated overburden is a nescessary precursor for the development of a cover-collapse sinkhole.

we wished to examine. These were then simulated by layering the sediments in the modelling tank and adding water, as appropriate. The modelling trial was initiated by removing the plug from the desired opening and then observing, photographing, and recording the development of the resulting sinkhole. The results were then analyzed to attempt to define general principles which might be useful in predicting areas and conditions prone to sinkhole collapse.

The Modelling Trials and Results

Fourteen modelling trials were conducted with a simple, one sediment overburden, either sand, silty sand, or sandy clay. Nine trials were conducted using mixed strata of sand and clay. Thirteen of the trials were conducted under dry conditions while ten were conducted under saturated or semi-saturated conditions.

Trials with Homogeneous Overburden

Six trials were duplicated using only dry, medium-fine sand; this sediment is very loose, non-cohesive, and has a small angle of internal friction. The results were all similar. When the plug was removed, small circular depressions appeared on the sand surface almost immediately. In a short time they enlarged to become inverted-cone shaped sinkholes. Sand movement appeared to be restricted to the cylindrical area between the hole and the surface because the settlement of the sand grains was not detectable until the depression grew large enough to intercept the plexiglass (see Fig. 3). The final slope of the depression formed an angle of 30-40°. The final diameter of the sinkholes ranged from 4"-10", depending on the size of the bottom opening and the thickness of the sediment. When holes of different diameters were opened, the corresponding depressions enlarged at diffrent rates (Fig. 4).

Two trials were conducted with dry, silty sand. The results were similar to above with the exception that the depressions were smaller with steeper slopes, approximately 90°. A shaft, or overhanging jar-shaped depression was created (Fig. 5). As in the previous trials, the depression enlarged more rapidly when the underlying opening was larger. The rate of sand erosion appeared to be proportional to the area of the opening which was unplugged, as would be expected.

One trial was attempted using dry sandy clay. No collapse occurred. The 0.5", 1.25", and 2.0" holes were all unplugged, but after 24 hours the sediment was still stable.

One trial was conducted with moist sand; the sand was mixed with water prior to placing it in the tank. A 0.5" silty sand stratum was placed in the middle of the sand. The 0.5", 1.0" and 2.0" holes were unplugged for 24 hours, but no movement of the sand occurred. The surface of the sand was loaded, but still no sinkhole developed. Obviously the moisture film had a significant enough effect to stabilize the sand, in contrast to the dry condition.

Three trials were conducted with saturated sand, simulating a phreatic aquifer. When a 0.25" hole was opened, the water drained rapidly, but the sand only sagged slowly circa 2" and then stopped. When a 0.75" hole was unplugged the sand eroded rapidly. A small conduit stoped upward and then horizontally to become a jar-shaped soil cave (Fig. 6). The cave was then filled by collapsing surface sediment and a funnel-shaped sinkhole with steep lower sides resulted.

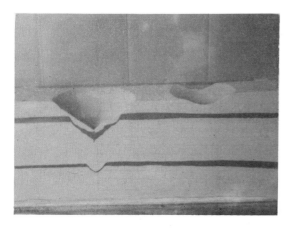

Figure 3: Modelling trial in homogeneous sand; stratum are simply dyed markers. Note, on the right, that the initial collapse effects only a cylindrical zone over the "drain" until the surface depression widens, as on the left.

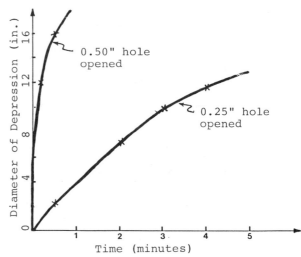

Figure 4: Graph showing the rate of enlargement of the surface depression for two different sized openings into the "limestone". The sediment was homogeneous medium-fine sand.

One trial was conducted using saturated sandy clay. When the 0.25" hole was opened, a small amount of water and clay drained. The sediment moved radially into the draining hole and caused a shallow depression in the surface, but little settlement of the entire layer occurred. When the 0.75" hole was unplugged, water and clay drained quickly through the opening forming a classic sinkhole with wide, gentle upper slopes and steep, near vertical walls below (Fig. 8). The diameter of the shaft-like portion was 6.5".

Trials with Stratified Overburden
        Three trials were conducted under dry conditions with a clayey stratum midway in the sand. When the holes were unplugged, the lower clay/sand interface quickly showed infinitesimal cracks. Sand ran out the hole forming small soil caves below the clay (Fig. 8). These voids grew wider until the clay began to collapse. Cracks first appeared over the two ends of the cavity. In one trial the clay stratum partially collapsed when the void was 3.1" in diameter, but the surface remained stable. In another trial the clay completely collapsed when the soil cave was 1.9" in diameter and a gentle depression occurred at the surface (Fig. 9). When this had stabilized the two holes 0.25" and 0.75" were unplugged. Both of these drained sand, but in the end only one depression occurred on the surface centered over the larger opening. The sagging of the clay layer which was caused by the smaller opening simply became part of the broader action resulting from the larger opening. In a third trial the clay layer collapsed differentially, so the migration of sand occurred on one side only and the shape of the surface depression was assymetrical (Fig. 10); the slope of one side was 40° and the other 24°.

        All of these trials were of a similar nature: The development of the sinkhole occurred in stages. First widening soil caves undermined the stable clay until it collapsed. If the collapsing clay blocked the hole, the subsidence ceased and little, or no, surface expression resulted. If the collapsing clay did not obstruct the hole, sand migration continued to form a surface sinkhole.

        Two trials were modelled under saturated conditions with a single silty sand stratum midway through medium-fine sand. In the first trial the silty sand was 0.5" thick. When the 0.75" hole was opened. The lower sand drained immediately and the silty sand layer above the hole dropped downward creating a shallow surface depression. Two minutes later the drainage of the hole became plugged. Precipitation was simulated and although it was only slight it quickly seeped down and opened the plug. Sediment continued to drain forming a typical sinkhole with a 75° slope and 9.1" diameter. In the second trial the silty sand stratum was 1.0" thick. When the 0.75" hole was opened a small cave formed briefly, but the silty sand collapsed and plugged the bottom hole. When precipitation was added, the trial continued as before.

        Three trials were conducted with two sandy clay layers, 0.5" and 0.7" thick, spaced 1/3 and 2/3 up through medium-fine sand. Marker beds were also used. As before, when the bottom holes were opened, cracks appeared in the lower clay/sand interface and a soil cave

Figure 6 (right): Modelling trial in saturated sand. The cylindrical soil cave later collapsed to form a typical sinkhole.

Figure 5 (above): Modelling trial in dry silty sand, as viewed from above. Note vertical sides.

Figure 7 (below): Modelling trial in saturated sandy clay. Normal stoping erosion formed a typical sinkhole as shown below.

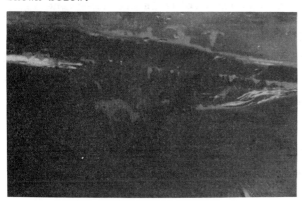

Figure 8 (right): Modelling trial with clay stratum (dark) in sand. A soil cave formed below the clay due to lateral erosion, then finally collapsed.

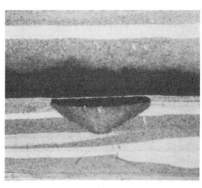

Figure 9 (below): As in Fig. 8. When the soil cave below the clay collapsed, the cylindrical column above settled producing only a shallow depression in the surface.

Figure 10 (below): As in Fig. 8. If the clay did not plug the opening, sand erosion continued above the clay forming a typical surface sinkhole.

Figure 11 (below): Modelling trial with two clay strata in sand. When the soil cave below the lower clay collapsed, the area above settled vertically causing a shallow depression as in Fig. 9.

formed below the clay. The cave grew laterally until the clay collapsed and a surface depression formed. The settling ceased because the bottom holes became plugged. The surface depression was approximately 9" in diameter while the collapsed hole in the clay was 4.5" in diameter and the underlying hole draining the sand was only 0.25" wide (Fig. 11). The total thickness of the sediment was approximately 12". In one trial a soil cave formed 4.5" in diameter and the settling process stopped. The thickness of soil above the cave was 8.3".

One trial was conducted with a similar arrangement of two clay strata between sand, but under saturated conditions. When the 0.5" hole was opened a small cavity 2" in diameter formed below the lower clay layer. When this cavity drained of water the clay collapsed and the upper clay subsided slightly while water continuously drained from the model. A depression 6" in diameter and 4" deep formed on the surface before the hole plugged with clay and subsidence stopped. After two hours the hole was manually opened; the subsidence grew to 8", but then plugged again and remained stable.

Analysis and Conclusions
We were pleased to be able to observe a series of sinkholes-in-progress without any danger. These observations provided some insight into the processes and conditions of sinkhole formation which are normally hidden underground. Scaled dimensions for the strata thickness and the cavity/sinkhole dimensions were recorded. The physical and mechanical properties of the overburden were not quantitatively measured. Qualitative indications of the moisture content are given by the terms dry, semi-saturated (moist), or saturated. While these are still preliminary experiments, the results were very informative and indicate that a continuation of similar research can be productive. The principle conclusions from our work are enumerated below.

1. Gravitational movement of overburden sediment down into voids in the limestone is a major factor in sinkhole development. This causes soil cavities which stope toward the surface. The flow of water is not necessary to cause the development of a sinkhole. However, it must be kept in mind that under natural conditions it may be the build-up of a hydrologic head which "pulls the plug" and opens up a formerly plugged void in the limestone down which the overburden can migrate.

2. Very small openings in the limestone surface can create sinkholes tens of times larger if there is sufficient void space within the limestone to accept the subsided sediment.

3. The interfaces of multi-layered stratigraphy affect the stoping collapse process causing substantial horizontal enlargement below more cohesive strata. This validates a similar conclusion reached by Newton (1987) from field data. Soil cavities beneath such strata may be stable for a long time. It is not unreasonable to speculate that, under some conditions, they could be permanent on a scale comparable to man's lifespan.

4. The interfaces of multi-layered stratigraphy may change the course of upward migration of soil voids. Therefore, the surface expression of the collapse is not always directly over the causative opening and voids. Newton (1987) has shown an unusual example of this principle from field investigations in Alabama.

5. The nature of the overburden immediately overlying the carbonate rock is a critical factor in sinkhole occurrence. Clayey strata may be stable, depending on the size of the opening. Dry sands collapse easily, flowing downward under the force of gravity. Water saturated sands also collapse easily, but somewhat less readily than in the dry state. Moist, or semi-saturated, sands collapse least easily and support near vertical slopes.

6. Several adjacent openings at depth may create one large depression in the surface.

7. Sinkholes collapse more rapidly where the overburden sediments are thinner.

8. The surface collapse occurs more rapidly and is larger if the opening to the limestone is larger. The surface depression may be many times larger than the opening to the limestone.

9. Infiltrating recharge aids in the collapse process. It may also cause a reactivation of a collapse which has been temporarily stopped.

10. Surface loading, especially slab-type loading, does not appear to cause additional collapse into a small opening at depth. There does not appear to be any reason to remove a load in an attempt to avoid sinkhole formation.

11. During the collapse process the hydrologic head above the opening to the limestone drops discontinuously with the surrounding water table, creating a hydraulic jump. As a

result the water flow in this area is vertically downward with a great scouring effect. Whereas, at a distance from the sinking point the groundwater flow in the overburden is primarily horizontal.

12. Time is still an unknown factor in sinkhole development. Most of these modelling trials were simulated within a scale of minutes or seconds. However, the rate of sinkhole development will vary with the size of the opening to the limestone, the nature and moisture content of the overburden sediment, and the thickness and stratification of the same. Even in areas where weak layers, karstic cavities, or even soil voids have been detected by geophysical surveys or drilling, one cannot assume that there is danger of imminent collapse. The time span of upward migration may be many tens of years. Of course, under some conditions the process may be very rapid.

Acknowledgements
We would like to thank Dr. Robert Lindquist and Scott Holm for allowing us to use the model which they constructed.

References

Beck, B.F. (ed.), 1984, Sinkholes: their geology, engineering, and environmental impact--Proceedings of the First Multidisciplinary Conf. on Sinkholes: Rotterdam, Netherlands, A.A. Balkema Publishers, 429 p.

-----, 1988, Evnironmental and engineering effects of sinkholes--the processes behind the problems: Environ. Geol. and Water Sci., V. 12, no. 2, p. 71-78.

Beck, B.F., and Wilson, W.L. (ed's), 1987, Karst hydrogeology: engineering and environmental applications--Proceedings of the 2nd Multidisciplinary Conference on Sinkholes and the Environmental Impacts of Karst: Rotterdam, Netherlands, A.A. Balkema Publishers, 467 p.

Holm, Scott R., 1986, The geology and karst processes of the southern Lake Wales Ridge: Unpub. M.S. thesis, Geol. Dept., U. of Fl., 128 p.

Jennings, J.N., 1971, Karst: Cambridge, Mass., M.I.T. Press, 252 p.

Newton, J.G., 1987, Development of sinkholes resulting from man's activities in the Eastern United States: U.S. Geol. Survey Circular 968, 54 p.

NWWA (Nat. Water Well Assoc.), 1986, Proceedings of the environmental problems in karst terranes and their solutions conference: Dublin, OH, Water Well Journal Pub. Co., 525 p.

-----, 1988, Proceedings of the 2nd conference on environmental problems in karst terranes and their solutions: Dublin, OH, Water Well Journal Pub. Co., 441 p.

Sitar, N. (ed.), 1988, Geotechnical aspects of karst terrains: exploration, foundation design and performance, and remedial measures: Am. Soc. Civil Eng., Geotech. Spec. Pub. No. 14, 166 p.

Sowers, G.F., 1975, Failures in limestones in humid subtropics: J. Geotech. Eng. Div. Am. Soc. Civil Eng., p. 771-787.

Sinkhole distribution and characteristics in Pasco County, Florida

JULIA L.CURRIN* *Mainland Senior High School, Daytona Beach, Fla., USA*
BRIAN L.BARFUS *Florida Sinkhole Research Institute, University of Central Florida, Fla., USA*

ABSTRACT

A statistical analysis of size, depth, date, frequency, and location was conducted of 168 reported new sinkholes in Pasco County. An average of twenty-four new sinkholes can be expected to form each year in Pasco County. The average sinkhole size is 10.23 feet in length, 9.15 feet in width, and 7.80 feet in depth. Seventy-five percent of these sinkholes will range from 0 to 10 feet in length providing a threshold target for foundation design engineers to limit sinkhole damage.

Sediment cover thickness is the major factor controlling sinkhole development in Pasco County. Sinkholes are most common in the western half of Pasco County on the Gulf Coastal Plain where cover sediment thickness ranges from 0 to 100 feet.

Most sinkhole occurrences in Pasco County form between March and May, the dry season, and, July and August, the rainy season. During the dry season the potentiometric surface level is lower, inducing increased flow of water from the surficial water table into the limestone of the Floridan Aquifer at depth. This increased flow of water washes sediment into cracks in the limestone forming cavities at the sediment/limestone interface. These cavities can potentially migrate to the surface causing sinkholes. During the rainy season, increased quantities of water deposited on the surface can induce greater flow of water into the limestone causing sinkholes in a similar manner.

INTRODUCTION

Throughout recent years, the occurrence of large numbers of sinkholes in counties along the central west coast of Florida have become a major concern for area residents. Among the most sinkhole-prone counties along the Florida west coast is Pasco County, which suffered 168 sinkhole collapses between 1970 and 1988. The main purpose of this paper is to utilize sinkhole data obtained from the Florida Sinkhole Research Institute at the University of Central Florida to assess the probability of sinkhole occurrences in Pasco County. The data will also be analyzed to determine various statistical parameters related to size, depth, date, and frequency of sinkholes in Pasco County.

GENERAL INFORMATION

Karst can be found in many areas of the United States, including Florida. Florida, however, is unique as one of the only areas of the United States that is almost 50% based upon karst terrain. In effect, Florida is an area where rainfall and other sources of water predominantly infiltrate into the ground instead of draining across the surface in normal, surface drainage basins. Florida possesses one of the largest underground confined water supplies or aquifers in the United States within highly permeable and porous limestones just beneath the surface.

The majority of Florida's residents live upon karst terrain and 92% of the residents obtain drinking water from groundwater supplies. Florida is highly susceptible to potential sinkhole damage and groundwater pollution due to high surface infiltration rates, the close proximity of limestone to the surface, and high groundwater withdrawal rates.

Figure 1 illustrates the general hydrogeologic conditions in Florida and depicts the surficial water table and the potentiometric surface when located in normal positions. The greater the separation between the surface water table and the potentiometric surface, the greater the head driving water downward and the greater the likelihood of a sinkhole occurrence.

*This project was the first-place winner in the Earth and Planetary Sciences category of The Florida State Science and Engineering Fair competition for Ms. Currin. Mr. Barfus contributed substantially in preparing this article from that project.

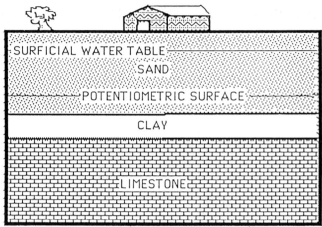

Figure 1: Generalized illustration of surficial
water table and potentiometric surface

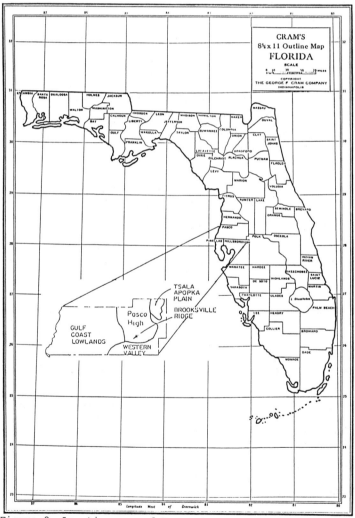

Figure 2: Location map of Pasco County, Florida
illustrating physiographic province boundaries

There are three major triggering mechanisms associated with sinkhole development in most regions of Florida. The first includes a significant lowering in the level of the potentiometric surface due to the heavy pumping of water from wells at an unusually rapid pace or drought conditions over a lengthy time period. Lowering the potentiometric surface induces greater movement of surface water to the subsurface washing sediment into cracks and cavities in the limestone. The resulting void could migrate to the surface forming a cover collapse sinkhole. The second includes the effects of heavy inflow from a drastic storm , a series of drastic storms, or a long duration lower intensity storm. Storms of these types add significant amounts of water to the surface water table causing a pressure build-up on the confining layer overlying the limestone. The pressure build-up also induces flow of surface water into the subsurface eroding the confining layer and washing sediment into the limestone resulting in a possible sinkhole. Finally both mechanisms can act in combination to cause a sinkhole in the manners described above.

There are two basic types of sinkholes forming presently in Pasco County. Cover collapse sinkholes are the most common type. Cover collapse sinkholes form rapidly by the sudden movement of sediment downward into voids at the sediment/limestone interface. They are characterized by steeply sloping sides. Cover subsidence sinkholes form as the entire block of cover sediment overlying the limestone moves downward as sediment is washed slowly into cracks in the limestone. Cover subsidence sinkholes form in a similar manner, but very slowly and are characterized by gently sloping sides.

STUDY AREA

Pasco County is located on the west central Florida Gulf Coast just north of Tampa (Figure 2). The County has a subtropical and humid climate. August is the warmest month and January is the coldest month, with mean temperatures of 82°F and 60°F respectively.

A detailed description of Pasco County geology and hydrogeology is shown in Table 1 (Fretwell, 1987). Pasco County is covered by a thin layer of Recent shallow marine and clay

Table 1: Generalized hydrogeologic column (modified from Fretwell, in press)

SERIES	STRATIGRAPHIC UNIT	HYDROGEOLOGIC UNIT	THICKNESS (FEET)	LITHOLOGY	WATER-PRODUCING CHARACTERISTICS
Holocene and Pleistocene	Undifferentiated surficial deposits	surficial aquifer system	0-100	Soil, sand, and clay of marine and estuarine terraces, alluvial, lake, and windblown deposits.	Wells generally yield less than 20 gal/min.
Early to Late Miocene	Hawthorn Formations	Intermediate confining unit	0-100	Predominantly clay, some grayish-green, waxy, some interbedded sand and limestone, phosphatic clay, marl, calcareous sandstone, limestone residuum.	Generally not a source of water due to extremely low hydraulic conductivity.
	Tampa Formation	Floridan aquifer system	0-60	Limestone, sandy, fossiliferous, sand and clay residuum in upper part in some areas.	Many domestic and irrigation wells produce water from the lower part.
Oligocene	Suwannee Formation		0-250	Limestone, cream to tan colored, fine-grained, grainstone, fossiliferous, thin-bedded to massive, porous.	
Eocene	Ocala Group		70-250	Limestone, white to tan, fossiliferous, massive, soft to hard, porous.	Yields large quantities of water from wells completed above evaporites.
	Avon Park Formation	Sub-Floridan confining unit	200-800	Limestone and dolomite. Limestone is light - to dark-brown, highly fossiliferous, and porosity is variable in lower part. Dolomite is gray to dark-brown, very fine to micro-crystalline and contains porous fossil molds, thin beds of carbonaceous material, and peat fragments. Formation generally contains evaporites in lower part.	Generally, not a source of water due to low permeability and poor water quality.

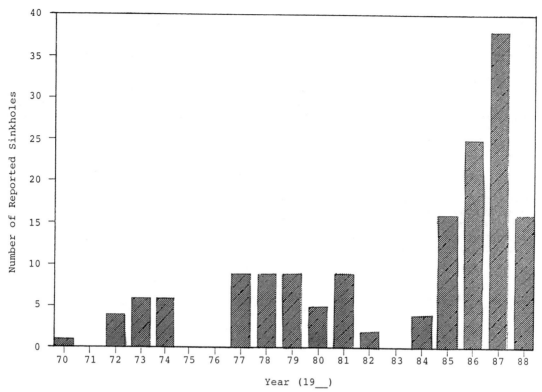

Figure 3: Number of new sinkholes per year in Pasco County, Florida.

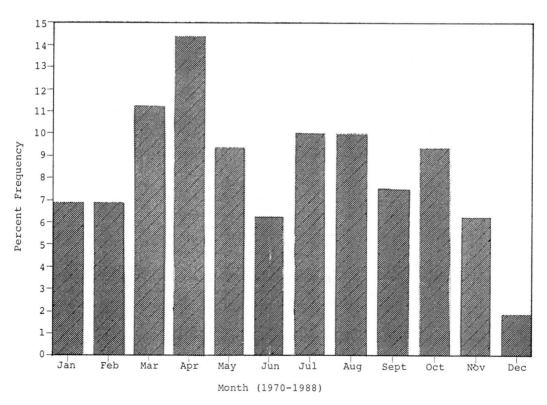

Figure 4: Frequency of reported new sinkholes per month in Pasco County, Florida (1970-1988).

deposits ranging from 10 to 100 feet in thickness. The county is underlain by a thick layer of Tertiary Limestones ranging in thickness from 700 to 1000 feet. These limestones form the main water bearing strata of the Floridan Aquifer and the dissolution of these limestones and their close proximity to the surface is directly or indirectly responsible for the preponderance of sinkhole activity in Pasco County.

White (1970) divided the county into four physiographic regions based on topographic relief and soil characteristics including the Coastal Swamps, the Gulf Coastal Lowlands, the Brooksville Ridge, and the Tsala Apopka Plain/Western Valley (Figure 2). The Coastal Swamps and the Gulf Coastal Lowlands of western Pasco County appear to contain the highest number of recent sinkholes. Surface elevations range from 10 to 100 feet above NGVD. The Brooksville Ridge, located in central Pasco County, forms the dominant topographic feature and ranges from 75 to 300 feet above NGVD in elevation. Few sinkholes form on the Brooksville Ridge. The Tsala Apopka Plain/Western Valley of eastern Pasco County contains a moderate number of sinkholes and ranges in elevation from 75 to 85 feet above NGVD.

STATISTICAL METHODOLOGY

Sinkhole reports from Pasco County, Florida were provided by the Florida Sinkhole Research Institute, University of Central Florida. These reports include data on location and date, length, width, depth, elevation, depth to the water table, potentiometric surface level, depth to the Floridan Aquifer (limestone), and various geologic and climatological parameters.

Standard statistical techniques as described in Wilson, and others, 1987, were used to analyze the sinkhole dimensional data to determine the minimum, maximum, mean, and, median values for the length, width, and depth of sinkholes in Pasco County. Histograms are utilized to illustrate the frequency of each sinkhole dimension. Histograms are also used to illustrate the frequencies of reported new sinkholes per month and per year.

Residual and influence analyses were conducted to determine the effect of each variable on all other variables and to determine if any of the variables exhibit a strong correlation with sinkhole development using a computer program called SOLO (Hintze, 1988). Correlation coefficients were determined as a measure of the degree of correlation between pairs of variables. Statistical tests were conducted to determine the confidence in the results. Finally, a sinkhole density map was constructed for Pasco County illustrating the regions or zones of high and low sinkhole probability. The sinkhole density map was compared to the map of physiographic regions to determine if there is any correlation between sinkhole occurrences and elevation.

RESULTS

Figure 3 shows the total number of new sinkholes reported per year in Pasco County from 1970 to 1988. Total sinkhole numbers steadily rise throughout the entire period. This rise probably reflects the increasing public awareness of sinkhole activity and the increasing growth and development of Pasco County. The period of most complete records ranges from 1985 to 1988 and most likely reflects efforts by the Sinkhole Institute staff to collect sinkhole data. The average number of sinkholes occurring per year during the last four years is 24. This frequency is probably a minumum value for the actual frequency of sinkhole occurrence in Pasco County.

Figure 4 shows the total number of new sinkholes per month for the period 1970 through 1988. The highest sinkhole frequency occurred in May. Inspection of the histogram shows that there are two modal groups, one in March, April, and May, and one in July and August.

The first modal group in March, April, and May corresponds with the dry season in Florida when potentiometric surface levels and rainfall are lowest. Under these conditions, water from the surficial water table can flow more easily into the Floridan Aquifer and sediment is washed into the limestone at depth potentially causing sinkholes.

The second modal group occurs in July and August during the peak rainfall season of summer thunderstorms. During the peak rainfall season large amounts of water are flushed through the surface sediments and infiltrated into the ground, washing sediment into the limestone at depth again potentially causing sinkholes.

Statistical analyses of the length, width, and depth were used to determine the minimum, maximum, mean and median size of sinkholes in Pasco County (Table 2). Length was arbitrarily defined as the largest surface dimension. New sinkhole lengths ranged from 1 to 80 feet. The mean length was 10.23 feet and the median was 5 feet. New sinkhole widths ranged from 1 to 80 feet. The mean width was 9.15 feet and the median was 5.00 feet. New sinkhole depths ranged from 0.3 feet to 75 feet. The mean depth was 7.8 feet and the median was 4.00 feet.

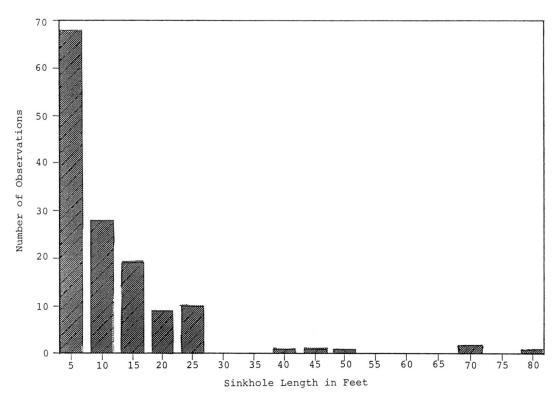

Figure 5: Histogram of new sinkhole lengths in Pasco
County, Florida.

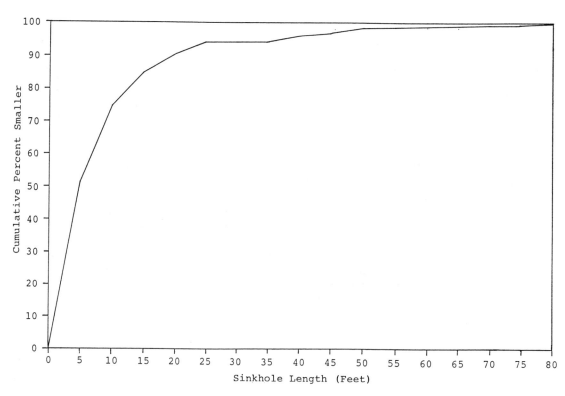

Figure 6: Cumulative frequency of new sinkhole lengths
in Pasco County, Florida.

Table 2: Results of standard statistical analysis on length, width,
and depth of new sinkholes in Pasco County, Florida

	Width(ft)	Length(ft)	Depth(ft)
Number of Observations	140	140	128
Minimum	1.00	1.00	0.30
Maximum	80.00	80.00	75.00
Mean	9.15	10.23	7.80
Median	5.00	5.00	4.00

Figure 5 shows a histogram depicting the frequency of new sinkhole lengths. The largest number of sinkholes fall in the 0-5 foot interval. The number of sinkholes decreases with increasing length in a log normal frequency distribution. Figure 6 shows the cumulative percent frequency of sinkhole lengths. The majority or seventy-five percent of all new sinkholes in Pasco County are between 0 and 10 feet in length.

Figure 7 shows a histogram depicting the frequency of new sinkhole widths. Again, the largest number of sinkholes fall in the 0-5 foot interval. The decrease in the number of sinkholes with width indicates a log normal frequency distribution. Figure 8 shows the cumulative percent frequency of sinkhole widths. Eighty percent of all new sinkholes in Pasco County are between 0 and 15 feet in width.

Figure 9 shows a histogram depicting the frequency of new sinkhole depths. The largest number of new sinkholes fall in the 0-5 foot interval. The decrease in the number of sinkholes with depth is a log normal frequency distribution. Figure 10 shows the cumulative percent frequency of new sinkholes versus depth. Eighty percent of all new sinkholes are between 0 and 10 feet in depth.

Results of residual and influence analyses show good correlation between the thickness of cover sediments and sinkhole occurrences. As expected the majority of new sinkholes in Pasco County form in areas where sediment cover is thin to nonexistent along the Gulf Coast. Good correlation was also exhibited between the level of the potentiometric surface and sinkhole occurrences. The potentiometric surface is at or near the topographic surface and the water table where most sinkholes form along the Gulf Coast of Pasco County. Sinkholes should form most often in areas where the potentiometric surface level is far below the water table and the topographic level. This discrepancy probably indicates that the cover sediment thickness is a much stronger factor in sinkhole development than the level of potentiometric surface in this area of Florida.

Figure 11 shows the density of reported new sinkholes in Pasco County for the entire period 1970-1988. Sinkhole densities ranged from 0 to 25 reported new sinkholes per 4 square miles during the period. The highest sinkhole densities occur in extreme western Pasco County. The overwhelming majority of all sinkholes in Pasco County occur in the western half of the county. The entire western half of Pasco County is located in the Gulf Coast Lowlands physiographic province (Figure 2) where cover sediment thickness over the limestone is known to be very thin. Limestones of the Floridan Aquifer are actually exposed along the Gulf Coast of Pasco County. Clearly, there is a strong correlation between cover sediment thickness, or the lack thereof, and the occurrence of sinkholes in Pasco County.

Several small zones of lower sinkhole density occur in eastern Pasco County on the higher elevations of the Brooksville Ridge and on the interior lowlands of the Tsala Apopka Plain and Western Valley. Sediment cover thickness over these regions is much greater than over the Gulf Coastal Lowlands. This indicates that sinkholes can occur in areas of greater cover sediment thickness; however, their density is very low compared to the Gulf Coastal Lowlands, a maximum of 5 reported new sinkholes per 4 square miles from 1970-1988, as compared to a maximum of 25 for the Gulf Coast Lowlands over the same period.

Conclusion

Sinkholes are becoming an ever increasing concern for the residents of Pasco County, Florida, as the population rises and development accelerates. A knowledge of the size, frequency, and density of sinkhole occurrences in Pasco County can be of immense value to engineers, developers, and county and city planners in their efforts to minimize the costs of sinkhole damage and limit pollution of the Floridan Aquifer.

An average number of 24 sinkholes can be expected to form each year in Pasco County. Seventy-five percent of these sinkholes will range from 0 to 10 feet in length. Most sinkholes in Pasco County form between March and May, the dry season, and, July and August, the rainy season. Foundations of structures could be designed to bridge 10 foot gaps eliminating the damage from three quarters of the sinkholes in Pasco County.

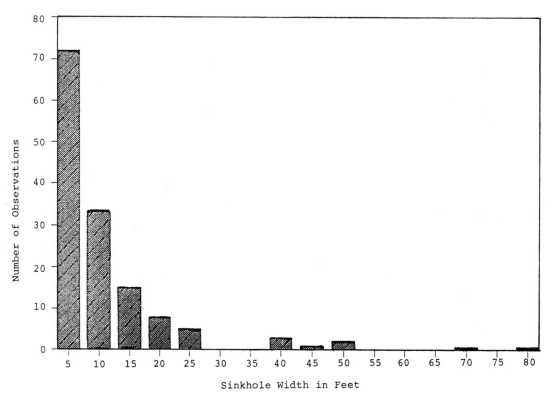

Figure 7: Histogram of new sinkhole widths in Pasco
County, Florida.

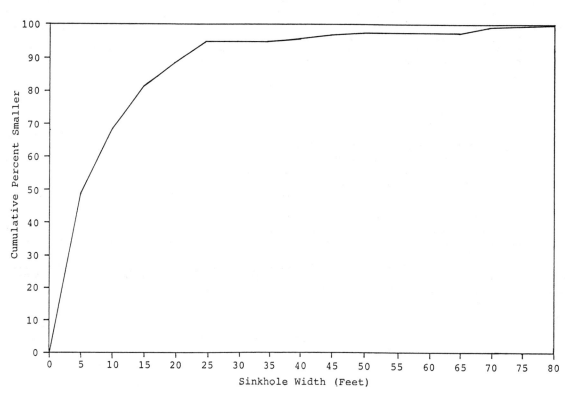

Figure 8: Cumulative frequency of new sinkhole widths
in Pasco County, Florida.

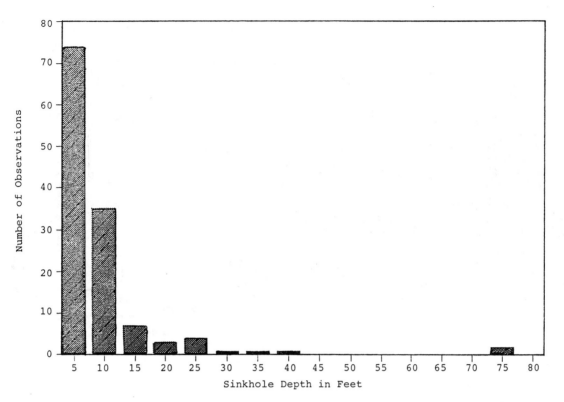

Figure 9:   Histogram of new sinkhole depths in Pasco
            County, Florida.

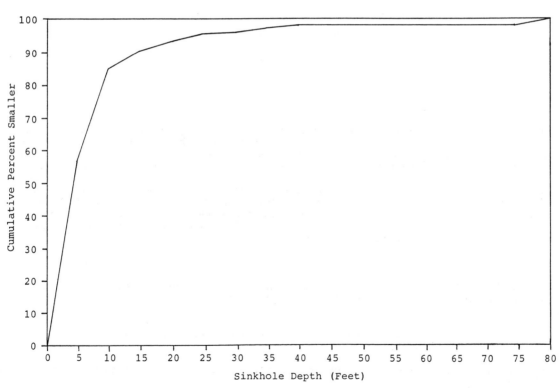

Figure 10: Cumulative frequency of new sinkhole depths in
            Pasco County, Florida.

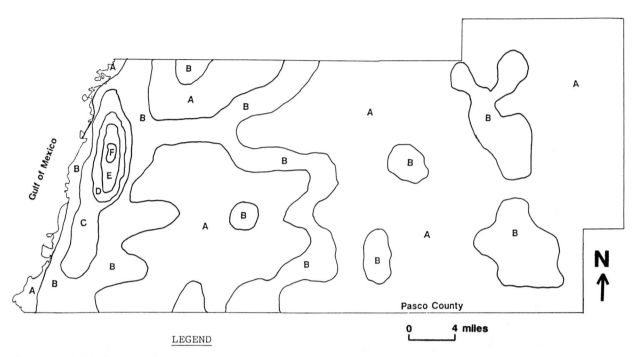

LEGEND

Zone   Sinkhole Density

A   No reported new sinkholes

B   1-5 reported new sinkholes per 4 square miles

C   6-10 reported new sinkholes per 4 square miles

D   11-15 reported new sinkholes per 4 square miles

E   16-20 reported new sinkholes per 4 square miles

F   21-25 reported new sinkholes per 4 square miles

Figure 11. Density of reported new sinkholes in Pasco County,
            Florida from 1970 to 1988.

        Sediment cover thickness is the main factor controlling the development of sinkholes in
Pasco County. The thinner the cover sediment thickness the greater the odds of sinkhole
development. Sinkholes are most common in the western half of Pasco County especially along
the Gulf Coast. Cover sediment thickness ranges from 0 to 100 feet in this area.

        Sinkhole density maps provide an excellent means of delineating areas of high sinkhole
probability. Sinkhole densities are highest along the Gulf Coast of Pasco County. Sinkhole
density maps should prove highly usefull to environmental engineers and county planners in
their efforts to locate landfills and industries in areas less sensitve to sinkholes and
potential pollution of the Florida Aquifer.

References

Fretwell, J. D., 1987, Water Resources and Effects of Ground-Water Development in Pasco
        County, Florida: U. S. Geological Survey Water Resources Investigations Report (in
        press).

Hintze, Jerry L., 1988, SOLO, Stastical Software, 1440 Sepulveda Blvd., Suite 316 Los
        Angeles, California 90025.

White, William A., 1970, Geomorphology of the Florida Peninsula, Florida Bureau of Geology,
        Bull. 51, 164p.

Wilson, William L., McDonald, Kathleen M., Barfus, Brian L., and Beck, Barry F., 1987,
        Hydrogeologic Factors associated with Recent Sinkhole Development in the Orlando Area,
        Florida, Florida Sinkhole Research Institute, University of Central Florida, Orlando,
        Rept. No 87-88-4, 109p.

Sinkholes and landuse in southwestern Wisconsin

MICHAEL DAY & PHILIP REEDER *University of Wisconsin-Milwaukee, Milwaukee, Wis., USA*

ABSTRACT

The karst developed in the Ordovician carbonates of southwestern Wisconsin contains several hundred sinkholes, in response to which landowners have adopted landuse strategies to minimize personal inconvenience and risk. Of 263 field-documented sinkholes, 75% are formed in the Oneonta Formation dolostones of the Prairie du Chien Group, and the remaining 25% are in the Platteville and Galena dolostones of the Sinnipee Group. Sinkhole densities are generally low, but often individuals are clustered on the tops and sides of interfluvial ridges between dry valleys. Although contemporary dissolution rates are low, sinkholes are currently forming at a rate of 5 to 10 per year. Most sinkholes are small, but locally they are a significant hazard to livestock and farm machinery. Contemporary landowner response includes avoidance, fencing and infilling. During the nineteenth century some sinkholes, particularly on the Platteville and Galena dolostones, were mined in search of lead ores.

Introduction

Sinkholes need be neither large nor active to have an appreciable impact on landuse practices. Even where sinkholes are small, scattered, inconspicuous and relatively benign, farmers and other landowners are acutely aware of them and adopt landuse strategies which minimize personal inconvenience and risk. Although locally in southwestern Wisconsin sinkhole development requires appropriate engineering practices, the broader environmental impact of sinkholes is manifested in subtle but widespread and traditional landuse adaptations. Generally landowners perceive sinkholes as a nuisance, but individual sinks have been used constructively for a variety of purposes.

The Study Area

The karst of southwestern Wisconsin is part of a larger karst area in the upper Mississippi drainage system, and it has developed in Paleozoic dolostones: the Ordovician-aged Prairie du Chien and Platteville-Galena Formations. The karst is not well known in the scientific literature, but it contains a wide array of caves, sinkholes, dry valleys and springs (Day, Reeder and Oh, 1989). There is evidence of pre-Cretaceous paleokarst (Hedges and Alexander, 1985). The karst is in the Driftless Area of southeastern Minnesota, southwestern Wisconsin, northeastern Iowa and northwestern Illinois (Figure 1) - an area about whose Quaternary history there has been much debate (Mickelson et al., 1982).

Geologic Setting

In contrast to the surrounding landscape, the Driftless Area of the Upper Mississippi Valley has not been covered by Pleistocene glacial deposits. Elsewhere in Wisconsin Pleistocene ice sheets obliterated or buried pre-glacial karst landscapes, but in the Driftless Area, Paleozoic dolostones outcrop at the surface, and it is in these that the sinkholes are developed.

The early Ordovician Prairie du Chien carbonates are medium-textured, light grey to buff dolostones, locally sandy, cherty and shaley (Heyl et al., 1970; Day, 1979, 1984). Quartz contents exceed 10% (Day, 1979), and clay contents average 2.1% (Frolking, 1982). Mean insoluble residue is 12.17% by weight (Day, 1979), with local range between 1.37% and 26.6% (Black, 1970). Porosity is about 10%, and in fossil fragments silica often has replaced the carbonate (Akers, 1965). The lowermost formation, the Oneonta Dolomite, is up to 30m thick and is "...a thick or poorly bedded, medium crystalline, saccharoidal dolostone with minor amounts of chert and shale, cavernous zones of poorly preserved algal stromatolites and...large secondary calcite crystals" (Barden, 1980, p.5).

The Middle Ordovician Platteville and Galena dolostones, often regarded as a single formation, are the most widely exposed carbonates in southwestern Wisconsin (Figures 2 and

Figure 1: Location of the Driftless Area.

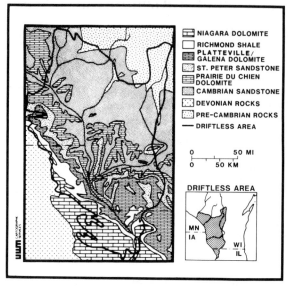

			NIAGARA DOLOMITE
			RICHMOND SHALE
			PLATTEVILLE/ GALENA DOLOMITE
			ST. PETER SANDSTONE
			PRAIRIE DU CHIEN DOLOMITE
			CAMBRIAN SANDSTONE
			DEVONIAN ROCKS
			PRE-CAMBRIAN ROCKS
			DRIFTLESS AREA

Figure 2: Generalized Geology of South-western Wisconsin.

System	Series	Formation or Group	Avg. Thickness in Meters
ORDOVICIAN	Upper	Richmond Shale	46
	Middle	Platteville/ Galena Dolomite	96
		St. Peter Sandstone	12
	Lower	Prairie du Chien Group	91

Figure 3: Simplified Ordovician Geologic Column for Southwestern Wisconsin.

3). The Platteville members are fine to medium crystalline, buff to blue-grey, fossiliferous dolostones and limestones (Barden, 1980) with a total thickness of about 25 m. The Galena members are thin to thick bedded, buff, shaley dolostones, locally with chert and shale bands. Total thickness is nearly 80 m (Barden, 1980).

The Platteville-Galena represents widespread, uniform carbonate sedimentation on a stable, shallow marine platform (Agnew, 1963). Reefs and bioherms are rare (Heyl et al., 1970), but Paleozoic tectonic stresses produced folding and fracturing; vertical and inclined joints are traceable for up to 90m vertically and 3km horizontally (Agnew, 1963). The carbonates are progressively more dolomitized eastward across the Wisconsin Arch, and many of the original structures have been accentuated by dissolution and by slumping during mineralization (Agnew, 1963).

Paleozoic lead ores (primarily galena) were deposited by rising hydrothermal solutions moving through dissolutionally-enlarged fractures which may have been formed by an aggressive pre-ore phase of ascending hydrothermal fluids (Behre et al., 1935; Hedges and Alexander, 1985). When meteoric water was drawn in subsequently the hydrothermal solutions were diluted and the brines were flushed out, leaving the ore deposits (Heyl et al., 1970). Lead deposits are most common in the upper Galena dolostones.

Geomorphology of the Karst

South of the Wisconsin River most karst is formed in the Platteville-Galena dolostones (Deckert, 1980); north of the river the karst is developed primarily in the Prairie du Chien group (Figure 2). In both areas the carbonates are expressed as interfluvial ridges, sometimes with carbonate cliff faces producing mesa-like forms (Barden, 1980). Much of the karst is covered by loess and colluvial deposits; it contains a combination of karstic and fluvial landforms, and thus is best categorized as fluviokarst (Sweeting, 1972).

The karst is part of an upland dissected by dendritic drainage systems. Drainage patterns and interfluvial ridge orientations are controlled by fracture orientations and by broad folds that range from 30km to 50km in length, 5km to 10km in width and 30m to 60m in amplitude (Heyl et al., 1959; Knox, 1982). The main valleys are deep, broad and alluviated, with local relief exceeding 100m. Maximum valley incision occurred during the Pleistocene, although the drainage configuration was probably established prior to then. Nebraskan deposits in the Wisconsin River Valley indicate entrenchment of some 85m since the Kansan (Mickelson et al., 1982).

Typically, valley side slopes are steep, up to 25 degrees, and are mantled by colluvium and talus. Contemporary hillslope process rates are high (Black, 1969, 1970) and were probably even higher in the late Pleistocene (Knox, 1982). Loess blankets much of the landscape, although much of it has been washed from the hillslopes into valley bottoms (Trimble, 1983). Dolostone ridge tops, particularly on the cherty lower Galena Dolomite, often are mantled by red clays, which Frolking (1982) concluded had been formed by clay illuviation in the zone of carbonate dissolution.

Contemporary dissolution of the dolomites is sluggish. Weathering rates for 80 weight-loss tablets of Prairie du Chien dolomite from 1979 to 1984 ranged from 1.02 to 3.18 mg. $10^{-3}/day/cm^2$ - up to 30% lower than rates for limestones (Day, 1984). Although dissolution rates are low, mechanical weathering, particularly freeze-thaw, is highly effective (Day, 1984).

The carbonate aquifers are dominantly of the diffuse flow type (White, 1969, 1977), which is typical of impure limestones and dolostones. Recharge is primarily through infiltration, and transmission velocities are low. Some parts of the Prairie du Chien aquifer appear to be of the free flow type with higher velocity, turbulent flow in integrated conduits. Paleoconduits, abandoned as the regional potentiometric surface was lowered, exist in several caves in both the Prairie du Chien and the Platteville-Galena (Day, 1986) (Figure 4). Recharge to active conduits may be via sinkholes or sinking streams but catchment areas are small. Water in the vadose zone achieves up to 80% of springwater hardness (cp. Smith and Atkinson, 1976), but the limited catchments inhibit sinkhole development.

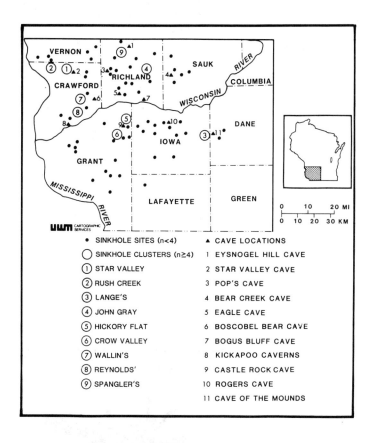

Figure 4: Selected Sinkhole and Cave Locations in Southwestern Wisconsin.

The predominant features of the karst landscape are tributary valleys which contain only ephemeral streams, having become dry as a result of incision of the main valleys and progressive development of the underground drainage system. Following the general model (Smith, 1975), the upper portions of surface drainage systems have become abandoned as dissolution has increased the subsurface drainage capacity.

Sinkholes

Sinkholes in southwestern Wisconsin were identified by early writers including Lapham (1873) and Martin (1932), who reported more than 70 sinkholes in Vernon, Crawford and Richland Counties. Since 1984 field investigation has revealed and documented 263 sinkholes: 197 (75%) in the Prairie du Chien Group dolostones and 66 (25%) in the Platteville-Galena dolostones. Counties with significant numbers of sinkholes include Crawford, Vernon, Richland, Grant and Iowa (Figure 4).

The 263 sinkholes which have been identified so far in southwestern Wisconsin represent only a portion of the total sinkhole population. Sinkholes are not visible on air-photographs or satellite imagery, and they are depicted only rarely on topographic maps. In this study sinkholes have been detected by field inquiry and investigation in areas of known sinkhole occurence. One hundred eighty-seven of the sinkholes are located in Woodland tracts, making their detection difficult even on foot. The remaining 76 are in cropped fields or livestock pasture;

62 of these sinks are marked by clumps of trees, although in 85% of cases trees alone do not indicate sinkholes. By rough approximation of carbonate areas and detection rates, the 263 sinkholes in this study probably represent less than 25% of the total southwestern Wisconsin population. In addition, it is possile that many former sinkholes are filled with windblown loess or with colluvium.

To date 174 individual landowners in known or suspected sinkhole vicinities have been asked whether there were sinkholes on their property. Of these, 88 replied in the affirmative, and an additional 36 were aware of sinkholes on adjacent or nearby property. These figures suggest that, at least in these areas, sinkholes are not uncommon and that landowners are well aware of them.

Overall, sinkhole densities are low, less than 0.5/km2, but there are several clusters with over 20 sinkholes/km2 (Figure 4). Most sinkholes are small - mean diameter is 6.8m and mean depth is 1.1m - but individuals are up to 50m wide and 6m deep. Seventy-eight percent of all sinkholes are less than 10m in diameter, and 51% are less than 5m wide. Width/length ratios indicate that on both the Prairie du Chien and Platteville-Galena sinkholes approximate circularity and that few individuals are markedly elongated along structural lines (Figure 5). There are also no statistically-significant (Chi-square, 0.05 level preferred long axis orientations, although visually a WSW-ENE trend is apparent (Figure 6). Long axis orientations correspond poorly with orientations of major joints and lead-bearing crevices (Grant, 1906) (Figure 7).

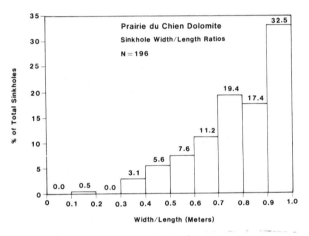

Figure 5a: Sinkhole Width/Length Ratios, Prairie du Chien.

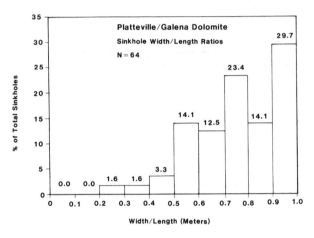

Figure 5b: Sinkhole Width/Length Ratios, Platteville-Galena.

Figure 6a: Sinkhole Long Axis Orientations, Prairie du Chien.

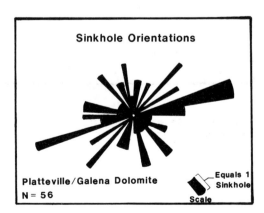

Figure 6b: Sinkhole Long Axis Orientations, Platteville-Galena.

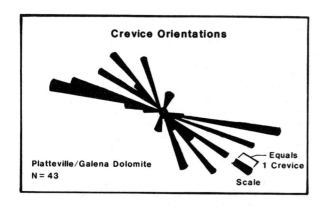

Figure 7a: Orientations of Lead-bearing Crevices (after Grant, 1906).

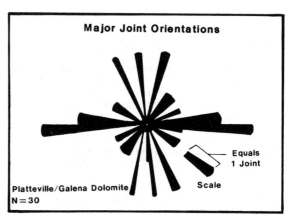

Figure 7b: Orientations of Major Joints (after Grant, 1906).

Two hundred forty-nine (95%) of the documented sinkholes have developed on the tops (119) and sides (130) of interfluvial ridges; only 14 (5%) are near slope bases, where bedrock is masked by colluvium. Surface collapse and the suffosion of regolith into bedrock cavities are important formative mechanisms, and it is estimated on the basis of personal obseration and communication with landowners that sinkholes are currently forming regionally at a rate of between 5 and 10 per year. Preliminary analysis of natural sinkhole fills suggests that these are derived both from bedrock weathering and from Wisconsinan loess (J. Oh, pers. comm.) and some are similar in appearance to the patchy upland red clays which Frolking (1982) attributed to clay illuviation in the zone of dolomite dissolution.

At least ten sinkholes lead to open cave passage beneath, but currently only two function as sinks via which surface streams are diverted underground. Because they are on ridges, most sinkholes have limited catchment areas, typically less than 500m2.

Sinkholes and Landuse

Agriculture is the dominant landuse in southwestern Wisconsin, and farmland (cropland and pasture) typically accounts for between 70% and 80% of county landcover. The relationship between agriculture and the karst landscape is generally harmonious but farmers and other landowners have adopted the following complementary tactics to minimize the inconvenience and risk associated with sinkholes:
1. Avoidance of existing sinkholes during plowing
2. Encouragement of tree growth in existing sinkholes
3. Fencing of sinkhole margins to prevent livestock access
4. Infilling of existing and developing sinkholes.

Avoidance is the primary response of landowners to southwestern Wisconsin sinkholes. Although 67 of those documented have been used for some purpose, the remaining 196 (75%) appear to have been left alone. This is most evident in farm fields, where in only three cases have broad, shallow sinkholes been plowed. The remaining 73 sinkholes in farm fields are avoided during plowing, planting and harvesting, primarily because the sinkhole slopes are too steep for farm equipment to negotiate and because of the perceived danger of ground surface failure.

All of the sinkholes which are avoided by landowners have trees growing in or immediately around them. In wooded tracts, many of which are on slopes deemed too steep for agricultural use, trees in sinkholes are broadly the same age and species as those surrounding them, suggesting that there is no special response to the sinkholes themselves. Where sinkholes are in farm fields tree growth is encouraged both to serve as an indicator of sinkhole position, hence avoiding equipment damage, and to hinder livestock entry.

Forty-three of the sinkholes (16%) have been fenced, using barbed or other wire, to prevent livestock (and human) access and injury. Of these, 28 are in pastures and the remaining 15 are in woodland. Sinkholes which have particularly steep (greater that 20 degrees) internal slopes or which open into caves account for 49% of those which are fenced. Vulnerability to falls or entrapment in sinkholes is perceived by farmers as greatest among young cattle and sheep.

Of the 67 sinkholes which have been used for some purpose, 62 (93%) have been used for disposal of fieldstones, other agricultural and domestic refuse. These activities serve two related purposes. First, the sinkholes provide convenient although imprudent dumping sites for waste; second, the waste serves to infill the sinkholes and may enable them to be reclaimed for agricultural use. According to farmers, at least 53 former sinkholes and at least 19 new, developing sinkholes have been completely filled because they represented an agricultural inconvenience, and perhaps a threat to livestock and farm machinery. While infilling of sinkholes with boulders and soil masks their presence, and is not guaranteed to prevent further ground failure, the use of sinkholes for waste disposal poses a potential threat to water supplies, since contaminants may move rapidly from the sinkhole to the aquifer via cave systems and along lines of preferential dissolution. To date, evidence of such pollution is limited to the immediate vicinity of a single sinkhole; regional groundwater contamination arises from nonpoint sources, primarily field application of agricultural chemicals.

Other contemporary uses of sinkholes are temporary and inconsequential, but historically many sinkholes were investigated as possible sources of lead. During the nineteenth century, especially between 1825 and 1870, the karst of southwestern Wisconsin was one of the most important sources of galena (lead sulfide) in the World. Lead mining began around 1815, and focussed on ore excavation from sinkholes, crevices and caves (Reeder and Day, 1989). The earliest mining was of surface "float" deposits, and it is sometimes difficult to distinguish natural sinkholes from borrow pits created by nineteenth century lead miners. The latter are usually surrounded by distinct mounds or ridges of excavated material, and they usually occur as high density clusters, but it is not unusual for diggings to follow joints and other lines of weakness, mimicking the development of sinks along lines of preferential dissolution. In many cases the lead miners excavated natural sinkholes, seeking surface deposits or access to underground ores. Since most lead mining was in the Platteville-Galena Formation, it is sinkholes in this geologic group that are most suspect, but the lack of correspondence between sinkhole and lead-bearing crevice orientations (Figures 6 and 7) suggests that the sinkholes considered here are not mining pits.

Conclusions

Although they are small, scattered and inconspicuous, sinkholes are not uncommon in southwestern Wisconsin. Moreover, farmers and other landowners are aware of them and have adopted a set of landuse strategies which minimize risk and inconvenience. Although avoidance is the primary contemporary tactic, sinkholes are treated in a variety of ways and they may have played a significant role in indicating lead deposits to nineteeth century miners.

References

Agnew, A.F., 1963. Geology of the Platteville Quadrangle, Wisconsin. U.S. Geological Survey Bulletin 1123-E, 245-277.

Akers, R.H., 1965. Unusual Surficial Deposits in the Driftless Area of Wisconsin. University of Wisconsin PhD Thesis, 169 p.

Barden, M.J., 1980. Geology Field Trip Guide. Wisconsin Speleological Society Hodag Hunt Guidebook, 1980, Section 2, 29p.

Behre, C.H., Scott, E.R. and Banfield, A.E., 1935. The Wisconsin Lead-Zinc District: A Preliminary Paper. Economic Geology, 20, 783-809.

Black, R.F., 1969. Slopes in Southwestern Wisconsin, U.S.A., Periglacial or Temperate? Biuletyn Peryglacjalny, 18, 69-81.

Black, R.F., 1970. Residuum and Ancient Soils of Southwestern Wisconsin. In: Pleistocene Geology of Southern Wisconsin, Wisconsin Geological and Natural History Survey Information Circular 15, I1-I12.

Day, M.J., 1979. Preliminary Results of an Investigation of Current Rates of Erosion in the Wisconsin Karst. The Wisconsin Speleologist, 16 (2), 11-23.

Day, M.J., 1984. Carbonate Erosion Rates in Southwestern Wisconsin. Physical Geography, 5 (2), 142-149.

Day, M.J., 1986. Caves of the Driftless Area of Southwestern Wisconsin. The Wisconsin Geographer, 2, 42-51.

Day, M.J., Reeder, P.P. and Oh, J.W., 1989. Dolostone Karst in Southwestern Wisconsin. The Wisconsin Geographer, in press.

Deckert, G., 1980. Geology of Southwestern Wisconsin. In: E.C. Alexander (Editor), An Introduction to Caves of Minnesota, Iowa and Wisconsin. National Speleological Society Convention Guidebook 21, 190 p.

Frolking, T.A., 1982. The Genesis and Distribution of Upland Red Clays in Wisconsin's Driftless Area. In: Quaternary History of the Driftless Area, Wisconsin Geological and Natural History Survey Field Trip Guide Book 5, 88-97.

Grant, U.S., 1906. Lead and Zinc Deposits Atlas. Wisconsin Geological and Natural History Survey Bulletin 14.

Hedges, J. and Alexander, E.C., 1985. Karst Related Features of the Upper Mississippi Valley Region. Studies in Speleology, 6, 41-49.

Heyl, A.V., Agnew, A.F., Lyons, E.J. and Behr, C.H., 1959. The Geology of the Upper Mississippi Valley Lead-Zinc District. U.S. Geological Survey Professional Paper 309, 310p.

Heyl, A.V., Broughton, W.A. and West, W.A., 1970. Guidebook to the Upper Mississippi Valley Base Metal District. Wisconsin Geological and Natural History Survey Information Circular 16, 49p.

Knox, J.C., 1982. Quaternary History of the Kickapoo and Lower Wisconsin River Valleys, Wisconsin. In: Quaternary History of the Driftless Area. Wisconsin Geological and Natural History Survey Field Trip Guidebook 5, 1-66.

Lapham, I.A., 1873. A Geological, Topographical and Subterranean Map of the Blue Mounds and Brigham Lead Mines.

Martin, L., 1932. The Physical Geography of Wisconsin. 2nd Edition, University of Wisconsin Press, 608p.

Mickelson, D.M., Knox, J.C. and Clayton, L., 1982. Glaciation of the Driftless Area: An Evaluation of the Evidence. In: Quaternary History of the Driftless Area, Wisconsin Geological and Natural History Survey Field Trip Guidebook 5, 155-169.

Reeder, P.P. and Day, M.J., 1989. Lead Mining of Caves in the Driftless Area of Southwest Wisconsin and Northwest Illinois. Proceedings 10th International Speleological Congress, in press.

Smith, D.I., 1975. The Problem of Limestone Dry Valleys - Implications of Recent Work in Limestone Hydrology. In: Processes in Physical and Human Geography, Peel, R.F., Chisholm, M. and Haggett, P., (Editors). Heinemann, London, 130-147.

Smith, D.I. and Atkinson, T.C., 1976. Process, Landforms and Climate in Limestone Regions. In: Geomorphology and Climate, Edited E. Derbyshire, London, Wiley, 367-409.

Sweeting, M.M., 1972. Karst Landforms. Macmillan, 362p.

Trimble, S.W., 1983. A Sediment Budget for Coon Creek Basin in the Driftless Area, Wisconsin, 1853-1977. American Journal of Science, 283, 454-474.

White, W.B., 1969. Conceptual Models for Carbonate Aquifers. Ground Water, 7, 15-21.

White, W.B.. 1977. Conceptual Models for Carbonate Aquifers: Revisited. In: Hydrologic Problems in Karst Regions, R.R. Dilamarter and S.C. Csallany (Editors), Western Kentucky University, Bowling Green, 176-187.

Salt dissolution, interstratal karst, and ground subsidence in the northern part of the Texas Panhandle

KENNETH S.JOHNSON   *Oklahoma Geological Survey, Norman, Okla., USA*

ABSTRACT

Natural dissolution of Permian salt beds in the northern Texas Panhandle during late Cenozoic time has resulted in interstratal karst, wherein sheet-like dissolution occurs in extensive areas beneath covering layers of nonkarstic rocks. A result of such interstratal karst is the disturbance of overlying strata as they subside or collapse into dissolution cavities. A study of interstratal karst in the Permian Flowerpot salt in the vicinity of the proposed Palo Duro dam and reservoir was conducted to determine whether karst development is an ongoing process and whether any special construction was needed to accommodate the interstratal karst. The Flowerpot salt is 0-107 m (0-352 ft) thick; it is at a depth of 180-335 m (600-1,100 ft), but most of it already has been dissolved from most of the study area. The Flowerpot salt is totally absent from the well just northeast of the dam. Strata beneath the Flowerpot salt are essentially flat-lying and undisturbed, whereas Permian and Tertiary strata overlying the interstratal-karst areas in the Flowerpot salt are chaotic and are structurally low.

The principal outcrops in the study area are horizontal sands and gravels in the Ogallala Formation, of Miocene-Pliocene age. The elevation of the base of the Ogallala is highly irregular, due to widespread dissolution of salt and resultant subsidence during and after Ogallala time. Calcrete, which forms a horizontal caprock near the top of the Ogallala, is disturbed in two large subsidence basins; thus, salt dissolution and subsidence continued locally after development of the caprock (after 3.5 to 2.4 million years ago). Salt dissolution and subsidence probably ceased at least several hundred thousand years ago, but one can only be certain that it ended more than 10,000 years ago, prior to development of modern stream valleys and alluvium. Therefore, inasmuch as interstratal karst is not an ongoing process in the project area, no special construction design is needed at the site to accommodate the karst.

Introduction

The current study of salt dissolution, interstratal karst, and resultant ground subsidence was undertaken in the vicinity of the proposed Palo Duro Dam and Reservoir in the northern part of the Texas Panhandle (Figure 1). Initial reviews of available information for this part of northeastern Hansford County showed a lack of specific data on salt deposits and salt dissolution in the project area. The lack of such data led to uncertainty as to whether shallow salt deposits still existed beneath the area or whether they had been removed long ago by natural dissolution; also, there was uncertainty as to whether dissolution of any remaining shallow salts is an ongoing process beneath the project area. Such information is critical in making final decisions concerning dam design and also in providing guidance for plugging of wells that penetrate salt within the reservoir area.

The focus of this report is on evidence and results of "interstratal karst," a term used for dissolution of soluble rocks beneath a covering layer of younger, nonkarstic rocks. Fresh water percolates down through overlying rocks and dissolves the salt, forming a system of cavities. As a cavity enlarges, the roof eventually can no longer be supported, and overlying rocks subside or collapse into the cavity.

Methods of Study

Investigations consisted of two principal phases: (1) a preliminary study of published data on regional geology and salt dissolution, and (2) a detailed study of the surface and subsurface geology at and near the proposed dam site and reservoir. The preliminary study consisted mainly of a review of principal studies of Permian evaporites and salt dissolution in the western Anadarko Basin region (Jordan and Vosburg, 1963; Johnson and Gonzales, 1978; Gustavson et al., 1980, 1982; Johnson, 1981; Simpkins et al., 1981; Gustavson and Finley, 1985; Gustavson, 1986); these data provided the framework for the detailed studies.

Detailed or site-specific studies consisted of three phases. (1) Evaluation of geophysical logs for 22 petroleum tests and driller's logs for 12 water wells drilled near the proposed reservoir. From these logs a series of maps and cross sections was prepared to show geologic structure and the distribution and thickness of salt deposits. (2) Stereoscopic study of aerial photos covering the proposed reservoir enabled determination of surface geologic structure and identification of two major collapse features or subsidence basins. (3) A site visit to evaluate and confirm the results of the first two detailed studies.

Site Geology, Salt Dissolution, Interstratal Karst, and Ground Subsidence

Outcropping rocks and sediments along Palo Duro Creek in northeastern Hansford County and the surrounding region are the Tertiary Ogallala Formation and the younger Quaternary cover sands, terrace deposits, and alluvium. The Ogallala rests unconformably upon several thousand feet of Permian red beds and evaporites (salt, gypsum, and anhydrite) deposited near the western end of the Anadarko Basin. The stratigraphic column for the area is presented in Figure 2. Permian and Ogallala strata overlying the interstratal karst zone in the Flowerpot salt are structurally complex and disrupted, whereas underlying strata, such as the Cimarron anhydrite, are essentially horizontal throughout the county.

The oldest stratigraphic unit of interest in this study is the Cimarron anhydrite, which is an excellent marker bed throughout the Anadarko Basin and nearby areas (Jordan and Vosburg, 1963). The unit typically is about 12 m (40 ft) thick and consists of several anhydrite beds interbedded with shale. Structure of the Cimarron anhydrite is quite simple (Figure 3D): the unit dips gently to the east and southeast at 1-4 m/km (5-20 ft/mi) and it lacks any of the chaotic structures present in younger strata. The Cimarron anhydrite underlies the zone of interstratal karst; thus, it shows the true structure of Permian strata in the area before dissolution of younger salt beds caused disruption of post-Flowerpot-salt strata.

The Upper Cimarron salt is a thick sequence of salt and salty shale that directly overlies the Cimarron anhydrite (Figure 2). This salt unit underlies all parts of the Texas Panhandle (Jordan and Vosburg, 1963; Johnson and Gonzales, 1978), and in the study area it ranges from 41-126 m (137-413 ft) thick. Some of the local thinning of the Upper Cimarron salt is undoubtedly due to dissolution (Figure 4), but its impact upon ground subsidence is relatively small, in comparison to the impact of Flowerpot-salt dissolution. The depth to the top of the Upper Cimarron salt ranges from 330-446 m (1,082-1,463 ft) below the land surface, and the salt is thus well below the depth at which salt commonly is now being actively dissolved (typically at depths of less than 150 m, or 500 ft).

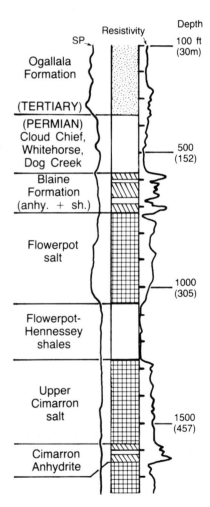

Figure 2. Geophysical log, lithology, and subsurface stratigraphy in the study area (R. H. Fulton, Lasater No. 3, H&TC, Block 45, sec. 72).

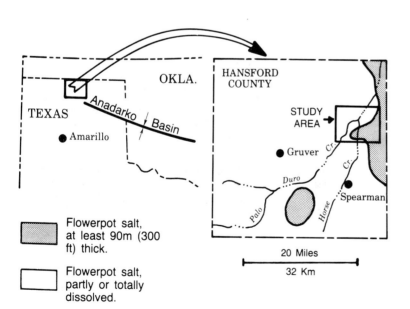

Figure 1. Location map showing study area in northeastern Hansford County, Texas.

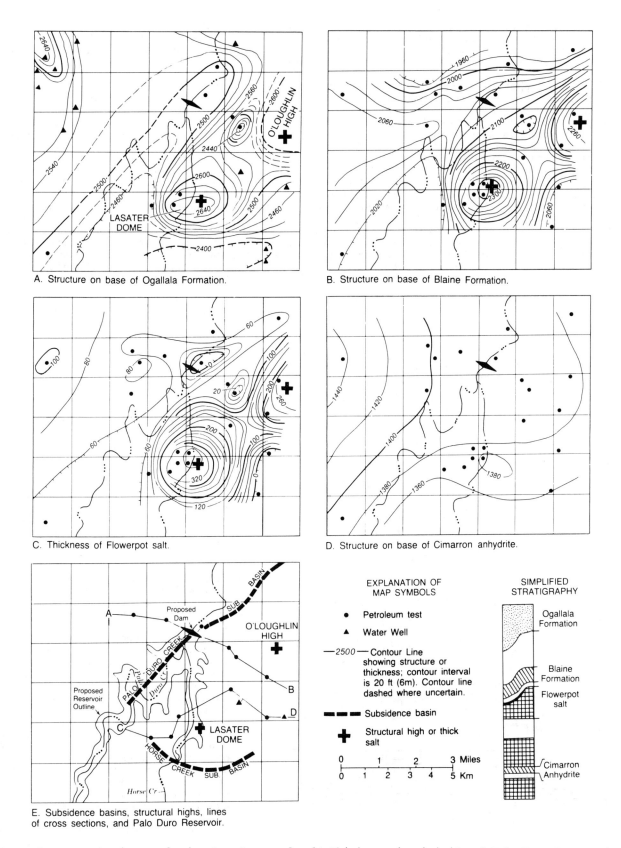

A. Structure on base of Ogallala Formation.

B. Structure on base of Blaine Formation.

C. Thickness of Flowerpot salt.

D. Structure on base of Cimarron anhydrite.

E. Subsidence basins, structural highs, lines of cross sections, and Palo Duro Reservoir.

EXPLANATION OF MAP SYMBOLS

• Petroleum test

▲ Water Well

—2500— Contour Line showing structure or thickness; contour interval is 20 ft (6m). Contour line dashed where uncertain.

▬ ▬ ▬ Subsidence basin

✚ Structural high or thick salt

SIMPLIFIED STRATIGRAPHY

Ogallala Formation

Blaine Formation

Flowerpot salt

Cimarron Anhydrite

Figure 3. Maps showing geologic structure and salt thickness in vicinity of Palo Duro Reservoir.

Nonsalty strata between the Upper Cimarron and Flowerpot salts are referred to as the Flowerpot-Hennessey shales (Jordan and Vosburg, 1963). These strata consist of 60-75 m (200-250 ft) of red-brown shales and siltstones, and their low permeability has probably helped minimize the amount of dissolution in the underlying salt.

The Flowerpot salt, the major unit involved in interstratal karst in the area, consists of interbedded salt and red-brown shale. The salt is thick and widespread in the eastern part of the Texas Panhandle, but has been largely or totally dissolved in most parts of Hansford County and farther to the west (Jordan and Vosburg, 1963; Johnson and Gonzales, 1978) (Figure 1). Where the salt has been dissolved, the remaining equivalent rock consists of insoluble red-brown shales and siltstones that are disrupted due to salt removal; where salt is missing, the strata are called the Flowerpot Shale.

The thickness of the Flowerpot salt varies sharply in the study area (Figure 3C). It is up to 107 m (352 ft) thick beneath the Lasater dome, and is 82 m (268 ft) thick beneath the O'Loughlin high; within short distances of these places the Flowerpot salt thins sharply to less than 30 m (100 ft). All the Flowerpot salt is missing in the well just northeast of the proposed dam, and also in two wells in the southeast part of the area. The main area of thin salt is beneath Palo Duro Creek, and this coincides with a near-surface collapse structure (in the Ogallala Formation) called the Palo Duro Creek subsidence basin. Inasmuch as the Flowerpot salt is absent or thin in the vicinity of the proposed dam, it appears that the dissolution process has already acted extensively on the Flowerpot salt at and near the dam site.

Based upon regional and local studies, it is clear that the irregular thickness pattern of the Flowerpot salt in the study area results from partial or total dissolution of the salt through the process of interstratal karst. As a result, overlying strata have subsided or collapsed into the dissolution cavities, and their structure conforms closely to the areas of thick and thin salt (Figures 3 and 4).

The depth to the top of the Flowerpot salt ranges from 180-335 m (600-1,100 ft) below the current land surface. In other parts of the Anadarko Basin, various salt units are now being dissolved locally where they are within 15-150 m (50-500 ft) of the land surface. The partial or total absence of Flowerpot salt at depths as great as 240-335 m (800-1,100 ft) in the project area results from earlier episodes of salt dissolution when the land surface was much closer to the top of the salt. The pre-Ogallala land surface was eroded down to a level within 90-150 m (300-500 ft) of the top of the Flowerpot salt during Late Cretaceous through late Tertiary time, and some of the salt dissolution may have occurred during this long interval of pre-Ogallala erosion. However, the highly irregular structure at the base of the Ogallala indicates that much of the salt dissolution and collapse in the project area occurred at the onset of, or during, Ogallala deposition.

Above the Flowerpot salt is the Blaine Formation, which consists of about 45 m (150 ft) of interbedded anhydrite, gypsum, dolomite, and shale (Figure 2). Locally the

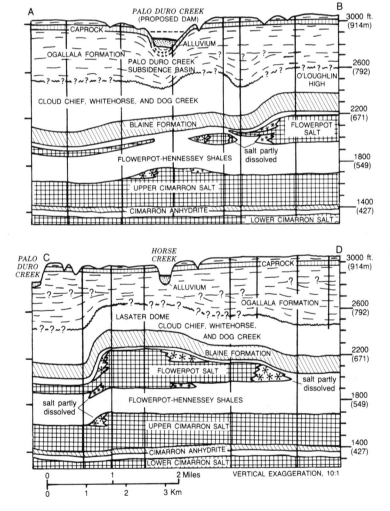

Figure 4. Cross sections showing geologic structure and collapse features related to interstratal dissolution of underlying salt. See Figure 3E for location.

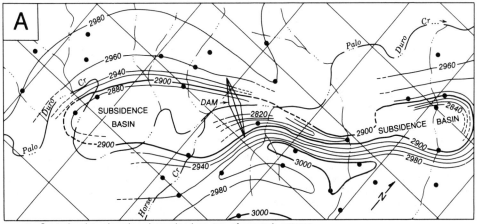

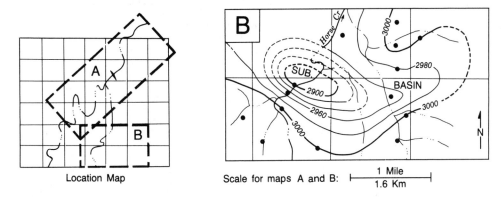

- Control point; elevation determined on top of caprock outcrop.
—2900— Structure contour line on top of caprock; contour interval is 20 ft (6m). Dashed where uncertain.

Location Map

Scale for maps A and B: |— 1 Mile —|
1.6 Km

Figure 5. Structure-contour maps on top of outcropping Ogallala caprock, showing Palo Duro Creek subsidence basin (map A) and Horse Creek subsidence basin (map B).

formation is thinner where some of the anhydrite or gyspum beds are dissolved or have collapsed due to dissolution of the underlying salt. Structure at the base of the Blaine is quite chaotic (Figures 3B and 4) as a result of partial or total dissolution of the Flowerpot salt and subsidence of the Blaine. The structurally highest areas are the Lasater dome and O'Loughlin high (herein named after principal landowners), which directly overlie the thick remnants of Flowerpot salt (compare Figures 3B and 3C). Structural lows include the southeast part of the study area and a modest syncline along Palo Duro Creek; both of these features coincide with areas of thin or missing Flowerpot salt.

Above the Blaine Formation are the youngest Permian units, the Dog Creek Shale, Whitehorse Group, and Cloud Chief Formation (ascending order) (Figure 2). These strata consist of red-brown shales interbedded with red-brown, fine-grained sandstone, and with several layers of white gypsum. The various units are not readily differentiated in the study area, and because of their color they are commonly called "red beds" by drillers. These strata range from 60-120 m (200-400 ft) thick in the area. Their structure, as well as the elevation of both their lower and upper surfaces, is highly irregular due to subsidence and collapse.

The youngest major unit to be studied, the Ogallala Formation of Miocene-Pliocene age, crops out extensively in Hansford County and rests unconformably upon the Permian red beds in the study area. The Ogallala consists chiefly of light-colored sand, silt, clay, and gravel deposited as interbedded fluvial and windblown sediments upon an erosional surface cut into the Permian (Seni, 1980; Gustavson and Holliday, 1985). The Ogallala is 105-180 m (350-600 ft) thick in the study area.

The elevation of the base of the Ogallala is highly irregular (Figure 3A), and this appears to result mainly from collapse or subsidence of the Ogallala base at the onset of, or during, Ogallala deposition (similar findings were reported by Gustavson and Holliday, 1985, and Gustavson, 1986, elsewhere in the Texas Panhandle). Such subsidence basins, resulting from collapse into underlying, interstratal-karst cavities, were then filled in by continued fluvial, eolian, and/or lacustrine sedimentation during later Ogallala time. The structural highs and lows at the base of the Ogallala coincide closely with those described earlier and appear to be related to interstratal-karst development in the Flowerpot salt.

Ogallala deposition apparently ended between 3.5 and 2.4 mya (million years ago), in the late Pliocene (Gustavson and Finley, 1985). At about the end of Ogallala time, distinctive and indurated beds of calcrete or silcrete (commonly called caliche or caprock) formed at or just below what was then the land surface. The caprock typically is 3-6 m (10-20 ft) thick and it caps conspicuous bluffs where it has been intersected by downcutting streams. In some areas, Pleistocene windblown sands and thin alluvial and lacustrine sediments were laid down upon the caprock.

In most parts of the study area the caprock and other outcropping Ogallala strata are horizontal or dip only 4-8 m/km (20-40 ft/mi), less than 0.5 degree, thus indicating that most salt dissolution occurred more than several million years ago (prior to development of the caprock 3.5-2.4 mya). However, at two places the caprock is sharply folded down into troughs or basins; these features, herein named the Horse Creek and Palo Duro Creek subsidence basins, were first identified during the aerial-photo study, and then were further examined during a field visit (Figure 5).

In the Horse Creek subsidence basin the main part of the caprock plunges down into the basin about 30 m (100 ft) within 450 m (1,500 ft) horizontally, or about 4 degrees (Figure 5); locally the dip reaches 10 degrees. Directly above and dipping parallel to the caprock are sands and silts that were in place at the time of subsidence, or were deposited during the subsidence. A thin, subhorizontal upper caprock unit extends across the top of the basin and rests upon these sands and silts. Salt dissolution and collapse occurred near the end of or after Ogallala time (less than 3.5-2.4 mya), but ceased before development of the thin, upper caprock unit. Inasmuch as the undisturbed, upper caprock probably took at least several hundred thousand years to form, according to estimates of the rate of calcrete formation by Machette (1985) and Gustavson and Holliday (1985), the subsidence would have ceased some time between the end of the Ogallala and several hundred thousand years ago.

In the Palo Duro Creek subsidence basin, a feature about 10 km (6 mi) long and 900-1,800 m (3,000-6,000 ft) wide, the caprock plunges at least 60 m (200 ft) down into the trough (Figure 5); the dip is locally as much as 15 degrees. In the northeast end of the basin, overlying strata dip into the basin the same as the caprock; these overlying strata are 30 m (100 ft) of windblown and lacustrine sediments. On the left (northwest) abutment of the dam the caprock splits into a major lower unit and a thin upper unit, similar to the Horse Creek basin. However, at the dam the lower bed dips 5-15 degrees, and the thin upper bed also dips 1-5 degrees; this indicates there may have been continued dissolution and subsidence after formation of the upper bed, perhaps in the past several hundred thousand years.

Several other lines of evidence indicate that salt dissolution and subsidence have not been active for a long period of time in the Palo Duro Creek basin. Basin-fill sediments now stand nearly 60 m (200 ft) above Palo Duro Creek, showing that much downcutting has occurred since the last episode of basin filling; Gustavson et al. (1980) estimate that the 60 m (200 ft) of downcutting by the nearby Canadian River in the Texas Panhandle required about 600,000 years. Also, the maturity of the Palo Duro Creek valley, which lacks any signs of sinkholes, depressions, fractures, or other evidence of recent disruption of the alluvium, indicates a lack of dissolution and subsidence for at least the past 10,000 years (the assumed approximate age of alluvial material in the area, based upon dating of paleosols in alluvium and low terraces elsewhere in the Texas Panhandle by Gustavson and Finley, 1985). In addition, there are no brine emissions, salt springs, or other evidence of present-day salt dissolution anywhere within Hansford County or Ochiltree County to the east. Therefore, it appears that salt dissolution and subsidence of the Palo Duro Creek basin began at or near the end of Ogallala time; it continued until some time after development of the upper caprock bed and ended before mature development of the Palo Duro Creek valley. Karst development and subsidence were probably completed at least several hundred thousand years ago, and certainly by 10,000 years ago.

Summary and Conclusions

Salt dissolution has occurred in the past at many places in the Texas Panhandle region, and it has occurred specifically beneath, and in the vicinity of, the proposed dam and reservoir. Dissolution is verified by subsurface mapping of the salt units, and by recognition of subsidence and/or collapse features above areas of abnormally thin salt. It is not possible to precisely date the various episodes of dissolution and subsidence in the study area, but it is certain that dissolution is not going on now and that subsidence has not occurred within the past 10,000 years or more.

Chaotic structures and subsidence features are present in the study area. The Blaine Formation and the base of the Ogallala are structurally high above thick remnants of Flowerpot salt (such as the Lasater dome and O'Loughlin high), and are structurally low in all other

120

areas where the salt is thin or absent. The fact that the complex structure of the Blaine Formation is not reflected in the deeper Cimarron anhydrite shows that the Blaine structure is due to interstratal karst and removal of salts between the Blaine and the Cimarron anhydrite. Further evidence of salt dissolution and collapse are the Horse Creek and Palo Duro Creek subsidence basins, where late Ogallala and/or post-Ogallala dissolution and subsidence have locally affected the caprock.

The highly irregular elevation of the base of the Ogallala Formation indicates that a major episode of Flowerpot salt dissolution occurred at the onset of, or sometime after the onset of, Ogalalla deposition; that is, some time during or after the Late Miocene, beginning some 12 to 10 mya. However, inasmuch as the Ogalalla caprock is flat-lying and undisturbed in most parts of the study area, it shows that dissolution and subsidence had ceased in most areas by the end of Ogallala time, some 3.5 to 2.4 mya. This Ogallala episode of dissolution and subsidence (from 12-10 to 3.5-2.4 mya) was undoubtedly the major period for removal of Flowerpot salt beneath the study area, although salt still remained beneath the Lasater dome, the O'Loughlin high, and areas that later were to become the Horse Creek and Palo Duro Creek subsidence basins.

The Horse Creek and Palo Duro Creek subsidence basins formed due to local salt dissolution and collapse in late Ogallala and post-Ogallala times, and probably ceased development at least several hundred thousand years ago. In the Palo Duro Creek basin, we can only be certain, however, that subsidence ended more than 10,000 years ago, before mature development of the stream valleys and alluvial deposits that cross the two subsidence basins. There is no evidence of ongoing or recent dissolution in the area, and thus there is no need for special construction design to accommodate modern karst activities beneath the dam or reservoir. However, trenches and other excavations will be monitored for evidence of additional subsidence features.

References

Gustavson, T. C., 1986, Geomorphic development of the Canadian River valley, Texas Panhandle: an example of regional salt dissolution and subsidence: Geol. Soc. of Amer. Bull., v. 97, p. 459-472.

Gustavson, T. C., and Finley, R. J., 1985, Late Cenozoic geomorphic evolution of the Texas Panhandle and northeastern New Mexico--case studies of structural controls on regional drainage development: Texas Bur. of Econ. Geology Rpt. of Inv. 148, 42 p.

Gustavson, T. C., Finley, R. J., and McGillis, K. A., 1980, Regional dissolution of Permian salt in the Anadarko, Dalhart, and Palo Duro Basins of the Texas Panhandle: Texas Bur. of Econ. Geology Rpt. of Inv. 106, 40 p.

Gustavson, T. C., and Holliday, V. T., 1985, Depositional architecture of the Quaternary Blackwater Draw and Tertiary Ogallala Formations, Texas Panhandle and eastern New Mexico: Texas Bur. of Econ. Geology Open-File Rpt. OF-WTWI-1985-23, 60 p.

Gustavson, T. C., Simpkins, W. W., Alhades, A., and Hoadley, A., 1982, Evaporite dissolution and development of karst features on the Rolling Plains of the Texas Panhandle: Earth Surface Processes and Landforms, v. 7, p. 545-563.

Johnson, K. S., 1981, Dissolution of salt on the east flank of the Permian Basin in the southwestern U.S.A.: Jour. of Hydrology, v. 54, p. 75-93.

Johnson, K. S., and Gonzales, Serge, 1978, Salt deposits of the United States and regional geologic characteristics important for storage of radioactive wastes: report to Union Carbide Corporation, Office of Waste Isolation, Oak Ridge National Laboratories, Y/OWI/SUB-7414/1, 188 p.

Jordan, Louise, and Vosburg, D. L., 1963, Permian salt and associated evaporites in the Anadarko Basin of the western Oklahoma-Texas Panhandle region: Oklahoma Geol. Survey Bull. 102, 76 p.

Machette, M. N., 1985, Calcic soils of the southwestern United States, in Weide, D. L., ed., Soils and Quaternary geology of the southwestern United States: Geol. Soc. of Amer. Special Paper 203, p. 1-21.

Seni, S. J., 1980, Sand-body geometry and depositional systems, Ogallala Formation, Texas: Texas Bur. of Econ. Geology Rpt. of Inv. 105, 36 p.

Simpkins, W. W., Gustavson, T. C., Alhades, A. B., and Hoadley, A. D., 1981, Impact of evaporite dissolution and collapse on highways and other cultural features in the Texas Panhandle and eastern New Mexico: Texas Bur. of Econ. Geology Circ. 81-4, 23 p.

Sinkhole hazards in Tasmania

KEVIN KIERNAN   *Forestry Commission, Hobart, Tasmania, Australia*

ABSTRACT

Subsidence and collapse have occurred in surficial mantles of glacial, glaciofluvial, periglacial and alluvial origin that are widespread overlying karstified rocks in Tasmania. Sinkholes have formed in several areas within a few years of forests being cleared for pasture or by the timber industry. These sinkholes have probably resulted from increased water movement through the regolith due to decreased transpiration, coupled with the rotting of tree roots that formerly helped bind the regolith. Other sinkholes have formed at some road margins, particularly where culvert design is poor, and some of these collapses have damaged roads. Sinkholes that are probably the result of groundwater pumping have formed adjacent to a large limestone quarry. Because economic losses associated with individual sinkhole incidents have generally been comparatively minor, and because the incidents have been geographically dispersed, they have attracted little official attention. However, the individuals affected may suffer significant personal loss and the combined cost of karst-related land surface instability in Tasmania is probably significant.

Introduction

Tasmania's karst occurs primarily in uninhabited or sparsely populated areas. There are no large-scale urban centres on the karst. Nevertheless, there are indications of a significant sinkhole hazard in some of the karsts. This paper presents a reconnaissance level survey of sinkhole problems in Tasmania. It describes some Tasmanian sinkholes, reviews their origin and assesses their impact.

The island of Tasmania lies at 41º-43º south latitude in the westerly wind regime of the Roaring Forties, and receives annual rainfall totals that range from over 3300mm in the west to around 500mm further east. Western Tasmania is largely forested, scrubland or sedgeland terrain characterised by fold mountain chains formed of pre-Carboniferous rocks. Here conglomerate and quartzite uplands rise above valleys eroded in phyllites and carbonate rocks. Eastern Tasmania is characterised by flat-lying and faulted post-Carboniferous sedimentary rocks and extensive sills of Jurassic dolerite. Here the vegetation structure is more open. The topographic relief of the carbonate rocks is highest close to the boundary between the fold and fault provinces where the carbonate has been protected from erosion on the flanks of dolerite-capped massifs.

The karst is formed primarily in outcrops of dolomite and limestone of Upper Precambrian to lower Cambrian age, and in limestone of Ordovician age. Minor karst also occurs in limestones of younger age. These rocks occur over an area of about 27,000 hectares (figure 1). Although some very striking karst has developed in the Precambrian dolomites, probably the most fully developed range of karst phenomena, and the greatest density of sinkholes, has formed in the Ordovician limestone. Impounded fluviokarsts predominate, but glaciokarsts and coastal karsts are also present. Quaternary surficial sediments that consist primarily of allogenic rock types overlie the carbonate bedrock in most areas. These sediments include tills, glaciofluvial gravels and sands, diamicton sheets that were probably formed by periglacial solifluction, and alluvial fans. Only in a few areas are the soils in the karsts derived from in situ weathering of the carbonate bedrock.

Recent sinkhole formation

There is little evidence of recent sinkhole formation through the collapse of cavities formed in carbonate bedrock. Evidence of recent collapse or subsidence of cavities formed in regolith is also extremely rare under undisturbed natural conditions -

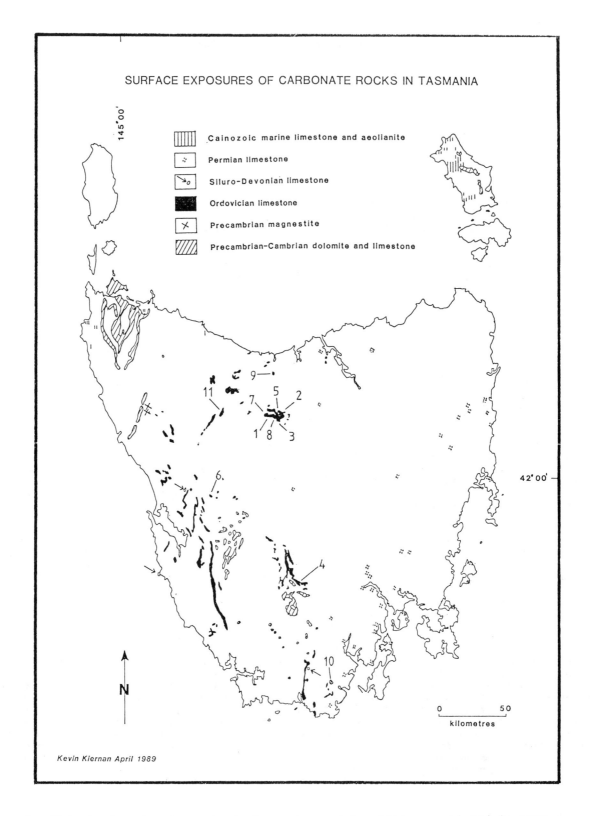

Figure 1: Extent of carbonate rocks in Tasmania, and localities mentioned in text.
1. Loatta; 2. Caveside; 3. Sassafras Creek; 4. Junee-Florentine; 5. Mole Creek; 6. Bubs
Hill; 7. Liena; 8. Mayberry; 9. Railton; 10. Hastings; 11. Vale of Belvoir.

the writer has seen only a handful of examples in 20 years. However, recent regolith instability is evident in some areas where timber cutting, pasture development, road construction, settlement and limestone quarrying have extended onto karst. These collapses are due to enhanced groundwater transport of regolith components into solution cavities formed in the bedrock. Three principal factors appear to have given rise to this situation. They are removal of the natural vegetation cover, drainage changes associated with road construction, and fluctuations in epikarstic water levels.

A. Enhanced downward percolation and root rotting following vegetation removal.

In several areas of Tasmania sinkholes have developed where natural forest vegetation has been removed to facilitate the development of pasture, or, less commonly, where clearfelling has occurred as part of forestry operations. That removal of the vegetation is the principle cause of this instability is suggested by the absence of fresh collapse in adjacent areas of uncut forest growing on similar regolith materials. The factor that seems most likely to predispose a deforested site to collapse or progressive subsidence is flushing of regolith components into cavities in the limestone as a result of increased free water in the regolith due to decreased transpiration following removal of the forest. Increased moisture also adds to the weight of the regolith while the rotting of the tree roots decreases regolith strength (Kiernan 1988).

Figure 2: Collapse sinkhole that formed overnight during a period of heavy rain in the Loatta polje, winter 1984.

In the Loatta area, north-central Tasmania, vertical sided collapse sinkholes up to 4-5m deep and 10-15m wide occasionally form overnight in pasture developed on periglacial alluvial sediments that overlie Ordovician limestone (figure 2). The collapses generally occur during periods of heavy rain. Most of the collapses appear to have occurred in pre-existing depressions or on the floor of shallow dry valleys. Additional stress may be imposed on the regolith and transport of soil particles enhanced by the import of minor quantities of irrigation water that is distributed by sprinkler on some of the pasture in the Loatta area (Kiernan, 1984). Subsidence at a more gradual rate is evident several kilometres further to the east at Caveside. This subsidence has

locally depressed pasture by up to 8m over a period of 20 years. The depressions are formed close to the edge of a glaciofluvial fan. Glacial meltwater was previously decanted from this fan into channels formed against the margin of a limestone ridge, thereby promoting the development of large vadose invasion caves (Jennings and Sweeting, 1959). A more recent anabranch of the stream in an adjacent cave may pass almost beneath the site of the depressions at a depth of about 10m (Jennings, 1967). The presence of this stream may have aided subsidence through permitting efficient evacuation of particles washed downward from the regolith. Seasonal flooding of this progressively subsiding site is reducing the area of pasture available during winter.

In the Sassafras Creek valley, also in central-northern Tasmania, collapses up to 2m deep and 10-15m wide occurred following clearfelling of an area of privately owned eucalypt forest. The collapses were first noted about 10 years after timber harvesting but their exact age is uncertain. The regolith comprises several metres of solifluctate in which clasts of sandstone, mudstone and dolerite occur in a sandy-silty matrix. Little effort was made to stimulate forest regeneration after cutting and to date only scrub has recolonised the logging coupe. More effective efforts to achieve forest regeneration may have prevented the sinkhole problem arising. However, the relationship between the time taken for tree roots to rot out, the time taken for a subsequent tree crop to dry out and bind the regolith, and the return period for rainfall events sufficient to trigger collapse is not known. Similar collapses have not been recorded from successfully regenerated logging coupes, although this may be simply an artefact of poor visibility where regrowth is vigourous.

B. Road Construction
(a) Concentration of diffuse runoff from road surfaces.
    In several areas collapse and/or subsidence appears to have been stimulated where natural diffuse infiltration of rainfall into the ground surface has been replaced by the concentration of runoff on road surfaces and its deflection into depressions neighbouring the road.

    At Loatta, collapse has occurred adjacent to a sealed road constructed across a periglacial alluvial fan deposited on Ordovician limestone. Discharge from a culvert under the road was directed into a shallow depression at the base of a low limestone hill. Subsequent collapse of the depression floor in the mid 1970's permitted access into a cavity in the regolith 4mx3mx3m high. The owner of the property covered the entrance to this cavity using large timbers to prevent livestock falling into the hole, but progressive enlargement appears to be continuing. An additional collapse into a soil chamber slightly larger in size occurred about 40m further west along a structural lineament in 1987. This collapse was located about 3m above the base of the limestone hill. It was filled in several months later, but signs of collapse were again evident in mid 1989. It would appear that subsurface drainage along the lineament is permitting collapse to be propagated some distance from the edge of the road.

    Progressive failure of a fill embankment around the margin of a sinkhole has led to partial collapse of a logging road in the Junee area, south-central Tasmania. At least 3m of dolerite-rich solifluctate overlies Ordovician limestone. There is little bedrock outcrop in the immediate area. In this case, failure appears to be occurring at the interface between the sloping side of the sinkhole and the fill material, in response to deepening of the sinkhole which has removed toe support from the fill. The forest on either side of the road was clearfelled in the early 1970's and failure of the road dates from sometime in the early 1980's. Removal of the forest may have contributed.

    At Mole Creek, central-northern Tasmania, small scale active collapse is evident on the inside of a right angled corner in a sealed road near a large spring known as Scotts Rising. Active soil collapse is focussed between small limestone outcrops in a very small sinkhole. Local residents have responded by filling the collapses with domestic and farm refuse, probably to the detriment of groundwater quality. The camber of the road deflects runoff from the road surface towards the apex of the corner. The bedrock outcrops probably promote the removal of sediments by further focussing the infiltration of runoff from the road.

(b) Gross interference with subsurface drainage.
    In the Junee area another unsealed logging road near Cave Hill has been partly engulfed by a sinkhole about 4m deep and 10m in diameter. This sinkhole is formed in a dolerite-rich solifluction sheet that was forested prior to clearfelling in the late 1960's. The collapse occurred in the late 1970's or early 1980's. The sinkhole is located on the upslope side of the road where it crosses a shallow dry valley close to the margin of a limestone hill. No evidence of fresh collapse has been recorded down valley from the road. No culvert was installed beneath the road at the time it was constructed, presumably because the dry valley did not appear to carry water. The collapse was probably caused by a progressive loss of components from the regolith beneath the floor of the dry valley due to the derangement of natural subsurface drainage in the regolith. Increased runoff and decreased transpiration after the forest was clearfelled may have been a contributing factor. Sinkholes that have formed in the surface of a sealed road through the pastures of Loatta have forced temporary closures of the road from time to time. Once again, collapse has occurred at a point where the road crosses a shallow dry valley. Some 200m upslope of the road a small stream sinks into the alluvial fan across which the road has been constructed.

    A different, but nevertheless related problem, appears to exist near Bubs Hill in western Tasmania. Here a highway traverses a small area of Ordovician limestone which is overlain locally by colluvial and alluvial sediments up to several metres thick. Partial progressive failure of a fill slope across which the highway was constructed has occurred on the margin of a small sinkhole (Houshold and Clarke, 1988). Subsequent inspection revealed the presence of a cave in bedrock limestone 100m from the road, in which a small stream was transporting foundation material eroded from beneath the road. No culvert discharged into the sinkhole beside the road. Table drains in the area were unsealed. Piping of material from previously filled karst cavities has produced very small collapses in the cut batter on the upslope side of the road. The highway is soon to be slightly widened by further partial filling of the sinkhole. The local road engineer contends that resurfacing the road will sufficiently strengthen it to arrest collapse. It would appear there has been gross interference with drainage in the epikarst with a significant role probably also being played by infiltration in the unsealed table drains and into the coarser subsurface horizons of the regolith exposed during excavation of the batters.

Subsidence can be expected to continue unless appropriate drainage and construction techniques are employed.

Significant interference with subsurface drainage also appears to have been involved in the formation of a collapse sinkhole 6m deep and 10m in diameter in the surface of a sealed road near Liena in north-central Tasmania. The sinkhole was formed in glaciofluvial gravels and sands that locally are overlain by basalt-rich colluvium and alluvial silts. It formed at a break of slope where a shallow valley discharged onto the glaciofluvial terraces beside the Mersey River. Older sinkholes are present downslope of the road. The impact of the road, the collapse, and subsequent remedial measures upon a neighbouring property owner became the subject of considerable acrimony, and the precise techniques used in road construction are not totally clear. Local reports suggest that a spring was partly filled during construction of the road. A major landslip occurred during heavy rain in September 1970 from the forested slopes above the road, and debris was spread across the pasture below it, demolishing a barn and narrowly missing a house. The spring that previously served the property ceased to flow and new springs emerged from the slip and adjacent areas to the south. There was vigourous discharge of clay-laden water from one spring in the bed of the adjacent Mersey River. The road formation remained largely intact until its collapse during a period of very heavy rain three years later in April 1974. Following this collapse a small stream disappeared into the sinkhole. The sinkhole was subsequently filled and drainage redirected. A broad depression about 10cm deep in the sealed road surface was again evident by 1987. The initial landslip may have been contributed to by increased soil water pressure due to impeded egress of drainage from the spring. This pressure may have been relieved catastrophically in a manner that removed toe support from the colluvium upslope. However, the magnitude and configuration of the landslide suggests that non karstic subsurface drainage in the regolith higher on the slope may have played at least as significant a role. The subsequent formation of the sinkhole involved a small landslip event that was propagated a short distance upslope.

C. Fluctuations in groundwater level.

Fluctuation of epikarstic water levels can flush components from the regolith, while major draw-down removes buoyant support, thereby increasing vertical stress on the regolith. Both processes have been responsible for sinkhole problems in Tasmania.

At Mayberry, central-northern Tasmania, the floor of a karst margin polje several kilometres in extent has been converted to pasture. Alluvial sediments several metres thick overlie the limestone. A gorge cut through an anticlinal core of non-carbonate rocks that underlie the limestone on the northern margin of the polje is active during times of flood, but for most of the year the polje is drained via underground channels that extend eastward to bypass the core of the plunging anticline. A sinkhole problem exists at the eastern end of the polje. In 1978 a local property owner reported the loss of 10 cattle during the previous 12 months due to the ground collapsing under their feet or the animals having fallen into newly formed sinkholes up to 7m deep. The most likely cause of the problem is pronounced natural fluctuations in the level of the watertable due to the incapacity of the underground outlets to evacuate the peak flows. Water levels rise in the regolith until the surface overflow channel becomes operative. These fluctuations occur over a fairly narrow vertical amplitude and facilitate removal of the finer components from the regolith and the formation of cavities within it. Collapses that have occurred a few kilometres to the north near Dogs Head Hill probably have a similar origin. Here a limestone strath that contains deep water-filled shafts, some of which are seasonally active springs, is overlain by glaciofluvial gravels and sands of variable thickness. Collapse of the pasture is again sometimes triggered by the passage of a heavy beast. The behaviour of the springs indicates fluctuating water levels in the regolith. Here, as at Mayberry, a reduction in the strength of the regolith following removal of the forest cover is likely to have played a significant role.

Artificial drawdown is the probable cause of a sinkhole problem at Railton in northwestern Tasmania. Here limestone is quarried from a deep pit on the floor of a broad valley beneath about 20m of overburden. Some of the overburden is authigenic. Prior to the advent of quarrying there was little evidence of karst topography. During the early years of the operation neighbouring farmers benefitted from the drawdown because it diminished seasonal flooding of parts of their properties. Local anecdotes suggest minor sinkhole problems subsequently arose including the formation of a small collapse between the sleepers on the local railway line. Instability of the land surface increased during the early-mid 1980's when a new bench was developed in the quarry, deepening the pit by 15-20m. A nearby waterfilled quarry that had been abandoned some years previously drained rapidly. Sinkholes appeared in pasture close to the quarry and in the backyards of at least two village dwellings. The town sewage main was ruptured by one sinkhole. One householder was subject to three separate collapse incidents. Exposures in the quarry reveal that the limestone surface beneath the overburden consists of a maze of pinnacles with a relief of 10-15m. At least two small caves and one major spring have been encountered at depth in the quarry. The sinkholes appear to occur within a cone of watertable depression around the quarry. Artificial lowering of the watertable due to the quarrying together with differential settlement of the overburden between the limestone pinnacles seem the most likely causes of the problem. However, inadequate drainage of runoff from the roofs of houses and outbuildings contributed to at least one collapse.

Discussion

Economic losses associated with individual sinkhole incidents have generally been comparatively minor. The incidents have been geographically dispersed and have attracted little official attention. The logging roads that have failed have generally done so after the tree crop was harvested, hence, these failures have had little economic impact on the forestry operations in the short term. On the other hand, private owners who suffer property damage or devaluation, or who lose livestock, may be faced with a cost that is very significant to them as individuals. There may also be a less readily quantified cost that flows from the uncertainty and anxiety with which householders may have to live if they perceive their homes or their safety to be under threat, whether real or imagined. The costs are likely to become more significant still if motorists are injured, compensation claims arise, expensive ameliorative requirements are enforced or enterprises are forced to cease operations.

Effective management of sinkhole problems in Tasmania is at present inhibited by the inexperience of local land managers with karst problems, coupled with a somewhat frontier mentality that sees brute force or recurrent repairs as the easiest response to a poorly understood problem. As more development extends onto Tasmania's karst sinkhole problems can be expected to increase. On the positive side, special requirements for timber harvesting operations in karst areas have now been enacted, and these may reduce the likelihood of sinkhole problems (Forestry Commission, 1987). Another hopeful sign is the apparent abandonment of a proposal for a series of sewage ponds in a potential sinkhole problem area at Hastings in southern Tasmania. On the negative side, a major new highway development through highly suspect terrain in the Vale of Belvoir in northwestern Tasmania still exhibits construction and drainage techniques that may be inadequate to prevent problems arising. In this locality epikarstic water level fluctuations occur in till deposits that overlie conspicuously karstified limestone. It remains to be seen to what extent lessons already waiting to be learnt in Tasmania's karst can be translated into more effective management of sinkhole-prone terrain.

References

Forestry Commission, 1987. Forest Practices Code. Hobart, Tasmania. Tasmanian Government Printer.

Houshold, I. & Clarke, A. 1988. Bubs Hill Karst Area - Resource Inventory and Management Recommendations. Unpublished report to Department of Lands, Parks & Wildlife, Tasmania. 80pp.

Jennings, J.N., 1967. Some karst areas of Australia. pp256-292 [in] J.N. Jennings and J.A. Mabbutt (eds). Landform Studies from Australia and New Guinea. Canberra. Australian National University Press.

Jennings, J.N. & Sweeting, M.M., 1959. Underground breach of a divide at Mole Creek, Tasmania. Australian Journal of Science, 21 pp261-262.

Kiernan, K., 1984. Landuse in Karst Areas: Forestry Operations and the Mole Creek Caves. Canberra. Australian Heritage Commission Library. 320pp.

Kiernan, K. 1987. Land development and slope stability in karst. pp65-66[in]. Geography and Public Policy. Programme Abstracts, Institute of Australian Geographers 22nd Conference. Canberra. Australian Defence Force Academy.

Kiernan, K., 1988. The Management of Soluble Rock Landscapes - An Australian Perspective. Sydney. Speleological Research Council Ltd. 61pp.

Ground subsidence associated with doline formation in chalk areas of Southern England

PETER MCDOWELL   *Department of Geology, Portsmouth Polytechnic, UK*

ABSTRACT

Many of the closed depressions found on, or near, the outcrop of the Chalk formation in England are dolines, or sink-holes, and have a significant impact upon engineering projects and the environment. These features potentially provide direct access of pollutants to a major groundwater aquifer, i.e. the Upper Chalk, and constitute a hazard to buildings and civil engineering works. In recent years, there have been many recorded cases of ground subsidence in Southern England, producing either steep-sided holes, up to 4 metres in diameter and depth, or shallow circular depressions with lower slope angles. These are considered to be associated with dissolution of chalk and the ongoing process of doline formation. Investigation of three sites in West Sussex, between Chichester and Arundel, has provided an understanding of the subsidence mechanisms. There is strong evidence to suggest that voids exist within the Slindon Sand formation, resulting from the washing of the silty sand into solution pipes and fissures in the underlying chalk. Collapse of the overlying gravels into these voids has been accelerated by construction activity and localised water ingress from soakaways and other sources. Ground profiling radar and dynamic probing were found to be the most effective methods for locating areas with near surface voids and loosened gravel. At other locations in Southern England, the greatest concentrations of dolines occur where the Upper Chalk is covered by a thin cover of Tertiary age sands, and/or argillaceous formations with sandy members. The subsidence mechanisms are thought to be basically the same as in West Sussex, but larger dolines can develop and a different combination of ground investigation methods is recommended.

Introduction

Approximately 15% of the land surface of England has either outcrops of Chalk, or a thin cover of post-Cretaceous deposits over Chalk. The Chalk is of Upper Cretaceous age and is mainly to be found in the Southern and Eastern counties of the country. In South East England, the Lower Chalk contains an appreciable proportion of clay minerals, but the Middle and Upper Chalk formations are generally remarkably pure, with over 90% calcium carbonate. The Upper Chalk has thin bands of marl, chalk-rock and flints, but is mainly a weak porous limestone, susceptible to dissolution. An essentially orthogonal discontinuity pattern allows groundwater to move vertically and horizontally, particularly within the Upper Chalk, which is a major aquifer. Dissolution has increased the permeability of this aquifer by widening fissures and, in some areas, forming very narrow and irregular cave systems, usually less than one square metre in cross-sectional area. Infilled hollows, or vertical cylindrical pipes, are often observed at the surface of the Chalk, as there have been many periods since the Upper Cretaceous uplift (70 Ma) when karstification could have been active. (Jones, 1981). Dolines, however, have probably mainly developed during the latter part of the Quaternary in areas where the Upper Chalk has a thin cover of sediments and/or drift deposits. Tracer tests carried out between dolines and natural springs in South Hampshire (Atkinson & Smith, 1974) demonstrate the potential for pollution of the Upper Chalk aquifer, but it is the possibility of ground subsidence which is considered in this paper.

Ground subsidence related to karst development in Chalk areas of England has been reviewed by Edmonds, Green and Higginbottom, (1987). This refers to earlier work by Edmonds (1983), which describes how a data base consisting of 2600 dolines, solution pipes and swallow holes, from more than 740 locations, was used to produce a map of the density of solution features upon the Chalk outcrop. Part of this map is reproduced in Fig.1. Although this is a valuable source of information for planning purposes, the spatial distribution density values must be considered to be conservative and can be expected to increase as more research is carried out and as new dolines are formed. In recent years, there have been many recorded cases of ground subsidence in Southern England, which are related to doline formation. Generally, there has been sudden collapse of the ground, producing circular steep-sided holes, up to 4 metres in diameter and depth, or the development of larger diameter depressions with lower slope angles.

The formation of these features is closely related to topographical, geological and hydrogeological conditions. A characteristic of all of the sites described is the presence of a thin, generally less than 10 metres thick, cover of Tertiary and/or Quaternary, age deposits containing a layer of sand and unconformably overlying the Upper Chalk. Evidence has been obtained, from several of these sites, that sand has been washed down into solution pipes and fissures in the chalk, producing voids within the sand horizon. The process of void enlargement is probably similar to that demonstrated by laboratory model studies (Hodek et al., 1984) and, eventually, there is collapse of the sand above the void. Where the sand is capped by a layer of cohesive soil, the mechanisms leading to subsidence may include gradual void migration and compaction, sudden shear failure or bending, (Duncan & Devane 1984). An appreciation of the local geology and the subsidence mechanism is essential for the design of appropriate ground investigations and remedial measures.

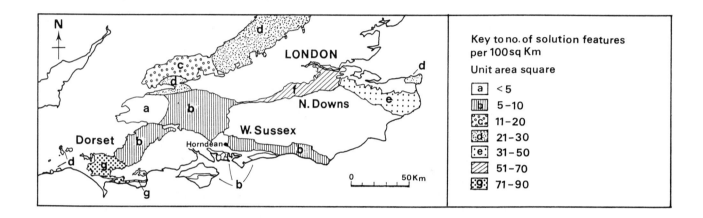

Figure 1: Outcrop of chalk in Southern England (after Edmonds, 1983)

Ground Subsidence in West Sussex

Although West Sussex is a small part of Southern England, the geological environment is very interesting and there have been several recent examples of ground subsidence near the outcrop of the chalk. Three sites are considered, one in a suburb of Chichester, known as Summersdale, and two others close to Fontwell and Slindon, near Arundel (Fig.2.). At all three locations, borehole investigations have proved the presence of marine sands of the Slindon Sand formation resting upon an irregular chalk surface, which can be shown to be part of a raised beach cut into the Upper Chalk (Fig.3.). The sand is overlain by head deposits, usually referred to as Coombe Deposits or Head Gravels, and/or brickearth of Quaternary age. These have been transported into the area by solifluction, wind or floodwaters in a periglacial environment. This succession is fully exposed in the gravel pits at Eartham, (Fig.2.), and frequent inspection of the worked faces has revealed void formation and slumping of the sands. The results of sieving analysis of a large number of samples has shown that the Slindon Sand is usually a fine sand, the Brickearth is a silty clay, and the Head Gravels are well graded. The sands and gravels have a much higher permeability than the brickearth and the clay-with-flints, which is occasionally present on the chalk surface, hence the distribution of these deposits is probably very significant in the development of dolines.

In 1976, ground subsidence at Summersdale caused damage to a road, adjacent houses, a water main and a gas pipe. The incident followed a period of heavy rainfall, which, in turn, followed an unusually hot and dry summer. The sink-hole was a wide shallow circular depression, suggesting compaction of the gravels following void migration. This interpretation is supported by the observation of loose gravel and voids in two of the holes subsequently drilled within the depression (Fig.4.). It is also interesting to note the absence of the sand layer in Borehole 6, located at the centre of the sink-hole; the sand presumably having been washed down into the chalk in earlier times. Almost exactly the same ground conditions have been recently found at an adjacent construction site where an expensive vibrocompaction and gravel placement process has been used to improve the bearing capacity of the ground. A variety of geophysical techniques, including electromagnetic ground conductivity and ground profiling radar were used at this site, to locate voids or loosened gravel, but with limited success. The most cost-effective method was found to be dynamic probing with a small portable system, known as the Borros Probe.

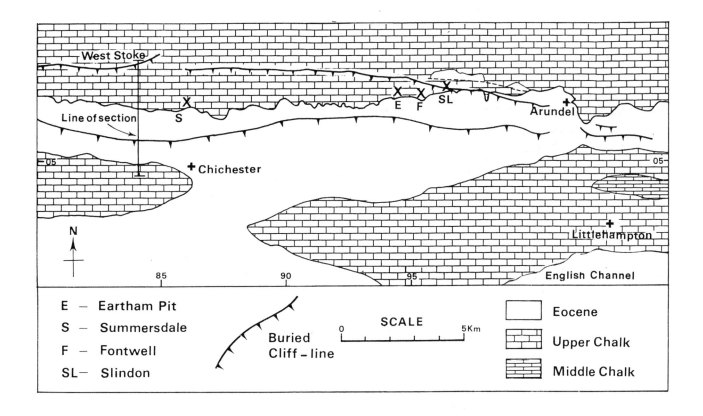

Figure 2: Sites of Ground Subsidence in West Sussex

In November 1985, a dramatic incident occurred at Fontwell (Fig.2.), involving ground subsidence along a road and on adjacent properties. Approximately 70 sink-holes developed within a circular area of about 90 metres in diameter. The holes were mainly cylindrical, up to 4 metres across at the ground surface and up to 3 metres in depth, or in the form of bell pits. Head gravels were exposed in all of the holes and it seems likely that gravel had collapsed into voids formed within the underlying Slindon Sand formation by the washing of sand down into pre-existing pipes and fissures in the chalk. Geophysical surveying provided interesting results at this site, with areas of anomalously low conductivity, obtained by traversing with the Geonics E.M. 31 ground conductivity meter, possibly representing loose gravel, and characteristic arcuate and planar reflection traces from ground profiling radar indicating small near-surface voids and shear planes within the gravels, respectively. The depth of investigation with these methods was, unfortunately, limited to the 4 metre thickness of the gravel layer, because the high clay and moisture content increased the ground conductivity of the gravels at this location.

In February 1988, several similar small diameter collapse dolines developed at Slindon (Fig.2.) and became news-worthy because a cow was killed. The borehole investigation proved once again a sequence of head gravels, silty sand and chalk. There were in this case, however, none of the subsidence triggers associated with urbanisation, such as soakaways and road traffic, and very little rainfall for a week before the event. Electrical resistivity and electromagnetic traversing did not, in this case, indicate the presence of any voids or loose ground, but clearly showed the areas where Lower Tertiary (Eocene) age clays were present above the chalk. This indirectly is advantageous, as several older and larger dolines, with gentler side slopes, are concentrated along the feather edge of the Eocene clays.

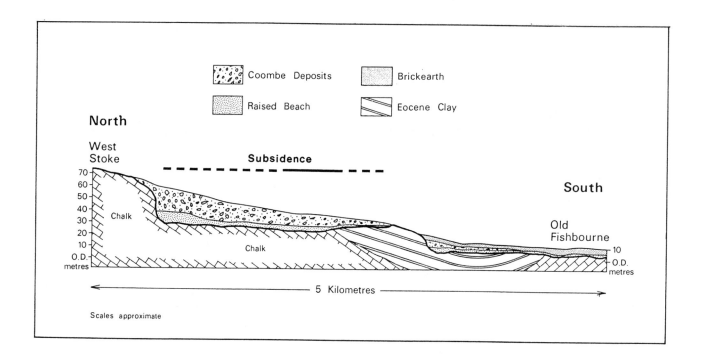

Figure 3:   Geological Cross-Section for West Sussex (after Mottershead, 1976)

Subsidence in Chalk Areas with Tertiary Cover

      In West Sussex and many other parts of Southern England, dolines have developed where deposits of Tertiary age overlie the Upper Chalk. The highest concentrations have been recorded in Dorset and along the edge of the North Downs (Fig.1.), where the Lower Tertiary, usually Eocene, formations are either predominantly sand or argillaceous deposits with sandy horizons or lenses. A very thorough geomorphological analysis of the Dorset dolines by Sperling et al, (1975) indicated approximately 500 dolines in an area of 16 square kilometres, with a maximum spatial distribution density of 99 per kilometre on Puddletown Heath. Most of the dolines were between 2 and 4 metres in depth and had side slopes of approximately 30°, but recent ground subsidence has produced smaller diameter holes with vertical sides, or widening with depth. Presumably, in time, the side slopes of these dolines will also degrade to 30°. As in West Sussex, the water table is within the chalk, and rapid localised downward percolation of water through vertical pipes and fissures has probably produced voids in the overlying sands.

      Although not previously described or recorded, there is also a concentration of dolines in parts of Hampshire, particularly around the village of Horndean (Fig.1.). In a small area of one square kilometre, around a major road interchange, there are approximately 20 holes; mostly dolines, but some being swallow-holes which have not been produced by subsidence. The tracer tests, already mentioned, were carried out at one of the largest of the dolines, which was a steep sided hollow, 30 metres in diameter and 10 metres in depth, at the interchange. Excavation of the side slopes of this doline, during construction, revealed a 2 metre thick layer of sand below 8 metres of clay and immediately above the chalk. Electrical resistivity sounding, in this area, did not clearly differentiate the sand layer from the underlying chalk, but a combination of electrical resistivity traversing and ground conductivity measurements, with the Geonics E.M.31, enabled the chalk surface to be profiled and sand lenses or channels within the clay to be mapped (McDowell, 1981).

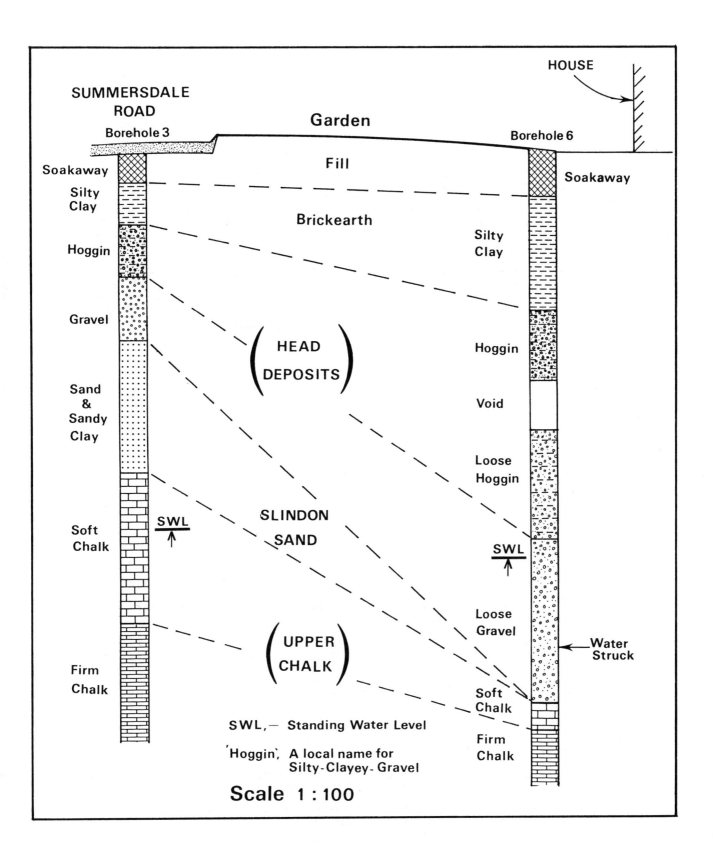

Figure 4:   Sink-hole Section at Summersdale, Chichester

Conclusions

Ground subsidence associated with doline formation in Chalk areas of Southern England is a greater problem than indicated by previously published data. This is an ongoing process which is accelerated by urban development and, even without vibration from road traffic and localised ingress of water, the ground can suddenly collapse; particularly where the geological environment is favourable for doline development.

The mechanisms producing ground subsidence are not always clearly understood and probably vary with location. It seems most unlikely that voids several metres in diameter could develop within the weathered and, in many cases, periglacially disturbed chalk, but there is ample evidence of sand having been washed down into fissures and pipes within the chalk. Voids up to 4 metres in height and 8 metres in diameter could conceivably develop in the sand layer, particularly where there is a capping layer of clay or clay-bound gravel. Theoretically, for collapse dolines, which are akin to crown-holes over mine workings, subsidence could occur when the cohesive capping layer is about 40 metres thick, assuming a 10% bulking factor, but there is no evidence that this has happened. The possibility of periglacial disturbance of the Chalk producing hydraulic continuity between sand members within clay and the Chalk also requires further research.

Ground investigations need to be carefully planned, bearing in mind the local geology and the anticipated subsidence mechanism. Geophysical methods have been found to be useful for defining the lithology of the deposits overlying the chalk and the variation in depth to the chalk surface, but of limited value for void location. Ground profiling radar has indicated, at some sites, the presence of disturbed ground, with small voids and shear planes, within 4 metres, or so, of the ground surface and may be more effective in dry periods and in areas with minimal amounts of clay in the surface layer.

Acknowledgements

The author gratefully accepts the support provided by Portsmouth Polytechnic and the Institution of Mining and Metallurgy, through the Bosworth Smith Trust Fund, for attendance at the conference.

References

Atkinson, T.C. & Smith, D.I., 1974. Rapid groundwater flow in fissures in the Chalk, an example from South Hampshire. Q.Jl.Engng.Geol., 7, 197-205.

Duncan, N., Devane, M.A., and Hickmott, J.M., 1984. The analysis of localised ground subsidence. Third Int. Conf. on Ground Movements and Structures. Univ. Wales, Cardiff, 9-12 July 1984, 1-16.

Edmonds, C.N., 1983. Towards the prediction of subsidence risk upon the Chalk outcrop. Q.Jl.Engng.Geol., 16, 4, 261-266.

Edmonds, C.N., Green, C.P. and Higginbottom, I.E., 1987. Subsidence hazard prediction for limestone terrains, as applied to the English Cretaceous Chalk. From Culshaw, M.G., Bell, F.G., Cripps, J.C. and O'Hara, M. (eds)1987. Planning and Engineering Geology, Geol. Soc. Engng. Gr. Spec. Publ., 4, 315-333.

Hodek, R.J., Johnson, A.M. and Sandri, D.B., 1984. Soil cavities formed by piping. Proc. 1st Multidisciplinary Conf. on Sinkholes, Orlando, Florida, 15-17 October 1984, 249-254.

Jones, D.K.C., 1981. The geomorphology of the British Isles, Southeast and Southern England. Methuan, 322p.

McDowell, P.W., 1981. Recent developments in geophysical techniques for the rapid location of near-surface anomalous ground conditions by geophysical methods. Ground Engineering, 14, 20-23.

Mottershead, D.N., 1976. The Quaternary History of the Portsmouth region. Portsmouth Geographical Essays. Dept. of Geography, Portsmouth Polytechnic, 21p.

Sperling, C.H.B., Goudie, A.S., Stoddart, D.R. and Poole, G.G., 1975. Dolines of the Dorset Chalklands and other areas of southern Britain, 205-223.

Morphometric analysis of sinkholes and caves in Tennessee comparing the Eastern Highland Rim and Valley and Ridge physiographic provinces

ALBERT E.OGDEN, WILLIAM A.CURRY & JAMES L.CUMMINGS   *Department of Earth Sciences, Tennessee Technological University, Cookeville, Tenn., USA*

ABSTRACT

To better understand the origin and growth of sinkholes, a morphometric analysis of sinkholes and caves was made of two different physiographic provinces in Tennessee. On the Eastern Highland Rim of central Tennessee, flat-lying Mississippian carbonates crop out producing broad sinkhole plains. An analysis of the percent area in sinkholes versus geologic formation was made. The St. Louis Limestone has the largest percentage land area in sinkholes followed by the Warsaw, Fort Payne, and Monteagle limestones. Slope appears to be the dominant factor with lithologic character being secondary. Orientation diagrams of 677 sinkholes axes and 978 straight cave segments were made. Cave passages show a very marked trend in a north/northeast direction as do the long axes of the sinkholes. The short axes of the sinkholes show little correlation to cave passage orientation.

A similar analysis of dolines was made for caves and sinkholes in eastern Tennessee which occur within the intensely folded and faulted Ordovician carbonates of the Valley and Ridge Province. Due to the steep dips in the area, caves were strongly oriented due north and between N30-50E. The long axes of sinkholes strongly correlate with cave orientations. The short axes of sinkholes were oriented nearly perpendicular to the long axes. Few cave orientations matched the short axis directions of sinkholes.

The resultant conclusion of the research is that where rocks are flat-lying, sinkholes and caves form along a wide range of orientations by dissolution along joints and fractures. In steeply dipping rocks, most cave passages and sinkholes are aligned along strike where dissolution has occurred along bedding planes within favorable lithologies. This enables the prediction of sinkhole formation and growth much easier in the Valley and Ridge Province compared to the Eastern Highland Rim.

Introduction

Understanding the origin, geologic distribution and morphology of sinkholes and caves helps provide geologists the ability to prevent engineering design problems that are common in karst terranes. This paper will compare the orientations of joints, sinkholes, photo-lineaments, and straight cave passage segments in flat-lying Mississippian limestones and folded Ordovician carbonates to determine if significant differences occur. The research results presented in this paper surround the basic concept of stratigraphic and structural controls on the origin of sinkholes and caves. The hypothesis that is being tested is whether the orientations of caves and sinkholes compare to zones of rock weakness as expressed in joint and photo-lineament directions. In addition, sinkhole density was calculated versus geologic formation to determine relative potential for sinkhole formation and growth.

Morphometric analysis of sinkholes is a means of quantifying the size, shape, distribution, and hydrogeologic controls on the development of closed depressions in karst terranes. Quantitative morphometric techniques applied to sinkholes were introduced by Cramer (1941) who first measured depression density. LaValle (1968) created the elongation ratio (length/width) as a means of quantifying sinkhole shapes. Williams (1966 and 1971) proposed a large number of morphometric parameters to describe karst terranes and was the first to apply Horton's (1945) and Strahler's (1957) ordering of river basins to sinkholes and spring water basins. William's work became the inspiration for a number of researchers to begin quantifying landforms in karst terranes. The structural controls affecting the elongation of sinkholes have been investigated by LaValle (1967), Kemmerly (1976), Ogden and Reger (1977), Soto and Morales (1984),

Littlefield et al. (1984), Ogden (1988), and Nelson (1988). A statistical comparison of cave passage and photo-lineament orientations was made by Ogden (1974) for the karst of Monroe County, West Virginia. Barlow and Ogden (1982) found little correlation between joints and photo-lineaments in the flat-lying carbonates of northwest Arkansas, but they did find a strong statistical correlation between photo-lineament and straight cave segment orientations. Kemmerly (1980, 1982, and 1989) has performed a detailed analysis of some 25,000 dolines from 42 quadrangles in the Western Highland Rim of Tennessee and the Pennyroyal Plain of Kentucky which are underlain by the St. Louis and St. Genevieve limestones of Mississippian age. He found a strong correlation between the orientations of depression long-axes and the systematic joints. The effect of highway construction on doline development in east Tennessee has been discussed by Moore (1976, 1981, and 1987). In the Eastern Highland Rim Province of central Tennessee, Mills et al. (1982) and Mills and Starnes (1983) have discussed sinkhole morphometry to aid in predicting sinkhole flooding in Cookeville. A recent study of sinkhole flooding around Cookeville has utilized the GIS and SWMM model to better predict the effect of storms (George et al., 1984).

Location and Geology

The study area is located within two geologically different physiographic provinces (Figure 1). The area between Sparta and Cookeville occurs on the Eastern Highland Rim Province which is underlain by flat-lying Mississippian limestones (Figure 2). Streams

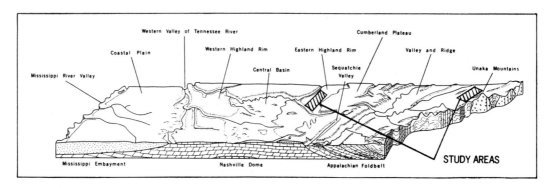

Figure 1: Location of the study areas (modified from Miller, 1974).

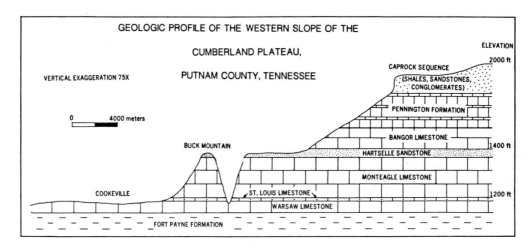

Figure 2: Geologic profile of the western slope of the Cumberland Plateau, Putnam County, Tennessee.

originating on the Cumberland Plateau east of Cookeville flow over shales and sandstones until underlying limestone beds are intersected. Some streams sink when they reach the Bangor Limestone while others sink into the lower Monteagle, St. Louis, or Warsaw limestones. Subterranean water moves through caves, pits, and solution enlarged fractures until emerging as spring flow. Deep, pit-type vertical caves form in the Bangor and Monteagle limestones, whereas long, horizontal caves usually occur in the St. Louis and Warsaw limestones. The underlying Fort Payne Formation is a dense silica-rich limestone. The lower Warsaw acts as an aquiclude. Stream sinkpoints are either open cave entrances which can be explored or sediment-filled sinkholes and fractures along stream bottoms. Descriptions of the caves have been made by Barr (1961), Matthews (1971), Faulkerson and Mills (1981) and the Tennessee Cave Survey. To date, over 200 caves are known in the study area with several having over one mile of passage. Unfortunately, most of the caves have not been mapped. Table 1 converts state-wide data published by Barr (1961) to show the stratigraphic distribution of caves in central Tennessee. Although Table 1 shows that the number of caves are greater in the Monteagle (St. Genevieve-Gasper) Limestone, substantially more cave footage occurs in the St. Louis and Warsaw limestones.

Table 1. Stratigraphic Distribution of Caves in Mississippian -aged Rocks of Central Tennessee (modified from Barr, 1961).

Formation	% of Total Number of Caves
Bangor	9
Monteagle	45.5
St. Louis	16.4
Warsaw	18.2
Fort Payne	10.9

The second study area for this investigation lies within Washington and Sullivan counties of eastern Tennessee within the Valley and Ridge Province of the Appalachians (Figure 1). The rocks are of Cambrian and Ordovician age and are strongly folded and faulted. Table 2 shows the stratigraphic distribution of Cambro-Ordovician caves as modified from Barr (1961). A vast majority of the caves occur in the Knox Group due to it being over 3,000 feet thick and having a large outcrop area. The Knox Group is composed primarily of dolomitic formations. Long caves and deep pits are not as abundant in eastern Tennessee as compared to the Eastern Highland Rim Province due to rock lithology and lower relief of carbonate rock exposure.

Table 2. Stratigraphic Distribution of Caves in Cambrian and Ordovician Rocks of Eastern Tennessee (modified from Barr, 1961).

Formation	% of Total Number of Caves
Chicamauga Limestone and equivalent rocks (Ordovician)	15.6
Knox Group (Ordovician and Cambrian)	65.6
Conasauga Group (Middle and Upper Cambrian)	7.8
Shady Dolomite (Lower Cambrian)	5.5
Other	5.5

Methods

Straight cave segments were delineated from cave maps provided by the Tennessee Cave Survey. These were divided into 50 foot segments and 100 foot segments in an attempt to distinguish between cave passages orientated along minor and major fractures, respectively. Sinkholes were randomly chosen from 1:24,000, U.S.G.S. topographic maps within the two study areas to delineate the orientations of their long and short axes. Joint orientation measurements were made using a Brunton compass. Photo-lineaments were delineated from 1:20,000, BXW, stereo aerial photographs. A rosette of straight stream segments was also made from a portion of the U.S.G.S. 1:100,000 Cookeville topographic map. The orientations of the straight cave segments, sinkhole axes, joints, and photo-lineaments were then placed in 10 degree classes to produce the rosette diagrams for each study area.

Sinkhole density and percent sinkhole area divided by area of formation were determined by placing 12 randomly selected one square mile mylar grided squares over the 1:24,000 geologic quadrangles. The number and area of sinkholes versus geologic formation were then plotted on bar graphs. Sinkholes less than 50 feet in diameter were probably not shown on the maps.

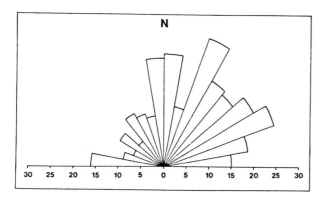

Figure 3: Rosette of 302 straight cave passage segment orientations of 100 foot length in the Cookeville area.

Figures 3 and 4 show the orientations of straight cave segments delineated from mapped caves in the study area around Cookeville. Cave passages are primarily developed in a north/northeast direction. This corresponds well to the mapped photo-lineaments (Figure 5) although weaker linear trends are seen in a northwest direction. There are two dominant joint directions seen in Figure 6, and these correspond very well to the photo-lineaments. The northeast trend follows that of the Appalachian Mountains. It is interesting to note that the orientations of straight stream segments delineated from the topographic map (Figure 7) do not correspond closely to the orientations of the photo-lineaments, joints, or major cave passages. This suggests that the streams

are not well adjusted to the geologic structure as are the caves. Major cave passages trend along the joints oriented northeast while the shorter passages occur along the joints oriented around N 45° W.

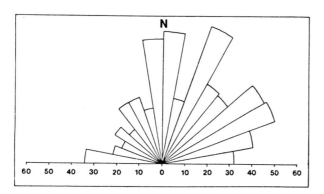

Figure 4: Rosette of 676 straight cave passage segment orientations of 50 foot length in the Cookeville area.

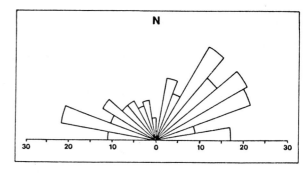

Figure 5: Rosette of 242 photo-lineament orientations in the Cookeville area.

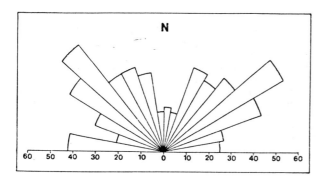

Figure 6: Rosette of 638 joint orientations in the Cookeville area.

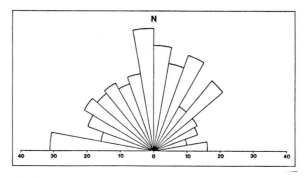

Figure 7: Rosette of 396 straight stream segment (topographic) orientations in the Cookeville area.

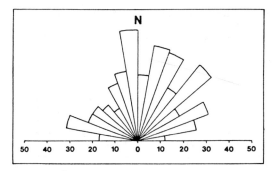

Figure 8: Rosette of 477 long sinkhole axis orientations in the Cookeville area.

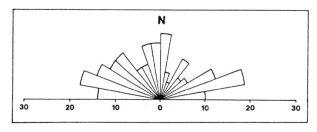

Figure 9: Rosette of 200 short sinkhole axis orientations in the Cookeville area.

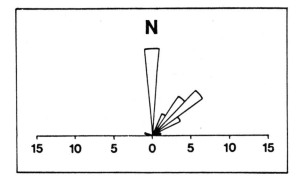

Figure 10: Rosette of 36 straight cave passage segment orientations of 100 foot length in east Tennessee.

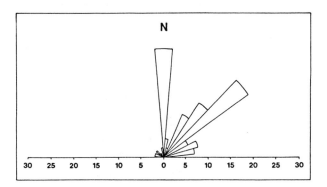

Figure 11: Rosette of 104 straight cave segment orientations of 50 foot length in east Tennessee.

Figures 8 and 9 show the orientations of sinkholes around Cookeville. Many sinkholes have only one significant axis of orientation which is usually in a northeast direction. Where short axes can be found, they are oriented in a northwest direction corresponding only slightly to some of the joint and photo-lineament trends. The long axes of the sinkholes are remarkably similar to the directions of the cave passages, demonstrating that dissolution is occurring preferentially along the joints oriented in north/northeast directions. This also suggests a correlation between sinkhole occurrence and the collapse of cave roofs, and perhaps more importantly, provides a tool to predict growth of sinkholes and their future occurrence.

Figures 10 and 11 show the orientations of straight cave segments in east Tennessee delineated from mapped caves. Cave passages show two strong orientations: due north and N 45° E. The N 45° E trend closely follows that of the strike of the rocks and the trend of the Appalachians in the study area. The orientations of the caves do not show a significant correlation to the measured joints (Figure 12). There is a significant difference in the spread of the data compared to the flat-lying rocks of the Eastern Highland Rim around Cookeville. Where the rocks are flat-lying, there is greater likelihood of cave development occurring along all dominant joint directions. In the steeply dipping rocks of east Tennessee, ground water moves rapidly down dip until reaching the water table, and then moves along strike within lithologically favorable beds. As a result, the cave patterns in east Tennessee show a strong linear or angulate pattern whereas in central Tennessee cave patterns are sinuous or maze-like. White (1988) has found similar results for the caves in Pennsylvania and West Virginia.

The orientations of the long and short axes of the dolines are shown in Figures 13 and 14. The long axes are oriented in a northeast direction similar to many of the straight cave segment orientations. The short axes are oriented primarily in a northwest

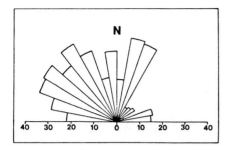

Figure 12: Rosette of 406 joint orientations in east Tennessee (Brown, FTDD, written communication).

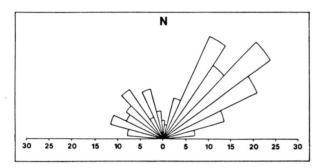

Figure 13: Rosette of 205 long sinkhole axis orientations in east Tennessee.

direction which is similar to the joints. Sinkhole and cave development and growth occur primarily along stratigraphic strike with less control by the joint orientations measured in rock outcrops.

Sinkhole density as expressed in number of dolines per square mile of formation was found to be greater in most of the formations of central Tennessee than in east Tennessee (Figures 15 and 16), although based on area (Figures 17 and 18), there is little

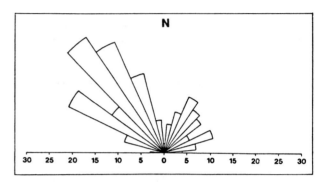

Figure 14: Rosette of 205 short sinkhole axis orientations in east Tennessee.

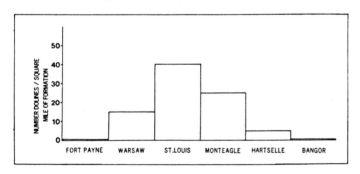

Figure 15: Stratigraphic distribution of dolines based on number in the Cookeville area.

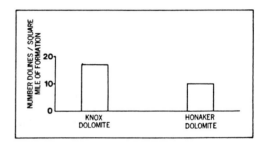

Figure 16: Stratigraphic distribution of dolines based on number in east Tennessee.

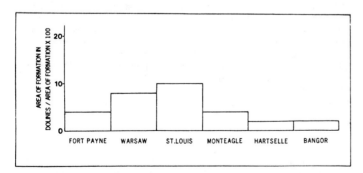

Figure 17: Stratigraphic distribution of dolines based on area in the Cookeville area.

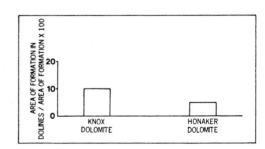

Figure 18: Stratigraphic distribution of dolines based on area in east Tennessee.

difference between the two physiographic provinces. The St. Louis Limestone has the greatest number of dolines per square mile of formation, but larger sinkholes are found in the other formations causing them to be more equal when expressed as an area measurement. In both study areas, slope and lithologic purity appear to be the primary factors in controlling sinkhole density. The St. Louis Limestone and Knox Dolomite occur on valley floors whereas the Honaker Dolomite and Warsaw and Monteagle limestones commonly occur on valley sides and are somewhat less pure.

Conclusions

Cave passages and sinkholes form along a wide-range of orientations in the flat-lying rocks of the Eastern Highland Rim corresponding to joint and photo-lineament trends. In the folded rocks of east Tennessee, most caves and sinkholes are strongly aligned along stratigraphic strike within bedding planes of favorable lithologies. The research for this paper has demonstrated the relative amount of structural control on cave and sinkhole development in the two physiographic provinces and shows that sinkhole occurrence and growth can be better predicted in the Valley and Ridge Province compared to the Eastern Highland Rim Province.

References

Barlow, C.A. and A.E. Ogden, 1982, A statistical comparison of joint, straight cave segment, and photo-lineament orientations: Bull. Nat. Speleol. Soc., v. 44, pp. 107-110.

Barr, T.C., 1961, Caves of Tennessee: TN Dept. Conservation, Division of Geology, Nashville, TN, Bull. 64, 567 p.

Cramer, H. 1941, Die systematik der karst dolinen: Neves Jb. Miner. Geol. Palaont., v. 85, pp. 293-382.

Faulkerson, J. and H.H. Mills, 1981, Karst hydrology, morphology and water quality in the vicinity of Cookeville, Tennessee: Unpublished Report for NSF Grant No. SPI-8004009, 67 p.

George, D.B., N. Taylor, T.E. Pride, and A.E. Ogden, 1988, Problem assessment of Cookeville's karst stormwater drainage system and spring water quality: Report to the City of Cookeville, TN (in press).

Horton, R.E., 1945, Erosion development of streams and their drainage basins: Bull. Geol. Soc. Am., v. 56, pp. 275-370.

Kemmerly, P.R., 1976, Definitive doline characteristics in the Clarksville quadrangle, Tennessee: Bull. Geol. Soc. Am., v. 87, pp, 42-46.

_____, 1980, A time-distribution study of doline collapse: framework for prediction: Environmental Geology, v. 3, pp. 123-130.

_____, 1982, Spatial analysis of a karst depression population: clues to genesis: G.S.A. Bull., v. 93, pp. 1078-1086.

_____, 1989, The karst contagion model: synopsis and environment implications: Environ. Geol. Water Sci., v. 13, no. 2, pp. 137-143.

LaValle, P., 1967, Some aspects of linear karst depression development in south-central Kentucky: Assoc. Am. Geographics Annals, v. 57, pp. 49-71.

Littlefield, J.R., M.A. Culbreth, S.B. Upchurch and M.T. Stewart, 1984, Relationship of modern sinkhole development to large scale photo-linear features: Proc. First Interdisciplinary Conf. on Sinkholes, Florida Sinkhole Research Institute, Orlando, FL, pp. 198-200.

Matthews, L.E., 1971, Descriptions of Tennessee Caves: TN Dept. of Conservation, Divsion of Geology, Nashville, TN, Bull. 69, 150 p.

Mills, H.H., D.D. Starnes, and K.D. Burden, 1982, Predicting sinkhole flooding in Cookeville: The Tennessee Tech Journal, v. 17, p. 1-20.

Mills, H.H., and D.D. Starnes, 1983, Sinkhole morphometry in a fluviokarst region: eastern Highland Rim, Tennessee, U.S.A.: Z. Geomorph. N.F., v. 17, pp. 39-54.

Moore, H.L., 1976, Drainage problems in carbonate terrane of east Tennessee: TN Dept. of Transportation Conf., Nashville, TN, pp. 111-130.

_____, 1981, Karst problems along Tennessee highways: an overview: Proc. 31st Highway Geology Symposium, Texas Bureau of Economic Geology, Austin, TX, pp. 1-28.

_____, 1987, Sinkhole development along "untreated" highway ditchlines in east Tennessee: Proc. 2nd Multidis. Conf. on Sinkholes and the Environmental Impacts of Karst, Florida Sinkhole Res. Institute, Orlando, FL, pp. 305-310.

Nelson, J.W., 1988, Influence of joints and fractures upon the hydrogeology of a karst terrane: Franklin, Co., Alabama: Proc. Int. Conf. on Fluid Flow in Fractured Rocks, U.S.G.S., pp. 240-261.

Ogden, A.E., 1974, The relationship of cave passages to lineaments and stratigraphic strike in central Monroe County, W. Virginia: Proc. 4th Conf. on Karst Geology and Hydrology, West Virginia Univ., pp. 29-32.

_____, 1988, A morphometric analysis of the sinkholes in the Greenbrier Limestone of West Virginia: Proc. 2nd Conf. on Environmental Problems in Karst Terranes and Their Solutions, NWWA, Dublin, OH, pp. 29-49.

Ogden A.E. and J.P. Reger, 1977, Use of morphometric analysis of dolines and cavern distribution for predicting ground subsidence in central Monroe, Co., W. VA: Proc. of the Int. Symposium on "Hydrologic Problems in Karst Regions: Western Kentucky University, pp. 130-139.

Soto, A.E. and W. Morales, 1984, Collapse sinkholes in the blanket sands of the Puerto Rico Karst belt: Proc. First Interdisciplinary Conf. on Sinkholes, Florida Sinkhole Research Institute, Orlando, FL, pp. 143-151.

Strahler, A.N., 1957, Quantitative analysis of watershed geomorphology: Trans. Am. Geophys. Un., v. 38, pp. 913-920.

White, W.B., 1988, Geomorphology and hydrology of karst terrain: Oxford University Press, 464 p.

Williams, P.W., 1966, Morphometric analysis of temperature karst landforms: Irish Speleol., v. 1, pp. 23-31.

Williams, P.W., 1981, Illustrating morphometric analysis of karst with examples from New Guinea: Zeits. Geomorph., v. 15, pp. 40-61.

Surface deformation of residual soil over cavitose bedrock

J.A.SCARBOROUGH, J.BEN-HASSINE, W.F.KANE & E.C.DRUMM   *Department of Civil Engineering, University of Tennessee, Knoxville, USA*

R.H.KETELLE   *Martin Marietta Energy Systems, Inc., Oak Ridge, Tenn., USA*

ABSTRACT

This paper presents a site specific model to predict the lateral and vertical extent of sinkhole subsidence. The deformation of the surface is studied using a hybrid approach of numerical and empirical analysis. Two-dimensional, nonlinear, finite element analysis is conducted for a thick, residual, clay soil and soil cavity above a discontinuity in rigid bedrock. This numerical analysis quantifies relationships in the subsurface geometry that drive surface deformation. Empirical profile functions are used to completely describe a continuous profile for a sinkhole subsidence basin. The vertical displacement, slope and curvature of the profile are major factors contributing to the damage of surface structures.

INTRODUCTION

The lateral and vertical extent, of surface subsidence or collapse, of sinkholes can contribute to the failure of surface structures, such as waste disposal liner systems, located in karst terrain. Site characterization is made difficult by the seemingly random nature of karst terrain, the poor predictability of future subsidence locations. The failure of a landfill liner in such a region is exacerbated by the geohydrological characteristics of karst, although potential consequences of failure can be mitigated by a sufficient liner system. Damage to surface structures due to subsidence is related to the curvature, slope and displacement of the surface profile (Karmis et al., 1987).

This study is limited to subsidence where the profile of the deformed surface, or basin, is continuous. Regional surveys of karst activity in eastern Tennessee suggest that collapse, resulting in a discontinuous profile, is a more likely type of failure (Newton and Tanner, 1986). Subsidence, however, has been noted as a precursor to collapse (Newton, 1976).

The subsidence of the surface due to karst activity was examined using both empirical curve fitting and numerical finite element analysis. The two-dimensional, numerical approach utilized a nonlinear material model for the strength-deformation characteristics of the residual, clay soil overlying cavitose bedrock. The empirical method develops the fit of a mathematical function to field profiles. Resulting constants control the shape of predicted basins. Sixteen profiles from four adjacent basins compose the field subsidence data for this study.

SITE DESCRIPTION

The site for which this analysis of sinkhole subsidence was performed is located at Oak Ridge, Tennessee near the western edge of the Valley and Ridge Province. It is underlain by silty clay soils and dolostone bedrock of the Knox group. The topography is hilly with parallel ridges, valleys, and elongate knobs. Regional bedrock structure causes bedding at the site to dip at an attitude of 35 to 45 degrees to the southeast. Variable weathering resistance and soil erodibility of the different stratigraphic zones has resulted in the parallel alignment of ridges and valleys. Effects of karst processes and erosion have combined in development of a rectangular surface drainage pattern. The karst system includes areas of doline karst on upland slopes, knobs and ridge crests with fluviokarst in the incised valleys.

Soils include ancient alluvium, loess, colluvium, residuum and saprolite ranging in thickness from 2 m to more than 40 m. Soils are predominantly residual silty clays with variable amounts of chert as boulders, nodules and gravel. Because of their fine texture, site soils have a high moisture retention. High natural moisture content, variable chert content and consolidation cause soils to range from very soft to very stiff.

Dolines occur in all five stratigraphic formations at the site, tending to align parallel to strike in some areas, along and near the trace of a prominent photolinear feature in one area, and along possible joint sets. Site investigation has included drilling, soil sampling, and testing within and outside visible karst features to obtain soil properties for use in subsidence analyses. The occurrence of ancient alluvium and loess on the upland landforms of the Knox is an indicator of long term stability of the soil mass. The analyses reported here were performed to quantify deformation characteristics of the site soils for consideration in planning potential site use.

Field Subsidence Basins

Ten subsidence basins and one collapse feature were measured using conventional plane surveying techniques. Due to similarities in location, material properties and geometry, four of these were selected for the study described herein. The basins were generally circular, but did exhibit elongated or elliptical bottoms. Basin diameters ranged from 22 m to 62 m. Measured vertical displacements ranged from 0.2 m to 5 m. Sloping topography, 5 to 10% slopes, and adjacent cut and fill, complicated determination of the lateral extent of subsidence. Profiles were drawn from 1:120 scale, 0.3 m contour maps using two criteria: 1) the profiles were oriented along and perpendicular to the major axis, and 2) the profiles were orthogonal to contour lines.

EMPIRICAL ANALYSIS

One method for empirical subsidence prediction is the use of profile functions. Profile functions have been developed to predict subsidence induced by longwall coal mining, but their use may be suitable in karst terrain, as well. Their application involves fitting a mathematical function to measured profiles of subsidence basins. The constants determined have been used to predict local subsidence with a knowledge of the subsurface geometry (Karmis et al., 1987; Karmis, 1984; Peng & Chaing, 1984; and Chen & Peng, 1981).

The negative exponential function suggested by Chen and Peng (1981) was used. This function was chosen because it does not force symmetry about the inflection point, and because it accurately represents the flat bottoms and steeply sloping sides of the measured profiles.

$$S(x) = S_o e^{-\alpha \left[\frac{x}{L}\right]^{\beta}} \qquad [1]$$

where: $S(x)$ = vertical displacement, S_o = maximum measured vertical displacement, X = horizontal distance from the origin, L = half-width of the subsidence basin, and α & β are the empirical parameters.

Figure 1 shows a generic subsidence profile. The origin is located on the centerline of the basin at the point of maximum vertical displacement. The half-width of the basin is taken as the horizontal distance from the origin to a point at 5% of the maximum vertical displacement. It is common practice to normalize vertical displacement by its maximum value. Horizontal distances may also be shown normalized to the half-width, L. It is worth noting that the slope at any point is the first derivative of the profile function, and the curvature is the second derivative. Vertical displacement, slope and curvature are associated with structural distress.

Figure 2 shows normalized field data from sixteen profiles and the associated best fit curve. The value of the empirical parameters, α = 2.75 and β = 3.91, were determined using least squares estimates from nonlinear regression analysis (STSC,

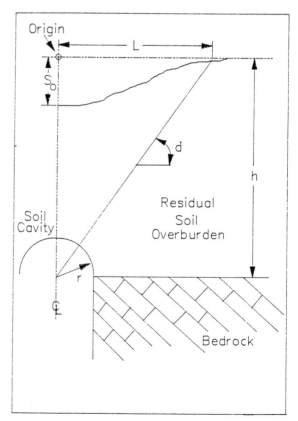

Figure 1: Idealized profile and definition of terms

1988). In contrast, Chen and Peng (1981) averaged parameters from each profile.

NUMERICAL ANALYSIS

Because ground water seepage tends to be drawn along the bedrock surface, washing of the residual soils tends to form elongated soil cavities that are likely to be larger than discontinuities in the bedrock (Kemmerly, 1980). The cavity-residual soil system can be idealized as a two-dimensional plane strain problem and finite element analysis utilized. For simplicity, the cross section of these cavities is assumed to be circular and hydraulic forces are neglected. Due to symmetry, only half of the domain is discretized, as shown in Figure 3. Eight-node, isoparametric, quadrilateral elements have been used to discretize the domain (Ben-Hassine, 1987).

Material Behavior and the Solution Procedure

The pre-peak, drained behavior of the residual soil can be adequately represented by a hyperbolic stress-strain model (Duncan and Chang, 1970). This model was chosen since it replicates the behavior of a soil more closely than a linear elastic model. In the hyperbolic model, Poisson's ratio is constant while the tangent modulus is a function of the stress state and is expressed as:

$$E_t = \left[1 - \frac{R_f (1 - \sin\phi)(\sigma_1 - \sigma_3)}{2c\cos\phi + 2\sigma_3\sin\phi} \right]^2 K_h P_a \left[\frac{\sigma_3}{P_a} \right]^n \quad [2]$$

where: E_t = tangent modulus, σ_1 = major principal stress, σ_3 = minor principal stress, P_a = atmospheric pressure, R_f = failure ratio, n = modulus ratio, K_h = modulus number for loading, c = cohesion, and ϕ = angle of internal friction. The values of the parameters used are given in Table 1.

An incremental-iterative Newton-Raphson procedure is utilized in the solution of the nonlinear problem. A mid-point Runge-Kutta procedure is adopted in the sense that tangent moduli are based on the old total stresses plus half the incremental stresses in order to further accelerate convergence. Nodal loads equivalent to the weight of the residual soil are applied incrementally in five steps. At

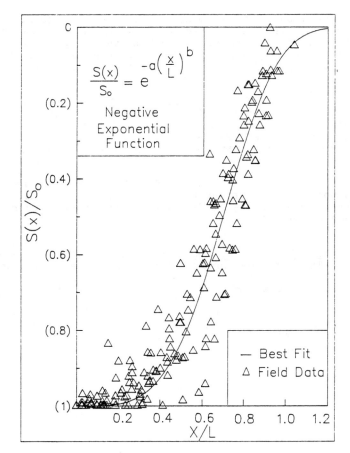

Figure 2: Normalized field profiles and best fit of the negative exponential function

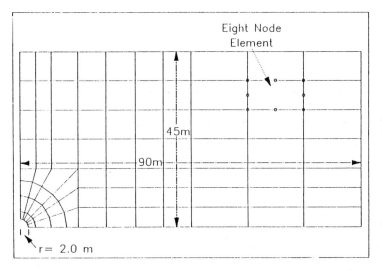

Figure 3: Finite element mesh for r = 2.0 m and h = 45.0 m

every load step, as many iterations as required to achieve convergence are performed. Convergence is monitored by comparing a norm based on the residual unbalanced forces in the system to the norm based on the original applied nodal forces with a tolerance of 1%. This approach is similar to previous analysis (Ketelle et al., 1987 and Drumm et al., 1989). Convergence of the non-linear problem was consistently achieved.

Results

A total of 25 finite element analyses were performed. The soil cavity radii considered were 0.3, 0.6, 1.0, 2.0, and 4.0 m. The thickness of overburden considered was 15.0, 22.5, 30.0, 37.5, and 45.0 m. These values of cavity radius and overburden thickness cover the range of values found at the site. The depth to bedrock, that is the depth to refusal for borings, made in and adjacent to Sinkhole 04, is known to be 41 m.

Figure 4 shows that for h = 45 m, the observed magnitude of the vertical displacement is bracketed by numerical predictions with radii of 2 m and 4 m, although the corresponding basin half-widths exceed observed values. The implications of this are discussed below.

INTEGRATION OF NUMERICAL RESULTS IN AN EMPIRICAL ANALYSIS

The numerical analysis provides a means to examine an unknown cavity radius, a known depth of overburden, and given soil properties in terms directly related to the profile of the deformed surface.

Vertical displacement is controlled by the radius of the soil cavity. This relationship is quantified by regression on the results of the numerical analyses. An exponential relationship, with the square of the correlation coefficient, R^2, equal to 0.933, for maximum vertical displacement in terms of the cavity radius was determined:

$$S_o = e^{-5.46 + 2.04\,r} \qquad [3]$$

where So = the maximum vertical displacement, and r = the radius of the soil cavity. The units of both variables are in meters.

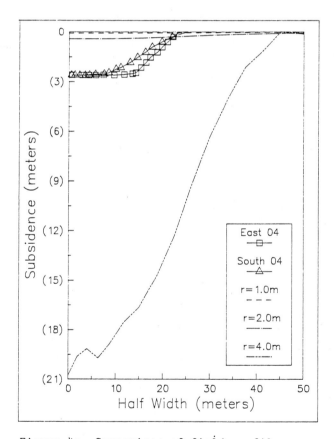

Figure 4: Comparison of field profiles with numerical results

Table 1: Material model parameters, basic soil properties from Woodward-Clyde (1984)	
Parameter	Value
Unit Weight, γ	18.8 kN/m3
Initial Tangent Moduli, E_t	1.006 E5 kPa
Poisson's Ratio, ν	0.35
Angle of Internal Friction, ϕ	23°
Cohesion, c	28.7 kPa
Failure Ratio, R_f	0.9
Modulus Exponent, n	0.5
Modulus Number, K_h	972.0
Atmospheric Pressure, P_a	103.5 kPa

The angle of draw relates the lateral extent of subsidence at the surface to the depth of overburden. It is measured from the horizontal to a line connecting the centerline of the basin at bedrock to the half-width of the basin at the surface, as shown in Figure 1. The depth of overburden is easily measured; and a known angle of draw allows estimation of the basin half-width.

Measuring the angle of draw from the centerline at bedrock, instead of from the outer edge of the cavity at bedrock as in mining, underestimates the lateral extent of the basin. For a given value of vertical displacement, an underestimated basin half-width will increase the slope and curvature of the profile. This definition of angle of draw was necessitated by the fact that the actual cavity radius in the field can rarely be determined. The relationship defining the angle of draw can be expressed as:

$$tan\ \delta = \frac{h}{L} \hspace{4cm} [4]$$

This relationship, for the numerical analysis, is virtually constant at $\delta = 31.9°$, with an $R^2 = 0.980$, as shown in Figure 5.

Considering only the single case of r = 4 m and h = 45 m, the magnitude of δ increases to 47.3°, reducing the half-width. However, when a larger cavity radius, r = 8.0 m, and a depth of overburden of 45.0 m were considered, convergent results were obtained up to 40% of the loading. In subsequent load increments, convergence was not achieved for any number of iterations suggesting total collapse of the domain.

In an axisymmetric analysis of slip surfaces about an open karst pipe, Yoon (1987) showed:

$$\delta = 45° + \phi/2 \hspace{3cm} [5]$$

where ϕ = the angle of internal friction for the clay soil. Incorporating this angle for $\phi = 23°$ and h = 41 m yields a predicted half-width of 27.1 m. This value compares favorably with observed values for Sinkhole 04.

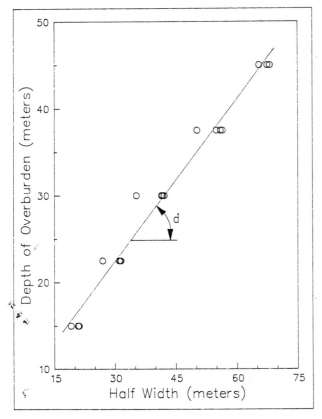

Figure 5: Angle of draw from the numerical analysis

147

SINKHOLE SUBSIDENCE PREDICTION

Direct substitution, of equations [3], [4], and [5] with φ = 23°, into the profile function [1] yields an expression for the vertical displacement at any point:

$$S(x) = e^{-5.46 + 2.04 r} \; e^{-\alpha \left[\frac{X}{0.66h} \right]^{\beta}} \qquad [6]$$

where: α = 2.75 and β = 3.91 are site specific empirical parameters, and r and h define geometry. Figure 6 is a comparison of the results from equation [6] for r = 3.0 m, r = 3.14 m and r = 3.5 m for constant h = 41 m with field profiles. The field and predicted curves compare favorably.

No distinct relationship was discerned in the numerical data for varying ratios of cavity radius to the depth of overburden (r/h) with L, the half-width of the basin, or So, the maximum subsidence. This ratio is significant in mining-induced subsidence (Karmis et al., 1987; Karmis, 1984; Peng & Chaing, 1984; and Chen & Peng, 1981). The absence of a significant r/h relationship prevents determination of an expression only in terms of the easily determined depth of overburden. Improved methods of geophysical exploration may allow routine determination of cavity sizes for use in the model. At present, estimates based on experience or probabilistic values can be used.

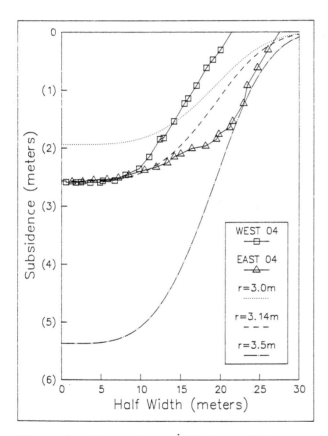

Figure 6: Comparison of field profiles with predicted results

CONCLUSIONS

Subsidence prediction requires knowledge of a relationship between the lateral and vertical extent of deformation and its driving force(s). The irregular and inaccessible nature of the bedrock surface in karst terrain necessitates the use of an idealized analysis to quantify relationships between unknowns.

Use of the finite element method can, with an adequate material model, provide reasonable estimates of the maximum vertical subsidence. Empirical analysis can then provide an estimation of the lateral extent of subsidence consistent with observed field conditions. Integration of the two in a hybrid approach provides a prediction tool for the complete subsidence basin profile. This is critical in the determination of the slope and curvature of the profile necessary for the damage assessment of structural components such as clay or geotextile landfill liner systems.

REFERENCES

Ben-Hassine, J., (1987). "Finite Element Applications in the Analysis of Nonlinear Problems in Geomechanics," M.E. Thesis, University of Tennessee, Knoxville

Chen, D and Peng, S.S., (1981). "Analysis of Surface Subsidence Parameters due to Underground Longwall Mining in the Northern Appalachian Coalfield," Technical Report 81-1, Department of Mining Engineering, West Virginia University

Drumm, E.C., Kane, W.F., Scarborough, J.A., and Ben-Hassine, J., (1989). "Strength and Deformation Characteristics of Residual Soils over Cavitose Bedrock," a report to Martin Marietta Energy Systems, Inc, Oak Ridge, Tennessee under contract # RO1133275, in preparation.

Duncan, J.M. and Chang, C.Y., (1970). "Nonlinear Analysis of Stress and Strain in Soils," Journal of Soil Mechanics, Foundation Division, ASCE, 96(SM5), pg. 1629 - 1653

Karmis, M., Goodman, G. and Hasenfus G., (1984). "Subsidence Prediction Techniques for Longwall and Room and Pillar Panels in Appalachia," Proceedings of the 2nd Conference on Stability in Underground Mining, pg. 541 - 553

Karmis, M., editor, (1987). "Prediction of Ground Movements due to Underground Mining in the Eastern United States Coalfields, Vol 1: Development of Prediction Techniques," Department of Mining and Minerals Engineering, Virginia Polytechnical Institute and State University

Kemmerly, P.R., (1980). "Sinkhole Collapse in Montgomery County Tennessee," Tennessee Division of Geology, Environmental Geology Series No. 6

Ketelle, R.H., Drumm E.C., Ben-Hassine, J. and Manrod, W.E., (1987). "Soil Mechanics Analysis of Plastic Soil Deformation over a Bedrock Cavity," Proceedings of the 2nd Multidisciplinary Conference on Sinkholes and the Environmental Impacts of Karst, pg. 383 - 387

Newton, J.G., (1976). "Early Detection and Correction of Sinkhole Problems in Alabama, with a Preliminary Evaluation of Remote Sensing Applications," Alabama Highway Department

Newton, J.G. and Tanner J.M., (1986). "Regional Inventory of Karst Activity in the Valley and Ridge Province, Eastern Tennessee, Phase I: ORNL/Sub/11-78911/1, Oak Ridge National Laboratory

Peng, S.S. and Chaing, H.S., (1984). "Longwall Mining," John Wiley and Sons

STSC, (1988). "STATGRAPHICS, Statistical Graphics System User's Guide, University Edition" Statistical Graphics Corporation, 2115 East Jefferson Street, Rockville, MD 20852

Woodward-Clyde Consultants, (1984). "Subsurface Characterization and Geohydrologic Evaluation, West Chestnut Ridge Site," ORNL/Sub/83-647/641/v1-2, Oak Ridge National Laboratory

Yoon, C.J. (1987). "Slip Surfaces in Sinkholes: An Analysis of the Slip Line of an Axisymmetric Open Karst Pipe," M.S. Thesis, University of Tennessee, Knoxville

3. Ground water quality and pollution in karst terranes

Physical parameters and microbial indicator organisms of a carbonate island karst aquifer, San Salvador Island, Bahamas

WILLIAM BALCERZAK & JOHN MYLROIE *Department of Geology and Geography, Mississippi State University, Miss., USA*

GERARD S.PABST *Department of Biological Sciences, Mississippi State University, Miss., USA*

ABSTRACT

San Salvador Island is located on the eastern edge of the Bahamian Archipelago and is one of 29 inhabited islands which are part of the island nation of the Bahamas. The physical isolation of the island, inland saline lakes, karst topography, and vertical and lateral variations in bedrock permeability make it difficult to characterize the groundwater environment. Previous investigations have shown that the freshwater lens is partitioned by coastline irregularities, inland lakes, and eolian dunes into a number of distinct aquifers. Lens volume, chemistry, and shape are highly variable. The present study is an assessment of water quality and distribution on San Salvador.

Water samples were collected from wells and catchments, and a base line study was conducted to establish various physical and microbial parameters. Measurements were made of conductivity, pH, dissolved oxygen, and temperature. Microbial analyses were made of total coliform, fecal coliform, and fecal streptococci. Average conductivity of the wells was found to be 2302 microsiemens/cm, corresponding to a total dissolved solids (TDS) value of 1472 mg/l, while conductivity in catchments was 227 microsiemens/cm or 135 mg/l TDS. Extremely high TDS values (up to 14,900 mg/l) were found in a few wells and possibly mark the site of a conduit flow between the ocean and hypersaline inland lakes. Delineation of individual lenses appears to be possible by similarity of physical parameters and the location of the aquifer in relation to lakes, shoreline, and topography. Microbial counts show very high fecal streptococci levels and little to no fecal and total coliforms, indicating recent animal contamination from an avian or ruminant animal source. The ubiquitous presence of fecal streptococci in all aquifers (counts were above 1,000 colonies/plate for 90% of the samples) measured from open, capped, abandoned, and active wells indicates a very high rate of flushing of these short-lived bacteria from the surface to subsurface. Karst processes may be allowing the rapid subsurface dispersal of bacterial contaminants in what would otherwise be a fine-grained porous aquifer. In a few locations higher total coliform indicates occasional aquifer contamination by human waste.

Introduction

San Salvador Island is located on an isolated carbonate bank of the Bahamian Archipelago 600 km (380 mi) east-southeast of Miami, Florida (Figure 1). The island is composed of shallow water marine and eolian limestones of Pleistocene and Holocene age. The island interior contains numerous saline lakes, most of which occupy sea level swales between dunes as high as 40 m in (64 ft) elevation. Karst features such as pits, sinkholes, and caves are well developed. There is no surface drainage.

Potable water on San Salvador Island is supplied by either rainwater catchments or groundwater. However, only a small portion of this water is provided by catchments because of the high cost of building materials. The government maintains two well fields on the island which supply 65% of potable water. The remainder is provided by shallow hand-dug wells located near each settlement. Karst processes may contribute to groundwater contamination by allowing direct recharge before natural filtration and adsorption can occur.

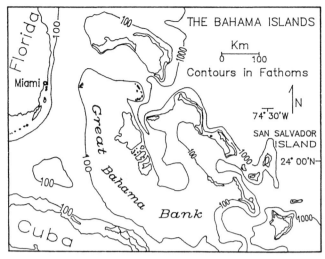

Figure 1. Location of study area.

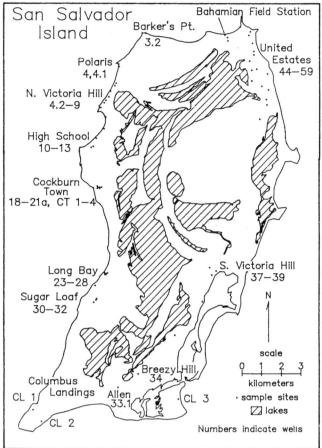

Figure 2. San Salvador Island showing sample locations.

Geologic and Hydrogeologic Setting

San Salvador Island is composed of Holocene and Pleistocene fossil coral reefs, subtidal deposits, beach rock, and eolian calcarenites. The relatively short duration and high amplitude of Quaternary eustatic oscillations has left its imprint on the geology of the island (Carew and Mylroie, 1985). Dune ridges, which reach to a maximum of 40 m (64 ft) above present sea level, dominate the topography. Many of the swales between the dune ridges contain saline and hypersaline lakes. Paleosols and other subaerial features were produced during periods of low sea level when the island was exposed to weathering. Pits, sinkholes, and caves have formed throughout the geologic history of the island (Mylroie, 1988).

Aquifers of San Salvador are partitioned by coastline, lake, and topographic features into numerous distinct lenses (Davis and Johnson, 1988). In addition, they are bounded by lateral and vertical variations in bedrock permeability. Depth to the top of the groundwater surface within shallow coastline wells ranges between 1.5 (4.9 ft) and 4.5 m (14.8 ft). Cavernous horizons have been reported at depths of 6.7 m (22 ft) to 7.9 m (26 ft), 10.6 m (35 ft) to 12.2 (40 ft), and at 18.3 m (60 ft) (Klein and others, 1958).

Purpose and Methods

The primary purpose of this study was to assess the water quality and distribution of ground water on San Salvador. A total of 86 water samples were collected from 67 wells and catchments. Sample sites were located primarily along the island perimeter where most settlements are located (Figure 2). In situ measurements were made with a YSI conductivity-salinity-temperature meter and a YSI dissolved oxygen meter and included conductivity, pH, dissolved oxygen, and temperature. Although salinity was recorded, its measurement is based on the standard composition of sea water. The values, therefore, are distorted by the high calcium ion concentration of the ground water. Colony counts were taken of total coliform, fecal coliform, and fecal streptococci using the membrane filter technique. These parameters were chosen because they are easily measured in isolated areas such as San Salvador and they give a good indication of the character and quality of ground water.

Physical Parameters

Readings obtained from wells and catchments along with minimum, maximum, and average values are shown in Table 1. Total dissolved solids were estimated using a factor of 0.65 to convert from microsiemens/cm to mg/l. Numbering of sample sites begins in the northeast end of the island at the Bahamian Field Station and increases counterclockwise around the island (Figure 2).

Generally, TDS increases toward the boundary of each village and decreases toward the center. Similarly, dissolved oxygen tends to be low near the boundaries and higher toward the center. Settlements are located along the perimeter of the island near potable water sources where the depth to the aquifer is minimal. For these reasons, individual aquifers can be assumed to be located at or near each village. Two areas will be looked at in more detail: the northwest portion of the island consisting of Barker's Point, Polaris, and North Victoria Hill; and the Long Bay/Sugar Loaf area (Figure 2).

Barker's Point/Polaris/North Victoria Hill

Well number 3.2 at Barker's Point shows TDS of approximately 3,600 mg/l and dissolved oxygen content of 1.0 mg/l. Wells 4 to 4.2 show fairly low TDS and high dissolved oxygen. TDS increases sharply in wells 4.3, 7, and 8 while dissolved oxygen increases slightly in wells 4.3 and 7 and decreases sharply in well 8. The southern boundary of the North Victoria Hill aquifer can be drawn somewhere between wells 8 and 10 due to a rise in elevation and an increase in TDS.

A possible explanation for the high TDS found in wells 4.3, 7, and 8 is salt water intrusion along the coast (average ocean salinity is 35,000 mg/l.). However, these wells are located only a few meters from wells low in TDS and the dissolved oxygen content of wells 4.3 and 7 approach levels expected of a lake. If salt water intrusion were occurring, a more gradual increase in TDS in adjacent wells and lower dissolved oxygen values (dissolved oxygen levels are lower in sea water) would be expected (Hitchman, 1978). In addition, these wells are located in a slight topographic depression where the

Table 1. Physical Parameter Data

well	C µS/cm	TDS mg/l	pH	DO mg/l	depth m (ft)	Temp °C (°F)
3.2	5500	3575	6.6	1.0	1.5 (4.9)	27 (81)
4	1350	878	6.6	3.8	2.8 (9.2)	28 (82)
4.1	1250	813	6.6	3.8	3.3 (10.8)	28 (82)
4.2	1200	780	6.3	5.8	2.6 (8.5)	29 (84)
4.3	3150	2048	6.6	6.4	3.0 (9.8)	29 (84)
7	6500	4225	7.2	7.4	3.4 (11.2)	28 (82)
8	23000	14950	6.6	0.1	3.2 (10.5)	28 (82)
8.a	900	585	7.2	4.1	2.4 (7.9)	28 (82)
9	1500	975	6.8	4.6	2.8 (9.2)	28 (82)
10	2600	1690	8.0	5.0	3.3 (10.8)	28 (82)
11	1500	975	6.7	2.0	2.1 (6.9)	29 (84)
11.1	1100	715	6.4	5.0	2.3 (7.5)	28 (82)
11.2	650	423	6.6	6.8	2.1 (6.9)	28 (82)
11.3	400	260	6.7	5.4	1.5 (4.9)	28 (82)
13	1400	910	6.6	1.8	--- ---	33 (91)
18	750	488	---	1.2	3.4 (11.2)	27 (81)
21	900	585	---	4.2	2.9 (9.5)	26 (79)
21a	1700	1105	6.3	4.2	--- ---	28 (82)
CT1	2000	1300	6.6	2.8	--- ---	28 (82)
CT2	2000	1300	7.1	4.8	--- ---	28 (82)
CT3	2050	1333	7.1	5.2	--- ---	28 (82)
CT4	1950	1268	7.1	3.0	--- ---	28 (82)
23	5000	3250	7.6	1.0	2.7 (8.9)	28 (82)
24	1700	1105	6.8	2.2	--- ---	28 (82)
25	1700	1105	6.4	2.0	--- ---	27 (81)
25.1	1600	1040	6.8	2.4	--- ---	27 (81)
27	800	520	6.7	2.4	--- ---	27 (81)
28	900	585	7.1	2.0	2.9 (9.5)	27 (81)
30	1000	650	7.1	4.0	--- ---	27 (81)
31	650	423	7.0	1.0	4.5 (14.8)	27 (81)
32	850	553	7.0	6.2	--- ---	28 (82)
CL1	1300	845	8.3	6.2	--- ---	28 (82)
CL2	1950	1268	7.6	8.0	--- ---	28 (82)
33.1	2000	1300	7.2	2.2	--- ---	28 (82)
CL3	750	488	6.9	5.6	--- ---	28 (82)
34	550	358	6.5	0.6	4.5 (14.8)	28 (82)

well	C µS/cm	TDS mg/l	pH	DO mg/l	depth m (ft)	Temp °C (°F)
35	6000	3900	2.6	2.4	--- ---	27 (81)
37	6000	3900	7.6	1.4	1.0 (3.3)	28 (82)
38	480	312	7.2	1.0	2.4 (7.9)	27 (81)
39	5000	3250	7.8	2.6	1.1 (3.6)	28 (82)
43	600	390	5.8	1.8	4.7 (15.4)	28 (82)
44	380	247	6.4	0.4	0.8 (2.6)	30 (86)
48	1350	878	7.0	1.6	1.6 (5.2)	27 (81)
51	850	553	6.8	3.0	--- ---	28 (82)
52	3000	1950	---	8.4	1.5 (4.9)	28 (82)
54	1900	1235	6.0	0.2	1.0 (3.3)	29 (84)
55	1200	780	7.4	7.2	0.8 (2.6)	29 (84)
57.1	1350	878	6.9	4.2	--- ---	27 (81)
59	600	390	7.5	6.3	1.9 (6.2)	27 (81)
min	380	247	2.6	0.1	0.8 (2.6)	26 (79)
max	2302	14950	8.3	8.4	4.7 (15.4)	33 (91)
avg	2302	1496	6.8	3.6	2.5 (8.2)	28 (82)

Catchments

Site	C µS/cm	TDS mg/l	pH	DO mg/l	Temp °C (°F)
1	150	98	6.4	4.6	29 (84)
3	160	104	9.4	10.0	30 (86)
36	700	455	3.3	3.4	28 (82)
46.1	290	189	---	2.6	28 (82)
47	125	81	6.5	4.4	26 (79)
50	120	78	---	3.0	28 (82)
58	90	59	7.2	6.6	28 (82)
58.1	190	124	8.5	5.1	27 (81)
31.3	190	124	7.4	4.8	28 (82)
min	70	46	3.3	2.6	26 (79)
max	700	455	9.4	10.0	31 (88)
avg	227	148	6.9	4.8	28 (82)

C = conductivity in microsiemens/cm (µS/cm)
TDS = total dissolved solids
DO = dissolved oxygen

155

distance between the coastline and an inland saline lake is minimal. The sharp increase in TDS, high dissolved oxygen, and location of these wells may indicate conduit flow between the ocean and adjacent saline lake.

Long Bay/Sugar Loaf

Long Bay and Sugar Loaf villages are located in the southwest portion of San Salvador and are bordered on the east by a 12 m (40 ft) high dune ridge. Wells 23 to 25.1 and well 30 are located along the north and west borders of the villages and show relatively high TDS. Wells 27, 28, 31, and 32 are lower in TDS and are located farther east near the base of the ridge. The southern extent of this aquifer could not be determined because of a lack of wells in this area. However, approximately 4.5 km (2.8 mi) south, at the base of the same ridge, well CL 1 indicates a TDS of 845 mg/l. Based on these wells and other data (Davis and Johnson, 1988), it is believed that a thicker fresh water lens exists within the dune ridge with groundwater flow to the west.

Indicator Organisms

Waste from warm-blooded animals can contaminate water sources with a number of bacterial genera and species (APHA, 1983). These include the coliform group and the genus Streptococcus, both of which are normally present in the intestinal tract of humans and animals. In addition, pathogenic microorganisms can be contributed by diseased individuals. The population of pathogenic bacteria tends to be low and the variety high since they are contributed by individuals with specific enteric illnesses while coliform and streptococci bacteria are contributed by the entire population (Viessman and Hammer, 1985). The survival rate of coliforms outside the intestinal tract is also much greater than that of pathogenic bacteria. For these reasons total coliform, fecal coliform, and fecal streptococci are widely used as indicators of microbial pollution.

Table 2. Pollution source indicated by FC/FS ratios.

human	4.0
duck	0.6
sheep	0.4
chicken	0.4
pig	0.4
cow	0.2
turkey	0.1

The ratio of fecal coliform (FC) to fecal streptococci (FS) may be used to determine possible sources of pollution. Fecal coliform are more prevalent in humans whereas fecal streptococci are more common in the intestinal tract of animals. Ratios of FC/FS (Table 2) have proven to be valid provided the samples are collected close to the source area. This is due to the lower survival rate of fecal streptococci relative to fecal coliform in the environment (APHA, 1985).

Ratios greater than 4.1 are considered indicative of domestic human pollution while ratios less than 0.7 are indicative of recent nonhuman pollution. Ratios between 4.1 and 0.7 suggest a mixture of waste from both animal and human sources (APHA, 1985).

Table 3. Microbial Data.

Wells

Well	TC	FC	FS	Well	TC	FC	FS	Well	TC	FC	FS
3.1	0	0	305	21*	0	0	2760	33.1	0	0	TNTC
3.2	0	0	TNTC	21a	24	0	TNTC	CL 3	148	0	TNTC
4	0	1	TNTC	CT1	0	0	---	34	216	0	TNTC
4.1	0	64	TNTC	CT2	0	0	TNTC	35	52	0	346
4.2	0	0	TNTC	CT3	30	0	TNTC	37	106	0	TNTC
4.3	0	63	TNTC	CT4	0	0	TNTC	38	8	0	TNTC
7	0	0	TNTC	23	56	0	TNTC	39	98	0	TNTC
8	1	0	TNTC	24	32	0	48	43	0	0	TNTC
8.a	0	3	TNTC	25	66	0	TNTC	44	0	0	TNTC
9	0	TNTC	TNTC	25.1	402	0	TNTC	48	0	5	TNTC
10	0	0	TNTC	27	132	2	TNTC	51	0	3	TNTC
11	0	0	TNTC	28	352	0	TNTC	52	0	0	TNTC
11.1	1	12	TNTC	30	292	2	TNTC	54	0	0	TNTC
11.2	0	0	TNTC	31	78	0	TNTC	55	22	0	TNTC
11.3	0	0	TNTC	31.1	62	0	TNTC	57.1	0	0	3
13	0	0	TNTC	32	0	0	TNTC	59	0	0	132

Catchments

Site	TC	FC	FS
1	0	0	TNTC
3	0	0	TNTC
36	104	0	TNTC
46.1	20	0	TNTC
47	0	1	TNTC
50	0	0	TNTC
58	0	0	51
58.1	0	0	68

TC = total coliform, FC = fecal coliform
FS = fecal streptococci
TNTC = too numerous to count (>1000 colonies/count)
All values in colonies/100ml
*Well 21 showed 276 colonies/10 ml sample (2760/100 ml)

Microbial data for both the wells and catchments are shown in Table 3. A full 90% of all samples processed showed fecal Streptococci counts greater than 1,000 colonies per plate. Fecal coliform counts were 0 to 79% of the samples and total coliform counts were 0 for 62% of the samples. The same results were obtained from both catchments and wells whether they were fresh, saline, open, capped, or abandoned and were not affected by dilution ratios of the sample water. The high fecal streptococci counts resulted in a ratio of 0.7 or less for all samples. This indicates a recent avian or ruminant source. These results are somewhat surprising since most residences have an outhouse near the home. There are no public waste disposal systems on the island. A few notable exceptions to the low coliform counts are found at the Long Bay/Sugar Loaf area (wells 23 to 32) and the eastern side of the island from Allen settlement to South Victoria Hill (wells 33.1 to 39). Here total coliform counts were as high as 200 colonies per plate. This may indicate past or occasional contamination by humans.

Conclusions

Results from the physical parameters measured agree with those obtained from more extensive investigations (Johnson and Davis, 1988). Individual aquifers can be delineated through geologic and geomorphic constraints and TDS and dissolved oxygen values. Microbial counts show high animal (avian and ruminant) contamination and low human contamination. The presence of extremely high fecal streptococci counts may be the result of secondary porosity produced by karst processes, allowing these short-lived bacteria to be quickly dispersed despite the sandy nature of the subsurface. The high TDS concentration found in most wells indicates a relatively low groundwater quality (the upper limit for fresh water can be considered as 1,500 mg/1) (Tchobanoglous and Schroeder, 1985). The potential for waterborne epidemics in these untreated waters is indicated by the high animal contamination.

Acknowledgements

The author would like to thank Dr. Donald T. Gerace and the Bahamian Field Station for financial and logistical support. Additional funding and materials were provided by Dr. Lewis R. Brown of the Department of Biological Sciences and the Department of Geology and Geography, Mississippi State University.

References

American Public Health Association (APHA), American Water Works Association, and Water Pollution Control Federation, 1985, Standard methods for the examination of water and wastewater, 16th ed.: Washington, D.C., American Public Health Association, 1200 p.

Carew, J. L. and Mylroie, J. E., 1985, The Pleistocene and Holocene stratigraphy of San Salvador Island, Bahamas, with reference to marine and terrestrial lithofacies at French Bay: in Curren, H. A., ed., Pleistocene and Holocene carbonate environments on San Salvador Island, Bahamas - Guidebook for Geological Society of America, Orlando annual meeting field trip: Ft. Lauderdale, Florida, CCFL Bahamian Field Station, p. 11-61.

Davis, R. L. and Johnson, C. R.., Jr., 1988, The karst hydrology of San Salvador Island, Bahamas [abs.]: Program of the 10th annual Friends of Karst meeting, San Salvador Island, Bahamas, Feb. 11-15, 1988.

Hitchman, M. L., 1978, Measurement of dissolved oxygen, John Wiley, New York, 225 p.

Johnson, C. R., Jr., and Davis, R. L., 1988, The use of Ca^{++} and Mg^{++} as indicators of ground and surface water source on San Salvador, Bahamas [abs.]: Program of the 10th annual Friends of Karst meeting, San Salvador Island, Bahamas, Feb. 11-15, 1988.

Klein, H., Hoy, N. D., and Sherwood, C. B., 1958, Geology and groundwater resources in the vicinity of the auxiliary air force bases, British West Indies, U.S. Geological Society, Open file report.

Mylroie, J. E., 1988, Karst of San Salvador: in Mylroie, J. E., ed., Field guide to the karst geology of San Salvador Island, Bahamas: College Center of the Finger Lakes Bahamian Field Station, San Salvador, Bahamas, p. 17-44.

Tchobanoglous, G. and Schroeder, E. D., 1985, Water quality: characteristics, modeling, modification, Addison-Wesley, Reading, Mass., 768 p.

Viessman, W. and Hammer, M. J., 1985, Water supply and pollution control, Fourth Edition, Harper & Row, New York, 797 p.

The impact of agricultural practices on water quality in karst regions

WILLIAM S.BERRYHILL JR.   *Water Quality Group, Biological and Agricultural Engineering Department, North Carolina State University, Raleigh, N.C., USA*

ABSTRACT

Agricultural activities have been identified as major nonpoint sources of pollution of water resources. Karst topography and drainage can present special concerns to farmers and other rural ground water consumers. Cropping and livestock operations can cause direct pollutant entry through sinkholes, or diffuse entry by leaching of nutrients, pesticides, bacteria, and other contaminants into shallow carbonate-rock aquifers, which are vulnerable to NPS contamination. Best Management Practices for cropland NPS control include conservation tillage, crop rotation, management of fertilizers and manure, and Integrated Pest Management. Practices that tend to increase infiltration, however, may be detrimental in karst areas. For livestock and poultry operations, planned grazing, livestock exclusion, and hygienic management of waste from confinement facilities are beneficial. Land-use changes by urbanization of existing farmland presents a special challenge for those concerned for karst ground water quality.

Keywords: agricultural BMPs, karst, NPS pollution, nitrates, pesticides, manure, sinkholes.

INTRODUCTION

Nonpoint source (NPS) pollution from agricultural activities and other sources has been identified as a high priority water quality problem in the U.S. (EPA, 1988). The U.S. Environmental Protection Agency and the U.S. Department of Agriculture have cooperated in efforts that include the Rural Clean Water Program (RCWP) to evaluate NPS pollution abatement progress. Two RCWP projects with karst concerns were sources of information on the scope of agricultural water pollution as well as appropriate Best Management Practices (BMPs) to alleviate the problem. Other sources included the Big Springs Basin study in Iowa and published research works from the literature. Except for brief mention, I have not addressed other rural pollution threats such as deep-well waste injection, highway and parking-lot runoff, oil and gas production, construction, mining, landfills (except sinkhole dumps), underground tanks, spills, and industrial wastes. Land-use change (development) is identified as a serious threat since in displacing farms it introduces pollution from the above-mentioned sources which may be more damaging to a karst environment than agricultural NPS pollution.

Crawford (1988) has reported that shallow carbonate-rock aquifers are especially vulnerable to NPS contamination. Contaminants may percolate through thin soils with little or no filtration (diffuse entry) or they may enter the ground water system directly through karst features like sinkholes, swallets, cave entrances, and fracture zones. Once underground, they can travel fast and far in slugs driven by storm surges through solution-enlarged passages that are analogues of non-karst surface drainage systems (Crawford, 1988; White, 1988; many other sources). The distinction between surface and ground water in karst becomes blurred with multiple sinkings and resurgences; however, contamination of water underground has been emphasized in this discussion. Analysis of soil qualities and topographical features such as slope, sinkhole density, and location upgradient from swallets may be helpful in determining potential for NPS karst contamination (Crawford, 1988). Although these factors were developed for the south central Kentucky karst, they are applicable elsewhere as well.

ORIGINS OF AGRICULTURAL NONPOINT SOURCE POLLUTION

Following is a listing of the sources of the major agricultural NPS pollutants in karst. We have emphasized nitrogen nutrients, pesticides, and bacteria, with brief mention of other contaminants. White (1988) provided a list (Table 1) of sources of 6 karst ground water pollutants, with sources other than agriculture included for the sake of perspective. Pollutant loadings are sometimes interrelated, for instance activities that cause increased erosion and soil loss can lead to elevated loadings of nutrients and pesticides.

Nutrients

Sources of contamination include fertilizers and manure applied to fields for crop production, manure storage facilities, feedlots, dairy parlors, poultry and hog houses, and

Table 1. Sources of Groundwater Pollutants[a]

	Oxygen demand	Nitrogen Phosphates	Chlorides	Heavy metals	Hydrocarbons Organics	Bacteria Viruses
Agriculture						
Barnyard waste	XXX	XXX				XX
Fertilizer		XXX				
Pesticides					XXX	
Domestic waste						
Septic tanks	XXX	XX				XXX
Outhouses	XXX	XX				XXX
Landfills	XXX	X		XX		
Sinkhole dumps	X		X	XX	XX	XX
Construction and mining			XXX	X	XX	
Industrial waste			X	XXX	XXX	

The number of Xs indicate, very roughly, the degree of pollution threat.

[a]Adapted from White, 1988.

pastures. Activities that may exacerbate nutrient losses include excess fertilizer and manure application, improper timing of application, poor manure storage management, and failure to supervise areas where livestock may concentrate. Other activities include inappropriate tillage practices such as up-and-down-hill plowing just before heavy rainfall, and allowing livestock access to sinking streams, cave entrances, and sinkholes. Conservation practices may also increase nutrient loss. Reduced tillage with residue cover can increase infiltration and cause loss of soluble nutrients to ground water. Level terraces can more than double ground water nitrate loading compared with contour farming (Johnson et al., 1982b).

Nitrate is recognized as a karst ground water pollutant with adverse health effects. In infants, nitrate can be reduced to nitrite, causing methemoglobinemia, the "blue-baby" syndrome (Grow, 1986). Levels of nitrate above the US EPA drinking water standard of 10 mg/l have been documented in karst ground water in Minnesota (Grow 1986; Smolen et al., 1989), Pennsylvania (Kastrinos and White, 1986), and Iowa (Mitchem et al., 1988). Kastrinos and White (1986) showed a linear relationship between mean nitrate concentration in karst spring water and agricultural land-use coverage. In the Conestoga Headwaters RCWP project area, about a third of which is underlain by carbonate rocks, wells in the carbonate-rock portion showed higher nitrate levels than non-carbonate wells (USDA, 1988).

Phosphorus, on the other hand, is not considered a ground water pollutant since it has no adverse health effect and does not migrate far into soils before it is immobilized (White, 1988). However, if excessive loadings of P (usually from manure) exceed the adsorptive capacity of a soil, leaching to ground water is possible (Johnson et al., 1982b) and direct entry and passage of P through conduit drainage to resurgences may occur. The major adverse environmental effect of high P levels is eutrophication of surface waters (Johnson et al., 1982b; other sources).

Pesticides
Since World War II agricultural pesticide use has greatly increased, especially in the last 15 years with widespread application of no-till and reduced-till practices that depend heavily on pesticides to kill cover crops or to control weeds, insects, and disease (Berryhill et al, 1989). Most modern pesticides move in the dissolved phase. In the Iowa karst, 55 to 85 percent of the atrazine found in ground water got there by infiltration through the soil (Mitchem et al.,1988). Maas et al.,(1984) identified the following primary routes of pesticide transport to aquatic systems: (1) By direct application to water surface. (2) In runoff: either dissolved, granular, or adsorbed onto soil particles. (3) By aerial drift. (4) Through volatilization and subsequent atmospheric deposition. (5) Through uptake by biota and subsequent movement through the food web. Sources of pesticide contamination in karst include excess or improper application, poor timing, wrong choice of chemicals, and improper cleanup and disposal practices.
Pesticides in ground water not only affect drinking water wells and springs adjacent to the point of use, but they may also kill fish (Maas et al., 1984) and affect municipal water supplies and surface water bodies downstream. Hallberg et al., (1985) reported year-round concentrations of atrazine in Iowa karst ground water of 0.2 to 0.5 ug/l. The level of atrazine reached a maximum in 1984-85 but recent cleanup efforts have caused the level to drop (Klein, 1989).

Bacteria

Bacteria and other microorganisms can enter karst water systems from animal wastes, outhouses and faulty septic systems, dead animals, and wastes thrown into sinkholes. Analysis of fecal coliform/streptococcal ratios can distinguish between human and animal waste sources (Sworobuk, 1984). Between 1925 and 1973, 75 epidemics of typhoid-paratyphoid fevers affected the populations of karst regions in Yugoslavia, facilitated by inadequate spring and well protection and poor hygienic conditions. This was perhaps a worst-case example of pathogenic water contamination (Pokrajcic, 1976).

Sediment

Erosion and soil loss in karst can result from bad tillage practices, feedlots, and continuously grazed pastures. Farmland is recognized as the largest contributor of sediment to U.S. waters with over 6.4 billion tons of topsoil eroded each year (Johnson et al, 1982c). Over the short term, sediment can transport nutrients, pesticides, bacteria, and other wastes to karst as well as surface waterways. It can fill solution passages and alter drainage patterns, clog wells, and cause turbidity in wells and springs. In the long run, erosion can cause the irreversible creation of karst deserts. In the limestone mountains of Mediterranean lands, centuries of overgrazing by goats has removed protective vegetation and caused soil loss to solution cavities. Slash-and-burn farming by the Maya in the karst Yucatan Peninsula has had the same effect. In both regions, soils and vegetation are not being replenished (White, 1988).

Other Problems

Agricultural operations can cause contamination by other substances than those discussed above. Fuels, oils, and other organic materials escape from leaking fuel tanks, discarded equipment (junk), chemical wastes, and spills. Sources of salts in ground water include animal feed and irrigation.

Farming activities in karst lands can in turn be adversely affected by non-agricultural pollution from off-site sources. The markets for a West Virginia dairy farmer's products were threatened when chemical leaks from a nearby industrial plant contaminated the well that provided water for his herd. In another case, a farmer was forced to sell his entire herd of beef cattle due to excessive chloride levels in his well. A hastily-drilled, badly cased gas well on adjacent property had caused brine to contaminate the farmer's fresh-water aquifer (Author's observation, 1986).

BEST MANAGEMENT PRACTICES FOR CONTROLLING NPS POLLUTION IN KARST

The volume of research literature in agricultural NPS control is quite large. Many of the conservation practices developed in recent years are applicable in karst, especially those for management of fertilizer, pesticides, animal wastes, and erosion. Those that increase infiltration are contraindicated if they also increase ground water loading of nutrients and pesticides.

Although carbonate rock-derived soils in karst areas are rich (White, 1988) and in proper topographical settings like parts of the Midwest make excellent cropland, in other areas soils are thin and rocky and fields are small and steep. Small farms raising livestock and poultry are more prevalent in the mountainous karst of the eastern U.S. Following are discussions of some BMPs that may be helpful on cropland and for livestock and poultry operations.

Cropland

Nonpoint source pollution from croplands can be reduced by such recognized practices (SCS, 1897) as conservation tillage with residue cover (Berryhill et al., 1989) crop rotation (Johnson et al., 1982b) reduced input agriculture (Odum, 1987), and Integrated Pest Management (IPM) (Maas, et al., 1984).

With respect to nutrient losses with chemical fertilizer use, Johnson et al., (1982b) reported that in a 3-year crop rotation of corn-soybeans-wheat, use of crop residue and cover crop resulted in 80 percent reduction in nitrate loss and 48 percent reduction in soluble phosphorus loss, compared with no residue or cover crop. They also reported comparable nutrient loss reductions due to other conservation practices. Regular soil testing and crop needs analysis are necessary to determine optimum fertilizer application rates. Their research showed that to minimize leaching losses further, applications should be made during periods of rapid plant uptake during spring and summer. Split nitrogen applications are especially beneficial. Applying fertilizer in advance of crop needs results in large nutrient losses.

Pesticide control in karst is especially important because of the potential for adverse effects far from the site of application. Although pesticides, like other contaminants, may enter karst aquifers directly and move rapidly in slugs, infiltration (leaching) is a major pathway causing chronic pesticide presence in ground water (Mitchem et al., 1988). Pesticides with high leaching potential and toxic properties include alachlor, aldicarb, atrazine, bromacil, carbofuran, cyanazine, dinoseb, disulfoton, methomyl, metolachlor, oxamyl, and simazine. (Water Quality Group, 1987)

Fortunately, information on BMPs for pesticide control is readily available from the US Department of Agriculture Extension Service, from land-grant universities, and from county agricultural extension agents. Maas et al. (1984) summarized pesticide BMPs into the following classes: (1) Application efficiency improvement (restricting aerial spraying, avoiding application when heavy rains are predicted, applying on windless days and late in the day, and others); (2) IPM (a combination of methods such as scouting, timing, biological controls, crop rotations, and others); (3) Soil and Water Conservation Practices (SWCPs) that affect runoff and leaching losses (Conservation tillage is a special case that, as mentioned before, can increase leaching losses depending on rainfall timing and other factors.); and (4) Substitution of less toxic, less persistent, less mobile, or more selective pesticides.

An example of coordinated BMP implementation in a karst area is the Garvin Brook Rural Clean Water Program (RCWP) Project in southeastern Minnesota, which is affected by ground water pollution from dairy and cash grain farm sources. About 17 BMPs were implemented, including sealing abandoned wells, filling sinkholes with clay, planting vegetative filter strips, split N applications, IPM, contour cropping, and building runoff diversions (Smolen et al, 1989).

Livestock and Poultry Operations
    Water quality BMPs for livestock and poultry operations fall into 3 categories: manure management, grazing systems, and high-density animal confinement facilities.

Manure Management
    Animal waste control is a major element in all livestock operations. Established practices include land application for fertilizer and soil conditioner, reuse of liquids for flushing, dry storage for later use, and export. Proper design and management of roofs, gutters, and other structural runoff diversions is also very important. Excess manure is a major concern in the Conestoga Headwaters RCWP, which has a large number of livestock operations. Alternatives to judicious management of manure as a fertilizer include export (hauling it away) and reducing the livestock population (USDA, 1988).

Land application on cropland or pastures is attractive for agronomic as well as economic reasons. (Johnson et al.,1982a) As with chemical fertilizer, manure application method, rate, and timing should be planned to minimize leaching. Surface broadcasting followed by disking minimizes nutrient losses and speeds up decomposition, however care must be exercised when applying manure on thin soils or near sinkholes and waterways. Crop needs, annual soil test results, and manure nutrient analysis should determine the rate of application. If one nutrient (usually phosphorus) is limiting, it is better to use a chemical fertilizer supplement than to risk overapplication of other nutrients. Manure should not be spread on snow or frozen ground. If winter application is unavoidable, it should be done on cover crops. The optimum times for application are shortly before planting and during the summer growth period (for top dressings onto standing crops not sensitive to ammonia). The goal in manure land application should be to minimize leaching below the root zone.

Grazing Systems
    Allowing livestock continuous and unrestricted access to sensitive parts of a pasture area, especially ponds, wetlands, streams, and other waterways can cause increased loadings of nutrients, sediment, and bacteria to ground water in karst. In addition, animals can wander into cave entrances seeking shade and water or fall into sinkholes, get trapped and die, and cause a high potential for ground water contamination (Author's observations, 1964-89).

Two BMPs that may apply especially to karst include Planned Grazing Systems and Livestock Exclusion (SCS, 1987). Planned Grazing refers to rotation of pastures with rest periods to allow ground cover to regenerate. The efficacy of planned grazing with respect to improvement of ground water quality was demonstrated by Chichester et al. (1979) and Van Keuren et al. (1979) who compared the effects of rotational summer grazing and continuous winter grazing by a herd of beef cattle in a 2 year study. There was no measurable soil loss from the 3 summer pastures, although about 1600 kg/ha per year was eroded from the 1 winter grazing and feeding area. Average annual surface runoff from the summer areas (13.2 mm) was reduced by 85% from the winter value (90.4 mm). Infiltration was slightly greater in summer under the rotational-grazed system (407 mm compared to 388 mm) as was subsurface transport of nitrogen and soluble salts. This study took place in a non-karst region. More research is needed on the effects of different grazing systems on karst soils and waters.

Livestock Exclusion is the practice of fencing animals away from water bodies and areas subject to erosion. In karst, sinkholes and cave entrances should be declared off-limits. Laurel Creek Cave in Monroe County, West Virginia was an obvious case where livestock exclusion was not practiced. Several years ago the low, wide entrance was a favorite loafing area for cattle since it was cool, dark, and moist there. The ground was a sea of mud and cow flops. During storms, however, the cave took large volumes of water, as evidenced by a highway bridge that had been washed deep inside by a flood (Author's observation, 1976). The potential for pollution of the aquifer recharged by this cave entrance could have been greatly reduced by a few feet of fencing.

High-Density Animal Operations

Many modern livestock and poultry production facilities depend on high-density confinement, such as beef feedlots and total-confinement indoor swine and poultry rearing facilities. (I have also included dairy milking parlors as sources, even though dairy herds are usually kept on pasture.) Large operations are considered point sources which must obtain NPDES permits and comply with discharge restrictions. Small facilities, however, are considered nonpoint sources. BMPs for these sources include paved animal living spaces, means for flushing or scraping daily accumulations of manure, leak-proof feed and manure storage facilities, and sanitary means to dispose of sick and dead animals. They also include runoff control structures and vegetative filter strips for erosion and nutrient control. In the Garvin Brook, Minnesota Rural Clean Water Project (RCWP) these measures were used to keep feedlot runoff from entering sinkholes (Smolen et al, 1989). Some operations rely for manure storage on unlined lagoons and pits. These should not be considered BMPs in karst unless local soils and underlying bedrock have been tested and found to be acceptably impermeable. The worst possible case would be a swine waste lagoon over cavernous limestone in an area of active sinkhole collapse.

Other Measures

A number of water resource management options can be beneficial in karst areas. Practices discussed previously such as runoff diversions, fencing, and vegetative filter strips can help prevent pollution of farm ponds, springs, fish hatcheries, and other surfaces receiving waters in karst as well as non-karst areas. Large sinkholes can be cleaned up and using them as trash dumps prohibited. Small ones can be filled, although this may not be a permanent fix. The US Department of Agriculture's Conservation Reserve Program (CRP) offers rental payments to farmers for up to 10 years for taking marginal land out of agricultural production and for planting filter strips in order to protect streams, lakes, and wetlands from nonpoint pollution. Although this policy does not presently apply to sinkholes and other karst water entry points, they may be included in the 1990 Farm Bill which as of this writing has not yet come up for debate in the U.S. Congress (Adler, 1989).

One of the biggest threats to karst as well as to family farms is encroaching development which introduces an entirely new set of challenges. These include upsets in land values and incentives to farmers to sell out; new urban pollutant loads from road salts, parking lot runoff, pet manure, underground tank leaks, and the like; and depressed water tables with increased probability of collapsing sinkholes. Discussion and evaluation of these problems is outside the scope of this paper; however, recognizing them is one of this author's aims in identifying impacts on water quality in rural and agricultural areas.

There are no engineering methods for dealing with development pressures. In spite of known foundation engineering problems in karst (White, 1988) and litigation problems (Turk, 1976) involving structural failure, insurance claims, and water pollution, the developers proceed apace. "Way-of-life" arguments by rural residents who like things the way they are seldom succeed in convincing country commissions and other authorities who also see prospects of increasing land values and an improved tax base. More research is needed in fields like applied rural sociology in developing practical methods of confronting developmental forces that intend to convert karst farmland and wilderness into suburbs and industrial "parks."

CONCLUSION

Major threats to water quality from agricultural activity in karst areas include nitrate, pesticides, bacteria, sediment, and other materials that are washed into ground water systems from croplands and animal-rearing operations. Although most recognized BMPs are universally applicable, some (manure-management practices, for instance) are especially beneficial in karst while a few are to be discouraged due to their infiltration effects.

Most conservation-minded farmers, equipped with an awareness of the special problems imposed by karst drainage, knowledge of a few practical management options, and considerable common sense, can keep nonpoint source pollution under control. Confronted with the threat of development, the rules change and farmers lose this control. The farmers need more clout when the time comes to "Just Say No" to the developers.

References

Adler, K.J., US Environmental Protection Agency. 1989. Personal communication.

Berryhill, W.S. Jr., Lanier, A.L., and Smolen, M.D. 1989. The Impact of Conservation Tillage and Pesticide Use on Water Quality: Research Needs. In: Proceedings of National Research Conference, Pesticides in Terrestrial and Aquatic Environments, Richmond, May 11-12, 1989. Virginia Water Resources Research Center, Blacksburg, Va. (in press)

Chichester, F.W., Van Keuren, R.W., and McGuinness, J.L. 1979. Hydrology and Chemical Quality of Flow from Small Pastured Watersheds: II. Chemical Quality. J Environ Qual 8(2):167-171.

Crawford, N.C. 1988. Karst Hydrologic Problems of South Central Kentucky: Ground Water Contamination, Sinkhole Flooding, and Sinkhole Collapse. Field Trip Guidebook, Second Conference on Environmental Problems in Karst Terranes and Their Solutions, Nov. 16-18, 1988. Nashville, Tenn.

EPA, 1988. Nonpoint Sources: Agenda for the Future. U. S. Environmental Protection Agency, Washington, D.C.

Grow, S.R. 1986. Water Quality in the Forestville Creek Karst Basin of Southeastern Minnesota. Master of Science thesis, University of Minnesota, Minneapolis, Minn. (unpublished).

Johnson, D.D., Kreglow, J.M., Dressing, S.A., Maas, R.P., Koehler, F.A., Humenik, F.J., Snyder, W.K., and Christensen, L. 1982a. Best Management Practices for Agricultural Nonpoint Source Control. I. Animal Waste. North Carolina Agricultural Extension Service, Biological and Agricultural Engineering Department, N.C. State University, Raleigh, N.C.

Johnson, D.D., Kreglow, J.M., Dressing, S.A., Maas, R.P., Koehler, F.A., Humenik, F.J., Snyder, W.K., and Christensen, L. 1982b. Best Management Practices for Agricultural Nonpoint Source Control. II. Commercial Fertilizer. North Carolina Agricultural Extension Service, Biological and Agricultural Engineering Department, N.C. State University, Raleigh, N.C.

Johnson, D.D., Kreglow, J.M., Dressing, S.A., Maas, R.P., Koehler, F.A., Humenik, F.J., Snyder, W.K., and Christensen, L. 1982c. Best Management Practices for Agricultural Nonpoint Source Control. III. Sediment. North Carolina Agricultural Extension Service, Biological and Agricultural Engineering Department, N.C. State University, Raleigh, N.C.

Klein, A.J. Monsanto Co. 1989. Personal communication.

Maas, R.P., Dressing, S.A., Spooner, J., Smolen, M.D., and Humenik, F.J. 1984. Best Management Practices for Agricultural Nonpoint Source Control. IV. Pesticides. North Carolina Agricultural Extension Service, Biological and Agricultural Engineering Department, N.C. State University, Raleigh, N.C

Mitchem, P.S., Hallberg, G.R., Hoyer, B.E., and Libra, R.D. 1988. Ground-Water Contamination and Land Management in the Karst Area of Northeastern Iowa. In: Ground-Water Contamination: Field Methods, ASTM STP 963, A.G. Collins and A.I. Johnson, eds, Am Soc for Testing and Materials, Philadelphia, Pa: 442-458.

Odum, E.P. 1987. Reduced-Input Agriculture Reduces Nonpoint Pollution. J. Soil and Water Cons. Nov-Dec 89:412-414.

Pokrajcic, B. 1976. Hydric Epidemics in Karst Areas of Yugoslavia, Caused by Spring Water Contaminations. In: Proceedings of US-Yugoslavian Symposium, Karst Hydrology and Water Resources, v.2. Dubrovnik, June 2-7, 1975. Fort Collins, Colorado: Water Resources Publications.

SCS. 1987. National Handbook of Conservation Practices. U.S. Department of Agriculture, Soil Conservation Service. Washington, D.C.

Smolen, M.D., Humenik, F.J., Brichford, S.L., Spooner, J., Bennett, T.B., Lanier, A.L., Coffey, S.W., and Adler, K.J. 1989. NWQEP 1988 Annual Report: Status of Agricultural Nonpoint Source Projects. Biological and Agricultural Engineering Department, N.C. Agricultural Extension Service, N.C. State University, Raleigh, N.C.

Sworobuk, J.E. 1984. Bacterial Contamination of Rural Drinking Water Supplies. Master of Science thesis, West Virginia University, Morgantown, W.Va. (unpublished)

Turk, L.J. 1976. Predicting the Environmental Impact of Urban Development in a Karst Area. In: Proceedings of US-Yugoslavian Symposium, Karst Hydrology and Water Resources, v.2. Dubrovnik, June 2-7. 1975. Fort Collins, Colorado: Water Resources Publications. USDA, 1988. Conestoga Headwaters Rural Clean Water Program 1987 Annual Progress Report. US Department of Agriculture, Agricultural Stabilization and Conservation Service, Harrisburg, Pennsylvania.

Van Keuren, R.W., McGuinness, J.L., and Chichester, F.W. 1979. Hydrology and Chemical Quality of Flow from Small Pastured Watersheds: I. Hydrology. J. Environ Qual 8(2):162-166.

Water Quality Group. 1987. Protecting Water - Protecting Crops: Pesticides and Water Quality. Biological and Agricultural Engineering Department, N.C. State University, 615 Oberlin Road, Raleigh, N.C., 27605. (919) 737-3723. Available as slides or videotape.

White, W.B. 1988. Geomorphology and Hydrology of Karst Terrains. New York and Oxford: Oxford University Press.

Waste disposal on karstified carboniferous limestone aquifers of England and Wales

A.J.EDWARDS & P.L.SMART   *Department of Geography, University of Bristol, UK*

ABSTRACT

Within England and Wales, there are 148 licensed landfill sites on the Carboniferous Limestone of which approximately 50% are abandoned quarries. The licensing of landfill sites which handle controlled waste was introduced in 1976 following the Control of Pollution Act (1974). The licence, regarded as the main tool of regulation by the legislative bodies, stipulates the perceived safe quantities and types of controlled wastes allowed at a site, the necessary site preparation and the mode of operation during disposal.

Of the 148 landfill sites 66 are licensed to receive "difficult" wastes, 19 for "household" wastes, 6 for "semi-inert" material and 56 for "inert" waste. The majority of the sites receiving "household" and "difficult" waste were licensed in 1976/7 with the introduction of the legislation. Some of these sites had been operational for up to 30 years, and site preparation at these sites had been minimal. It was not until the mid-1980s that the majority of new "difficult" and "household" sites were lined. Lining usually consists of infill by inert material up to the level of the highest water table recorded prior to disposal, followed by 1 m of clay, but variations in practice occur between different regional regulatory authorities. There has been an increase in total waste capacity with time, accounted for by an increase in both number and size for "inert" sites, but by size alone for "difficult" and "household" sites. This trend will continue, with particular interest in potential sites with rail links in Southern England.

Significant groundwater contamination has only been reported from "difficult" and "household" waste sites. In some cases, rapid and gross contamination of adjacent springs occurred, while in others migration of contaminants appears to be by diffuse flow. In both situations regional geological structure is a strong controlling factor but aquifer heterogeneity is a major problem in prediction of contaminant movement. At present there are no national guidelines or requirements for site monitoring and this has lead to confusion and differences in practice. Boreholes and springs are employed at new high risk sites, but for the majority of sites, regular monitoring has only taken place after a pollution event has occurred. Tracing studies have been few and generally inconclusive.

Introduction

The Carboniferous Limestone is widely distributed in England and Wales (Figure 1), and provides approximately 2% of the total annual groundwater abstracted (Newson, 1973). It is a thick dense, massively bedded karstified aquifer with low primary porosity and specific yield of about 1% (predominantly fissures and fractures) (Atkinson, 1977). In this account we have also included Devonian Limestones, which have similiar character. Conduit flow is well developed in the Carboniferous Limestone, and gives the potential for rapid unattenuated pollutant transmission (Atkinson, 1971; Quinlain & Aley, 1987). For this reason waste disposal on the karstified limestones has always been regarded as problematic (Dodge, 1985). However, Water Authority aquifer protection polices include both physical hydrology and the extent of resource utilization. Thus in some areas of England and Wales, despite the potential for groundwater contamination, relatively low levels of aquifer protection are operative, as abstraction is very limited.

In this paper we review the development and current status of waste disposal on the Carboniferous Limestone aquifer. Information has been collected from Waste Disposal Authorities, Water Authorities, site operators and our own work. We have concentrated on the period since 1976 when site licensing was first introduced, as prior to this time, relatively little information is available. The British Geological Survey undertook a desk study, brief details of which were published by Gray, Mather & Harrison (1974) and The Department of the Environment (1978) however access to the more comprehensive information is controlled by the Department of the Environment.

Legislation and The Regulatory System
There are two principal Acts of Parliament which regulate disposal of waste by landfill in England and Wales, although at least 48 other Acts have some relevance. The Town and Country Planning Act (1971) covers the general principles of site suitability within the context of the County Structure Plan. This includes consideration of access, public health, water pollution, need for disposal and site suitability in relation to local land-use. Once planning permission has been obtained Part 1 of the Control of Pollution Act (1974), which is concerned with collection and disposal of waste, becomes applicable. This act requires all sites to be licensed by the County (or District in Wales) Waste Disposal Authority (WDA), and is regarded as the main tool of regulation by legislative bodies.

The WDA can only reject an application on the grounds of danger to public health or risk of water pollution, and has a statutory obligation to consult the local Water Authority (WA). Information concerning site development and operation must be supplied to the WDA, who may also require a detailed hydrogeological site investigation. The WA has the power to veto any application if the proposed site is considered to present undue risk. Evidence for such action must be well founded. If there is disagreement between the operator, the WDA and the WA, they may appeal to the Secretary of State for the Environment. Here an assessment of the evidence occurs, and the subsequent decision determines the final path of the application.

In addition to the issuing of licences, the WDA is responsible for monitoring operations at licensed sites and enforcement of regulations. Throughout England and Wales, the activities of the WDA are monitored by the Air, Water and Waste arm of Her Majesty's Inspectorate of Pollution (HMIP). Guidence is given through the publication of Waste Management Papers, which reflect the current thinking on various aspects of waste disposal. The present issue is, for instance, concerned with the control of landfill gas. (Her Majesty's Inspectorate of Pollution, 1989a). In practice however, there has been a lack of central guidence due to the smallness of the HMIP. Until 1988, only 6 field inspectors were employed to cover the 5000 licensed facilities in England and Wales. In the period of 5 years, 3 WDA never recieved a visit, over 50% were visited once and no one WDA ever recieved a visit in consecutive years (The House of Commons Environment Committee, 1989).

The majority of WDA are also small; of the 39 that exist in England and Wales, 18 possess 4 or less staff. This means that adequate scientific knowledge and experience is often lacking and that duties are often shared within a tight schedule. Furthermore the low political profile that is attached to waste disposal means that resources are often limited, and the Waste Disposal Officers are of low grade and status.

Because of these problems, site licence conditions are, in general, not well enforced and the licence thus fails to satisfactorily regulate operations. Indeed, the Conservative Chairman of the Commons Environmental Select Committee stated that "Britain had come perilously close to suffering a disaster from its appalling waste disposal system" (The Guardian, March 7th, 1989). There are also inadequacies in the legislation (Khan, 1987), and prosecution is difficult because:

i. Non-compliance with the licence conditions is not a statutory offence.

ii. The presence of prohibited waste on a licensed site is not sufficient for prosecution; the act of disposal must be witnessed

iii.The penalties for any offence do not adequately reflect the degree of potential or actual environmental damage (Smith, 1988), or the profits that may derive.

iv. Waste disposal licences do not remain in force once the landfill is completed, thus adequate aftercare is not ensured.

v. Loopholes in the Control of Pollution Act (1974) enable site operators to surrender their site licence at any time, terminating their responsibility for the site.

vi. During appeals to the Department of the Environment concerning alteration to or revoaction of site licences, which normally take 2 years to reach a decision, the operator can continue operations.

At present no "Duty of Care" system operates within England and Wales. Waste producers are under no legal obligation to hold precise records of their controlled waste and, they do not either bear the responsibility for ensuring that the wastes are properly handled and disposed of. Thus environmentally unsound or illegal disposal methods may be used because they are cheapest.

<u>Landfill Sites Upon the Carboniferous Limestone</u>
Potentially harmful wastes can be referred to as "Controlled", "Special", "Hazardous", and "Toxic", depending on the relevant Acts of Parliament. However the Department of the Environment has issued other guidelines which have been generally used by WDA, WA and operators (The Department of Environment, 1986; Her Majesty's Inspectorate of Pollution, 1989b). Classification within this paper is based upon the Register of Facilities for the Disposal of Controlled Wastes in England and Wales (Aspinwall & Co., 1987):

i. Inert Wastes - Wastes that will not physically react or biodegrade and do not pollute under normal circumstances

ii. Semi-inert Wastes - Generally inert material which incorporate small amounts of more difficult waste, which may be capable of biodegrading or reacting to produce a polluting leachate.

iii. Household Wastes - Waste arising from private dwellings and general wastes from commercial and industrial premises.

iv. Difficult Wastes - Wastes whose properties present a risk during handling, and which could be harmful in either the long or short term to the environment.

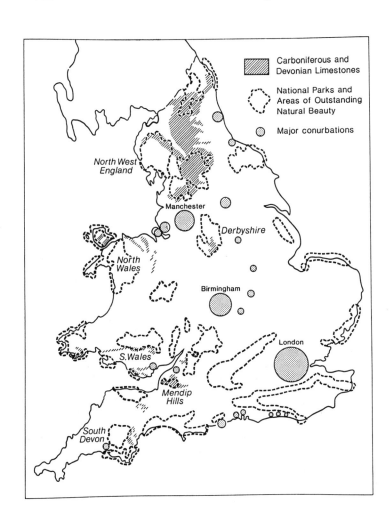

Figure 1. Distribution of the Carboniferous Limestone Aquifer in England and Wales in relation to major conurbations, National Parks and Areas of Outstanding Natural Beauty.

Although in the majority of cases sites licensed for difficult wastes are able to take all other categories, this is not always the case. Thus at specific sites taking predominantly inert or semi-inert materials, consent may be given to dispose of difficult waste at a specific and low loading. In many cases upgrading of an existing low category licence in this manner is often more favoured than the development of new sites.

Surprisingly, sites licensed for difficult waste comprise the largest category on the Carboniferous Limestone (Figure 2), and a larger proportion of Carboniferous Limestone sites than for the whole of England and Wales. The vast majority of these are in quarries. The second largest category comprises inert landfills, but these are a smaller proportion of the total for England and Wales as a whole as are the semi-inert sites. This may well be due to availability of alternative disposal sites being available nearer to the source of the inert waste.

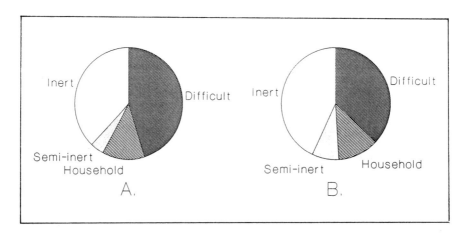

Figure 2. Comparison of the proportion of difficult, household, semi-inert and inert wastes landfilled on the Carboniferous Limestone (A) compared to that for all sites in England and Wales (B).

As shown in Figure 1, many of the Carboniferous and Devonian Limestone areas are generally distant from the major conurbations and are rural. It is therefore only in the case of difficult wastes, that the higher transport costs incurred in transit to distant sites can be justified. Furthermore, in rural areas, poor communication means that landfills must be provided to take household and difficult wastes from the local community and rural industry. Because there is a low waste input to such sites they have the potential for a long operational life, as exemplified by the low percentage (1-2%) of sites which have closed since 1976. This suggestion is supported by the higher proportion of publically owned landfill sites upon the Carboniferous Limestone (33%), compared to the England and Wales average (17%). However, a more jaundiced interpretation of this statistic is that WDAs are more lenient in licensing public than private sector sites in potentially difficult locations. This may result from the dual role forced upon WDA by legislation, which both requires them to provide sufficient local capacity for waste disposal, and to police the issue and compliance with site licences.

The number of landfill sites per km^2 is shown for the Carboniferous Limestone of England and Wales upon a county basis in Figure 3. High densities in South Wales, South West England and the Peak District reflect proximity to areas of high population density (Figure 1). Densities are much lower in Northern England. However, limestone upland areas provide an important recreational resource throughout England and Wales, as can be seen from the remarkably high proportion of the Carboniferous Limestone areas which correspond with National Parks and Areas of Outstanding Natural Beauty (Figure 1). Consequently, there is an intrinsic land-use conflict in these areas, which is particularly acute in areas with high site densities. Furthermore given the great shortage of suitable waste disposal sites in the populous south east of England, pressure for landfill development in areas such as the Mendip Hills can be expected to increase, particularly as there is a parallel high demand for quarried stone (Stanton, 1989). Indeed in the UK many of the larger waste disposal firms are subsidiaries of large mineral extraction conglomerates such as ARC, part of the ConsGold multi-national.

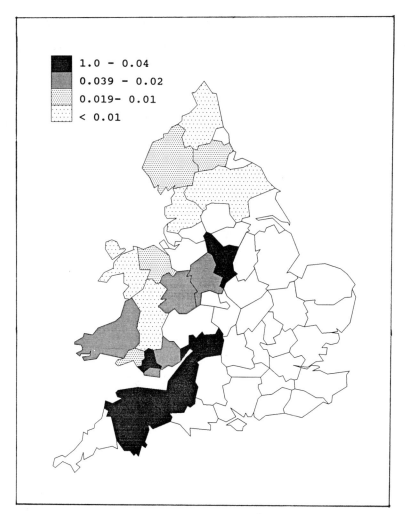

Figure 3. Density of landfill sites per km^2 of Carboniferous Limestone outcrop for different counties in England and Wales.

There are also regional differences in the proportions of sites taking difficult wastes (Table 1). These partially reflect the need to develop sites for disposal of difficult and household wastes in the areas where alternatives are not available, for instance North West England. Differences within these areas, may relate to the quantities of difficult waste produced (low in the Northumbrian Water Authority area for instance), or to differences in the policies of the county WDA. For instance some WDA do not specify the requirement to avoid groundwater contamination in the site licence. More important, however is the classification of the Carboniferous Limestone aquifer in the area adopted by the Water Authority. The classification is based on the vulnerability of the aquifer to pollution, and the WAs' dependence upon it for potable supplies. Thus Wessex Water Authority rely heavily on the Carboniferous Limestone aquifer for supply and accordingly rate most of the outcrop as Zone 1/2. In this zone all proposals for waste disposal are discouraged. In contrast in Yorkshire Water Authority, there is no significant abstraction from the limestone, and the aquifer is classed as 3/4. This policy has only been recently implemented, and there are therefore discrepancies which result from previously established sites, particularly those which were in operation (and therefore a _fait accompli_) prior to 1976 when licensing was introduced. Thus despite Severn-Trent WA rating the Carboniferous Limestone as Zone 2, there are a large proportion of sites taking difficult waste in this area. There are also contrasts in zoning between different outcrops within WA areas; for instance the North Bristol Coalfield area, where abstraction does not occur, accounts for a substantial proportion of the difficult category sites on the Carboniferous Limestone in the Wessex area.

Table 1. Aquifer protection zone classification for the Carboniferous Limestone Aquifer of England and Wales for different areas in relation to percentages of sites licensed for difficult, household, semi-inert and inert wastes. Figures in brackets show numbers of WDAs responsible for site licensing in each WA area.

LIMESTONE AREA	WATER AUTHORITY	AQUIFER PROTECTION ZONE	NUMBER OF SITES	% DIFFERENT OF LANDFILL TYPES			
				1.	2.	3.	4.
MENDIP HILLS	WESSEX (2)	1/2	21	43	5	10	43
PEAK DISTRICT	SEVERN TRENT (4)	2	29	34	4	10	52
WALES	WALES (10)	3/4	28	56	0	4	41
NORTH WEST ENGLAND	NORTHUMBRIAN (2)	N/A	21	43	10	38	10
	YORKSHIRE (1)	3/4	8	13	0	25	63
	NORTH WEST (2)	N/A	31	24	0	15	60
SOUTH DEVON	SOUTH WEST (1)	2	9	44	11	22	22
TOTAL LANDFILLS UPON THE CARBONIFEROUS LIMESTONE			148	38	4	13	45
TOTAL LANDFILLS IN ENGLAND & WALES			4088	43	8	12	37

N/A - Water Authority as yet has not implemented an official policy of aquifer protection.

1 - Inert Landfill Sites
2 - Semi-Inert Landfill Sites
3 - Household Landfill Sites
4 - Difficult Landfill Sites

The majority of difficult and household landfill sites were licensed with the introduction of the Control of Pollution Act (1974) in 1976/1977. Many of these sites had been operational for 30 years. Subsequently the number of new sites being licensed each year has decreased from 29 in 1977 to 2 in 1988 (Figure 4) The reluctance of the legislative bodies (predominantly the WA) to licence household and difficult sites accounts partially for the decline, but the increasing cost of site preparation on sensitive aquifers means that alternatives have been utilised. Those sites that have been developed have tended to be of large capacity in order to give economies of scale. Some Authorities have recently insisted upon landfills being smaller, and these are often used as trials for potentially larger developments. The number of licences being granted for landfills handling inert and semi-inert wastes has increased rapidly since 1976 (Figure 4). In 1988, they accounted for over 80 % of all new licences that were granted upon the Carboniferous Limestone. The majority of the sites are small, with low waste inputs from local producers or "in-house" wastes. Site preparation costs are also low, and in most cases, hydrological surveys would not be neccessary.

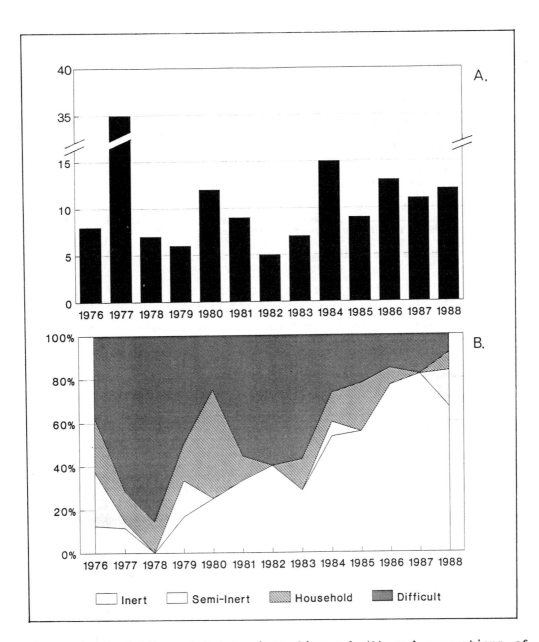

Figure 4. Variation of total sites licensed (A) and proportions of sites taking difficult wastes (B) for the Carboniferous Limestone with time.

Site Preparation

Many of the difficult and household sites licensed in 1976 or 1977 were operated upon the dilute and disperse philosophy (Swinnerton, 1984), and none of these sites were lined (Figure 5). This contrasts with the containment philosophy which has been widely adopted for recently developed sites, following the publication of Waste Management Paper Number 26 "Landfilling Wastes" (The Department of the Environment, 1986) which described the operation of the containment philosophy. Prior to this many WAs and WDAs relied upon the infilling of proposed household and difficult sites with inert materials to a level of the previous highest-recorded water table (Table 2). The infill material was of variable nature. 6" clean stone was utilized in a number of sites although the use of on-site quarry waste and inert demolition and construction waste was more common. Some WA who regarded the

Carboniferous Limestone aquifer as a major water resource operated the containment philosophy prior to 1986. Both Wessex and Severn Trent prior to 1986 stipulated that lining should be installed from approximately 1983, both WAs thus have a high percentage of lined difficult and household sites within their area (Table 2). Both still however have a legacy of dilute and disperse sites which still remain in operation.

Lining of sites consists usually of 1 m of natural materials such as clay, the permeability of which should be less than 1×10^{-7} cm/sec. For Carboniferous Limestone landfill sites, both the walls and the base of a quarry are lined. The quarry floor is initially blinded and then the lining materials are rolled on top. Quarry walls are lined during the disposal operations as the depth of the waste increases and different areas of the quarry are utilized. Lining materials are smeared onto the walls using the on-site heavy machinery. In some cases, lining materials are specially imported on site although the utilization of stripped overburden from previous quarrying activities is widespread as well as the segregation of inert material loads from the incoming waste that the landfill is receiving. A standard specification to cover the testing and placing of natural materials for basal liners, bunds and final caps has recently been proposed (North West Water, 1988). This would be based upon established civil engineering practices employed in the construction of highways and earthern dams. The integraty of natural liners is uncertain (Lundgren & Söderblom, 1985), although there is an acceptance that natural materials may not provide containment but allow a "delay and decay" philosophy (Fleming, 1988).

Given the uncertainty as to the integrity of natural liners, only one site has been developed using artifical liners. A butyl liner was utilized on the Carboniferous Limestone in the South West WA as early as 1978, and may have been in response to an unsuccessful conflict with the operator of an unlined difficult waste site on the Limestones taking liquid wastes. Subsequently, butyl liners have been avoided due to their susceptability to attack from hydrocarbon solvents and oils. Preference now lies in the high density polyethylene (HDPE) with which a site in North West England may be lined. Two sophisticated sites, which have been recently developed, have included underdrainage to a pumped sump to extract uncontaminated groundwater from beneath the site liner, and leachate extraction for treatment and disposal from above. As larger and hydrogeolgically more difficult sites are developed such technology will doubtless become the norm.

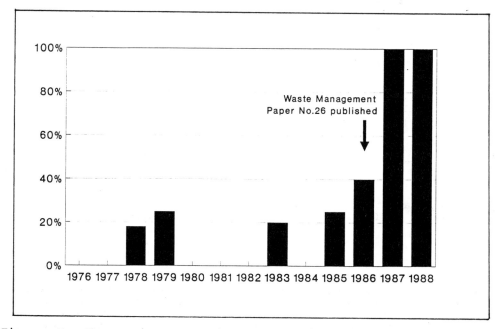

Figure 5. Change in proportion of difficult and household waste landfills which have been lined with clay or artificial liners.

Table 2. Differences in site preparation, monitoring and pollution between the WA for combined difficult and household landfills.

WATER AUTHORITY	% INFILLED	% LINED	% MONITORED	METHOD OF MONITORING PRECAUTIONARY			POST POLLUTION			% POLLUTION
				streams	wells	springs	streams	wells	springs	
WESSEX	18	27	46	✓	✓	✓	✓	✓	✓	36
SEVERN TRENT	17	44	33	✓	✓	✓	-	-	-	6
WALES	0	8	31	✓	✓	-	-	-	-	0
NORTH- UMBRIAN	60	0	50	✓	-	-	-	-	✓	10
YORKSHIRE	0	0	29	✓	-	-	✓	-	-	29
NORTH WEST	32	7	26	✓	-	-	-	✓	✓	22
SOUTH WEST	0	50	50	✓	✓	✓	-	-	-	50

* No Category 1 or 2 sites are lined, monitored or reported to have caused pollution

Site Monitoring.

Visual monitoring of all landfill sites takes place during site inspections, which are carried out by the WA and the WDA on a periodical basis. Such inspections only identify surface seepages of leachate, usually associated with failings in engineering such as insufficient bund thickness, inappropriate bund material and drainage pipes which have not been sealed.

There are no legal requirements for groundwater monitoring of landfill sites within England and Wales. A high ranking officer of the HMIP, however, stressed that routine monitoring is advisable because of the lack of understanding of the leachate response of landfills (Rae, 1985). The extent of groundwater monitoring is, therefore, very variable between different areas (Table 2). WAs who adopt a stringent approach to site preparation also employ precautionary monitoring in order to protect resources. This involves the utilization of boreholes specifically drilled around difficult and household landfills, and the abstracting boreholes and springs which are already in the vicinity. Springs are not employed for the specific monitoring of landfill sites unless proven connections have been found. Given the almost total lack of tracer tests prior to development, this usually entails the weekly monitoring of springs which have already become polluted. Figure 6 illustrates the monitoring network in a Carboniferous Limestone area with conflicting waste disposal, abstraction, quarrying and recreational interests. Broadfield Down comprises a plateau of low elevation underlain by massively karstified Carboniferous Limestone aquifer in the form of an East-West trending pericline. Impermeable materials surround the limestone, which is overlain in the east by younger interbedded limestone and shales. Groundwater springs rise around the margins of the upland, with Chelvey and Cold Bath Springs on the Northern Limb accounting for 44% of the calculated recharge to the Carboniferous Limestone aquifer. These springs are utilized by the local water supply company, which also operates a surface intake on the east of the pericline, which is partially fed by springs. A number of relatively small private licensed borehole abstractions are also in operation. The area is used extensively for recreation, includes a National Nature Reserve, and is of high amenity value. Quarrying of limestone is still active, with the Regional Structure Plan specifying substancial additional areas to provide aggregate for the next 40 years (Avon County Council Planning Department, 1978). There are 3 landfills within the area, and a number of abandoned quarries, which are currently disused.

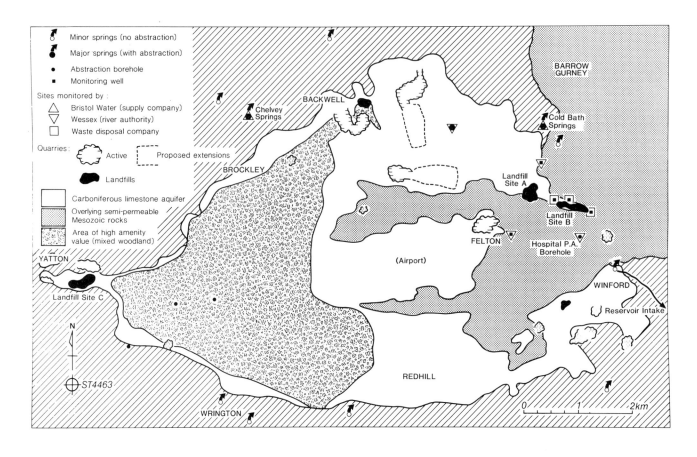

Figure 6. Hydrology of Broadfield Down north of the Mendip Hills, showing quarries, landfills and monitoring sites.

Landfill C has been closed for approximately 10 years but took difficult wastes. Although no abstraction points are in the vicinity of the area, intensive monitoring of this site by the WA was undertaken during its operation because seepages of leachate were thought to be polluting surface waters in the immediate surrounding area. Tracer was injected into a borehole drilled on the quarry floor although successive traces failed to identify connections between the landfill and the surrounding springs. Figure 6 illustrates that current monitoring is much less intense since the cessation of operations.

Landfill A started operation in 1979, and it initially took predominantly inert materials. An upgrading of the existing licence was then sought to permit disposal of more difficult wastes such as road sweepings, sawdust, paper and small quantities of PCB's. The WA required a tracer test be undertaken to predict the direction of possible leachate movement, and assess the potential for spring contamination. An injection borehole was drilled in the quarry floor, but would accept water only at a very small rate. Tracer was then injected with water flush direct on the floor, but after progressively larger injections, no tracer was recovered at any of the sampled sites. Nevertheless the licence was upgraded, but the more difficult wastes were to total less than 25% of the daily intake volume and PCB's were to be less than 10 mg/1 Kg. This site has recently closed and has been partially restored.

Landfill B intiated in 1983, is still operating and can dispose of difficult wastes such as pressed sewage cake, timber and paper. Like Landfill A, difficult wastes can only be handled provided they are a small proportion of the total inert wastes which were originally licensed. Observation boreholes exist to monitor Landfill B. They were specifically installed by the site operator as a condition of the upgraded licence, and monitored by a hydrogeological consultant, The analytical data are made available to the local WA. Infrequent monitoring of the water quality at two other observation boreholes, and at the private abstraction borehole, is carried out by the WA. Finally the main public abstraction sites are routinely monitored on a monthly basis by the supply company, Bristol Waterworks Ltd. This ad hoc arrangement is typical of those commonly adopted for site monitoring on the Carboniferous Limestone.

With increased emphasis upon containment and site preparation, there has been a movement to monitoring the level of leachate within the waste. Concrete sewer access pipes are installed prior to emplacement of waste, which is filled around them. This enables the monitoring of leachate production and may indicate groundwater inflow. Removal of leachate is also possible by pumping to a sewer or tankering. This prevents the build up of a significant leachate head which may give high hydraulic gradients across the lining materials causing their rupture, and limits total leachate released should failure occurs. Permissible standing levels of leachate vary between WAs, ranging from 1 m to 2.5 m, but in other cases the satisfactory level is unspecified and determined by site visits.

The Problems of Landfill Sites Upon Carboniferous Limestone

Four types of problems are reported for landfill sites upon the Carboniferous Limestone: ingress of water, groundwater contamination by leachate, fire, and the migration of landfill gas.

1. Water Ingress

Movement of groundwater into landfill sites will result in enhanced leachate production, the concentration of which will depend on the age and composition of the waste. Methanogenesis and production of landfill gas will be accelerated (Pacey & DeGier, 1986). Because large volumes of water may discharge through a conduit, groundwater inflow is a considerable potential problem for sites on the Carboniferous Limestone. At some sites, very rapid flooding has occurred. For instance, After 10 years of dry operation, the water level in an unpumped landfill site in North West England rose from 61 m AOD to 71.5 m AOD within one month, completely flooding the landfill. Two 6" pumps operating continuously for 2 weeks were needed to lower the water levels to 62 m AOD. Subsequently two 4" pumps were able to maintain this level, but when they were stopped the water levels rose at a rate of about 1 m a month, reestablishing flooded conditions. Investigations of the problem revealed that after torrential rain, a new swallet had opened in the bed of a stream some 6-800 m along the strike from the site, and was engulfing over 80 % of the flow. Cement and bentonite were used to seal the opening, and the site is now again dry.

Flooding of a previously dry site in the Peak District was less simple to remedy. Excavation showed a series of small springs issuing from the quarry floor, but the problem could not be related to a surface stream capture. Pump shafts were therefore placed through the waste, and the site has subsequently been dewatered using these. A difficult landfill site in North West England was developed in a disused quarry with no site preparation. During disposal operations, water emanating from springs within the site flowed across the quarry floor through the disposed waste and into an adjacent disused quarry. The resultant leachate lake then drained into the disused quarry walls via a series of fissures. Bunds to prevent leachate escape were emplaced although they proved inadequate due to the ponding of the water behind them (see below). Continual pumping to tanker therefore had to take place from a sump which was installed within the waste adjacent to the bunds. In both these cases water ingress was via diffuse flow, and was related not to an essentially random capture by a conduit, but rather was due to a general increase in the regional groundwater levels.

This is a particular problem for sites developed in deep quarries because these may function as large diameter wells, lowering water levels in the vicinity. Inflow of water may cause flooding, and installation of a pump may be neccessary to keep the workings dry. On abandonment, water levels will rise, but this may take place slowly as the diffuse permeability of the Carboniferous Limestone is very low, and large volumes of water are required to raise the water levels in the quarry void. Consequently, even if sites are infilled with inert material to the previously recorded high water level, this may be below the level that will be established over the longer term. Furthermore, as deposition of waste reduces the total void volume, recovery of water levels may be more rapid than prior to landfill.

The re-establishment of the water table in the area surrounding a lined landfill may create significant hydraulic gradients across the lining or bund material. This process is augmented by the current approach to landfill operation, which minimises direct recharge into the waste. Natural lining materials may prove ineffective if differing hydraulic heads existed either side of them. For instance Smart (1985) describes a site in Jurrassic Limestones where leakage of the clay bunds resulted from ponding of water moving downdip towards the landfill. Physical erosion of natural lining materials by concentrated inflow may also be a problem, as was the case at a recently installed site in the Peak District. To prevent the development of significant hydraulic gradients, pumping needs to proceed throughout the operational life of the landfill. thus in a lined, difficult waste landfill site near the Mendip Hills, pumping of clean groundwater occurs from a sump at the low point in a gravel blanket installed below the liner. Leachate levels within the waste are controlled by pumping from a second sump within the liner to balance the internal and external water levels, and prevent the build up of significant hydraulic gradients across the lining material.

2. Groundwater Contamination By Leachate.

As discussed above, despite the change to a containmant philosophy in newer sites, the majority of landfills taking difficult and household waste are unlined, even when liquid industrial wastes are deposited. The Carboniferous Limestone is exposed predominantly in the western part of Britain and in areas of high relief. Their distribution coincides with areas of high rainfall, and the potential for leachate migration from landfill sites is high. Although there is a considerable potential for spring pollution, due to rapid unattenuated movement of leachate through conduits, this problem has only been reported at 4 sites. At these the pollution generally occurred soon after site development and was associated with the onset of water ingress problems at 2 sites, and particularly heavy rainfall at another. Gross contamination of the springs occurred, with, in one case, the leachate travelling over 2 km. No prior hydrogeological or tracing work had been undertaken. Contaminated spring waters were treated with hydrogen peroxide to improve leachate oxidation at three of the four sites. Remedial measures included capping the landfill with low permeability materials to prevent leachate generation, and removal of leachate by tankering where groundwater inflow was the major cause of the problem. In all cases water quality improved rapidly, and had returned to pre-pollution levels within 6 months of the remedial works, suggesting that little significant storage of leachate occurred.

In the cases described above, no groundwater monitoring using boreholes was undertaken. Indeed Quinlain & Ewers (1985) have suggested that " observation wells drilled to monitor pollutants in most limestone terranes are ... a waste of time and money because it is extremely improbable that wells will intercept the conduits through which the pollutants move". They therefore place emphasis on the use of springs to monitor groundwater contamination from landfills. However, in the case of Broadfield Down, the monitoring strategy for which was described above, leachate migration from landfill B has clearly been detected in the boreholes adjacent to the site, whose base is just above the water table. No prior tracing to these monitoring wells was undertaken (cf Quinlain, 1988). At 5 other sites in South Devon, the Mendip Hills and North West England, leachate migration has been successfully detected in monitoring wells and has also been reported by Leonard, Gilbert & Kane (1986) for an unlined site on limestones in Huntsville, Alabama. These results suggest that some form of contaminant plume does occur within the diffuse flow zone, and that the case of Quinlain & Ewers (1985) may have been considerably overstated. Concentration of recharge in karstified aquifers occurs at a relatively shallow depth within the epikarst aquifer (Smart & Friederich, 1986). Thus where sites are established within this zone, movement of leachate will be predominantly into conduits in the manner envisaged by Quinlain & Ewers (1985). Indeed one of the four sites which displayed gross contamination was such a site located in a shallow quarry positioned on top of a hill, and with an extensive vadose zone underlying it. For landfills in deep quarries however, the focusing of recharge into conduits will be less prevalent; the zone for lateral movement in the epikarst has been removed; solution fissures within the quarry floor are choked with fine wastes, and are perched on topographic highs (Figure 7). Thus diffuse seepage predominates, and boreholes may be used successfully to monitor groundwater contamination. However, it is important to stress that there are considerable differences in the degree of anisotropy and heterogeneity of different limestone aquifers (Smart & Hobbs, 1986), which relate to the primary structure and geomorphic evolution of the area. Some boreholes may be poorly intergrated with the surrounding aquifer (Hobbs & Smart, 1988).

A second problem in the sole use of springs to monitor groundwater contamination, is that dilution and attenuation of leachate may occur during transmission through the aquifer. The latter is of most significance in the diffuse flow zone, while the former may be large where slow movement into a major conduit occurs. In the case of Broadfield Down, there has been a gradual but statistically significant increase in the levels of chloride and oxygen demand at Cold Bath Spring to the north west of landfills A and B (Figure 8). Note also that oxygen demand now shows a much higher variability than prior to landfill, with sporadic high peaks, raising questions as to the frequency with which monitoring should be undertaken (Quinlain & Alexander, 1987). Chloride concentrations in the spring are over 12 times lower than in the monitoring boreholes as a result of dilution, which would be greater still at the much larger Chelvey Spring. Leachate attenuation is also occurring. Ammonia concentrations decrease from 240 mg/l in the monitoring boreholes to below detectable limits over the distance of 1.5 km, well over an order of magnitude greater than simple dilution would explain. Significant attenuation in the unsaturated zone was also observed by Clark (1984)

The hydraulic gradient in Broadfield Down is from Landfill B towards the Hospital abstraction borehole, but unlike Cold Bath Spring, this shows no evidence of contamination. Movement of leachate appears to be strongly controlled by the regional structure, with downdip migration predominating as a result of strong anisotropy in diffuse permeability. This is also reported at Tythegston by Clark (1984). A plume of high density leachate is migrating downdip to the northwest at this site, even though flow to the south was expected from the regional hydrogeology (The Department of the Environment, 1978). At another site in South Devon, liquid wastes appear to be trapped beneath confining beds following downdip migration. It is however important to remember that groundwater flow direction in karst terrains is generally down the steepest hydraulic gradient or to the lowest outlet from the aquifer. The models of Palmer (1986) may therefore be of considerable value in initial desk studies of potential landfill sites.

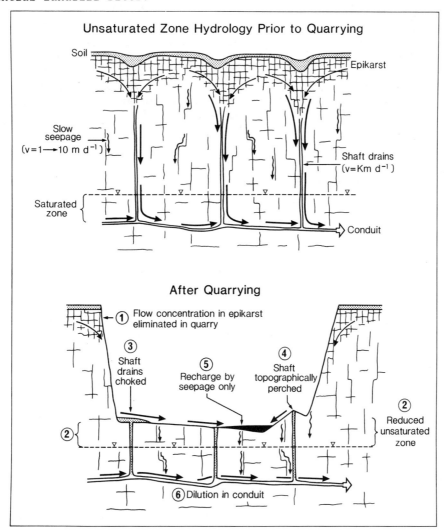

Figure 7. Cartoon contrasting the change in dominant recharge between natural condition (A) and that after quarrying (B).

3. Fires

Landfills A and B (figure 6) have both had several fires during their operation. Both had relatively high inputs of dry inert material, wood and timber products. As well as providing fuel, these are low density wastes permitting ready inflow of air for combustion. Fires in landfills taking putrescible wastes do occur, although the greater moisture and dominance of carbon dioxide in the gas phase greatly reduce the risk. A particularly intense fire occurred in 1986 at Landfill B. The core of the fire was difficult to find, and the waste had to be excavated and doused with water on a 24 hour basis for between 3 to 4 weeks. Such substantial additions of water considerably affect the water budget of the landfill, and this may explain why organic contamination has increased since 1986 at Cold Bath Springs (figure 8).

The fires have been more frequent during the winter months, possibly because the thermal gradient between the warm landfill and the cool atmosphere promotes the most active circulation of gases within the landfill at this time. The inflow of air is enhanced by the highly fractured rock walls of the landfill, which result from blasting during quarrying operations. Lining may inhibit this effect, although installation of stone or pipe ventilation chimneys between the waste and the liner to reduce gas migration may augment the problem of fire.

Landfill sites that have a high proportion of tyres are particularly susceptible to fire, in 1972 an extensive fire at Thythegston landfill site followed the previous disposal of over 2 million tyres (The Department of the Environment, 1978). Tyres have a large void space giving good ventilation. The pyrolitic distillation of the tyre rubber at high temperatures, generates a complex mixture of liquid and gaseous hydrocarbons including pollutants such as phenols, cresols and benzenes (Severn-Trent Water Authority, 1982). These pollutants are present in 2 forms: an insoluble, oily material which floats, and a soluble phase which may emulsify, cause discolouration of water (Lord, 1977) and migrate into the groundwater. Fires within a limestone quarry receiving this type of waste have occurred in the Peak District in 1970 and 1977. However, no pollution has subsequently been monitored. Best & Brooks (1981) document a fire which caused significant pollution to a surface stream draining a quarry on Carboniferous Limestone in Scotland. They did not investigate the effects of the pollutants upon groundwater quality. The persistence of the pollutants produced from tyre fires and their potential toxicity suggest significant potential for contamination.

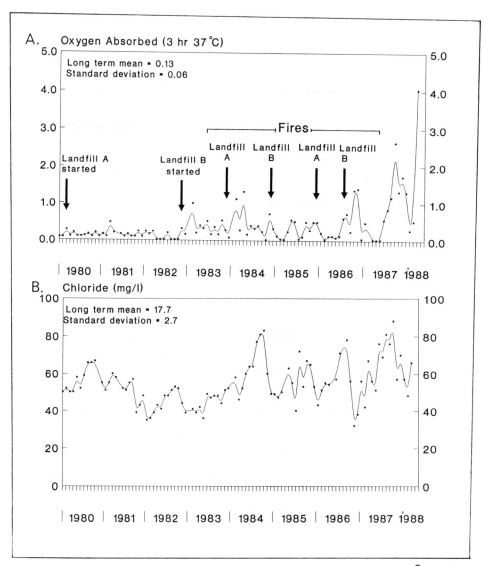

Figure 8. Variation of oxygen adsorbed in 3 hours at 37°C (A) and chloride (B) for Cold Bath Spring, Broadfield Down (Data kindly provided by Bristol Waterworks PLC). Note High initial chlorides are probably related to prior use of quarry A to store road de-icing salt.

4. Migration of Landfill Gas

There is great concern regarding the dangers associated with landfill gas, which can contain up to 65% methane. When the concentration range of methane in air is in the range 5-15%, then an explosion can occur if a source of ignition is present. Explosions have occurred within the UK (Parker, 1981), which were associated with high gas concentrations building up in confined, unventilated spaces such as roof voids, factory inspection pits, basements and beneath parked cars (Parsons & Smith, 1986).

The flow of methane out of a landfill is driven by the large gas pressures generated within the waste material during methanogenesis. Vertical migration of the produced gas is preferred as it reduces the danger to persons and property outside the confines of the landfill. Vertical migration is however inhibited by procedures which are employed, during and post landfill operation, to reduce direct recharge into the waste. During landfill operation, thin layers of waste material (2.5 m) are heavily compacted and covered daily by 6" of inert materials such as clay and soil. This covering and the perched water tables that it produces promotes horizontal gas migration out of the landfill into the surrounding strata. Horizontal gas migration is even further promoted after the completion of a landfill site, as the final layer of waste is domed off with a 1 m cap of low permeability.

The potential for horizontal gas migration within Carboniferous Limestone is very large due to the existence of fractures and fissures which can allow gases to migrate over long distances in a rapid and unpredictable manner. In 1973, 2 cavers within a party of 3 collapsed through asphyxia while exploring an old lead mine in the Peak District. The noxious gas was found later to be carbon monoxide. After extensive enquiries it was discovered that a large quarry blast had taken place several hours before hand, approximately 0.75 km away, and the prevailing wing had forced the noxious gases into the fractures and the fissures which connected through to the old lead mine (Tankard, 1986). Although this is unrelated to landfill operations, it does demonstrate the potential for gas migration within Carboniferous Limestone.

All landfill sites that have received biodegradable wastes and are within 250m of buildings must be monitored on a weekly basis. Due to the unpredicatable nature of the gas flow within Carboniferous Limestone, monitoring is very problematic, even so, gas monitoring boreholes are now being widely installed. The precise configuration of the gas monitoring boreholes varies between WDAs. For instance, in the Peak District 2" gas monitoring boreholes surround selected landfills at intervals not greater than 50 m around the landfill's circumference but within 5 m of the edge of the fill material. In North West England, however, 4" gas monitoring boreholes are being installed every 20m between the landfill and any buildings which are within a 250 m radius. Gas Monitoring at a landfill site must continue until gas production has fallen below being considered a risk. Monitoring can therefore only be stopped once the monitored concentration of methane by volume in air is less than 1% (Her Majesty's Inspectorate of Pollution, 1989a). Due to the very recent implementation of gas monitoring facilities, it is difficult to assess the effectiveness of boreholes in monitoring gas migration within Carboniferous Limestone. Limited monitoring to this date has also meant that the true extent of the problems associated with gas migration is as yet unknown.

Within operational sites, measures to prevent the lateral migration of gas to the surrounding strata have included the installation of vertical stone chimneys between the waste infill and the landfill liner (if present). These vertical chimneys are emplaced at intervals of 50 m around the edge of the waste infill and are approximately 2 m in diameter. Horizontal vents are installed every vertical 8 m to aid ventilation. These chimneys are employed to inhibit lateral gas migration and to divert gas movement vertically to the surface. Clean 6 " stone is utilized for this purpose as it is vital that a good ventilation is maintained to promote gas flow.

With completed and capped landfills, passive gas vents have been drilled through the waste material. These are to promote vertical gas flow and usually consist of boreholes containing slotted casing which is surrounded by approximately 25 mm of none-fine aggregate materials. Very few active extraction systems exist within England and Wales, unless the gas is being utilzed for energy production although no sites on the Carboniferous Limestone produce energy from their gases.

Conclusions

There are significant differences in both the proportions of landfill sites accepting different types of waste, and in site preparation between WDA and WA areas for the Carboniferous Limestone. These primarilly reflect the pressure for waste disposal, and the significance of the aquifer for water supply in the area, as there is general recognition that due to the karstified nature of this aquifer, there is a high potential for groundwater contamination.

There have been substantial changes in the philosophy and practice of landfill disposal of waste in Britain. Prior to the introduction of site licensing in 1976, many sites received no preparation, and a legacy of such sites still taking difficult wastes remains today. Site preparation practice has more recently been directed by HMIP, and now most difficult sites are lined with natural low permeability materials. As a result of this change, there has been a reduction in the proportion of sites on the Carboniferous Limestone taking difficult waste, and a corresponding increase in sites licensed for inert waste.

A specific problem with much inert and semi-inert waste is that it may contain a high proportion of combustible material, but may not develop an anaerobic atmosphere because the high porosity allows ventilation. Fires are thus a considerable hazard, particularly those that occur in sites taking tyres, and generate substantial volumes of poor quality leachate as a result of dowsing with water. Excess leachate generation also occurs when there is ingress of groundwater. This is rarely associated with catastrophic re-routing of conduit flow, but often results from incorrect estimation of the highest water level in quarries due to the recovery of the regional water table after cessation of pumping. Gross contamination of springs fed by conduits has occurred, but rarely affects sites in deep quarries because the natural recharge concentration function of the epikarst zone is inoperative. Remedial measures on-site give rapid improvement in quality. Many sites do however generate a contaminant plume within the saturated diffuse zone, which may eventually affect springs. There is evidence that chemical attenuation does occur in both the vadose and diffuse flow zones. Monitoring of sites occurs generally as a response to a pollution incident, and is not generally a requirement of the site licence. Water quality at abstraction sites is however monitored routinely. Experience suggests that the best monitoring strategy involves both the sampling of boreholes in the vicinity of the quarry and potentially affected springs. Neither alone is adequate, but the former are more sensitive indicators of leachate pollution than the latter, because considerable dilution occurs within the conduit system.

Further changes in waste disposal policies within England and Wales are imminent, with the much needed separation of the disposal and policing duties of WDAs, together with an increase in staffing to allow adequate inspection and enforcement of site licences. In the longer term "Duty of Care" legislation will be introduced to control waste producers. This may herald a new and more enlightened attitude to waste dispsoal problems.

Acknowledgements

A. J. Edwards is supported by a CASE Award from the Natural Environmental Research Council (Training Award GT4/88/AAPS 5), in cooperation with ARC.

References

Atkinson, T.C., 1971. The dangers of pollution of limestone aquifers with special reference to the Mendip Hills, Somerset. Proceedings of the University of Bristol Speleogical Society, v.12, pp. 281-290.

Atkinson, T.C., 1977. Diffuse flow and conduit flow in limestone terrain in the Mendip Hills, Somerset. Journal of Hydrology, v. 35, pp. 93-110.

Aspinwall & Company., 1987. Sitefile; A digest of authorised waste treatment and disposal sites in Britain. Aspinwall & Company, ISBN 09512878 0.

Avon County Council Planning Department, 1978. Mineral Workings in Avon, Avon County Council.

Best, G.A. & Brookes, B.I., 1981. Water pollution resulting from a fire at a tyre dump. Environmental Pollution (Series B.), v.2, pp. 59-67.

Clark, L., 1984. Groundwater development of the Chepstow Block: A study of the impact of domestic waste disposal on a karstic limestone aquifer in Gwent, South Wales. Proceedings of the International Symposium on Groundwater Sources - Utilization and Contaminant Hydrology, Montreal, May 1984, pp. 300-309.

Control of Pollution Act. 1974. HMSO

Dodge, E.D., 1985. Heterogeneity of permeability in karst aquifers and their vulnerability to pollution. Examples of 3 springs in the Causse Comtal (Avergon, France). Annales de la Societe Geologique de Belgique, t.108, pp. 43-47

Fleming, G., 1988. Landfill and water: A difficult mix, Hydrology of landfill sites. Proceedings of the SHG/BHS Symposium, Glasgow, 3rd November.

Gray, D.A., Mather, J.D. & Harrison, I.B., 1974. Review of groundwater pollution from waste disposal sites in England and Wales, with provisional guidelines for future site selection. Quarterly Journal of Engineering Geology, v.7, pp. 181-196.

Her Majesty's Inspectorate of Pollution, 1989a. Waste Management Paper Number 27; The Control of Landfill Gas, A technical memorandum on the monitoring and control of landfill gas, HMSO

Her Majesty's Inspectorate of Pollution, 1989b. Waste Management Paper Number 4; The Licensing of Waste Facilities, A revision of Waste Management Paper Number 4 (1976) to provide a technical memorandum on the licensing of waste facilities including a review of relevant legislation, HMSO.

Hobbs, S.L. & Smart, P.L., 1988. Heterogeneity in Carbonate aquifers - A case study from the Mendip Hills, England. Proceedings of the 21st Congress, Karst Hydrogeology and Karst Environment Protection, Geological Publishing House, Beijing, China, pp. 734-730.

Khan, A.Q., 1987. Problems with enforcement of good landfill practices in the UK. Proceedings of the Landfill Practices Symposium, Harwell, 1987.

Leonard, K.M., Gilbert, J.A. & Kane, W.F., 1986. Groundwater problems associated with a municipal landfill in the Nashville Dome region of Alabama. Proceedings of the Environmental Problems in Karst Terranes and their Solutions Conference, October 28-30th, Bowling Green, Kentucky, pp. 309-320.

Lord, S.P., 1978. Cowm 1975 - An incident. Journal Institute of Water Engineers and Scientists, v.32, pp. 463-467.

Lundgren, T. & Söderblom, R., 1985. Clay barriers - A not fully examined possibility. Engineering Geology, v.21, pp. 201-208.

Newson, M.T., 1973. The Carboniferous Limestone of the UK as an aquifer rock. Geography Journal, v. 139(2), pp. 294-305.

North West Water, 1988. Earthworks on landfill sites, Unpublished report.

Pacey, J.G. & DeGier, J.P., 1986. The factors influencing landfill gas production. Proceedings of the Energy from Landfill Gas Symposium, 28-31st October, Solihull, pp. 51-59.

Palmer, A.N., 1986. Prediction of contaminant paths in karstic aquifers. Proceedings of the Environmental Problems in Karst Terranes and their Solutions Conference, October 28-30th, Bowling Green, Kentucky, pp. 32-51.

Parker, A., 1981. Landfill gas problems - case histories. Proceedings of the Landfill Gas Symposium. Harwell, Paper 3.

Parsons, P.J. & Smith, A.J., 1986. The environmental impact of landfill gas. Proceedings of the Energy from Landfill Gas Symposium, 28-31st October, Solihull, pp. 17-21.

Quinlain, J.F., 1988. Special problems of groundwater monitoring in karst terranes. Symposium on Standards Development for Groundwater and Vadose zone Monitoring Investigations, ASTM STP 000, Neilson, Philadelphia, pp. 1-37.

Quinlain, J.F. & Aley, T., 1987. Discussion of "A new approach to the disposal of solid wastes on land". Ground Water, v.25, pp.615-616.

Quinlain, J.F. & Alexander, E.C., 1987. How often should samples be taken at relevant locations for reliable monitoring of pollutants from an agricultural, waste disposal, or spill site in a karst terrane ? A first approximation, in Beck, B.F. & Wilson, W.L. (Eds.), Karst Hydrogeology: engineering and environmental applications, A.A. Balkema, pp. 277-286.

Quinlain, J.F. & Ewers, R.O., 1985. Groundwater flow in limestone terranes: Strategy, rational and procedure for reliable efficient monitoring of groundwater quality in karst areas. Proceedings of the 5th National Symposium and Exposition on Aquifer Restoration and Groundwater Monitoring, Columbus, Ohio.

Rae, G.W., 1985. The need for monitoring. Proceedings of the Landfill Monitoring Symposium, Harwell, pp. 3-11.

Severn-Trent Water Authority, 1982. Control of Pollution Act 1974 Section 5(4): A Reference to the Secretary of State in respect of a site at Upper Oldhams Farm, Blakemoor, Derbyshire.

Smart, P.L., 1985. Applications of fluorescent dye tracers in the planning and hydrological appraisal of sanitary landfills. Quarterly Journal of Engineering Geology, v.18, pp. 275-286.

Smart, P.L. & Friederich, H., 1986. Water movement and storage in the unsaturated zone of a maturely karstified Carbonate aquifer Mendip Hills, England. Proceedings of the Environmental Problems in Karst Terranes and their Solutions Conference, October 28-30th, Bowling Green, Kentucky, pp. 59-87.

Smart, P.L. & Hobbs, S.L., 1986. Characterisation of Carbonate aquifers: A conceptual base. Proceedings of the Environmental Problems in Karst Terranes and their Solutions Conference, October 28-30th, Bowling Green, Kentucky, pp. 1-14.

Smith, H., 1988. Legislative aspects of landfill - a River Inspector's view. Hydrology of landfill sites. Proceedings of the SHG/BHS Symposium, Glasgow, 3rd November.

Stanton, W., 1989. Bleak prospects for limestone. New Scientist, 13th May.

Tankard, J.D., 1986. The hazards associated with landfill gas production. Proceedings of the Energy from Landfill Gas Symposium, 28-31st October, Solihull, pp.60-69.

The Department of The Environment, 1978. Cooperative programme of research on the behaviour of hazardous wastes in landfill sites, HMSO.

The Department of the Environment, 1986. Waste Management Paper Number 26; Landfilling Wastes, A technical memorandum for the disposal of wastes on landfill sites, HMSO

The Guardian. March 7th, 1989.

The House of Commons Environment Committee, 1989. Toxic Waste, Second Report of the Environmental Committee 1988-89, v.1, House of Commons Paper 22-1, HMSO

Town and Country Planning Act. HMSO. 1971.

Assessing ground water flow paths from pollution sources in the karst of Putnam County, Tennessee

ELWIN D.HANNAH   *Division of Superfund, UST Program, Department of Health and Environment, Nashville, Tenn., USA*
TOM E.PRIDE, ALBERT E.OGDEN & RANDY PAYLOR   *Center for the Management, Utilization and Protection of Water Resources, Tennessee Technological University, Cookeville, Tenn., USA*

ABSTRACT

Throughout the last two years, dye tracing techniques have been used to delineate the subterranean pathways between sinking streams and springs in Putnam County, Tennessee. A portion of Putnam County is located on a sinkhole plain underlain by nearly flat-lying Mississippian carbonates. Over 300 sinkholes have been delineated from topographic maps. An earlier study (Mills et al., 1982) showed that approximately 25% of the sinkholes contained solid waste, organic fill, or were receiving sewage. This alarming information prompted the present study aimed at delineating the recharge areas to the springs.

The three largest communities in the county (Cookeville, Algood, and Monterey) have experienced growth that has contributed to ground water contamination from a variety of sources such as leaky underground storage tanks, municipal storm water runoff, poorly treated and untreated wastewater, and acid mine drainage.

At least three corporations having leaky underground storage tanks contribute petroleum products to one sinking stream. The contaminated water in the sinking stream moves a distance of 3000 ft. to a spring in less than eight hours and then travels to the old Cookeville City Lake. The town of Algood injects its poorly treated wastewater into a sinkhole causing contamination of a spring and creek just 2000 ft. away. An old dump outside Algood occurs in a deep sinkhole. Runoff from the landfill area is believed to travel approximately one mile to a spring that usually remains turbid, likely due to the dumping of limestone dust into a sinkhole next to quarrying operations.

Ground water tracing has been conducted south of Interstate 40 from the industrialized portion of Cookeville. Petroleum products have been observed at the springs along Hudgens Creek, but the point sources of contamination have not yet been identified.

Coal mining operations around Monterey have caused acid mine drainage and the production of yellow-boy. Tracing has documented which springs are being impacted.

The end product of the research is a map delineating the recharge areas for the springs in the study area. This map can be utilized for planning purposes, contamination remediation, and for emergency response to spills along Interstate 40 and state highways within the county.

Introduction

During recent years, there has been increasing concern for protecting the quality of karst ground water systems. Attitudes concerning disposal of solid and liquid wastes in karst environments have changed with the realization of the dynamic relationship between surface water and subsurface cave and spring systems. In the southeastern United States, major karst regions are associated with Mississippian limestones of the Interior Low Plateaus and faulted and folded Ordovician limestones in the Appalachian Valley and Ridge Province. As municipalities located in these regions grow, new and innovative actions will be necessary to avoid contamination of the ground water systems.

During the last two years, a systematic analysis of ground water flow paths utilizing dye tracing techniques has been performed in Putnam County, Tennessee, between pollution sources and springs. The area is suffering growing pains that have resulted in ground water contamination from leaky underground storage tanks, untreated municipal sewage, garbage dumps, municipal storm water runoff, agricultural practices, and acid mine drainage. Dye tracing enables researchers to delineate surface and ground water basins that provide recharge to springs. This information can be utilized by health officials and city planners to determine sources of spring water pollution and to prevent further ground water contamination from occurring.

Location and Geology

A portion of Putnam County, Tennessee, is located on the sinkhole plain of the Eastern Highland Rim Province, and Monterey is located on the Cumberland Plateau Province (Figure 1). Streams originating on the plateau around Monterey flow over shales and sandstones until underlying Mississippian-aged limestone beds are intersected (Figure 2). Some streams sink when they reach the Bangor Limestone while others sink into the lower Monteagle Limestone. Subterranean water moves through caves, pits, and solution-enlarged fractures until emerging as spring flow near base level. Deep, pit-type vertical caves form in the Bangor and Monteagle limestones, whereas long, horizontal caves usually occur

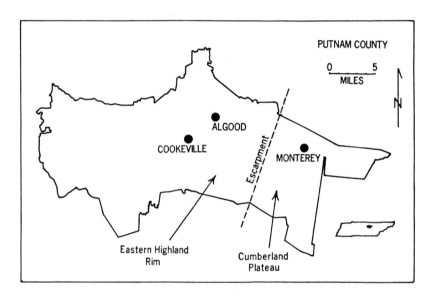

Figure 1: Location of the Study area.

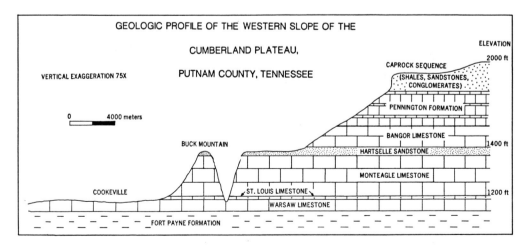

Figure 2: Generalized geologic profile of the Cumberland Escarpment.

in the St. Louis and Warsaw limestones. The underlying Fort Payne Formation is a dense, silica-rich limestone. The lower Warsaw acts as an aquiclude, and thus many springs occur perched above this member. The upper Warsaw is commonly arenaceous, but dissolution of the calcite cement enables cavern growth by corrasion downward from solution conducts developed in the lower St. Louis. Descriptions of the caves have been made by Barr (1961), Matthews (1971), Faulkerson and Mills (1981), and the Tennessee Cave Survey. Over 200 caves are known in Putnam County. To date, approximately 11 km of cave passage have been surveyed within the city limits of Cookeville. Most of these caves occur in the Warsaw Limestone and all are hydrologically active. Monterey is located on Pennsylvanian-aged rocks that are being mined for coal. Acid mine drainage from these mines eventually sinks downstream when the Mississippian-aged limestones are intersected.

Methods

Ground water tracing was conducted using fluorescein green dye. An activated charcoal dye adsorption packet was placed in the spring that was the expected emergence point of each dye trace. Blanks were placed at locations not believed to be hydrologically connected with the trace to eliminate chances of false positives. The packets were generally changed every week and then tested with a base elutriant.

Results

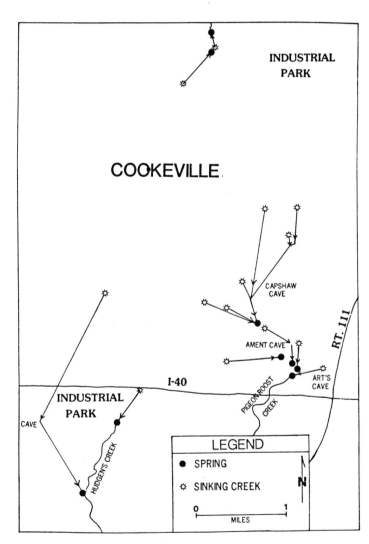

Figure 3: Dye trace results within the Cookeville city limits. Most of the traces north of I-40 were conducted by Faulkerson Mills (1981).

Physiochemical degradation of the subterranean stream system beneath most of Cookeville has been previously documented by Smithson (1975), Faulkerson and Mills (1981), Wilson (1985), and Pride et al. (1988). Also, Pride et al. (1989) reported the adverse effects of ground water contamination on the benthic-macro-invertebrate communities in cave streams beneath Cookeville. Faulkerson and Mills (1981) conducted several dye traces north of Interstate-40 within the city limits of Cookeville. These are included on Figure 3. The purpose of the present study is to document ground water flow paths south of Interstate 40 around Cookeville and areas east and north of Cookeville around the towns of Algood and Monterey.

Figure 3 shows dye traces conducted within the City of Cookeville and south of Interstate-40. Spring water is contaminated north of the interstate primarily by storm water runoff, leaky underground storage tanks, and occasional by-passes of untreated municipal sewage. The springs south of the interstate on Hudgens Creek smell of petroleum product. The drainage area for the springs includes part of the interstate as well as an industrial park with some industries located in sinkholes.

Figure 4 shows dye trace results east of Cookeville and in the town of Algood. City Spring drains a portion of the interstate's runoff. Elevated levels of nitrate in the spring are attributed to abundant cattle grazing in the recharge area (Pride et al., 1988). Kirby Spring is contaminated by petroleum products known to be leaking from at least three underground storage tanks including a large truck stop.

Hidden Hollow Springs is one of the largest resurgences in the area and is used as a commercial, recreational area. The recharge area includes hog and cattle farms, an industrial park, and many subdivisions. Significant water quality problems have arisen in the past due to occasional bypasses of untreated municipal sewage into a sinking stream that recharges the spring (Pride et al., 1988).

One spring north of Algood on Bear Creek has water quality problems from poorly treated municipal wastewater that is piped into a sinking stream. A nearby spring on Bear Creek is contaminated by storm water runoff from Route 111. Drain wells along the highway have intersected Indian Cave which contains the underground river that leads to the spring.

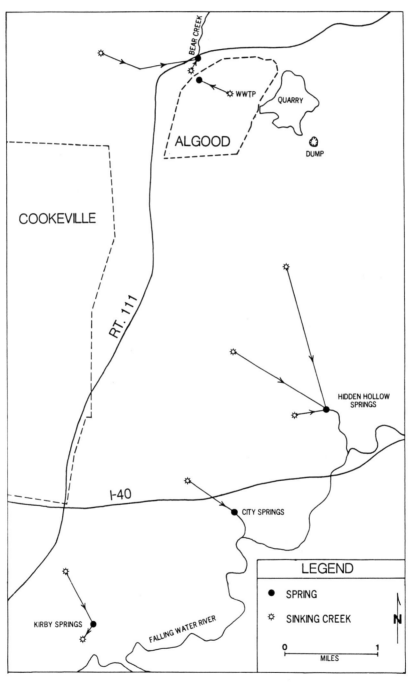

Figure 4: Dye trace results east of Cookeville.

A sinkhole east of Algood has been used as an unpermitted dump for years. Several attempts have been made to trace water entering this sinkhole, but all have been unsuccessful. A large limestone quarry occurs next to the dump and washed lime dust runs into an adjacent sinkhole. A spring along Bear Creek was located that always runs turbid. It is believed that the quarry runoff recharges this spring as does the dump, and the high turbidity absorbs the dye.

Figure 5 shows dye trace results around the City of Monterey which is located about fifteen miles east of Cookeville on Pennsylvanian-aged clastics of the Cumberland Plateau. Northeast of town, the recharge areas for springs at Spring Creek are primarily forest with some grazing on grasslands. As a result, spring water quality is relatively good. East of Monterey there are abandoned deep and surface coal mines. Acid mine drainage sinks downgradient when the surface water intersects the Mississippian-age carbonate rocks. Where the acid waters enter caves, yellow-boy encrusts the walls due to the neutralizing effect of the limestone. The West Fork and East Fork (not shown) of the Obey River have been severely contaminated by mine runoff. The effects of the mines on the Obey River would be even worse if not for the "natural" acid mitigation provided by the limestone caves. Unfortunately, blind cave fish and other troglobitic species have been severely impacted.

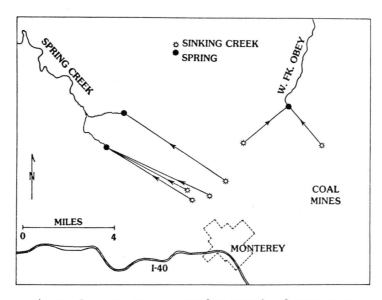

Figure 5: Dye trace results north of Monterey.

Conclusions

Dye tracing in Putnam County has been effective in defining the recharge areas of the springs and thus identifying present sources of pollution. Perhaps more importantly, this information, if used wisely, can provide a planning tool to lessen the impact of future development. In addition, if accidental spills of hazardous substances occur along the interstate south of Cookeville, its emergence point will be known. Much more dye tracing along the karst reaches of our country's interstate system is needed to enable immediate response to spills for better environmental protection.

References

Barr, T.C., 1961, Caves of Tennessee: TN Dept. of Conservation, Division of Geology, Nashville, TN, Bull. 64, 567 p.

Faulkerson J. and H.H. Mills, 1981, Karst hydrology, morphology and water quality in the vicinity of Cookeville, Tennessee: Unpublished Report for NSF Grant No. SPI-8004009, 67 p.

Matthews, L.E., 1971, Descriptions of Tennessee Caves: Tennessee Division of Geology, Nashville, TN, Bull. 69, 150 p.

Mills, H.H., D.D. Starnes, and K.D. Burden, 1982, Predicting sinkhole flooding in Cookeville: The Tennessee Tech Journal, v. 17, pp. 1-20.

Pride, T.E., A.E. Ogden, M. Harvey, and D. George, 1988, The effect of urban development on spring water quality in Cookeville, Tennessee: Proc. Natl. Water Well Assoc. Karst Conf., Nashville, Tennessee, pp. 97-120.

Pride, T.E., A.E. Ogden, and M. Harvey, 1989, Biology and water quality of caves receiving urban runoff in Cookeville, Tennessee, U.S.A.: Proc. 10th Int. Cong. of Speleology, Budapest, Hungary (in press).

Smithson, K.D., 1985, Some effects of sewage effluent on the ecology of a stream ecosystem, M.S. thesis: Tennessee Technological University, Cookeville, Tennessee, 74 p.

Wilson, D., 1985, The physical, chemical, and biological recovery of Pigeon Roost Creek following the closing of a sewage treatment plant: M.S. thesis, Tennessee Technological University, Cookeville, Tennessee, 73 p.

Anatomy of a hazardous waste site in a paleosink basin

DENNIS J.PRICE   *Suwannee River Water Management District, Live Oak, Fla., USA*

ABSTRACT

For 30 years the Brown Wood Preserving Company treated timber with creosote and pentachlorophenol, disposing waste to a three acre unlined lagoon in a paleosink basin. This basin has been naturally filled with reworked Hawthorn clays and a thin veneer of sand. The lagoon is dependent on rainfall for its water supply. A thick layer of creosote sludge had accumulated in the deepest part of the lagoon. Recent subsidence occurred along the west shore line of the lagoon. The upland areas are underlain by slightly sandy Hawthorn clays which contain extensive secondary fracturing. The Suwannee Limestone is encountered at variable depths and dips into the basin. The area is semi-artesian with regional groundwater flow to the south and southwest.

Monitor wells were installed and geophysical techniques employed to determine the local geologic conditions. The monitor wells and residential wells were sampled, no creosote products were found. Water level data were analyzed and mapped. This information indicated that the potentiometric surface was relatively flat with flow lines that indicated a slight bulge in the contours under the site. Analysis of data indicated anomalies in the ground water levels and the stratigraphic column. Subsequently, additional monitor wells were installed. Sampling indicated no water quality problems.

Because questions remained about groundwater flow direction, sinkhole development and recharge, another level of geologic investigation was initiated. Split spoon samples were taken and the geologic information was mapped and interpreted. Based on the new information additional monitor wells were installed. When sampled, one of the two new wells and one of the old wells was found to have creosote products present.

General Setting

The basin is typical of the geomorphology of the transition zone along the Cody Scarp which separates the Northern Highlands and the Gulf Coastal Lowlands (Ceryak, 1981). The method of the continued expansion of these paleosink basins is demonstrated in the Brown Wood Lagoon by the recent subsidence features. The recognition of these basins and their location along fracture traces is of particular importance to the well drilling industry and the study of sites in karstic terrain.

Humid, subtropical, climatic conditions exist in the study area. The average yearly temperature is 19 degrees C. Rainfall averages 137 centimeters per year, with highest rainfall rates occurring from June through August. The average evaporation transporation rate is 107 centimeters per year.

History

For thirty years the Brown Wood Preserving Plant was operated in Live Oak, Florida by at least two different companies. The plant quit operating in 1978 (fishbeck, thompson, carr & huber, 1986). The purpose of the plant was to treat poles and lumber with creosote and, to a lesser degree, pentachlorophenol.

Chemicals

Creosote has been reported to contain as many as 10,000 substances (EPA, 1984). The primary substances of environmental concern in creosote are the Polynuclear Aromatic Hydrocarbons (PAH). "PAH have little solubility in water and tend to sorb onto media such as soil" (fishbeck, thompson, carr & huber, 1986). Pentachlorophenol is produced by the chlorination of phenols. Pentachlorophenols readily degrade through photodegradation and biodegradation. It is slightly more soluble in water than creosote. Table 1 is a list of the chemicals found in the groundwater after extensive exploration and monitoring. They are constituents of creosote, not pentachlorophenol.

Table 1: List of chemicals in creosote detected in the groundwater.

DIBENZOFURAN
FLUORENE
PHENANTHRENE
ANTHRACENE
FLUORANTHENE
PYRENE
NAPHTHALENE
2-METHYLNAPHTHALENE
ACENAPHTHENE
FLUBRENE

Disposal

During the period the plant operated, all waste sludge and product was discharged from the plant site through a ravine and into a three-acre lagoon. The lagoon drains into a 74-acre basin. Through the years, the lagoon accumulated a layer of creosote product that was a couple of feet thick in the deepest portion of the lagoon. The shallow west side had a very thin layer of residuum that was deposited as a film when, in dry times, the water re-treated to the deepest part of the lagoon.

Site listing

Prior to 1983, a study was made that put the site on the United States Environmental Protection Agency's (USEPA) superfund list. An administrative order issued by the USEPA on September 14, 1983, required the cleanup. Two former owners were contacted as the parties responsible for the cleanup. Both companies voluntarily agreed to combine resources and begin the cleanup.

Focus of the report

This report is being written to describe the site, the methods used to investigate the site, the geology of the paleosink basin and the difficulty in performing site specific analysis of groundwater characteristics over a small area in Karstic terrain.

Study methods

The initial task was to study information that existed as published documents filed at the office of the Suwannee River Water Management District (SRWMD) and at the offices of the Department of Environmental Regulation (DER).

Sampling was initially begun in 1983 using shallow soil sampling procedures. Much of the information relating to the location and thickness of the gross contamination was discovered during this procedure and also through a later extensive sediment sampling program. Four monitor wells were installed in 1984. Another round of test holes was performed in early 1985. In August of 1985, two more monitor wells were installed to continue the exploration for a plume. Analysis again found no contaminants. The presence of two collapse features in the Northwest corner of the lagoon indicated a real possibility existed for groundwater contamination to be occurring. Due to this another series of soil borings was drilled in 1986, the geology reviewed and, based on all available information, two more monitor wells were installed in hopes of locating the plume, if one existed. All the monitor wells (Figure 1) were sampled and analyzed. The potentiometric surface was mapped on a regular basis for about two years; two examples of the surface are presented as Figures 2 and 3.

During this time several geophysical studies were performed. These included Ground Penetrating Radar (GPR), Neutron and Gamma Logs plus a video survey of a production well located on the site.

Early in the course of the study, the two collapse features, mentioned previously, appeared in the northwest end of the lagoon. These collapse features maintained a water level that was consistently lower than the water in the lagoon except when the lagoon was full. Because of their occurrence, a concern was raised about the possibility of heavy equipment causing a general collapse of the lagoon resulting in the lagoon sediment ending up in the aquifer.

Soil around the plant site was heavily contaminated with creosote. Because of this and also knowing that the site was underlined with the thick sequence of reworked Hawthorn clays, efforts were made to determine if the clays were attenuating the flow of contaminants. A backhoe was used to dig through the creosote and several feet into the clay and sandy-clay.

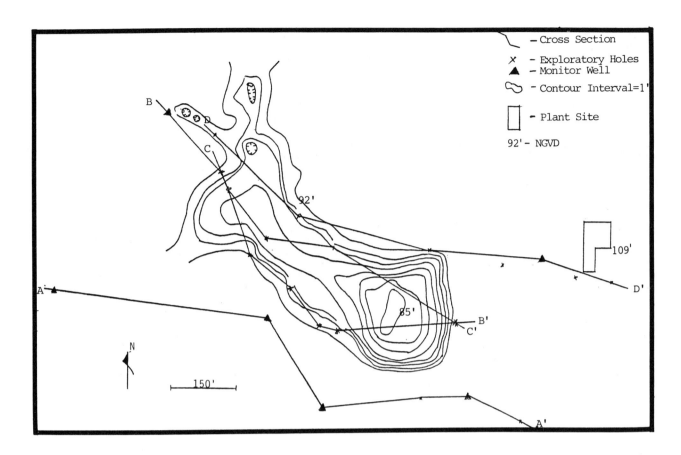

Figure 1: Location of sampling points, monitor wells, bathemetric survey and geologic cross-sections.

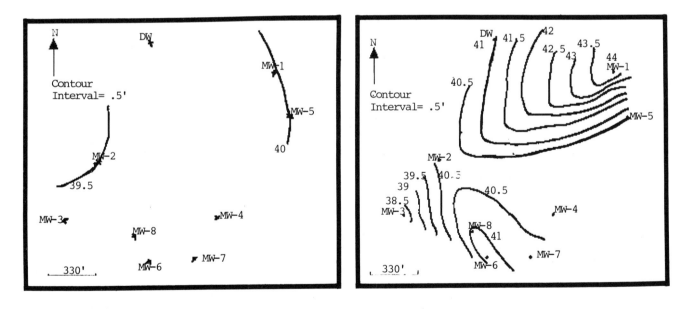

Figure 2: 9/17/85 contour of the potentiometric surface.

Figure 3: 11/06/89 contour of the potentiometric surface.

Geology and Hydrogeology

The Floridan Aquifer is the principal aquifer in the entire region. In the area of the Brown Wood site the aquifer is unconfined. Aquifer tests performed by the Suwannee River Water Management District have indicated a very high hydraulic transmissivity 9,000 - 17,000 square meters per day.

The following are the geologic units which affect, control, and limit Karst features in the study basin. The oldest unit is the Ocala Group which was described by Puri (1953). The Ocala Limestone was established as Eocene in age by Cooke (1915). The Suwannee Limestone unconformably overlies the Ocala group as established by Cooke and Mansfield (1936). The Suwannee Limestone is Oligocene in age and is overlain by the Hawthorn Formation in the study area. The Hawthorn Formation is phosphatic throughout and is composed of sands, clays and silty, dolomitic calcareous units. It is a confining unit in the area and restricts karst development. Elevations at the site range from near 33.2 meters, NGVD, around the plant site to near 26 meters, NGVD, in the deepest part of the lagoon.

The site sits in the Transition Zone that is the product of a retreating escarpment called the Cody Scarp (Ceryak, 1981). In this area, Pliocene and Miocene sediments overlie the Suwannee and Ocala limestones. The limestone formations are characterized by an abundance of solution channels. Beginning in the late Miocene, after the Hawthorn Formation, of Miocene age, was deposited, the Ocala Uplift occurred. This movement produced an anticlinal feature that fractured the Ocala and Suwannee limestones in Northeast/Southwest and Northwest/Southeast directions. The crest of the anticline runs through the vicinity of the Brownwood site.

Lineament studies of the area indicated that the lagoon is located at the intersection of two fracture traces. Borings on the site indicated that a thin veneer of sand existed at the surface at higher elevations around the plant. This was underlain by mostly reworked Hawthorn clays and sandy clays. The lagoon contained sludge, a veneer of sand, and below the sand, more clay and sandy clay. Reworked clay is a term used to try to describe the effects sinkhole activities have on the natural Hawthorn clay and sandy clay strata as they move, slump and erode, and redeposited into the sinkholes and sinkhole depressions as the scarp retreats and the basins expand.

A GPR survey over the lagoon site was performed by Dr. Berry Beck of the Florida Sinkhole Research Institute. The survey was inconclusive in that it failed to detect any features that might positively identify a method for surface waters to enter the groundwater. What it did confirm was the thin veneer of sand overlying a layer of clay and sandy clay. This information, when related to other geologic information, was very helpful in determining whether or not heavy equipment might cause further collapse when deployed in the basin.

Excavations around the plant site were performed in areas heavily contaminated with waste creosote. A backhoe was used to excavate about four meters down. Plastic dense clay and sandy clays were found to underlie the site. The clay contained an intricate network of micro-fine secondary fractures. At several locations, at the depth of two to four meters, creosote was found to be covering the sides of the fractures. It is probable that the fractures are desiccation features and the depth to which they extend is most certainly to the top of the first carbonate unit encountered in the plant area.

Downhole Gamma and Neutron geophysical surveys were conducted. The results showed that the Suwannee and Ocala formations were not homogeneous on the site. The logs showed the Suwannee to be slightly clayey and soft. The Ocala was alternating with hard and soft limestone, free of clay. A video survey of the old production well revealed cavernous zones in the indurated limestone at depths of over 61 meters (fishbeck, thompson, carr & huber, 1986). The lagoon normally holds water only in the deepest portion (Figure 1) nearest the plant. Borings showed that this portion of the lagoon was completely underlain by an impermeable plastic clay.

The decision was made that the presence of heavy equipment in the lagoon area would not pose a serious potential for further collapse of the area.

The water level information was discussed at length because of the apparent anomaly which occurred in Monitor Well 6 (MW-6). The data shows that the water table under the site is flat. In most cases, about 50 percent, MW-6 had a higher water table. Of this 50 percent, the difference above the mean was, in all cases but one, less than one third of a meter. The high water in MW-6 ranged from 2/10 of a meter to 6/10 of a meter higher than the lowest water level measured that date. The data indicates the basin may be recharging the aquifer in the vicinity of MW-6. From data gathered from the Suwannee River Water Management District, it was known that the regional ground water flow was to the south-southwest (Suwannee River Water Managmenet District).

192

Data from the drill log indicated that MW-6 was finished entirely into a firm, tan, silica limestone. This contrasted from the usual formation that the monitor wells were set into. Ordinarily, the wells were in hard, highly fractured limestone. The possibility existed that the resulting tighter pore space and the more clastic characteristics of the material might be artificially raising the water table. There were cases, however, when the high water table was encountered in other monitor wells. The difference was in centimeters and if the theory proposed above holds true, it should certainly hold true in cases where the water level in the high well and the water level in MW-6 were so close.

The water level information was reinvestigated for this report to relate it to rainfall data. Table 2 shows several months of rainfall data related to the differences in the amount of rise and fall of the water level as the aquifer reacted to the rain. The rainfall during the month is assumed to be reflected in the next months water level readings. The month of December was chosen as the starting point. The previous three months rainfall, which was nearly constant, averaged 5.72 centimeters per month. This allowed water levels to reach equilibrium and provide a reliable baseline to begin observations relating to the effect rainfall has on water levels under the site. MW-6 reacted the slowest to the increased rainfall for January and February. If the previously described formational characteristics exist at the well screen, the data is expected. The water level reacts slower because the material is retarding the movement of water through it.

Table 2: GROUNDWATER ELEVATION AND RAINFALL (R/F) DATA COMPARISONS

DATE	MW-1	MW-2	MW-3	MW-4	MW-5	MW-6	DW	R/F
12/10/85	40.0	39.7	39.7	38.0	40.0	40.0	39.9	3.75in
01/14/86	41.4	41.0	40.9	40.8	41.4	.41.2	40.7	7.57in
02/12/86	43.8	43.7	43.4	43.3	43.8	43.3	43.8	7.51in
03/12/86	48.7	48.8	48.5	48.2	48.7	48.9	48.6	2.93in
04/10/86	49.6	49.6	49.2	48.9	49.6	49.8	49.6	0.45in
05/13/86	48.7	48.5	48.0	47.8	48.7	48.7	48.5	1.88in
06/11/86	47.2	46.9	46.4	46.1	47.2	47.3	47.4	9.25in
07/10/86	46.0	45.6	45.0	44.8	46.0	46.3	46.2	1.70in

Water level fluctuation due to the next months continued heavy rains reflected quicker in MW-6. This is an unexpected reaction to the water level if the proposed theory is held to be correct. What must have happened, is the water in the lagoon has begun to recharge the aquifer in the vicinity of MW-6. The water level would be higher as the material in the formation impedes the flow of water reaching the more indurated, fractured, rock further out horizontally in the formation. Figure 4 shows the location of the well screens in the formation.

April's reaction to March's low rainfall (relative to the preceding two months) showed a slight rise. The rise in MW-6 was in the middle range of how the water level was adjusting in all the monitor wells. Rainfall was scarce during April and May and the water levels began to decline at a relatively even rate. A large rainfall event in June retarded the decline considerably in MW-6 and not so much in the others. This reflects recharge in the vicinity of MW-6.

Geologic data gathered to date indicated that the thin veneer of sand that covered the lagoon was thicker in the vicinity of TH-9, TH-10 and TH-5. Experience has shown that, at least in the sinkhole basins along the Transition Zone, most basins contain a vertical sand conduit that allowed drainage of the basin watershed into the aquifer (Figure 4).

Most of the holes drilled on site encountered carbonate material, the first carbonate unit encountered was soft and easily drilled. Below this, some of the borings encountered a hard indurated limestone. The soft limestone was generally described as; pale orange to white, intergranular-intercrystaline porosity, micrite and crystaline graintype, grainsize fine to medium, poorly indurated, mostly a non-fossilferous, friable limestone (Ceryak, personal conversation, 1989). The contour of the top of the hard limestone is presented in Figure 5. The contour of the top of the hard limestone is presented in Figure 6.

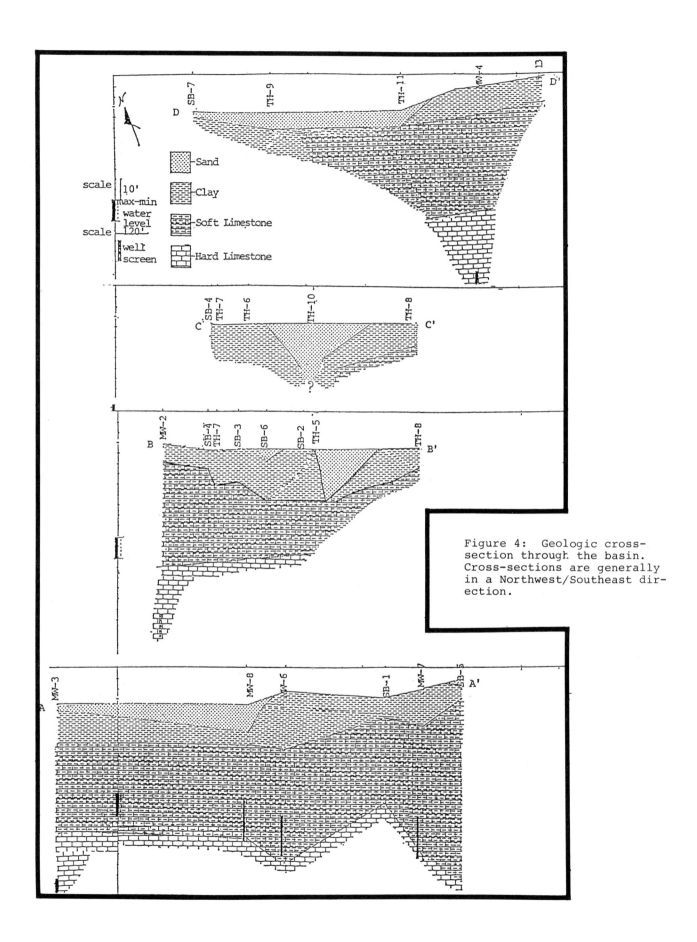

Figure 4: Geologic cross-section through the basin. Cross-sections are generally in a Northwest/Southeast direction.

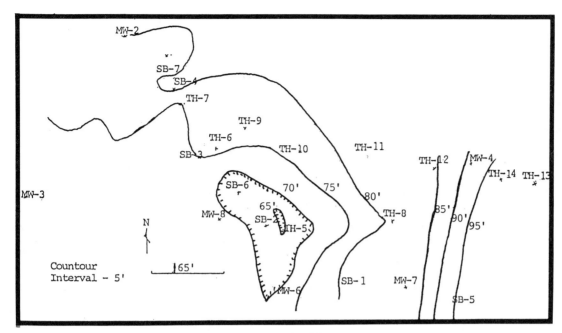

Figure 5: Contour of the top of the soft limestone.

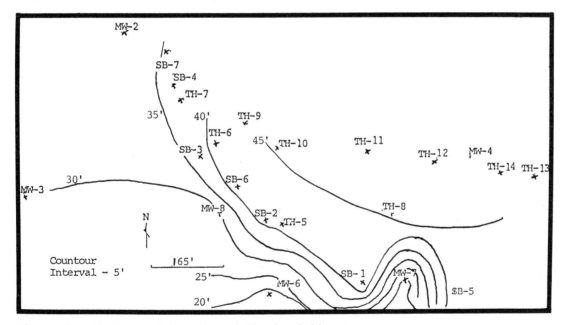

Figure 6: Contour of the top of the hard limestone.

When all the geologic data was available, the geology was mapped in an effort to determine the most likely place to intercept the flow of contaminants from the site, if the flow
existed. MW-7 was placed several hundred feet to the southeast of the lagoon. The possibility existed that conduit flow in the aquifer might be following that trace since that is
the fracture trace direction.

MW-8 was located based on the following information: the thicker sequence of sand
found in TH-5, TH-9 and TH-10; the anomaly seen in MW-6; and the evidence from the soft-hard
limestone contact. If this sand was the channel that contaminants were taking when leaving
the site, and all data seemed to indicate this, then it was important to monitor the aquifer
in the area. MW-6 was already located in a southerly direction so MW-8 was located in a
southwesterly direction. Also, MW-8 was crowded against the lagoon just out of the high
water level. On the next round of sampling, contaminants were picked up in MW-4 and MW-8.
The amounts were quite low.

Months later, during the excavation of the creosote from the lagoon, a sand pipe was encountered in the Northeast side of the lagoon in the vicinity of MW-4. Creosote had migrated in mass down this sand conduit. A backhoe was used to clean the pipe out and plug it. Gross creosote contamination had migrated 4.5-6 meters below the bottom of the lagoon. The pipe was about a meter in diameter.

Conclusion

An exploratory approach, in determining ground water characteristics and the placement of monitor wells, should be of primary concern when investigating plume movement in karst or highly fractured environments. The gathering of existing data from a variety of sources needs to be done. Begin the exploratory work by setting out to prove or disprove existing theories about the site. Decide if answering questions will help in the goal before you.

In karst terrain, anomalies in the geologic and hydrogeologic data have arisen shortly into the data gathering and exploratory program. All anomalies should be explored and explained. During the entire period of exploration, a detailed geologic map should be maintained, this is extremely important. This precludes wildcatting with test holes.

On a site-by-site basis, test holes can be used without having to adhere to the strict guidelines necessary when drilling and installing monitor wells. This ability can be determined almost immediately when the geologic investigation begins. The advantage gained by using test holes in this manner is the rapid accumulation of site-specific conditions.

At some point the hydrogeologic studies and the actual cleanup of the site should be two different issues. The initial geologic investigation should immediately allow the engineers to determine the potential of heavy equipment causing further collapse of the area, this same information should provide an idea of how the gross contamination might move during cleanup. From then on the cleanup plans can be implemented concurrently with trying to determine contaminate migration through the soils and groundwater.

In the case of the Brown Wood site, the drilling did not encounter any caverns in the limestone under the lagoon. The initial limestone formation encountered was soft, plastic and almost clastic in nature. This overlaid a hard indurated limestone. The hard limestone would contain the caverns which could cause collapse. The soft limestone was plastic enough that it would fill most of the caverns that might have been present. The soft limestone was overlain with a firm clay and sand base that would support heavy equipment.

The desiccation features in the clay provide a route for contamination to migrate downward. Due to the insolubility of the creosote, contaminants move in mass. Test holes TH-13 and TH-14 were bored at the plant site in the vicinity of where the desiccation features were first described. Carbonates were encountered at about 3 meters below ground surface. The secondary fractures extended to that depth in the plant area. The carbonates below the clay are soft and easily drilled. As described in this report, the soft carbonate unit probably does not contain fissures or fracture traces. These carbonate units would prevent the continued migration of contaminants after passing the clay unit.

When the lagoon fills and overtops the clay bowl it normally resides in, water is carried to the sand conduit located at the edge of the lagoon North of MW-6. Only trace amounts of contaminants are carried with the water (fishbeck, thompson, carr & huber, 1986). These trace amounts migrate downward and are further reduced in amount by adsorption onto the silts and clays in the carbonate unit. Only minor amounts reach the groundwater as indicated in the analysis of water from MW-6 and MW-4.

References

Ceryak, Ron, 1981, Significance of The Cody Scarp on the Hydrology of North Central Florida, Southeaster Geologic Society, Guidebook Number 23.

Cooke, C. Wythe, 1915, The Age of the Ocala Limestone, U.S. Geological Survey Professional Paper 95.

EPA, "Creosote" Special Review Position Document 2/3, August 1984.

fishbeck, thompson, carr & huber, engineers and scientists, 6090 East Fulton, P. O. Box 211, Ada, Michigan 49301, Report on the Remedial Investigation, Brown Wood Preserving Site, Live Oak, Florida, in cooperation with Environmental Engineering and Management, Ltd., Minnesota.

Puri, Harbins, 1953, Zonation of the Ocala Group in Peninsular Florida. (Abstract) Journal of Sedimentary Petrology, Volume 23.

Suwannee River Water Management District, Route 3, Box 64, Live Oak, Florida 32060

Potential groundwater contamination of an urban karst aquifer: Bowling Green, Kentucky

PHILIP P.REEDER   *Geography Department, University of Wisconsin-Milwaukee, Wis., USA*
NICHOLAS C.CRAWFORD   *Geography/Geology Department, W.Kentucky University, Bowling Green, Ky., USA*

ABSTRACT

Bowling Green, Kentucky is built upon a sinkhole plain and is almost entirely located within the Lost River Karst Groundwater Basin. Recharge of the aquifer predominantly occurs via sinkholes, sinking or losing streams, caves and storm water drainage wells. The existence of an urban center in this karst setting and the ease with which potential contaminants entered the aquifer led to concerns about groundwater contamination in Bowling Green. Fourteen contaminant levels and related parameters were monitored for fourteen months at four sites in and around Bowling Green. Composite and grab samples were collected at (1) the Lost River Rise, the final resurgence of the Lost River which is the master subsurface stream that flows under Bowling Green, (2) By-Pass Cave, a tributary to the Lost River located in a commercial area, (3) The Lost River Blue Hole, located just upstream from the city, and (4) Federal Well, located in a rural area three kilometers upstream from the Blue Hole. This investigation revealed that groundwater contamination is a more serious problem in the rural area south of Bowling Green near Federal Well and in the urban area near By-Pass Cave. Federal Well had the highest concentrations of cadmium, chromium, nitrates and total dissolved solids. By-Pass Cave had the highest recorded concentrations of silver, copper, lead, and microbiological constituents. All these contaminant levels were substantially lower at the Lost River Rise due to the more contaminated waters mixing with less contaminated waters, thus reducing overall concentrations. Therefore, the Bowling Green area has nonpoint source groundwater contamination problems at some specific locations. However, when the overall picture is viewed, contamination of the Lost River under Bowling Green is not that substantial, primarily because of the dilution of contaminants by water from the entire groundwater basin.

Introduction

Bowling Green is located in the central portion of the Pennyroyal Sinkhole Plain of Kentucky. Most of the city lies within the Lost River Karst Groundwater Basin which drains about 140 square kilometers of southcentral Warren County (Figure 1). Bowling Green, because of its karst setting and the impracticality of constructing a conventional storm sewer system on the rolling sinkhole plain, utilizes the karst aquifer for removal of storm water runoff. Runoff naturally enters the karst aquifer through sinkhole drains, swallets of sinking streams and infiltration through the soil. Due to urban expansion, many sinkholes have been filled and large areas covered by impervious surfaces, resulting in increased amounts of storm water runoff.

In order to reduce the flooding of homes, buildings and streets, Bowling Green and Warren County adopted a storm water management plan in 1976 that included a) sinkhole floodplain easements, b) storm water retention basins and c) the use of storm water drainage wells to increase the rate at which runoff is transmitted to the subsurface (Matheney, 1984). These drainage wells (referred to as Class 5 Injection Wells by EPA) are drilled or excavated to permit surface runoff to be transmitted into solutionally-enlarged bedding plane partings, joints and caves of the karst aquifer. Over 600 wells now exist, and the number is rapidly increasing as further development continues.

Pollution of the karst aquifer by contaminated urban storm water runoff is a real concern in Bowling Green. However, attempting to keep urban storm water runoff out of the caves under Bowling Green is not practical, nor possible, even though it carries pollutants into the underground streams. Crawford and Groves (1984) point out that accepting the use of caves as storm sewers may be difficult, but in reality most cave streams receive direct input from the surface and thus function as storm sewers for the karst landscape whether it be agricultural, forested or urban. Even a city-wide storm sewer system could not prevent storm water runoff from entering underground streams since runoff sinks naturally at thousands of locations throughout the area (Crawford and Groves, 1984).

Concern about pollution of the aquifer by urban storm water ruoff was the impetus for this study. Groundwater quality was monitored for 14 potential contaminants over a 14 month period at four locations in and around Bowling Green. Drinking water standards were used as a means of comparision between the four sample sites because these standards represent the most stringent available criteria.

Research Design and Methods
Contaminant levels and related parameters were monitored for 14 months between May 11, 1983 and July 27, 1984. The sample design consisted of grab and composite sampling. Grab samples were taken routinely during base flow and intensively during the rise and fall of the hydrograph during and following selected storm events. Composite samples were gathered during selected storm events by ISCO automatic water samplers.

The sampling locations were at four strategic sites within the Lost River Groundwater Basin. Station #1 was located at the Lost River Rise, with station #2 located at the Lost River Blue Hole. Station #3 was located at the entrance to By-Pass Cave in a commercial area of Bowling Green, and station #4 was located at Federal Well (formerly a water well) located three kilometers upstream from the Lost River Blue Hole (see Figure 1). These four sample locations yielded data pertaining to water quality: (1) in a rural area south of the city (Federal Well), (2) at a location on the edge of the city before the Lost River flows under Bowling Green (Lost River Blue Hole), (3) in a commercial section of Bowling Green (By-Pass Cave), and (4) at the final resurgence of the Lost River (Lost River Rise). By using this sampling program it was possible to assess groundwater quality in the Bowling Green area before, during and after interaction with potential contaminants associated with urbanization.

Schneider Robot Monitors located at the Lost River Blue Hole and at the Lost River Rise continuously recorded conductivity during the sampling period. Grab and composite water samples were tested for the presence of inorganic chemicals, microbiological constituents, and other related parameters. Inorganic chemicals were: (1) cadmium, (2) chromium, (3) nickel, (4) silver, (5) copper, (6) iron, (7) lead, (8) nitrates, and (9) total dissolved solids. Microbiological constituents were: (1) fecal coliform, and (2) fecal streptococci. Other parameters tested for included: (1) phosphorus, and (2) suspended solids.

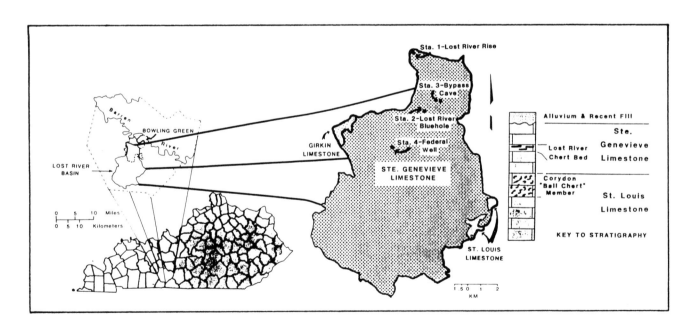

Figure 1 - Study area location, geology, and sampling site locations.

Discussion of Results

Groundwater quality is subject to degradation by chemical and biological pollutants with some typical sources of groundwater pollution being septic tanks, landfills, sewage-treatment facilities, agricultural chemicals, chemical spills, and contaminated urban storm water runoff. In order to properly regulate the quality of water in the United States, water quality standards were developed by the United States Environmental Protection Agency (EPA). The National Interium Primary Drinking Water Standards, which established the maximum contaminant level (MCL) for several inorganic and organic substances in public water supplies, as well as bacteriological standards. The authority to issue these standards was granted by the Safe Drinking Water Act and are based on known human toxicities of these substances at a consumption level of 2 liters/day (Fetter, 1980). Furthermore, each state was directed by Congress to establish quality standards for surface water bodies. In Kentucky this took the form of the Kentucky Public and Semipublic Water Supply Regulations (401 KAR 6-015) which were put into effect on July 6, 1983. The standards set forth in these regulations are similar to federal standards set forth in Public Law-93-523, the Safe Drinking Water Act.

Cadmium

Sources of cadmium include mine tailings and metal smelters. There is no human physiological need for cadmium, and it is toxic to almost all body systems. Cadmium can induce hypertension in humans (Thind, 1972), and chronic exposure can result in pulmonary edema, osteomalacia, and kidney disease (Kendrey and Roe, 1969). Cadmium is used in the metal plating industry, storage battery manufacture, and numerous other industrial processes. Industrial water releases of cadmium have resulted in contamination in both surface and groundwaters (McCaull, 1971). The drinking water standard for cadmium is 0.01 mg/l. Enriched cadmium soil (Thornton and Webb, 1980) occurs naturally in marine black shales of Carboniferous age which outcrop in portions of northwest England. Peak concentration of cadmium as high as 150 mg/l were measured in these rocks.

Depicted in Figure 2 are the mean, maximum, and minimum levels of cadmium detected at each of the four sample sites. The mean concentration of cadmium exceeded the maximum contaminant level of 0.01 mg/l in all instances and in the case of Federal Well (0.02 mg/l) the mean cadmium concentration was double the enforcement standard. The maximum concentration of cadmium also occurred at Federal Well (0.11 mg/l). Maximum values were lower at By-Pass Cave (0.085 mg/l), the Lost River Blue Hole (0.077 mg/l), and the Lost River Rise (0.070 mg/l). The mean cadmium concentrations at By-Pass Cave, the Lost River Blue Hole and Rise were 0.012, 0.02, and 0.014 mg/l respectively.

Examination of the data indicates that the possible source of cadmium was located in the rural area south of Bowling Green, hence the highest readings occurred at Federal Well. The potential source may be industrial wastewater improperly disposed from one of the small manufacturing plants located in the area. As the contaminated water moved toward and then under Bowling Green via the Lost River, it was sufficiently diluted by less contaminated water to the point that the maximum cadmium level was lowest at the Lost River Rise. The lowest mean concentration for cadmium was recorded at By-Pass Cave which indicated that the source of cadmium pollution was not urban storm water runoff. This was further substantiated by the fact that the Lost River Rise had the second lowest mean cadmium concentration.

Chromium

Although chromium is relatively abundant in the earth's crust, it is not present in very high concentrations in natural waters (Fetter, 1980). Chromium can exist in a number of valence states with trivalent (+3) the most common in nature and hexavalent (+6) found primarily in industrial applications. No harmful effects were found in laboratory animals ingesting trivalent chromium (U.S. EPA, 1976), but hexavalent chromium was found to be a systemic poison. Therefore, hexavalent chromium is a primary concern in drinking water with the standard for hexavalent chromium set at 0.05 mg/l. Chromium has been noted to oxidize from its trivalent state to the more toxic hexavalent state. Robertson (1975) noted this change in valence state in naturally occurring waters as sediments were oxidized as the result of alkaline groundwater and a positive Eh.

Samples collected at the four sample sites were analyzed for total chromium content only. Therefore, it was not possible to equate the collected values with water quality standards only concerned with hexavalent chromium.

The mean chromium content was greatest at Federal Well (0.10 mg/l) indicating a source from the rural area south of the city (Figure 2). Federal Well also had the highest maximum concentration of chromium at 0.41 mg/l. By-Pass Cave and the Lost River Blue Hole had similar mean chromium concentrations (approximately 0.75 mg/l) with the Blue Hole having a maximum concentration of 0.17 mg/l and By-Pass Cave having a maximum concentration of 0.12 mg/l. This indicates that the commercial area around By-Pass Cave was not a significant source of chromium contamination and that the groundwater flowing from the Federal Well area to the Lost River Blue Hole was significantly diluted by water less contaminated with chromium. The water which resurged at the Lost River Rise had a higher mean concentration of chromium than the Blue Hole (Rise 0.09 mg/l and Blue Hole 0.08 mg/l) and a higher maximum chromium concentration (Rise 0.29 mg/l and Blue Hole 0.15 mg/l). This indicated that the chromium contamination in the Lost River was increasing between the Blue Hole and the Rise. It was determined that By-Pass Cave was not the source of this contamination, although it may contribute some chromium contamination. In order to significantly increase the chromium content of the Lost River between the Blue Hole and the Rise, multiple urban sources of chromium are necessary.

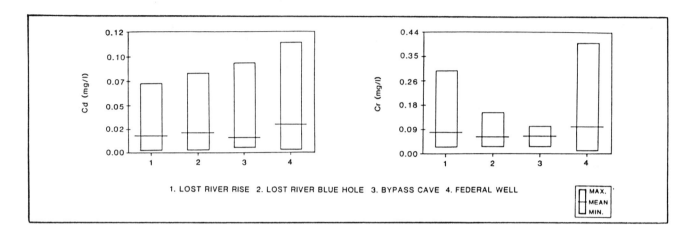

Figure 2 - Minimum, mean, and maximum cadmium and chromium concentrations at the four sampling sites.

Nickel

Nickel is a whitish metal that is relatively non-toxic to humans in some forms, but gaseous nickel carbonyl is highly toxic and a suspected carcinogen (Waldbott, 1973). Currently no U.S. National Interium Primary Drinking Water Standard exists for nickel. But from research in the aquatic environment using crustaceans and minnows, a water quality criterion of 0.01 mg/l has been proposed for both fresh and marine life (Fetter, 1980).

Data pertaining to nickel contamination indicated that Federal Well had the lowest mean and maximum concentrations of nickel at 0.07 and 0.29 mg/l respectively (Figure 3). The mean and maximum concentrations for nickel at By-Pass Cave, the Lost River Blue Hole, and the Lost River Rise were all similar (approximately 0.12 mg/l) with a maximum of about 0.50 mg/l. The disparity between the concentrations of nickel at Federal Well and the other three sample sites indicate that these concentrations are probably due to minor inputs of nickel within the city limits.

Silver

Silver occurs in both elemental form and as various salts. If ingested it accumulates in human skin, eyes and mucous membranes. Because of the tendency for silver to accumulate in the body, the drinking water standard was set at 0.05 mg/l.

Samples collected at the four sample stations indicated that there was substantially more silver at By-Pass Cave than at the other sample sites (Figure 3). The mean concentration of silver at By-pass Cave was 0.19 mg/l with a maximum concentration of 0.86 mg/l. At Federal Well, the Lost River Blue Hole and the Lost River Rise, the mean concentration of silver was approximately 0.05 mg/l (the enforcement standard), with the maximum of 0.28 mg/l occurring at Federal Well. The elevated values for By-Pass Cave indicate a source of silver from the commercial area around the cave. The silver concentrations from By-Pass Cave are diluted by water less contaminated with silver when the By-Pass Cave stream enters the Lost River. At the Lost River Rise the mean silver concentration was 0.05 mg/l which indicates dilution of the contaminated By-Pass Cave water to a point equal to the 0.05 mg/l drinking water standard.

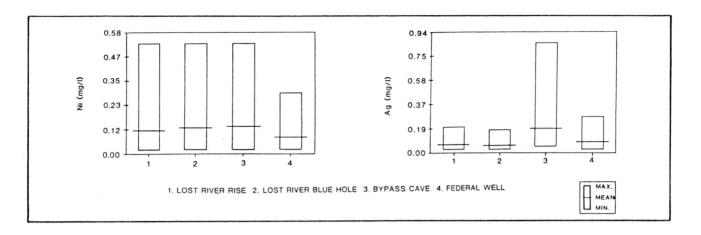

Figure 3 - Minimum, mean, and maximum nickel and silver concentrations at the four sampling sites.

Copper

Copper occurs in nature as a native metal and various copper minerals and salts. It is required in trace amounts in both vertebrate and invertebrate animals, but excessive amounts of copper can be toxic. Copper is found in relatively low concentrations in natural waters (U.S EPA, 1976), and copper sulfate is sometimes used as an aquatic herbicide. It was noted by Purves (1966) that soils in two Scottish towns contained four times as much copper as rural arable soils. This contamination was ascribed to gardening activities such as the use of ash or municipal composts as fertilizer and general fallout from open burning, car exhausts and industry. The drinking water standard for copper is 1.0 mg/l which is based on esthetics (higher concentrations give water a metallic taste), rather than toxicity.

All of the values obtained at the four sample sites fell well below the enforcement standard (Figure 4). By-Pass Cave had the highest mean concentration of copper at 0.19 mg/l and the highest maximum concentration at 0.62 mg/l. The similarity between mean copper concentrations obtained at Federal Well (0.11 mg/l), the Lost River Blue Hole (0.10 mg/l), and the Lost River Rise (0.11 mg/l) probably indicates the background concentration of copper in Bowling Green's groundwater. The maximum copper concentrations were also very similar for these three sample locations (approximately 0.40 mg/l).

The copper concentration at By-Pass Cave owes its origin to the urban nature of the area. Car exhaust is a possible source of the higher copper levels. These elevated levels were diluted upon joining the Lost River, hence the concentrations at the Lost River Rise are similar to the assumed background concentrations.

Iron

Iron is a common element found in rocks and soils and is an essential element for both plant and animal growth. However, iron has been noted to be toxic to some aquatic species at concentrations of 0.32 to 1.0 mg/l (Warnick and Bell, 1969). Iron in drinking water imparts a taste at concentrations of 1.8 mg/l (Cohen and Others, 1960) and soluble iron concentrations in excess of 0.3 mg/l can impart a brownish-color to laundry and porcelain (Hammer, 1986). A drinking water standard of 0.3 mg/l has been imposed to avoid the staining of plumbing fixtures, and for the protection of aquatic life, the maximum iron concentration is set at 1.0 mg/l for surface water.

The highest mean concentration of iron occurred at the Lost River Rise (2.2 mg/l) (Figure 4). The mean iron concentrations at By-Pass Cave and the Lost River Blue Hole were measured at 2.0 mg/l. The highest maximum concentrations occurred at the Lost River Blue Hole and Rise at 9.95 mg/l. By-Pass Cave had a maximum iron concentration of 4.4 mg/l. The lowest mean concentration of iron was 1.05 mg/l which was recorded at Federal Well, where the maximum concentration was 2.7 mg/l.

The source of elevated iron concentrations, all of which exceed the enforcement standard of 0.3 mg/l, is probably from leaching of iron from the soil zone by downward moving meteoric water.

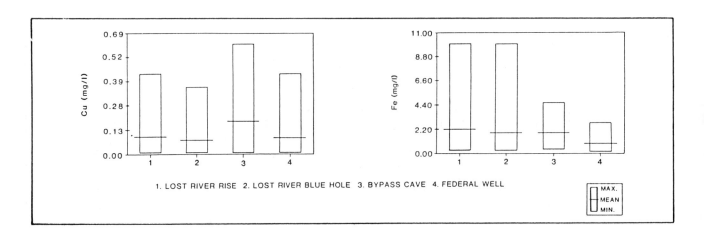

Figure 4 - Minimum, mean, and maximum copper and iron concentrations at the four sampling sites.

Lead

Lead is a toxic metal which fulfills no known physiological requirement and which has long been associated with environmental and occupational disease (Shukla and Le Land, 1973). Lead accumulates in the bones and soft tissue and inhibits the formation of hemoglobin leading to sypmtoms of anemia. Lead poisoning is known to cause mental retardation, cerebral palsy and optic nerve damage in children (Perlstein and Attala, 1966). Because of the toxicity of lead, the drinking water standard is 0.05 mg/l. Some of the major sources of environmental lead pollution are lead-based paints and car exhaust with the main sources of lead in drinking water being lead solder and pipes.

Collected data revealed one sample site to be much higher in lead concentrations than the other three (Figure 5). At By-Pass Cave the mean concentration of lead was 0.22 mg/l, with a maximum concentration of 0.98 mg/l. At Federal Well and the Lost River Blue Hole and Rise mean concentrations of lead were below the 0.5 mg/l drinking water standard and substantially less than By-Pass Cave.

The elevated lead levels at By-Pass Cave were a result of its commercial location. Rain water efficiently washes lead and other car exhaust contaminants from area streets and parking lots into By-Pass Cave. Federal Well and the Lost River Blue Hole had the lowest lead concentrations because they are upstream of the urbanized areas of Bowling Green and are thus isolated from the source of lead contamination. The Lost River Rise had higher maximum lead concentrations when compared to Federal Well and the Blue Hole because it is the eventual recipient of all lead contaminated water from the urban area of Bowling Green. However, lead concentrations at the Rise are lowered because of the diluting effect of the less contaminated water from south of the city.

Nitrates

Nitrates are inorganic salts of nitrogen formed as nitrogen passes through a number of valence states during biological reactions. Nitrate is reduced to nitrite in the gastrointestinal tract allowing nitrite to enter the bloodstream where it reacts with hemoglobin impairing the blood's ability to transport oxygen. This condition is known as methemoglobinemia and is known to be fatal in fetuses and infants under the age of three months (Fetter, 1980). Some common sources of nitrates in drinking water are nitrogen fixing bluegreen algae, lightning strikes which convert elemental nitrogen to inorganic forms, fertilizers, municipal and industrial wastewater, refuse dumps, animal feed lots, septic tanks, urban drainage and decaying plant debris.

The data indicate that nitrate pollution is not a substantial problem in the Bowling Green area (Figure 5). The highest nitrate values were recorded at Federal Well where the mean concentration of nitrates was 2.9 mg/l, with a maximum concentration of 8.9 mg/l, both of which are below the enforcement standard of 10 mg/l. Federal Well had the highest concentration of nitrates because it is located in a rural area where the heavy use of nitrogen fertilizer is common. At the other end of the spectrum is By-Pass Cave where agriculture is minimal. At By-Pass Cave the mean concentration of nitrates was 0.70 mg/l with a maximum of 4.0 mg/l. The Lost River Blue Hole and Rise had lower nitrate concentrations when compared to Federal Well (2.3 and 2.15 mg/l respectively). This is the result of the water from south of the city, which is higher in nitrate concentration, being diluted by water from the urban areas which is lower in nitrate concentration.

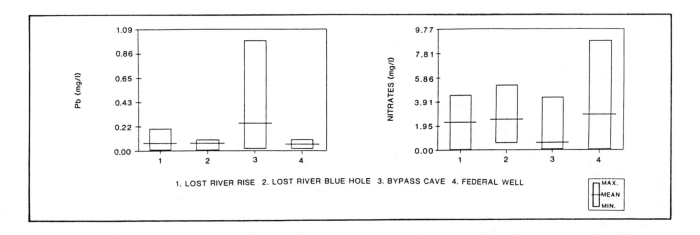

Figure 5 - Minimum, mean, and maximum lead and nitrate concentrations at the four sampling sites.

Total Dissolved Solids

The result of chemical and biochemical interactions between groundwater and the geological material through which it flows causes groundwater to contain a variety of dissolved inorganic constituents or total dissolved solids (TDS). The amount of total dissolved solids in water determines whether it is fresh or saline. Water with less than 1000 TDS is considered fresh. The enforcement standard for drinking water is no more than 500 mg/l.

The data indicate that Federal Well had the greatest amount of dissolved solids with a mean concentration of 410 mg/l and a maximum of 515 mg/l which is slightly above the enforcement standard (Figure 6). Federal Well exhibited the highest TDS levels because the well is recharged by diffuse rather than conduit flow, hence groundwater was in contact with the rock mass longer and was able to dissolve more chemical constituents of the rock mass. By-Pass Cave had the lowest mean TDS concentration of 72 mg/l because the majority of water sampled at this station was storm water runoff from area parking lots and streets. The Lost River Blue Hole and Rise had very similar mean TDS concentrations of approximately 226 mg/l. The Blue Hole had a maximum concentration of 500 mg/l and the Rise 390 mg/l. The Blue Hole had higher concentrations of TDS because it is recharged by groundwater from south of Bowling Green which had longer access to the rock mass. As the Lost River flows under Bowling Green its flow is diluted by tributaries similar to By-Pass Cave which have limited contact with the rock mass.

Conductivity

Groundwater can be viewed as an electrolyte solution because nearly all of its dissolved constituents are present in ionic form (Freeze and Cherry, 1979). The electrical conductivity therefore reflects the amount of dissolved material in the water. The electrolyte capacity of natural water is a function of the molar concentration of each ion and its electrical charge (Fetter, 1980). Freeze and Cherry (1979) however, point out that ions present in solution are a mix of charged and uncharged species, therefore conductance measurements cannot be used to obtain accurate measurements of ion concentration or total dissolved solids.

The examination of the collected data indicates that conductivity and TDS data gathered at the four sample sites were very similar, with Federal Well having the highest mean conductance (610 mhos/cm) and the highest maximum conductance (915 mhos/cm) (Figure 6). By-Pass Cave had the lowest mean and maximum conductance at 160 and 330 mhos/cm respectively. The Lost River Blue Hole and Rise had very similar mean and maximum conductance at approximately 375 (mean) and 540 mhos/cm (maximum). Comparsion of conductance with TDS for the Blue Hole and Rise indicate a strong similarity.

Microbiological Contamination

Coliform bacteria residing in the intestinal tract, occur in large numbers in the feces of humans and other warm blooded animals, averaging about 50 million coliform per gram (Hammer, 1986). Coliform bacteria are generally typified by Escherichia coli and fecal streptococci (enterococci). These are harmless bacteria which leave the bodies of humans and warm blooded animals in excrement. They usually cannot live for more than a few days outside the body. If even one organism is detected in a 100 ml water sample it means that the water sampled has recently been in contact with either human or animal feces. Therefore, they are useful indicators of potential pathogens that could also live in human or animal feces. After heavy rains surface runoff water flowing past cattle and other animal excrement in the fields flows into sinkhole drains and the swallets of sinking streams directly into cave streams. Also, percolating soil water is believed to flush septic tank effluent from absorption fields down into the karst aquifer. Many septic tanks in the Bowling Green area are discharging effluent directly into the aquifer with little, if any, treatment by the soil. The drinking water standard for coliform bacteria generally specifies that water shall contain no more than an average of one coliform organism per 100 ml. Hammer (1986) notes that for use in a public water supply (before treatment) 2000 fecal and 10,000 total coliform bacteria are acceptable. For water contact recreation, a mean of 1000 total coliform with 200 being fecal with no more than 10% of the samples exceeding 2000 total coliforms, 400 of which may be fecal, is acceptable. High numbers of coliform bacteria in drinking water may indicate pathogens which can cause gastrointestinal disorders and more serious diseases such as virus hepatitis A. Relatively recent outbreaks of hepatitis A have been implicated with improper disposal of human waste and contamination of water wells and springs in karst areas in three Kentucky counties. One outbreak, involving over 150 cases, occurred in Knott and neighboring Floyd counties. A similar outbreak in Meade County resulted in 110 cases with one fatality. Dye trace studies implicated a cave stream which resurges at Buttermilk Falls Spring, a local source of drinking water. Septic tank effluent is believed to be the source of the outbreak (Beaulac, 1984). High concentrations of coliform bacteria in recreation waters may indicate pathogens which can cause skin, eye, ear, nose and throat infections if these waters are used for swimming.

The review of the collected data for fecal coliform and fecal strep contamination yielded some surprising results (Figure 7). Mean fecal coliform and strep concentrations were highest at By-Pass Cave. This is surprising because By-Pass Cave is hydrologically isolated from heavy agricultural areas where coliform counts are expected to be high because of animal wastes. The primary reason that By-Pass Cave was higher than the other three monitoring sites was because virtually the entire discharge was storm water runoff while at Federal Well, Lost River Blue Hole and Lost River Rise, surface runoff constitutes only a portion of the total discharge. Therefore, it appears that feces of dogs, cats and birds are the source of the high levels at By-Pass Cave.

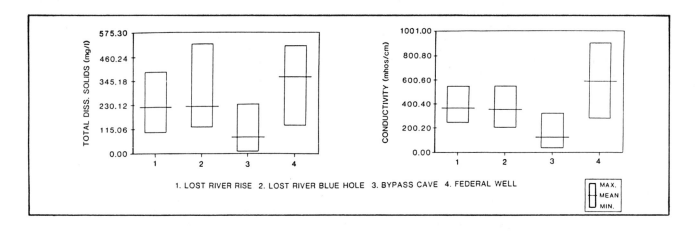

Figure 6 - Minimum, mean, and maximum total dissolved solids and conductivity at the four sampling sites.

Phosphorus

Phosphorus in organic, and inorganic phosphate form is the key nutrient in controlling the eutrophication of surface waters (Fetter, 1980). The natural distribution of phosphorus in groundwater is limited to areas of phosphate rock, but the greatest sources of phosphorus in natural waters are organic pollution and detergents containing phosphates. Excess phosphorus concentrations in raw water are associated with eutrophication problems, and because of the toxic nature of this process standards are established for water supplies. Fetter (1980) considers phosphorus levels above 0.01 mg/l potentially eutrophic.

The data revealed that phosphorus contamination at Federal Well, By-Pass Cave and the Lost River Blue Hole is insignificant (Figure 8). This fact can be attributed to a low population density upstream of Federal Well and the Blue Hole, hence less phosphate detergents enter the aquifer. At By-Pass Cave the water collected at the sample site was derived from streets and parking lot runoff which had minimal contact with detergents. The Lost River Rise had the highest mean concentration of phosphorus because of its location downstream from Bowling Green. The mean phosphorus level recorded at the Rise was 0.55 mg/l with the maximum being 10.35 mg/l. The disparity between the mean and maximum was due to a single isolated incident on April 20, 1984.

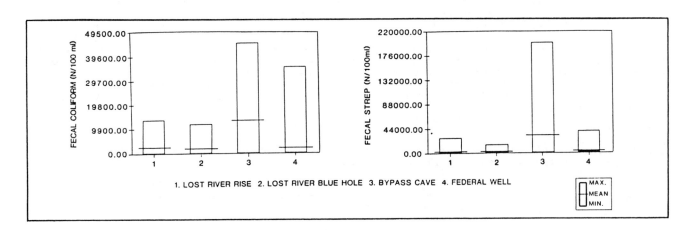

Figure 7 - Minimum, mean, and maximum fecal coliform and fecal strep counts at the four sampling sites.

Suspended Solids

Suspended solids are materials which are derived from soils and parent material which does not go into solution. The highest mean concentration of suspended solids (105.0 mg/l) occurred at By-Pass Cave, which reflects the fact that the sample site is only fed by storm water runoff (Figure 8). Federal Well had the lowest mean concentration of suspended solids because it is recharged by less turbulent waters, having less suspended solids. The Lost River Blue Hole and Rise had mean suspended solid concentrations of approximately 90 mg/l. The highest maximum suspended load concentrations (approximately 555 mg/l) occurred at the Blue Hole and Rise and represent soil erosion from the farmland south of Bowling Green during intense precipitation events.

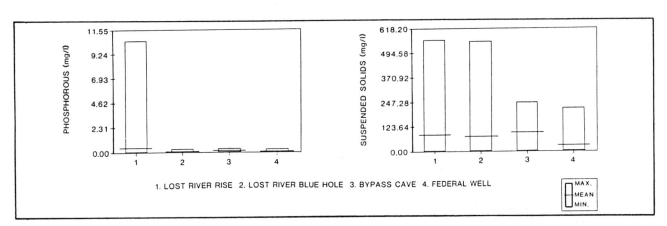

Figure 8 - Minimum, mean, and maximum phosphorus and suspended solid concentrations at the four sampling sites.

Conclusions

As previously noted, drinking water standards were used to compare the concentrations of chemical constituents at the four sampling sites in this study. By using drinking water standards, the worse case scenario was developed pertaining to groundwater contamination in and around Bowling Green. In actuality Bowling Green uses water from the Barren River (a surface stream) as a potable water supply, and has a municipal water distribution system. However, in the rural area south of the city (in the vicinity of Federal Well) groundwater is used almost exclusively for potable water supplies.

This investigation revealed that groundwater contamination is a more serious problem in the rural area south of Bowling Green near Federal Well and in the urban area near By-Pass Cave. Federal Well had the highest recorded concentrations of cadmium, chromium, nitrates and total dissolved solids and the second highest concentrations of microbiological contaminants. By-Pass Cave had the highest recorded concentrations of silver, copper, lead and microbiological contaminants. All these contaminant levels were substantially lower at the Lost River Rise due to dilution with less contaminated water, thus reducing overall concentrations. Therefore, the Bowling Green area has nonpoint source groundwater contamination problems at some specific locations. However, when the overall picture is viewed, the contamination of the Lost River under Bowling Green is not that substantial, primarily because of the dilution of contaminants by water from the entire groundwater basin.

References

Cohen, J.M., 1960, "Taste Threshhold Concentrations of Metals in Drinking Water," Journal of American Water Works Association. Volume 52, pp. 660-670.

Crawford, N.C. and Groves, C.G., 1984, "Storm Water Drainage Wells in the Karst Areas of Kentucky and Tennessee," U.S. Environmental Protection Agency Underground Water Source Protection Program Grant Report. 52p.

Fetter, C.W., 1980, Applied Hydrogeology. Columbus, Ohio: Merrill Publishing, 488 p.

Freeze, R.A. and Cherry, J.A., 1979, Groundwater. Englewood Cliffs: New Jersey, Prentice-Hall, 530p.

Hammer, M.J., 1986, Water and Wastewater Technology. New York: John Wiley and Sons, 550p.

Kendrey, G. and Roe, F.J.C., 1969, "Cadmium Toxicology," Lancet. Volume 1, pp. 1206-1207.

Matheney, J.B., 1984, Bowling Green and Warren County, Kentucky Storm Water Management: Program Review. Bowling Green, Kentucky: City/County Planning Commission. 104p.

McCaull, J., 1971, "Building a Shorter Life," Environment. Volume 13, pp. 31-34, 38-41.

Perlstien, N.A. and Attala, R., 1966, "Neurologic Sequel of Plumbism in Children," Clinical Pediatrics. Volume 5, pp. 292-298.

Purves, D., 1966, Nature. Volume 210, pp. 1077-1078.

Robertson, F.N., 1975, Hexavalent Chromium in Groundwater in Paradise Valley, Arizona," Groundwater. Volume 13, pp. 516-527.

Shukla, S.S. and Le Land, H.V., 1973, "Heavy Metals: A Review of Lead," Journal of the Water Pollution Federation. Volume 45, pp. 1319-1331.

Thind, G.S., 1972, "Role of Cadmium in Human and Experimental Hypertension," Journal of the Air Pollution Control Association. Volume 22, pp. 267-270.

Thorton, I. and Webb, J.S., 1980, "Regional Distribution of Trace Element Problems in Britain," In Applied Soil Trace Elements. Brian Davies (Editor), New York: John Wiley and Sons, pp. 381-439.

United States Environmental Protection Agency, 1976, Quality Criteria for Water. Washington D.C.: U.S. Government Printing Office, 256p.

Waldbott, G.L., 1973, Health Effects of Environmental Pollution. St. Louis: C.V. Mosby Company, 316 pp.

Warnick, S.L. and Bell, H.L., 1969, "The Acute Toxicity of Some Heavy Metals to Different Species of Aquatic Insects," Journal of the Water Pollution Control Federation. Volume 41, pp. 280-284.

Sinkhole dumps and the risk to ground water in Virginia's karst areas

D.W.SLIFER   *New Mexico Environmental Improvement Division, Sante Fe, N.Mex., USA*
R.A.ERCHUL   *Department of Civil Engineering, Virginia Military Institute, Lexington, Va., USA*

ABSTRACT

Virginia's Valley and Ridge Province consists of 24 counties on the western edge of the Commonwealth. Due to the karst/limestone geology that characterizes the region these counties share a common vulnerability to their ground water resources. Karst aquifers are among the most sensitive to disturbance and are readily contaminated. One of the most visible and obvious sources of contamination is the dumping of trash and waste into sinkholes. Botetourt and Rockbridge Counties, the areas which were the focus of the research, have thousands of sinkholes within their boundaries. The possibility of contamination to local ground water is great due to the presence of household hazardous waste in dumps. A total of 260 illegal dump sites were documented in the study area, with 75% of these existing in karst areas. Approximately 23% of the illegal dumps were in sinkholes. More than 90% of the population of the study area takes their water from wells and springs. The research has concentrated on an effective methodology of locating and assessing potentially dangerous sinkhole dump sites. The project has documented these techniques allowing other counties in karst terrain to obtain similar results. A slide show and video are also being produced to show local schools and citizen organizations the hazards associated with sinkhole dumping. The results of the research underscores the need for local governments to play an active role in the areas of waste disposal, land use, and protection of ground water quality in their critical karst areas.

Introduction

Virginia's Valley and Ridge Province consists of 24 counties on the western edge of the Commonwealth. Due to the karst/limestone geology that characterizes the region, these counties share a common vulnerability to their ground water resources. Karst aquifers are among the most sensitive to disturbance and are readily contaminated. One of the most visible and obvious sources of contamination is the dumping of trash and waste into sinkholes and caves. Such locations are the worst possible place to dispose of anything since they are direct conduits for the infiltration of surface water and contaminants into the underlying aquifer. While the harmful aspects of this practice may be obvious to the informed reader, they unfortunately are not obvious to the many rural residents who have traditionally used sinkholes for convenient dump sites. With the exception of dumping dead animals, the use of sinkhole dumps was probably relatively benign in the past. The advent of our modern "throw-away" society has made this practice much more dangerous however, due to the greatly increased volumes of waste, manufactured goods, and the thousands of chemicals that are now used in the home and on the farm. Despite the existence of sanitary landfills in each county there are still many illegal open dumps in use throughout rural Virginia, and in the Valley and Ridge Province a significant number of these dumps are located in sinkholes. Although existing state law (Sec. 10-150.14) "prohibits dumping refuse, garbage, dead animals, etc. in caves or sinkholes," it is fair to say that this has had little impact. Although the problem of illegal dumping in sinkholes has been recognized by state regulatory agencies, enforcement is lax and the practice continues throughout rural areas of Virginia. Local governments vary in their approach to the problems of illegal dumping, waste collection, and ground water protection. Many local governments and the general public remain ignorant of the health hazards inherent in karst areas. With the exception of Clarke County, local governments have not effectively addressed this aspect of land use. The key elements in trying to correct the problem are education and enforcement. If the public, local governments, and state agencies are made aware of the magnitude and nature of the problem it is likely that behavior patterns will change for some of the offending population and that more effective enforcement will be devised to help create and maintain an improved situation.

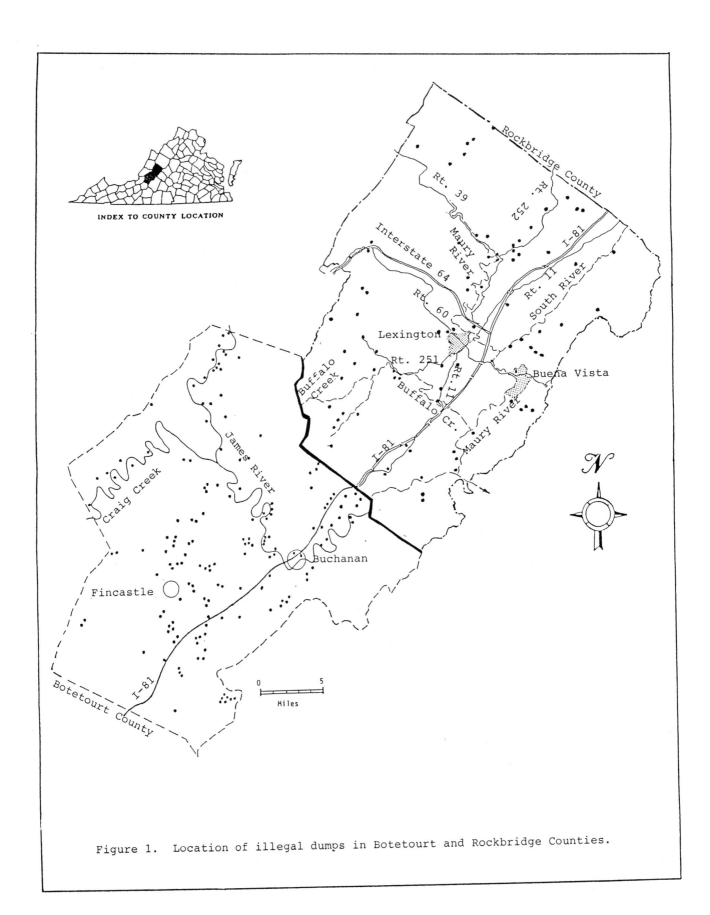

Figure 1. Location of illegal dumps in Botetourt and Rockbridge Counties.

The publication of "Rockbridge's Illegal Dumps" by the Virginia Water Resources Research Center in 1987 created widespread interest in the subject of illegal dumps. Numerous articles in local papers and segments on TV and radio followed. This attendant publicity was good but often failed to stress adequately the findings related to karst ground water and sinkhole dumping. This is the common tie that should relate Rockbridge County's dumps to the other 23 counties in the Valley and Ridge Province. It is important to expand this existing data base of one county to include at least one other. An accurate inventory of sinkhole dumps for another county further documents the scope of the problem by proving it is not unique to Rockbridge. Since many dump sites are not visible from roads, aerial survey data produce the most accurate inventory in the least amount of time. This technique is effective in mapping dump sites and represents a powerful tool for localities, planners, state regulatory agencies, and anyone else interested in working on the problems associated with illegal dumping.

Results

Botetourt County was selected as the study area for this project. Botetourt is adjacent to Rockbridge and is similar in size, geology, and population. Both counties are primarily rural in nature and each operates a single sanitary landfill. While Rockbridge maintains a solid waste collection system of dumpster boxes and trucks, Botetourt has no collection system other than private commercial haulers. For the survey of illegal dumps in Rockbridge in 1986, information was collected by mailing a letter and county map to approximately two dozen persons who are knowledgeable about the county. Reported sites were then field checked and evaluated. A similar approach was used in this study also and targeted to Botetourt residents, who reported 53 dump sites. In both cases the rate of response exceeded 50 percent. This technique is inexpensive, simple, and produces reasonably good data for those sites that are the most visible and that are usually located along road sides. However, it has been found that this questionnaire technique does not approximate the actual number of dump sites in a county because of the large number of sites located on private property and not visible from roads. Consequently, in the Botetourt study it was decided to supplement the questionnaire data by conducting aerial investigations of the county.

Because county-wide aerial photography of an appropriate scale was not available (and if it were, image interpretation can be subjective), it was decided to fly over Botetourt County in a small plane at altitudes low enough to identify most illegal dumps. The same technique has been used successfully by researchers with the Tennessee Valley Authority to inventory dump sites in that region. To survey Botetourt County two flights were made at altitudes ranging from 183 to 305 meters (600 to 1,000 feet), for a total of approximately 10 hours. A set of topographic quadrangle maps was used by a spotter to help guide the pilot to fly a series of parallel flight lines across the county, and to record the location of dump sites as they were observed. It was found that the most efficient way to record site locations was to note the latitude and longitude coordinates from the plane's digital LORAN-C display as the plane was directly above the site. These coordinates, when plotted on the topographic maps afterward, were usually accurate to within a thousand feet. A total of 106 dump sites were recorded during these two flights, and many of them were photographed. During the flights over northern Botetourt, a small area of southern Rockbridge was also surveyed for dump sites. Seven previously unknown sites were observed in a period of 15 minutes -- all of them had been undetected in the Rockbridge study because they are remote and are not visible from roads.

When the aerial data from Botetourt was combined with the list of sites reported by respondents to the questionnaire, some duplications were noted and the total data base was then verified by field checking each site that could be found. During the summer months of 1988 field personnel investigated a total of 140 dump sites. While most reported sites were verified in the field, some could not be located -- probably due to vegetation obscuring them. It is preferable to conduct this type of field work when vegetation is dormant; for aerial detection it is essential to fly when the leaves are off. In order to complete the inventory and compensate for any seasonal effect, a second period of field work was conducted in January and February 1989, bringing the total number of confirmed sites in Botetourt to 168. Figure 1 shows the location of these dump sites in Botetourt. An update of the 1986 survey of dumps in Rockbridge County has lead to documentation of 92 sites, which are shown in Figure 1.

During the field check each site was evaluated and many were photographed. A field check form was developed to insure a standardized evaluation. By assigning numerical values to factors such as dump size, site geology, topographic setting, public access, frequency of use, waste types, and proximity to water a ranked score was developed for each site. The range of ranked scores for the dumps evaluated in Botetourt County is from a low of 7 to a high of 64. Scores greater than 40 are considered to be a serious threat to ground water, while scores less than 16 are considered to be a low impact to the environment. While such rankings are approximate and are based on a certain amount of

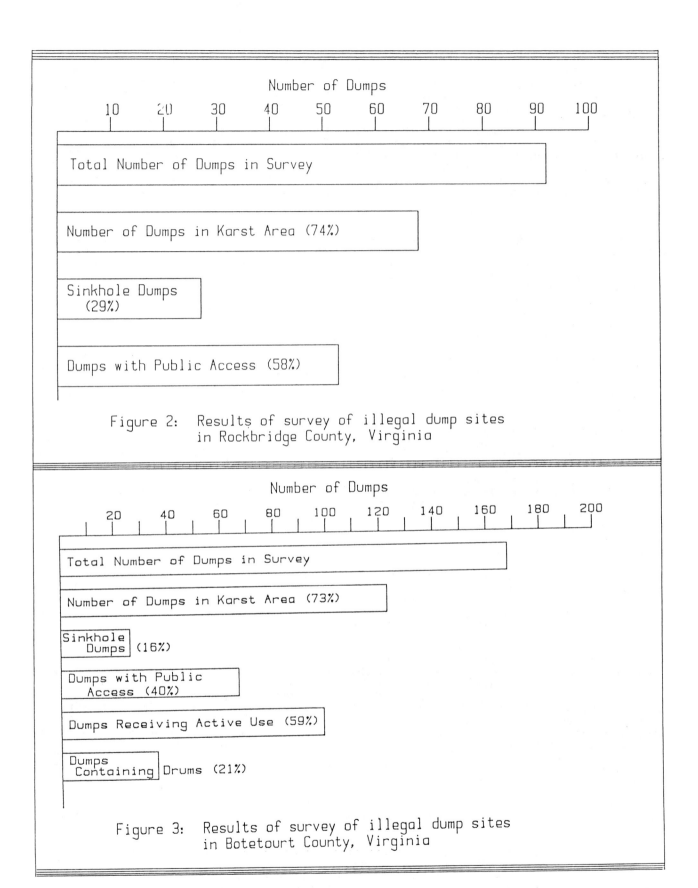

Figure 2: Results of survey of illegal dump sites
in Rockbridge County, Virginia

Figure 3: Results of survey of illegal dump sites
in Botetourt County, Virginia

judgement, they still represent the best available data and are reasonably good assessments of illegal dumping in the study area. The ranked scores have been assigned to four categories which describe the relative threat to ground water of each site. The categories were 1) Low Impact to Environment, 2) Moderate to Serious Threat, 3) Moderate Threat to Ground Water, 4) Serious Threat to Ground Water. For both counties those dumps in the two most serious categories comprise approximately 29% of the total. These dumps are believed to be actual or imminent hazards in terms of ground water contamination. Rockbridge has a relatively greater proportion of low impact sites than does Botetourt. Dumps in this category are primarily only aesthetic problems and represent little or no risk of environmental harm. They are often small, infrequently used, and contain inert debris, brush, rubble, etc. It is important to understand, however, that the ranking of any dump, no matter how apparently benign, can change at any time to a more severe category with a single act of dumping trash that contains hazardous material (eg. partially used containers of pesticide, herbicide, or other harmful chemical product). The ranking of many dumps, if monitored carefully, should fluctuate with patterns of usage and levels of household hazardous waste present as well as response to events such as significant and sustained enforcement policy by local government, presence or absence of a local landfill, and availability of a county-wide waste collection system. The absence of a county operated collection system in Botetourt may be a factor in explaining that only 10% of Botetourt's dumps are in the low impact category, as compared to 24% low impact sites seen in Rockbridge, which does have a collection system. Another possible factor may be that Botetourt's landfill is not centrally located in the county, as is Rockbridge's landfill.

Other aspects of illegal dumps are depicted in bar graphs on Figures 2 and 3. Twice as many sites are known in Botetourt, primarily because that county was inventoried by aerial survey. There is little doubt that a similar aerial survey for Rockbridge would probably double the number of dumps known there. It is interesting that the number of dumps located in karst areas is nearly identical at 73% for Botetourt and 74% for Rockbridge. Rockbridge has a greater share of sinkhole dumps (29%) than does Botetourt (16%). This is probably due to a greater density of sinkholes in the Rockbridge area. For the Botetourt study a count of sinkholes shown on topographic maps and aerial photographs indicated a total of approximately 580 sinkholes in that county. By contrast, 966 sinkholes have been counted in a single Rockbridge area topographic map (Lexington 7 1/2' Quadrangle) in a detailed survey by map, air photo, and field work. Because ground water is inherently at risk from dumps in sinkholes, the weighing of ranking factors tends to place most sinkhole dumps into the more serious categories. The 26 sinkhole dumps in the Botetourt study have ranked scores ranging from 26 to 42, and averaging 33 (in the moderate to serious threat category.

Figures 2 and 3 also indicate that 58% of the Rockbridge dumps have public access, compared to only 40% in Botetourt. This is a reflection of the survey methodologies; since the Rockbridge survey was not by airplane it tends to favor the more visible roadside sites which by definition have public access. In addition, the Botetourt survey addressed the "active use" factor. While difficult to quantify, a little experience inspecting these sites enables observers to judge how frequently a dump is used in a relative sense. The data suggests that 59% of the Botetourt sites receive "active use." By implication, the other 41% are either old abandoned dumps, or ones that are only occasionally used.

An important, apparent difference between Rockbridge and Botetourt dumps is the prevalence of 55-gallon drums observed in Botetourt sites. Drums are rare in Rockbridge sites but were counted in 21% of the Botetourt sites. Drums are of course often used to contain hazardous liquid chemicals or waste. In the dump sites where drums have been observed, many are now empty but some do still contain unknown substances. For instance, the field notes for dump site #DA-1 record: "Two 55 gallon drums with solvent retained heavy parafumic distillate in them, containing zinc dialkydithiophosphate, polyalkene succinicanhydride magnesium detergent, methylacylate copolymates." This list of chemicals was taken from labels on the drums, which may or may not contain the actual labeled contents in them. At another site, sinkhole dump #BU-1, a half dozen steel drums had rusted enough to now be empty, but a plastic drum still held residual photographic chemicals and was labeled as corrosive and poisonous. Since the field personnel evaluating the Botetourt sites were not trained to safely sample for hazardous materials, definitive data is not available to further characterize the contents of the drums observed. The interesting question is why do Botetourt dumps contain more chemical drums than Rockbridge dumps. Both counties have similar transportation networks, including Interstate 81, and both are predominantly rural, non-industrial areas. Perhaps Botetourt's relative proximity to the City of Roanoke and its numerous industrial and commercial facilities may explain the presence of drums dumped illegally in at least 36 sites.

With the intention of detecting possible contaminants, ground water samples were collected from nearby wells or springs at a number of dump sites in both counties. Results have been inconclusive, perhaps because the severe drought conditions encountered during both studies may have created adverse hydrologic conditions for detecting any contamination. It is difficult to detect ground water contamination in karst because the flow is confined to discrete channels or conduits which may not be connected to the particular well or spring being sampled. Timing of sampling is also critical, since ground water in karst flows at a high rate and one would have to sample at just the right time following a precipitation recharge event, thereby "flushing" a pulse of contaminated water through the sinkhole dump and into the underlying aquifer. In both studies the circumstances of severe drought and the limited resources available were factors that worked against successfully detecting actual contamination emanating from a dump site. Most likely, a long-term study of a few chosen sites would produce conclusive data.

Using the results of the study of these two counties, the total number of illegal dumps within the Valley and Ridge Province is projected to be approximately 6240. Of that total, about 4617 would be located in karst, and about 1372 would likely be sinkhole dumps. Based on an average of 10%, approximately 624 dumps would be in the most serious threat to ground water category. These are alarming statistics which portray an intolerable situation. The active participation of local governments is essential if the specter of illegal dumping, especially in karst areas, is to be eliminated. Without vigilant enforcement and a sustained program to increase public awareness, the problem could get even worse. Recent legislative changes in Virginia have created very stringent regulations governing landfills. Implementing these requirements may result in some counties having to close their landfills and participate in regional waste management programs. One possible effect of such changes could be an increase in illegal dumping if convenience decreases and costs of legal disposal increase. Such a scenario would make the data presented here valuable for documenting illegal dumps in the future following changes in waste management policy of state or local government.

To illustrate the need for a strong and unified approach to the illegal dump problem, consider the two counties studied in this research. These counties differ in their legal approach to the problem. Rockbridge lacks an ordinance that even recognizes or defines illegal dumps, and has therefore made essentially no effort at enforcement since there is no local law to enforce. Botetourt does have an adequate local ordinance banning illegal dumping, which empowers law officers as well as private citizens to prosecute violators based on direct observation or on names found on litter in illegal dumps. But until recently enforcement efforts have been haphazard at best. In Botetourt, as in most other counties, dumps are investigated by health department sanitarians, usually only in response to a complaint, rather than from a systematic program to identify and eliminate illegal dumps. Since enforcement of illegal dumping was an implied low priority, Botetourt decided in 1988 to assign this duty to two animal control officers (who were funded by a combination of state and local funds). Consequently, several articles in the Fincastle Herald in 1988 indicate that a more significant effort is being made to enforce illegal dumping regulations in Botetourt.

Conclusion
The data gathered in the study of sinkhole dumps in Botetourt County, the update of existing data for Rockbridge County dump sites, and the comparison of these two counties has produced a reasonable approximation of the magnitude of the illegal dump problem within the karst region of Virginia's Valley and Ridge Province. With the total number of illegal dumps in the Valley and Ridge Province projected to be near 6,240, and by using the information available in the Rockbridge and Botetourt studies of the percentage of dump sites existing in sinkholes, it is estimated that approximately 1,372 of these illegal dumps are likely to be in sinkholes. Because sinkholes in the 24 counties in Virginia's karst regions are assumed to be similar to those found in the study of Botetourt and Rockbridge Counties, one can assume that almost 400 sinkhole dumps exist and are causing a moderate to serious threat to ground water. Noting the polluted conditions of these sinkholes, and having the knowledge that contaminated water flows through sinkholes into underlying aquifers, one can see the very serious nature of this problem and the current threat to Virginia's ground water supply.

The importance of the role of local government in dealing with the problems associated with illegal dumping cannot be over-stressed. As shown previously, the study areas of Botetourt and Rockbridge Counties are very similar geographically, but they take differing approaches in response to illegal dumping. It is fair to say that a careful study of all 24 counties in the Valley and Ridge Province may find that there are 24 ways of attempting to cope with this difficult problem. It is increasingly recognized that the land use policies of localities represent a crucial, but presently fragmented, element in protecting ground water. It is hoped that the findings of this project will increase public and governmental awareness of the problems associated with illegal dumps in Virginia's karst areas. Only with a concerted effort by the public and by state and local governments is there hope of reversing the damage being done to Virginia's ground water supply.

Tennessee's superconducting super collider proposal: A proposed large scale surface/subsurface construction project located in a karst environment

PATRICIA J.THOMPSON   *Tennessee Department of Conservation, Division of Geology, Nashville, Tenn., USA*

NICHOLAS C.CRAWFORD & RANDY L.VILLA   *Center for Cave and Karst Studies, W. Kentucky University, Bowling Green, Ky., USA*

ABSTRACT

The Tennessee proposed Superconducting Super Collider Site was located in a regionally stable geologic environment with a shallow karst terrain. It was in close proximity to Snail Shell Cave, the largest and most intricate cave system in Middle Tennessee. Placement of the Collider Ring, Booster systems and construction of the Campus/Injector Complex, if not designed correctly, could have severely impacted the karst environment.

Regional investigations were conducted to determine the general depth to water around the SSC Site. Most of the water occurs within 61 meters (200 feet) of the surface where solution cavities are developed. The Collider Ring was placed below this at 107 meters (350 feet) MSL to avoid the shallow groundwater system.

An extensive hydrologic investigation was carried out to study the impacts on the shallow karst from construction and operation of the Campus/Injector Complex. The investigation included a karst hydrologic inventory, construction of a potentiometric map of the upper karst aquifer and seventeen dye tracer investigations.

Two main subsurface flow routes were identified in the upper part of the drainage basin. These cave streams merge downstream to form the subsurface/surface Overall Creek. A spill or leak of hazardous materials on the site could contaminate the groundwater for as much as 15.5 kilometers (9.7 miles).

A groundwater monitoring and emergency containment system was recommended. It included continuous monitoring instrumentation located in the cave stream under the Campus Complex. Groundwater contaminated by a spill would be detected and immediately pumped into a retention basin located on the surface for treatment before being released back into the groundwater system.

Introduction

The Department of Energy's (DOE) Superconducting Super Collider (SSC) is a high energy particle accelerator designed to "push" protons in opposite directions around a ring and collide them at combined velocities nearly twice speed of light. The entire collider system consists of an underground tunnel 85 kilometers (53 miles) in circumference (Collider Ring), three smaller tunnels (High, Medium and Low Energy Boosters), and a 203.3 hectare (500 acre) Campus Injector Complex on the surface.

DOE announced its intent to construct the SSC in early 1987 and invited site proposals from any interested party. The State of Tennessee, and 25 other states, submitted site proposals.

DOE specified certain site requirements, necessary to operate the SSC successfully, the two most important being the geology and hydrology of the location. An area of low seismic risk with a uniform geologic regime was essential. The tunnel had to be dry, therefore, the hydrologic environment was a key factor.

Regional Setting Of The Tennessee Site

After careful examination of the entire State, it was determined that Middle Tennessee possessed the best qualifications for a competitive site. The specific site selected was 48 kilometers (30 miles) southeast of Nashville, near Murfreesboro, Tennessee. The 85 kilometer (53 mile) tunnel ring, to be excavated by tunnel boring machines, was to extend through parts of four counties, Rutherford, Williamson, Marshall and Bedford. The Campus/Injector Complex, where the experiments take place, was to be located on the eastern side of the ring.

The proposed site was in the physiographic region known as the Central Basin, an elliptical basin formed by the erosion of the Nashville Dome (Figure 1). The topography consists

primarily of flat lying areas, a few rolling hills and scattered knobs. The most significant surface feature is the shallow karst characterized by the presence of springs, karst
windows, caves, sinkholes and sinking streams.

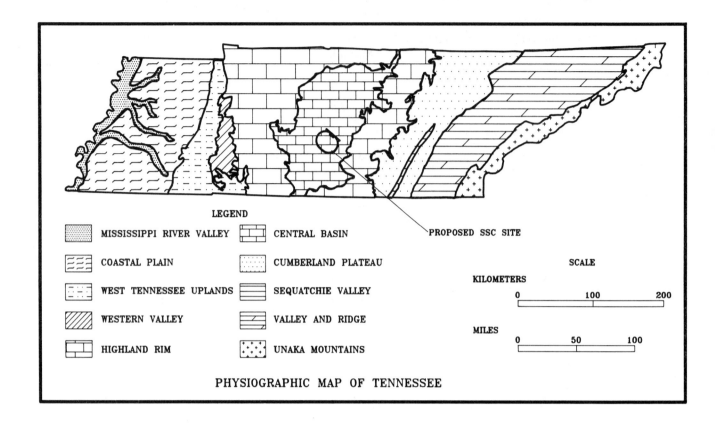

Figure 1: Location Map of the Proposed Location for the Tennessee Superconducting Super
Collider Site (SSC).

General Geology and Hydrology
    The geology of the region was fairly well understood, because the surface had been map-
ped by the Tennessee Division of Geology. Subsurface information was obtained from detailed
lithologic logs of core holes drilled by mining companies in search of zinc ore bodies. Core
and rotary holes were drilled during the site investigation and geophysical logs and hydrological tests were made to support existing data. All the information gathered was correlated and a detailed picture of the surface/subsurface geology was constructed.

    The rock units within the site are Ordovician age limestones of the Stones River and
Nashville groups (Figure 2). They are generally flat-lying, but gentle folds occur locally.
The formations comprising these groups alternate between fairly thick-bedded, pure carbon-
ates and thin-bedded limestones containing thin shale partings.

    The groundwater is primarily confined to solutionally enlarged joints, bedding planes
and caves in the low-porosity limestones. The aquifers are in the thick-bedded, purer
limestones, and the thin-bedded shaly limestones act as confining layers.

    The Ridley and Pierce Limestones are of particular importance in the development of the
karst system in the SSC area (Figure 3). The Ridley is a fairly pure, thick-bedded limestone about 34-40 meters (110-130 feet) thick. About 10 meters (33 feet) above the base
there is a thin-bedded, shaly member that acts as a confining layer (Ridley Confining
Layer). Most of the karst features, including Snail Shell Cave, are developed either above
or below the Ridley Confining Layer (Figure 3). The Pierce Limestone is a very thin-bedded,
argillaceous limestone about 10 meters (33 feet) thick and it acts as a confining layer,
thus protecting the underlying Murfreesboro Limestone from significant solution. Most of
the karst features bottom out near the Ridley-Pierce contact.

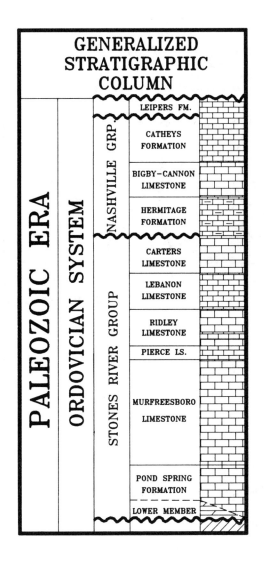

Figure 2: Generalized Strati-
graphic Column, Stones River and
Nashville Groups, Central Basin,
Tennessee.

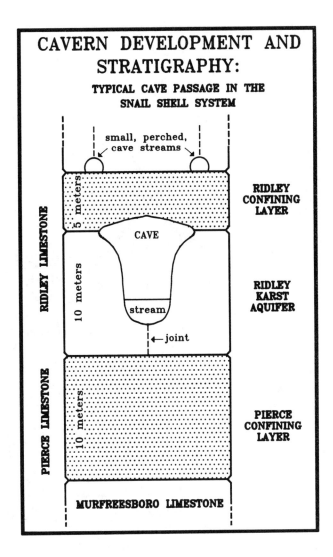

Figure 3: Profile View of a Typical Cave
Passage in the Snail Shell System.

Collider Ring

Placement of the Collider Ring was dependent on the depth of the groundwater in the
area. Regional water well data was compiled to determine the depth to water around the
site. It was found that most of the groundwater occurs within 61 meters (200 feet) of land
surface. Solution cavities and joints diminish with depth, consequently very little water
was found below 61 meters (200 feet). Site-specific hydrologic packer testing of several
test holes was undertaken to test the regional water well study. The results of the tests
supported the early findings. In order to avoid the groundwater in the karst aquifers
above the Pierce Confining Layer, the proposed tunnel was to be 107 meters (350 feet) above
mean sea level (MSL), from 91 to 182 meters (300 - 600 feet) below the irregular land surface. At this depth the tunnel would be below the shallow groundwater system and considered
environmentally safe.

Campus/Injector Complex

The shallow karst groundwater system could have been adversely affected by the construction and operation of the Campus/Injector Complex. Although known major karst features
were avoided in siting the Complex, Snail Shell Cave, the largest and most intricate in
Middle Tennessee, was close to the Complex. Concern about the possible threat to the
cave's environment prompted an intensive investigation of the area.

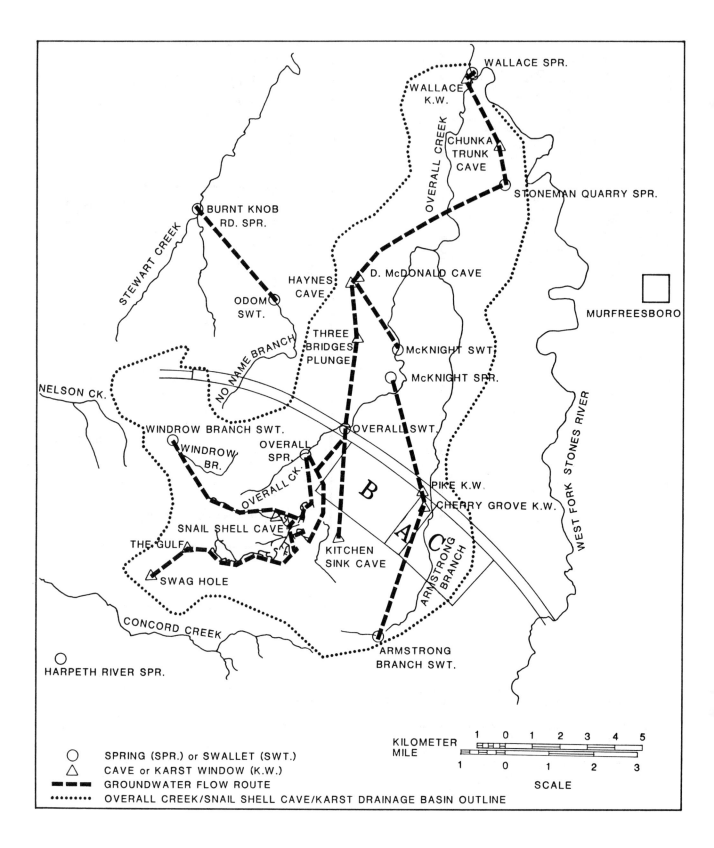

Figure 4: Groundwater Flow Routes, Overall Creek/Snail Shell Cave/Karst Drainage Basin Outline.

Karst Hydrology Inventory

An extensive field investigation was made to locate karst hydrologic features in the area of the Campus/Injector Complex, the Snail Shell Cave area, and the Overall Creek - West Fork of the Stones River area. The features examined included springs, caves, cave streams, karst windows, significant sinkholes, dry surface streams and sinking streams. The investigation resulted in the discovery of several new caves and springs, including Wallace Spring, the largest in the area (Figure 4).

Studies conducted by Dr. Nicholas Crawford in 1975 - 76 confirmed that the cave streams in the Snail Shell Cave system resurge at Overall Spring (Figure 4). The water normally flows on the surface as Overall Creek, but in dry weather it sinks below the surface at a point approximately 1.6 kilometers (1 mile) downstream from the spring at Overall Swallet. It was originally thought to resurge again at McKnight Spring farther downstream. Water flowing from McKnight Spring sinks below the surface approximately 1 kilometer (.62 miles) downstream at McKnight Swallet. Overall Creek then flows in the subsurface from McKnight Swallet to its confluence with the West Fork of the Stones River, approximately 11.8 kilometers (7.3 miles).

One of the most important discoveries was Wallace Spring, on the West Fork of Stones River about 300 meters (1000 feet) downstream from the confluence with Overall Creek (Figure 4). During low discharge, the entire flow of Overall Creek is through caves to a final resurgence at Wallace Spring. At such times, the 33 meter (100 foot) wide Overall Creek channel is dry even at its confluence with West Fork of Stones River. Wallace Spring had not been reported in the literature, although it is the largest spring in the area and the final resurgence for the subsurface Overall Creek.

Two karst windows, Cherry Grove and Pike, were located on areas A and B of the Campus/Injector Complex and are related to the subsurface flow of Armstrong Branch (Figure 4). Dye traces proved that the water from these karst windows, rather than water from Overall Creek, resurges at McKnight Spring.

Members of a group of speleogists calling themselves the Tennessee Cave Survey provided locations of several more caves in the area that were strategic in defining the subsurface flow routes.

Potentiometric Map Of The Uppermost Karst Aquifer

A potentiometric map of the uppermost karst aquifer was made by using static water levels from water well file data, new field measurements from wells near the Campus/Injector Complex, and base flow water elevations of springs, karst windows, streams and cave streams (Figure 5). Although karst aquifers are extremely inhomogeneous and anisotropic, a potentiometric map is very useful for determining general groundwater flow directions and locating groundwater basin divides.

Dye Trace Investigations

Dye trace studies were planned and carried out after completion of the karst hydrology inventory, construction of the potentiometric map and reviews of the literature of the area's groundwater conditions. The principal question to be answered was whether Snail Shell Cave and the Campus/Injector Complex are hydrologically connected. Over 50 sites in four drainage basins were "bugged" to detect the dyes.

A total of seventeen dye trace studies were made, four by Dr. Nicholas Crawford in 1975 - 76, ten in the summer of 1988, and three in the winter of 1988-89.

Four kinds of dye were used for the groundwater tracer investigations:

a: Fluorescein, C. I. Acid yellow 73
b: Rhodamine WT, C. I. Acid Red 388
c: Optical Brightener, C. I. Tinopol 5BM GX F.B.A. 22
d: Direct Yellow 96, C. I. Diphenyl Brilliant Flavine 7 GFF D. Y. 96

The following water tracing techniques were used:

a: Qualitative tracing with Fluorescein and Rhodamine WT. About 0.45 kilograms per 1.6 kilometers (1 pound per mile), up to a maximum of 2.2 kilograms (5 pounds) were used, and activated charcoal dye receptors were placed at all springs, karst windows, cave streams and some water wells.

b: Qualitative tracing with Optical Brightener and Direct Yellow 96. About 1.3 kilograms per 1.6 kilometers (3 pounds per mile) were used, and surgical cotton receptors were placed at all springs, karst windows, cave streams, and some water wells.

c: Quantitive tracing with Fluorescein and Rhodamine WT. An ISSO automatic water samp-

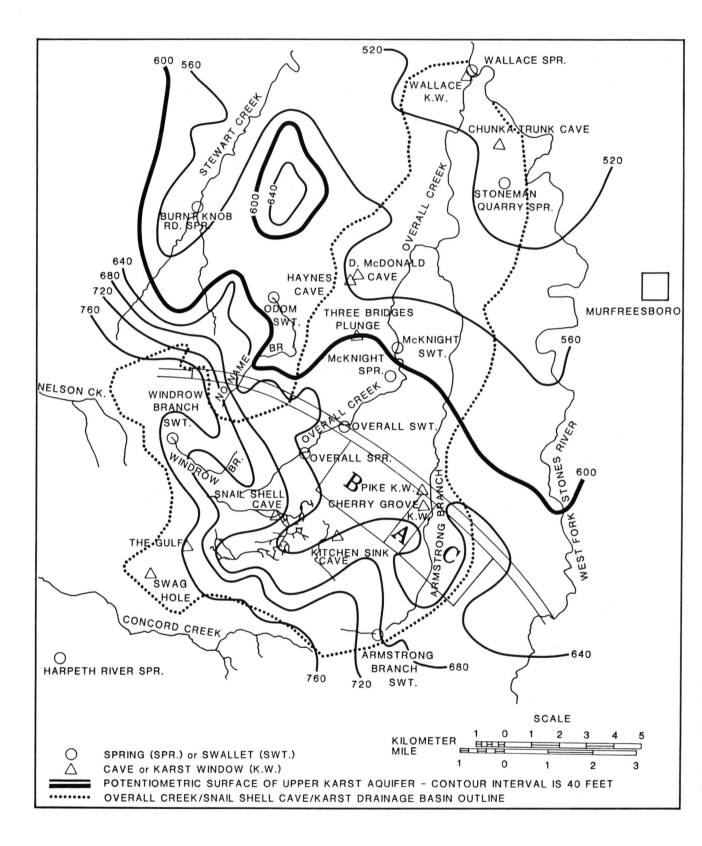

Figure 5: Potentiometric Map of the Uppermost Karst Aquifer.

ler was placed at Overall Spring and samples were collected at four hour intervals. The samples were analyzed for Rhodamine WT and Fluorescein on a Turner Fluorometer. Concentrations as low as 0.0001 ppm are detectable on the Fluorometer.

Summary Of Dye Traces

Dye Traces In The Snail Shell Karst, 1975 - 76. Four dye traces were performed by Dr. Nicholas Crawford during an earlier effort to investigate the cave streams in the Snail Shell Karst, and are briefly described as follows:

The Gulf: Dye was injected at "The Gulf," a large karst window near the beginning of the southwest cave stream that flows through the Snail Shell Karst. Dye flowed through Nanna Cave and Echo Cave and resurged at Overall Spring (Figure 4).

Snail Shell Cave System: Two traces were started in the cave system, one at Echo Cave and one in the Grand Canal Sump of Snail Shell Cave. Dye was detected at Overall Spring for both traces. McKnight Spring was checked for dye on the Grand Canal trace but results were negative (Figure 4).

Windrow Branch: Dye was injected into the farthest upstream swallet on Windrow Branch, which is the head of the northwest cave stream in the Snail Shell Karst. Dye was detected at the Snail Shell Sink entrance and Overall Spring (Figure 4).

Dye Traces To Define The Subsurface Drainage Of The Campus/Injector Complex. During the late summer of 1988, ten dye traces were conducted to define the subsurface drainage routes in the area of the SSC Campus/Injector site. Due to drought conditions, groundwater flow was extremely low.

McKnight Swallet: Fluorescein dye was injected at McKnight Swallet on Overall Creek and detected at several caves and springs, including Stoneman Quarry, before its final resurgence at Wallace Spring on the West Fork of Stones River. A second trace was started at McKnight Swallet after the discovery of Three Bridges Plunge Karst Window. However, dye was not detected at Three Bridges Plunge (Figure 4).

Grand Canal Sump: Two traces were started at the Grand Canal Sump in Snail Shell Cave. When the first trace was started the flow conditions in the cave were so low that no water was flowing down the Canal to the sump. Nearly six weeks later, dye was injected again after a channel was dug in the cave floor to start water flowing down the Canal to the sump. It had been established during the dye traces in 1975 - 76 that water flowing down the Grand Canal resurges at Overall Spring. It was thought possible that the subsurface route splits, with part of the water going to McKnight Spring. Results at McKnight were negative, but positive results elsewhere confirmed that all the water from Snail Shell Cave goes to Overall Spring and on to Three Bridges Plunge Karst Window to join the subsurface Overall Creek just upstream from Dennis McDonald Cave (Figure 4).

Snail Shell Cave Stream: Rhodamine WT was injected in the Snail Shell Cave Stream, which sumps at the back of the cave. Dye was detected at Overall Spring, but not at McKnight Spring (Figure 4).

Chunka Trunk Cave: Direct Yellow was injected at Chunka Trunk Cave Stream and was detected at West Fork Cave, Wallace Karst Window and Wallace Spring (Figure 4).

Overall Swallet: Two dye traces were started at Overall Swallet in an attempt to trace the flow to McKnight Spring. McKnight Spring was negative for both traces, but dye was detected at Three Bridges Plunge Karst Window during the second trace (Figure 4).

Cherry Grove Karst Window: Fluroescein dye was injected at Cherry Grove Karst Window, which is on the proposed Campus/Injector Complex. The dye was detected at Pike Karst Window, McKnight Spring and Stoneman Quarry (Figure 4).

Kitchen Sink Cave: Dye was injected into a stationary pool at the bottom of Kitchen Sink Cave and detected at Three Bridges Plunge Karst Window, but not at Overall Spring, McKnight Spring, Cherry Grove or Pike Karst Window (Figure 4).

Dye Traces Designed To Define The Limits Of The Overall Creek Drainage Basin. Three additional dye traces were conducted to define the limits of the Overall Creek/Snail Shell Cave drainage basin.

Armstrong Branch Swallet: Dye was injected into Armstrong Branch Swallet, which is close to the head of Armstrong Branch and upstream from the proposed Campus/Injector Complex. Dye was detected at Cherry Grove and Pike Karst Windows as well as McKnight Spring (Figure 4).

Swag Hole: Swag Hole Karst Window is midway between The Gulf and the Head of the Harpeth River Spring. It was thought that the water flowing in Swag Hole would go to the Harpeth River Spring but the only dye detected was at The Gulf and Overall Spring (Figure 4).

No-Name Branch - Odom Swallet: No-Name Branch originates north of Windrow Branch and during high flow travels north to Odom Swallet, where it sinks. Dye was injected in Odom Swallet and detected at two locations in Stewart Creek and at Burnt Knob Road Spring, which flows into Stewart Creek (Figure 4). This trace defined the northwest basin divide between Overall Creek and Stewart Creek.

The subsurface drainage in the area is indeed complicated, but it appears that there are two major subsurface drainage systems, one of which drains the streams in the Snail Shell Karst by way of Blue Sink, Horseshoe Cave Karst Window, Overall Spring, Overall Swallet, Three Bridges Plunge Karst Window, and Haynes Cave (Figure 4). Some of the water from the Snail Shell streams appears to flow past Overall Spring directly to Three Bridges Plunge Karst Window. During high flow following heavy rains almost all the discharge from the Snail Shell Karst flows from Overall Spring past Overall Swallet and on down Overall Creek. This water appears to flow over the top of the subsurface stream flowing to McKnight Spring.

The second system is the surface/subsurface Armstrong Branch which originates at a spring south of Area A on the campus and then sinks at Armstrong Branch Swallet, about 3.2 kilometers (2 miles) south of Area A (Figure 4). It flows in the subsurface through Cherry Grove Karst Window to Pike Karst Window and resurges at McKnight Spring. It sinks again at McKnight Swallet and becomes part of subsurface Overall Creek.

The two streams join somewhere between Haynes Cave and Dennis McDonald Cave and flow through subsurface Overall Creek to a final resurgence at Wallace Spring (Figure 4).

Groundwater Monitoring And Containment System

Construction in a karst environment can be a threat to a groundwater system and construction and operation of the proposed SSC was no exception. The hydrologic investigation proved that there are cave streams under the Campus/Injector Site and that they flow for several kilometers underground before finally resurging at Wallace Spring. A spill or leak of hazardous chemicals could quickly sink through the thin soils or flow as surface runoff directly into a sinkhole and into a cave stream and then contaminate the groundwater all the way to Wallace Spring.

Monitoring wells are used as a safety check to ensure that a leak or spill has not contaminated groundwater. Since it is extremely difficult to drill a well directly into a cave stream, the preferred method of monitoring in karst areas is at a spring, even though the spring may be off the site.

In the case of the SSC Campus/Injector Complex, a cave stream is located on site (Cherry Grove and Pike Karst Windows) and a monitoring station could be installed on the proposed SSC property. The Pike Karst Window is ideally located just inside the eastern boundary of the proposed Campus/Injector Complex. This is an outstanding location for a downstream monitoring station for the subsurface drainage of Areas A and C.

Due to the potential for rapid flow in karst aquifers, the cave stream would need to be continuously monitored. This could be done by installing a small pump in the Pike Karst Window that would continuously pump a small volume of water directly into the SSC laboratory and through continuous monitoring instruments.

The only way to ensure that a contaminant would not flow off site would be to install a recovery system before a spill or leak takes place. The following procedures were recommended for the SSC project. Install three pumps at the Pike Karst Window. Determine the drawdown necessary in order to prevent a discharge from flowing off site. (This could be determined by a dye trace from the Cherry Grove Karst Window. If dye does not reach McKnight Spring, the drawdown level is adequate). The pumps would have float controls adjusted to the water drawdown level such that if one pump could not handle the subsurface stream discharge, a second and if needed, a third pump could automatically kick on.

If a contaminant was detected by the continuous monitoring instrumentation, the pumping system could be turned on either automatically or manually to prevent any contaminants from flowing past Pike Karst Window. The pumps would discharge the contaminated water into a lined surface impoundment for later treatment. Once the contaminant slug had passed, the pumps would be turned off. The water pumped into the surface impoundment could then be treated before being released.

A surface impoundment could be made by placing an earth dam across the usally dry Armstrong Creek near the eastern boundary of the Campus/Injector Complex. The impoundment would need a plastic-rubberized liner, since most of the dry bed of Armstrong Branch is only about 6 meters (20 feet) above the water surface level of the cave stream below. During periods of high flow, the cave stream cannot handle all the discharge and the water level rises to the surface and supplies water to Armstrong Branch. Steel gates could be closed at the dam if there were any contaminants in either the groundwater or the surface water. A 3 to 4 meter (10 - 13 feet) earth dam is probably all that would be needed. It might be desirable to keep the water level at 1 to 2 meters (3 to 6 feet), creating a lake perhaps 122 meters (4000 feet) long that would add to the beauty of the Campus Area. Water from the reservoir could be monitored for contaminants before being released downstream. Storm drains from the campus area could be directed into the lake. If a surface spill or leak occurs, it would flow into the lake. Liquid sinking into the ground would be detected and pumped into the lake at the monitoring-pumping station at Pike Karst Window. This system would detect and then stop any contaminants in both groundwater and surface runoff from getting off site.

Conclusion

Hydrologic investigations in the area of the proposed Superconducting Super Collider showed that the karst aquifer system is shallow and complicated. Construction of a large scale surface/subsurface facility would require special considerations to ensure an environmentally safe and successful operation.

Most of the groundwater occurs within 61 meters (200 feet) of the surface, where dissolution of limestone has created cavities for groundwater movement and storage. The most significant karst features, such as caves, cave streams, and sinkholes, occur in the Ridley Limestone, within 15 - 24 meters (50 - 80 feet) of the surface.

The regional groundwater study, along with site-specific hydrologic packer testing, confirmed that placement of the Collider Ring tunnel at 107 meters (350 feet) MSL, in the lower Murfreesboro Limestone would be well below the Pierce Confining Layer and the karst aquifers in the overlying Ridley Limestone. Therefore, any impacts of the tunnel on the karst groundwater system would be negligible.

The hydrologic inventory and dye trace investigations showed the complexity of the karst system. Although most of the major caves and related karst features were found, there are undoubtedly others that would need to be located if the SSC were to be built in Tennessee.

The two major flow routes in the vicinity of the Campus/Injector Complex are: 1) cave streams draining the Snail Shell Karst and 2) subsurface Armstrong Branch. The two systems flow separately in the upper part of the Overall Creek Drainage Basin and merge between Haynes Cave and Dennis McDonald Cave. From that junction subsurface Overall Creek flows to a final resurgence at Wallace Spring, on West Fork of Stones River. During high discharge the cave cannot handle all the flow and several overflow springs direct the water into surface Overall Creek. This investigation demonstrated that the cave stream that drains the Snail Shell Karst area would not be impacted by the SSC project because it is upstream from the Campus/Injector Complex and on a different tributary of subsurface Overall Creek.

Karst aquifers are extremely vulnerable to contamination that can adversely affect a large area in a very short time. A monitoring and containment system as described above would be necessary to prevent contamination of groundwater down-gradient from the SSC Campus/Injector site.

It was recommended that a continuous monitoring and total containment system be installed before operation of the SSC began. The type recommended consists of a monitoring station on the cave stream that flows under the Campus, which would continuously monitor for contaminants. If a spill or leak occurred, pumps installed in the cave stream would be activated and the entire discharge of the relatively small cave stream would be pumped into a surface retention basin. The contaminated water would then be treated before being released back into the karst groundwater drainage system.

References

Crawford, N., and Barr, Thomas, Jr. 1989, Hydrology of the Snail Shell Cave - Overall Creek Drainage Basin and Ecology of the Snail Shell Cave System, Prepared for the Tennessee Department of Conservation, Nashville, Tennessee.

Crawford, N., 1982, Karst Hydrogeology of Tennessee, guidebook for U. S. Environmental Protection Agency Karst Hydrogeology Workshop, Nashville, Tennessee, 102 p.

Wilson, C. W., Jr., 1949, Pre-Chattanooga Stratigraphy in Central Tennessee, Tennessee Division of Geology, Bull. 56, 407 p.

4. Engineering geology of karst terranes

A volcanic breccia in the buried karst of Hong Kong

N.J.DARIGO   *Dames & Moore, Wanchai, Hong Kong*
Seconded to Hong Kong Geological Survey, Kowloon, Hong Kong

ABSTRACT

A subsurface mapping project has been undertaken by the Hong Kong Geological Survey to study buried karst problems associated with an area of intense building activity in the New Territories of Hong Kong. The project has led to the identification of a volcanic breccia composed largely of marble clasts, which has unique weathering and karst characteristics.

Steeply dipping breccia layers subcrop beneath Quaternary superficial deposits in a northeast-trending valley, which extends through northwest Hong Kong and into the Guangdong Province of China. Other rocks underlying the valley include Carboniferous marble, which is presumably the source of the clasts in the breccia. The breccia unit comprises the newly named Tin Shui Wai Member of the Upper Jurassic Tuen Mun Formation. It is composed of marble-bearing tuff-breccia and conglomerate interbedded with fine-ash tuff and siltstone. Much of the unit is metamorphosed, masking evidence of depositional environment. The degree of metamorphism is highest in the northern portion of the subcrop, where marble clasts have been severely stretched to give the appearance of banded marble.

Deep tropical weathering of the breccia sequence has created an extremely variable rockhead surface, which is overlain by completely weathered tuff composed of sandy silt. The variation in rockhead elevation may be the result of different rates of weathering between the tuff and the marble breccia layers. Cavities occur in rare instances in the northern part of the breccia where marble is the dominant clast type.

The identification of the Tin Shui Wai Member as a volcanic breccia and not banded marble, as was formerly believed, has eliminated a large area in northwest Hong Kong where cavitous marble was suspected. Foundation problems associated with the variation in weathering depth of the breccia sequence outweigh the minor occurrence of cavities.

Introduction

A low-lying region in the northwestern New Territories of Hong Kong (Figure 1) is the subject of an ongoing geologic study to locate and map buried karst. The study has been undertaken by the Hong Kong Geological Survey in an effort to identify areas which may have potential building foundation problems. The extremely dense population of Hong Kong combined with its mountainous topography have necessitated the use of all available flat-lying space, including reclaimed land, for high-rise housing. In the past decade, several new cities have been built in the New Territories to house the burgeoning population. During this period of intense building activity, buried cavitous marble was discovered in the town of Yuen Long underlying Quaternary alluvial deposits.

Subsequent to the discovery of marble and prior to the present study, the Geological Survey outlined a region in the New Territories, called the Designated Area, where buried marble was suspected or where the subsurface geology was completely unknown (Geotechnical Control Office, 1988a). The Designated Area comprises a region of plains and low hills composed of Carboniferous strata, which are roughly continuous with Carboniferous marble and limestone in the adjacent Guangdong Province of China. It is now known that much of the western Designated Area is composed of Jurassic volcanic rocks. This paper addresses a Jurassic volcaniclastic unit containing clasts derived from the Carboniferous marble, which was formerly thought to be banded marble. A related paper in this volume by D.V. Frost describes the Carboniferous cavitous marble in the central part of the Designated Area.

Methods

The subcrop map shown in Figure 2 was completed mainly through the study of approximately 1000 logs of boreholes which were drilled in the study area over the past 15 years. These

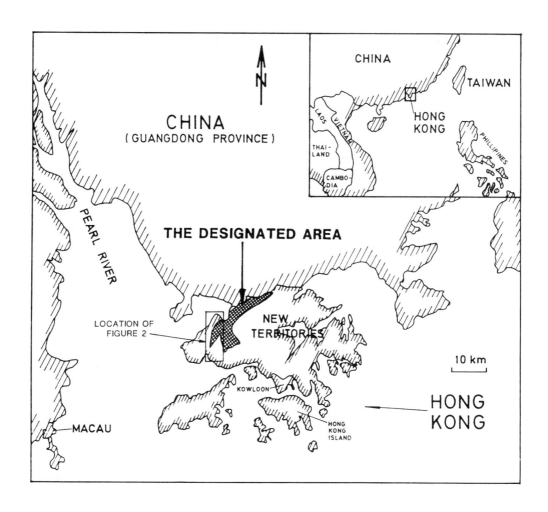

Figure 1. Location of the Designated Area - a region of suspected buried karst in Hong Kong.

data are stored by the Geotechnical Information Unit (GIU) of the Geotechnical Control Office, Hong Kong Government. Generally, most boreholes were continuously cored and were drilled to depths of less than 20 m below rockhead. Nearly all of the boreholes were drilled for building foundation purposes and were therefore logged according to accepted geotechnical engineering practices. As a result, many details of purely geological interest may have gone unrecorded. For many of the boreholes drilled within the last five years, the logs are accompanied by core photographs which proved invaluable for confirming and enhancing existing core descriptions. In rare instances, core was available for viewing by the author; however, most core drilled for geotechnical purposes in Hong Kong is not stored indefinitely.

In addition to the above data, about 40 deep boreholes have been specially commissioned by the Hong Kong Geological Survey to study the Designated Area. Seven of these are located within probable Jurassic strata in the western part of the Designated Area. These penetrated to depths of 40 to 130 m below rockhead providing accessible continuous core for more thorough geologic study and sampling.

Subsurface borehole information was correlated with isolated outcrops of Carboniferous and Jurassic strata, which have been mapped most recently by Arthurton et al.(1988), Lai (1987), and Langford et al.(1988). Solid rock is exposed mainly in the northwestern and southeastern parts of the subcrop map shown in Figure 2.

Regional Setting
The Designated Area follows a structurally controlled valley with an approximate northeasterly trend. Granite generally forms high relief topography on either side of the valley, and isolated patches of deeply weathered metasedimentary rocks form low hills adjacent to the granite. The remainder of the study area is covered with Quaternary superficial deposits. The complex structure of the area is dominated by northeast-trending and west-

226

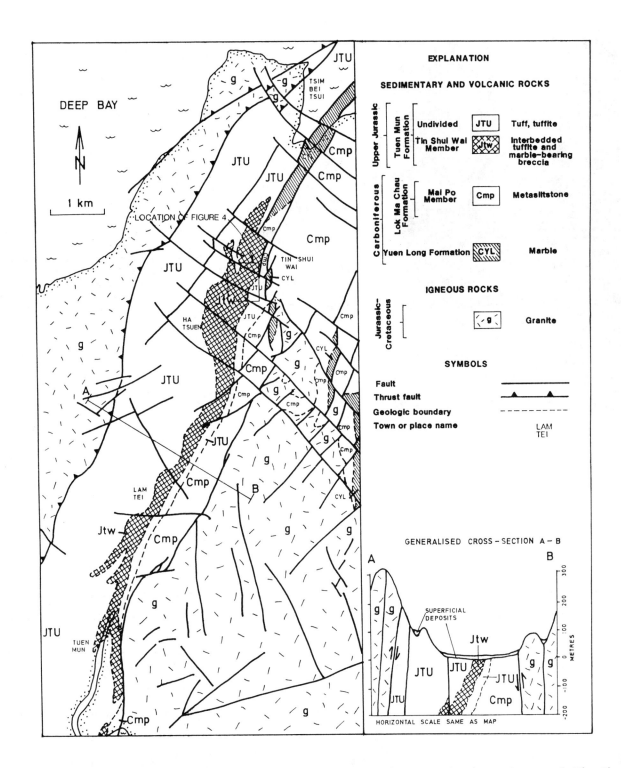

Figure 2. Generalised solid geology of western Designated Area, showing subcrop of Tin Shui Wai Member.

dipping thrust and reverse faults. More recent northwest-trending near-vertical cross faults appear to be common, but show only minor offsets. Strata identified beneath Quaternary deposits in the Designated Area include marble, metasiltstone sandstone, and conglomerate of the Carboniferous Yuen Long and Lok Ma Chau Formations (San Tin Group), and volcanic rocks of the Jurassic Repulse Bay Volcanic Group. Along the west side of the Designated Area (Figure 2), the volcanic rocks comprise the Tuen Mun Formation and the Tin Shui Wai Member breccia,

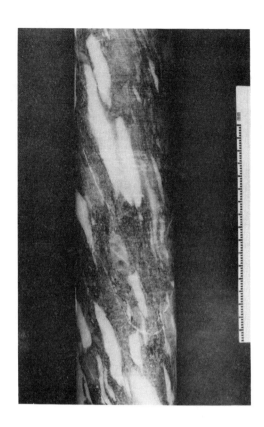

Figure 3. Drill core showing metamorphosed tuff-breccia with marble clasts, from Tin Shui Wai Member near Ha Tsuen (borehole BGS 18, 25.5 m Hong Kong grid reference 817446E, 834144N).

which are probably some of the oldest rocks in the Repulse Bay Volcanic Group. A similar marble-bearing volcanic breccia was discovered recently in the eastern New Territories. It was encountered beneath reclaimed land and in water tunnels near the town of Tai Po (S. W. Y. Ho, Acer/Freeman Fox Ltd., personal communication, 1989). This breccia may be part of the Upper Jurassic Yim Tin Tsai Formation, which is exposed in nearby outcrops and has been mapped as predominantly crystal tuff (Addison and Purser, 1986). Since the Yim Tin Tsai Formation is the oldest unit of the Repulse Bay Volcanic Group mapped in the eastern New Territories, it may be roughly equivalent to the Tuen Mun Formation of the western New Territories.

The Tuen Mun Formation probably lies unconformably on top of Carboniferous rocks in the southwestern part of the Designated Area. Towards the north, Jurassic strata are often faulted against the Carboniferous. Along the western edge of the Designated Area, a thrust- or reverse-faulted intrusive contact juxtaposes granite over Jurassic volcanic rocks (Langford et al., 1989).

The strata encountered in boreholes in the western Designated Area generally show steep metamorphic dips ranging from 40 to 90 degrees, with most dips around 60 to 70 degrees. Because of the shearing and metamorphism of most rocks, it is difficult to distinguish original sedimentary dip from metamorphic foliation. Some cores in the northern part of the Tin Shui Wai Member show probable bedding contacts dipping roughly 15 to 30 degrees shallower than the foliation. In many cores, however, the foliation appears to be roughly the same as bedding. Also, many cores and outcrops in the area show two different directional metamorphic events, as evidenced by secondary kinks or folds overprinting the dominant metamorphic dip.

Tin Shui Wai Member -Description

The Tin Shui Wai Member, which has been recently named as a result of this study, subcrops beneath Quaternary superficial deposits in the western Designated Area, from north of Tin Shui Wai to south of Tuen Mun (Figure 2). It comprises an interbedded sequence of volcaniclastic rocks, including tuff-breccia (Figure 3), fine- to coarse-ash andesitic tuff and tuffite, tuffaceous siltstone, lapilli tuff, and andesitic conglomerate. The thickness of layers in the Tin Shui Wai Member typically ranges from 0.1 to 5 m, although in the southern part of the subcrop, layers as thick as 10 m have been encountered in boreholes.

The Tin Shui Wai Member subcrop is bounded both to the east and west by undifferentiated andesite lava, tuff, and tuffaceous siltstone of the Tuen Mun Formation. Although original stratigraphic dip is masked by complex metamorphism, regional structural trends indicate that the stratigraphic base of the unit is to the east, and that on a gross scale, the whole formation dips westerly (cross-section A-B, Figure 2). In unmetamorphosed parts of the subcrop near Tuen Mun, the breccia is massive and shows no sharp contacts with adjacent fine-grained units which would indicate a stratigraphic dip. The contact between the Tin Shui Wai Member and the undifferentiated andesite and tuff appears to be gradational on both top and bottom. Likewise, at its northern and southern extent, the Tin Shui Wai Member pinches out, so that most of the sequence is tuff or siltstone with sporadic thin layers of tuff-breccia and conglomerate. In the main part of the Tin Shui Wai Member, breccia and/or conglomeratic layers make up 50 to 70% of the interbedded sequence. Towards the edges of the subcrop and in the undifferentiated Tuen Mun Formation rocks, coarse-grained units may constitute less than 5% of the sequence.

The Tin Shui Wai Member appears to undergo a facies change from primarily sedimentary waterlain rocks in the Tin Shui Wai area in the north, to more pyroclastic rocks to the south. Although depositional origin is difficult to identify due to metamorphism, fine-grained portions of the rock appear to be siltstones or tuffaceous siltstones in the north, while those to the south are tuff or tuffite. R.L.Langford (Hong Kong Geological Survey, personal communication, 1989) suggests that rocks in the southern part of the subcrop near Lam Tei

(Figure 2) form a vent facies, which originated by underground brecciation and mixing, a common mode of origin for andesitic volcanic breccias recognized by others (e.g. Parsons, 1969; Fisher and Schmincke, 1984). Rocks in the Lam Tei area were either emplaced underground, or were extruded along dikes and sills and deposited as massive pyroclastic flows. Rocks further to the north and south of the suggested vent source area were possibly deposited as laharic type flows, that is, water-lubricated volcanic mudflows in which brecciation occurred as a mixture of volcanic and sedimentary processes. In the region south of Ha Tsuen (Figure 2), occasional evidence of normal graded bedding is present in possible laharic rocks. Even further from source, volcanically derived material may have been reworked by sedimentary processes resulting in epiclastic breccias and waterlain siltstones and sandstones.

Clasts in the Tin Shui Wai Member are angular to subrounded, and are composed of white marble, quartzite, metasiltstone, and andesitic tuff. In the northern part of the subcrop, the clast type is dominantly marble; whereas to the south the clasts are more heterolithologic. Marble, quartzite, and siltstone clasts were probably derived from the existing country rock of the time; the Carboniferous Yuen Long and Lok Ma Chau Formations. Tuff clasts were probably derived from slightly older Upper Jurassic volcanic rocks or were formed contemporaneously with mixing and brecciation.

North of Tin Shui Wai, near Tsim Bei Tsui (Figure 2), heterolithologic green metaconglomerates are commonly interbedded with metasiltstone and metasandstone in the Jurassic rocks. Conglomerates comprise up to 40% of the section encountered in boreholes near Tsim Bei Tsui and thin southward, interfingering with siltstone and sandstone until they comprise about 5 to 15% of the section near the northern extent of the Tin Shui Wai Member. Clast types distinguish the Tin Shui Wai Member from the undifferentiated conglomerates. The conglomerates contain quartz, siltstone, and tuff clasts, but are notably lacking in marble clasts. The conglomerates often appear silicified and altered by fluidized metamorphism. It is possible that the conglomerates in this area, which were previously thought to be part of the Carboniferous Lok Ma Chau Formation (Allen and Stephens, 1971), are roughly equivalent to the Tin Shui Wai Member rocks, but are a sedimentary facies further removed from the volcanic source.

Metamorphism of the Tin Shui Wai Member is more pronounced in the northern part of the subcrop than to the south. In the south near Tuen Mun and Lam Tei, metamorphism occurs as

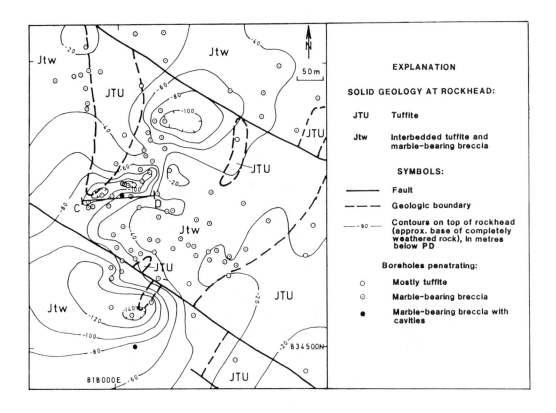

Figure 4. Contours on top of rockhead, northern Tin Shui Wai subcrop. Cross-section C-D is shown in Figure 5. Hong Kong Principal Datum (PD) is 1.2 m below Mean Sea Level.

narrow, shear zones several metres to tens of metres wide. which are separated by unmetamorphosed unsheared rock. From about 1 km north of Lam Tei northward, all cores show a directional metamorphic dip, but clasts are still recognizable. In the northern Tin Shui Wai area, marble clasts become so stretched by shearing as to lose all original clast appearance, thus appearing as marble bands or augen-like structures in a schistose metamorphic rock (Figure 3). Often a quartzite or marble clast appears to be preserved in its original shape while other marble clasts are sheared all around it. Chloritized and occasionally slickensided joints are also present in the rocks of the northern part of the subcrop.

Weathering and Karst Characteristics

Nearly all rocks in Hong Kong are affected by deep chemical weathering, which is enhanced by the subtropical climate of the area. Weathering has left a ubiquitous mantle of residual soil or completely weathered rock throughout the land surface of Hong Kong. Weathering grades in Hong Kong are most often described using a classification system which ranges from Grade I (fresh rock) to Grade VI (residual soil) (Geotechnical Control Office, 1988b). Some geotechnical engineering authors use the term rockhead to refer to the boundary between moderately and highly weathered rock (Grades III and IV). The base of the zone in which completely weathered (Grade V) material dominates, however, is generally the point at which a positive identification of rock type can be made and thus provides a more useful definition

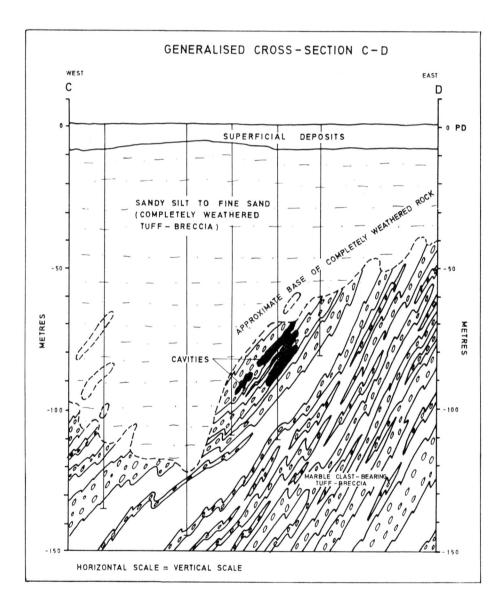

Figure 5. Cross-section C-D, showing structure and weathering of Tin Shui Wai Member in northern part of subcrop. Patterned layers are tuff-breccia; blank layers are siltstone.

of rockhead for geologic mapping purposes.

The completely weathered material which overlies rockhead in the area of the Tin Shui Wai Member subcrop is typically greenish grey sandy or clayey silt. In rare instances, a relict texture of breccia clasts is noted on borehole logs. Usually this identifying feature is absent and it is difficult to distinguish the volcanic origin of the Tin Shui Wai Member from completely weathered material alone, as it appears similar to residual layers overlying Carboniferous siltstones in other parts of the Designated Area.

The base of the completely weathered material is highly variable in the Tin Shui Wai Member, particularly in the northern part of the subcrop. In the southern part of the subcrop, rockhead elevation ranges from about 20 to 40 m below PD (approximately 25 to 50 m below ground level); while in the northern part of the subcrop, it varies from 20 m to more than 140 m below PD (25 to 150 m below ground level) over very short distances (Figures 4 and 5). Thus, in the northern area, the thickness of completely weathered material can range from 10 m to as much as 140 m.

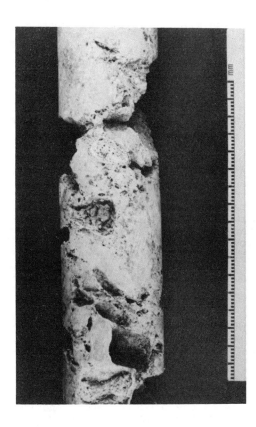

Figure 6. Core showing honeycomb-weathering pattern created by dissolution of marble clasts in breccia, from Tin Shui Wai Member near Lam Tei (Figure 2). Remaining rock consists of tuff and quartzite clasts in a tuff matrix.

This variation may be due to different rates of reaction to weathering between the fine-grained layers and the breccia layers. The tuffite and siltstone layers are likely to be more resistant to weathering than the marble breccia layers. Marble clasts in the breccia layers probably dissolve early in the weathering process, leaving a honeycomb-weathered structure composed of non-marble clasts and the remaining fine-grained matrix (Figure 6). This feature is commonly encountered in the top few metres of rock, particularly in the southern part of the Tin Shui Wai subcrop. Honeycomb weathering is not as common in the northern part of the subcrop, perhaps because of the high density of marble fragments. Dissolution of clasts in the northern area may not leave enough matrix to support the rock. Following complete dissolution of the marble fragments, and once the remaining rock is attacked by weathering until it reaches Grade V, the rock probably collapses or gradually compresses to fill voids left by marble clasts. Standard Penetration Tests (SPT's) performed during drilling through the completely weathered zone show N values (measured in blows/300mm(0.98ft)) of about 15 to 30 just below alluvium, ranging to typically 100 to 200 near rockhead. Quite often, however, the N values do not increase continuously with depth as might be expected, but fluctuate up and down with depth, indicating the presence of anomalous soft zones beneath more dense material. The soft zones are likely to be former breccia layers which have lost all evidence of clasts due to dissolution. This weathering pattern results in outliers or corestones of more solid highly weathered rock encased in completely weathered material (Figure 5). The pattern of fluctuating N values with depth exists throughout the Tin Shui Wai Member subcrop, but is more pronounced in the north, where marble is the dominant clast type.

The variation in weathering depth in the northern part of the Tin Shui Wai subcrop is probably enhanced by the steep dip of the layers, and by the fact that they are relatively thin. A flat-lying massive unit of the same rock type would probably show a more uniform weathering front. Ground water in the rock probably penetrates the individual upstanding layers of marble breccia more easily than the adjacent fine-grained layers, resulting in the convoluted weathering front suggested in Figure 5.

Metamorphism of the northern Tin Shui Wai rocks may also influence weathering depth. Fractures and joints add conduits allowing ground water access to the rock. Shearing alters the original packing arrangement of the clasts, effectively shortening the distance between clasts in at least one direction, so that there may be less matrix between marble clasts which has to weather before the rock collapses. Shearing may also increase the surface area of marble clasts, possibly making them more susceptible to dissolution.

Relatively large cavities have been encountered in only two boreholes located in the northern Tin Shui Wai Member subcrop area (Figure 4). Although cavities are rare, this area may be more susceptible to cavity formation because of the high density of marble fragments available for dissolution. The cavities range from about 0.6 to 8 m in vertical dimension, and occur at depths of 10 to 25 m below rockhead (about 80 to 105 m below ground level). According to borehole logs, no cavity infill material was recovered during the drilling process. As suggested in Figure 5, cavity shape may follow the dip and width of the marble breccia layers.

Other minor solution features are present sporadically throughout the Tin Shui Wai Member. Linear solution features sometimes occur along subvertical joint surfaces, and are typically 0.1 to 0.3 m in length. Such features may be the result of dissolution of secondary calcite veins, which are common in all rocks of the Designated Area.

Engineering Implications

The correct identification of the Tin Shui Wai Member rocks as volcanic breccia has eliminated a large part of the Designated Area which would formerly have been treated as cavitous marble for building foundation purposes. However, the unpredictable weathering pattern in the Tin Shui Wai Member does necessitate detailed geotechnical investigation, particularly in the northern part of the subcrop. Driven ungrouted H-piles are generally the favoured pile type in the Tin Shui Wai area (W.K.Pun, Geotechnical Control Office, personal communication, 1989). The cavities found in the Tin Shui Wai Member are not a great engineering concern. Because of the deep weathering and the depth of the cavities below rockhead, driven piles are expected to reach refusal well above the cavities.

Conclusions

The present study has resulted in the identification of an Upper Jurassic marble-bearing volcanic breccia called the Tin Shui Wai Member, which was previously mapped as Carboniferous banded marble. Although complex metamorphic and structural overprints have distorted the original depositional features of the rock beyond recognition in some places regional geologic relationships suggest that pyroclastic and volcaniclastic rocks dominate in the southern part of the subcrop and grade laterally to sedimentary rocks in the north.

Several factors have contributed to the convoluted weathering pattern in the Tin Shui Wai Member: 1) the difference in reaction to weathering between the breccia layers and the tuffite layers; 2) the steep dip of the sequence; 3) the thickness of the beds; and 4) the shearing and metamorphism of the rocks. The likelihood of cavity formation in the marble breccia is small. However, where they have been encountered in boreholes they measure up to 8 m high.

Acknowledgements

This study was performed for the Hong Kong Geological Survey and the Geotechnical Control Office of the Hong Kong Government. The paper is published with the approval of the Director of the Civil Engineering Services Department, Hong Kong Government. Dames & Moore-Hong Kong is gratefully acknowledged for funding presentation of the paper.

References Cited

Addison, R. and Purser, R.J., 1986, Solid and superficial geology of Sha Tin, 1:20,000 sheet 7: Hong Kong Geol. Svy., Geotech. Ctrl. Off., Hong Kong, Series HGM20, Ed. 1-1986.

Allen, P.M. and Stephens, E.A., 1971, Report on the geological survey of Hong Kong 1967-1969: J.R.Lee, Govt. Printer, Hong Kong, 107 pp.

Arthurton, R.S., Lai, K.W., and Shaw, R., 1988, Solid and superficial geology of Tsing Shan (Castle Peak), 1:20,000 sheet 5: Hong Kong Geol. Svy., Geotech. Ctrl. Off., Hong Kong, Series HGM20, Ed. 1-1988.

Fisher, P.V. and Schminke, H.U., 1984, Pyroclastic rocks: Springer-Verlag, New York, 472 pp.

Geotechnical Control Office, 1988a, Carboniferous solid geology of the northwest New Territories: 1:20,000 scale map, Drwg. No. GS-SP/713, Hong Kong Geol. Svy., Geotech. Ctrl. Off., Hong Kong.

Geotechnical Control Office, 1988b, Guide to rock and soil descriptions: Geoguide 3, Civil Eng. Services Dept. Hong Kong, 189 pp.

Lai, K.W., 1987, Solid and superficial geology, sheet 2: Unpubl. map at 1:10 000 scale, Hong Kong Geol. Svy., Geotech. Ctrl. Off., Hong Kong.

Langford, R.L., Lai, K.W., Arthurton, R.S., and Shaw, R. 1989, Geology of the western New Territories: Hong Kong Geol. Svy. Geological Memoir No. 3, Geotech. Ctrl. Off., Hong Kong, 139 pp.

Langford, R.L., Lai, K.W., and Shaw, R., 1989, Solid and superficial geology of Yuen Long, 1:20,000 sheet 6: Hong Kong Geol. Svy., Geotech. Ctrl. Off., Hong Kong, Series HGM20 Ed. 1-1989.

Parsons, W.H., 1969, Criteria for the recognition of volcanic breccias: review: In Igneous and metamorphic petrology, Geol. Soc. of Amer. Memoir 115, p. 263-304.

Practical concerns of Cambro-Ordovician karst sites

JOSEPH A.FISCHER & JOSEPH J.FISCHER   *Geoscience Services, Bernardsville, N.J., USA*
THOMAS C.GRAHAM   *Lanidex Corp., Parsippany, N.J., USA*
RICHARD W.GREENE   *Storch Engineers, Florham Park, N.J., USA*
ROBERT CANACE   *State of New Jersey, Department of Environmental Protection Division of Water Resources, Trenton, N.J., USA*

ABSTRACT

Solution-prone carbonates are found in many Appalachian valleys. Once relatively ignored, their significance in many rapidly developing areas requires a plan of investigation not in accordance with conventional soil mechanics concepts. Preliminary investigations should start well in advance of typical site development planning procedures. Conventional geophysical techniques are inapplicable, other than as an accessory to special test borings, test pits and often, carefully controlled pneumatic probes. An aware property owner, planner and designer are a necessity in completing a useful study.

Introduction

Construction in areas underlain by carbonate rocks can be environmentally unsound and economically infeasible unless the possible solutioning-related problems can be defined in advance of project planning, design and construction. In this paper, the authors will concentrate on the hard, crystalline, Cambro-Ordovician carbonates which extend from Alabama to New England, generally found within the valleys of the Appalachians. Both the investigatory techniques and the engineering solutions can be significantly different than those used in the more familiar, Recent carbonate deposits of the southern United States and the Caribbean.

In these older carbonates, our concern is essentially two fold; 1) the migration of relatively undiluted contaminants through voids and fractures in the residual soils and the underlying rock, and 2) the formation of collapse features as a result of a) the continuing erosion of soils into rock voids, particularly if exacerbated by the removal of overburden soils during construction operations, or b) the increased intensity of loadings above soil or rock cavities.

The prospective developer must be aware of the possible impacts of karst in his pre-acquisition, property analysis. This need for early information must be considered in light of the lack of funding generally available for a pre-purchase study.

Thus, the first step is to obtain and correlate the readily available geologic information. Next, using this information, is to plan and execute a geologic reconnaissance with experienced personnel. Subtle features, such as wide, bowl-like depressions, small swales, changes in vegetation, unplowed areas, and\or wooded areas in otherwise normally farmed lands, changes in soil moisture, and the ponding of water are all pieces in the karst puzzle. "Old timer" information is extremely valuable, particularly if the site has been farmed.

Most property sales have a "due diligence" period prior to final passage of title. The results of the Phase 1 study and its' impact on property values must be considered by both the property owner and the prospective buyer. The planning team for the prospective buyer should analyze the significance of the results of the preliminary geological studies on both land planning and the design of the project (including the relationship to other possible site constraints).

Should the site remain viable for its' intended use after the initial phase of study, the investigator should prepare a program of field explorations which incorporates a preliminary understanding of the nature of the planned project.

The results of this more intensive field investigation(s) will impact most civil engineering design issues, including storm water management, building layout, septic or sewer system design, potable water supply, earth moving during construction, foundation design, roadway layout, pavement design, and utility location and design.

In this paper, the authors principally address the investigation of large sites for major projects. As a result, a multi-phased study is presented. The first phase is appropriate for all site studies. The second phase may be subdivided into as many segments as necessary for the site in question, however, the investigative tools and techniques subsequently discussed in <u>Phase 2</u> are appropriate for the largest or smallest site study. The approach for different sized projects differs in scope, not in procedures.

<u>Geologic/Engineering Concerns</u>

Certainly, doline formation is one of the more spectacular effects produced by solution-prone carbonates below a site. However, many other aspects of solutioned rocks can be evaluated by the geological engineer charged with developing an appropriate foundation scheme for any proposed construction over carbonates. The very likely, great irregularity of the rock surface is a concern in both foundation design and excavation operations. The natural weathering process, in these materials, can produce a variety of shapes, and erratic and large changes in the physical properties of rock and residual soils over short, horizontal distances.

Much useful engineering/planning data for these rocks has been developed by the Geological Surveys in a number of the Appalachian states. Unfortunately, many of these sedimentary rocks, which were deposited under the same tectonic and environmental conditions, often receive different formation names. References are available to correlate some of these different formations (e.g. Fischer and Canace, 1989).

The mere existence of cavities in these rocks, in itself, may not necessarily be of concern to man's construction activities. The existing cavities in these rocks were created over long time periods, and in materials that are hundreds of millions of years old. Thus, the formation of new cavities, or the enlargement of existing ones by solutioning or erosion, is usually not a threat to engineered construction that has an economic life of some 50 to 100 years (unless the pH of the ground water is lowered). Furthermore, the available strength of these rocks, when unweathered, is quite high (the unconfined compressive strength of sound rock is on the order of 500 kPa). As a result, a relatively thin layer of sound rock over a cavity can support relatively large loads (e.g. The Natural Bridge of Virginia).

What is of major concern, however, is the formation of dolines beneath roadways, structures and utilities as a result of the migration of overlying soils into rock voids. The formation of a sinkhole in Cambro-Ordovician carbonate terrane occurs in a somewhat different manner than those in the well-publicized Florida occurrences. In these harder rocks, water and gravity forces enlarge soil voids formed above solutioned joints, fractures and channels in the underlying rocks. The result is an ever-growing soil or rock arch supporting the soils above. Increased erosion, additional loading and/or removal of a portion of the arch, by, for example, construction activity, can result in the formation of a sinkhole which is the surface manifestation of this soil collapse phenomenon. These events can occur much more rapidly than can solutioning of sound rock. They do, in fact, occur in time periods of concern to engineers and planners.

It is obvious that soil and rock voids do not provide the same absorption, adsorption or dilution opportunities for potential pollutants that conventional overburden materials do. Thus, the opportunity for slugs of contaminant to move rapidly to sources of potable water is a constraint that must be examined in any proposed development. The problem is certainly exacerbated in the many rural areas that utilize the fractured, solutioned carbonate formations for water supply because of their normally good aquifer characteristics.

<u>Phase 1 - Pre-Acquisition Property Analysis</u>
In a "limestone" investigation, the greatest return on investment is possible during the initial stage of any site study. The first step is to review the data available from Federal and State sources. Aerial photographs should be obtained from whatever time period is possible. The use of a series of photographs can be invaluable as karst features may develop slowly, over decades. Some features visible in early photographs can be obscured by later development. Persistent lineaments and circular shapes are particularly suspicious when they can be observed on a number of photographs, taken over a long time period.

The photographs are used as a basis for a geologic reconnaissance of the site and surrounding locale. Land forms are many times of assistance in assessing subsurface characteristics, except, perhaps, where glaciation has changed a pre-existing karst surface. Areas that a farmer will avoid during cultivation usually represent rock pinnacles (outcrops) or persistent sinkholes. Forested areas in otherwise fully farmed lands, many times indicate areas of shallow rock. Changes in vegetation sometimes indicate incipient sinkhole activity. Do not ignore subtle changes in texture or color on the photographs for not all karst features are large and\or obvious.

Old sinkholes are, many times, used as local garbage dumps and can be found below debris piles. Old mined areas can indicate mineralization along a fault zone. These fault zones are particularly susceptible to solutioning.

At this point in time, little investment has been incurred, but a great deal of information has been generated. The prospective owner/developer can thus evaluate the effect of the existence of a carbonate subsurface on property value and possible construction costs. Property acquisition costs may be reduced if negotiations are being conducted in good faith and with an understanding of the effects of the subsurface on facility layout, as well as design and construction costs.

<u>Phase 2 - Site Investigation</u>

Should the property in question continue in the development process, a conceptual design should be available to the owner's investigation team, prior to planning the site studies. For larger projects, it is usually advisable for the geotechnical consultant to acquire some site specific data, i.e. test borings, test pits, probes, to be used with the regional, Phase 1 data. These results can be used to develop the conceptual site plans to a greater degree than is possible with only regional data. From this site specific data, a generalized, subsurface conditions map should be developed. This map (not to be confused with the more common "constraints map") should provide information on the nature of the subsurface; approximate depths to rock, locations of doline prone areas, estimates of the permeability of subsurface soils and rock, estimated ground water depths, water quality, any anomalous subsurface conditions and similar. The project planners and designers, in concert with the geotechnical team member(s), can then ameliorate the possible deleterious effects of the carbonate subsurface.

These aforementioned aspects of the subsurface should be considered in site planning, as are other environmental constraints, e.g. wetlands, surface water flows, stands of trees, etc. Although, karst concerns often can be solved with an infusion of money, their economic impact can be lessened by appropriate planning.

The pneumatic probes, utilized as one of the investigative tools to develop the section presented on Figure 1, can be quite valuable, when used in conjunction with test borings, because of the air-trac rigs mobility, the speed of drilling and the relative economy of use. Although no core can be extracted, rock depths and qualitative competency can be estimated. With an experienced operator, it is possible to see or feel major changes in the stiffness of the overburden soils as well. Noting the depths that a loss in return air is experienced also aids in understanding the subsurface. It is particularly useful when the air loss in one probe hole returns from another.

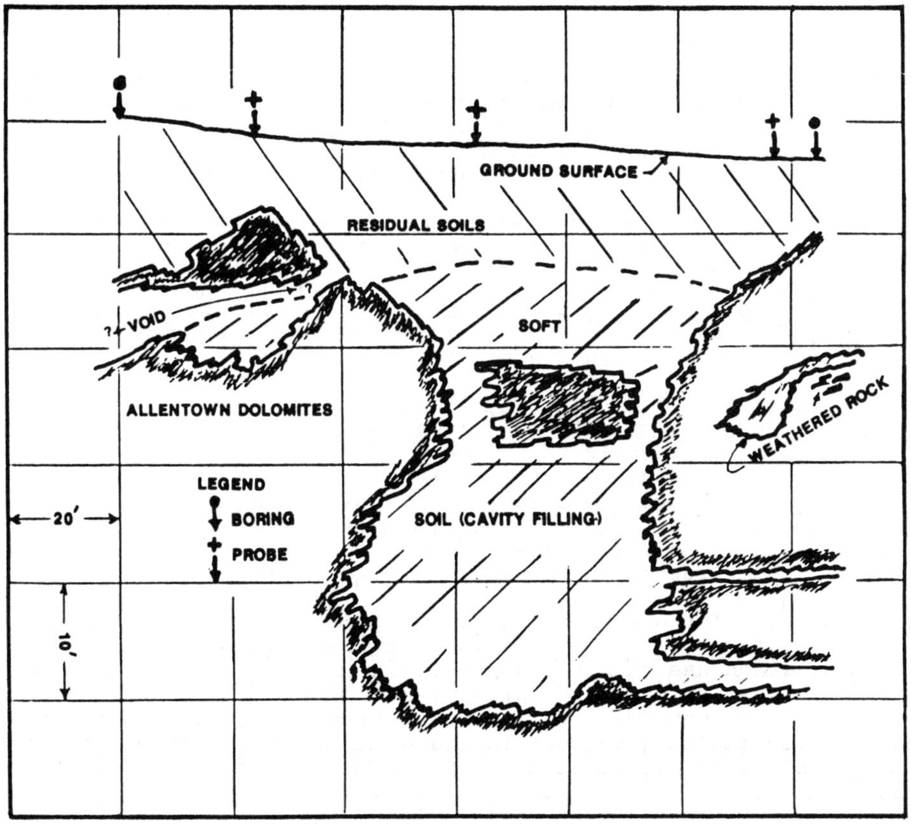

However, the authors believe that a judicious program of site exploration in Cambro-Ordovician aged carbonates, requires experienced personnel, wash boring drilling and double-tube, split-barrel coring. Increased knowledge at decreased costs can be developed from test pits, pneumatic probes and, in some

SECTION–ALLENTOWN FORMATION–CLINTON TOWNSHIP

FIGURE 1

instances, geophysical tools.

Many indirect procedures have been advanced for the detection of subsurface cavities, which have not yet become obvious as dolines. If we exclude the use of aerial photography and satellite imagery from our discussion, a host of geophysical techniques have been advanced as being suitable. These procedures include seismic reflection and refraction, electrical resistivity and conductivity, self potential, ground penetrating radar and gravity surveys.

If one is aware of the nature of the subsurface target, it is difficult to visualize how these geophysical prospecting procedures can be considered any more that marginally useful in carbonates of this nature. They function only in an auxiliary role to the performance of direct measurements of subsurface parameters.

Pinnacled rock, large boulders in place, the size and nature of soil voids and the lack of a coherent ground water table compromise the use of seismic surveys in most engineering applications in karst. Conductivity and resistivity are useful in evaluating soil conditions and can indicate variations in rock depth. However, often decomposed rock will look like sound rock on the records, 1/2 inch wide soil voids atop erratic rock surfaces cannot be inferred, inconsistencies in ground water levels can confuse the geophysicist, and cavities will not be seen unless they are of a very large size.

The use of ground penetrating radar (GPR) is also not recommended. The Cambro-Ordovician carbonates, with which the authors are familiar, usually weather to a clayey soil. During most of the year, the moisture content of these soils are quite high. Thus, conventional GPR is ineffective after a depth of a meter or two in these clayey, high moisture content, residual soils. However, in areas of shallow rock or sandy soils, the use of GPR may be of assistance. In most instances, the great variability in rock depths over a site and the need to better understand the pattern of variability makes the use of GPR uneconomical.

It is valuable to recognize that there are times in which these indirect methods can be effectively combined with direct methods of investigation, such as test borings or test pits. However, these geophysical tools cannot be substituted for carefully drilled test borings, qualified full time inspection, experienced and professional drillers and large diameter (Nx or better), double-tube, split-barrel coring.

There is invaluable information to be gained in examining the water softened soils immediately above the rock surface, noting the water losses at various depths, observing the clay filled-seams, stained joints, weathered cavity or joint sides in a full five foot run laid out in the core barrel, watching the drill rods fall through 10 or more feet of void, into soft cavity fillings, or angle off in one direction from contact with a rock pinnacle.

As an example of the variations that can be expected in a short distance, the section shown on Figure 1 was prepared from a composite of supervised test pits, test borings and pneumatic probes (with interpretations based on the authors' experience with similar subsurface conditions). A closer boring or probe spacing may have indicated even more changes in the subsurface topography and stratigraphy. Indirect measurements will not see a subsurface as variable as that shown on Figure 1.

Conclusions
There are many areas of the eastern U. S. that are underlain by solution-prone, Cambro-Ordovician carbonates. Many times the possible site development concerns in these areas are misunderstood or ignored.

Large cavities can be found in what are otherwise hard, crystalline rocks; voids form in the soil immediately above the rock; the rock surface is incredibly erratic, and large boulders can be found in an otherwise clayey soil matrix. Possible ground water pollution as a result of the open channels in soils and rock is of major significance. As the unweathered rock is competent, excavation operations can change from blasting (a task better off avoided so as not to disturb soil arches or soil-filled voids and channels), to backhoe removal of soft wet soils, within feet to tens of feet. Additionally, much of man's construction activity can seriously exacerbate karst-related problems, posing questions of safety and economy.

The procedures and techniques appropriate for multi-phased investigations are relatively simple, but predicated upon common sense and sound engineering judgment. Results are primarily based upon previous regional studies, aerial photography, test borings and experience. The organization of a successful study and the applicability of its' conclusions depend upon the interaction of a knowledgeable, multi-discipline acquisition, planning and design team.

References

Canace, R. & and R. Dalton, 1984, A Geological Surveys Cooperative Approach to Analyzing and Remedying a Sinkhole Related Disaster in an Urban Environment, in Sinkholes: Their Geology, Engineering and Environmental Impact, A.A. Balkema.

Fischer, J. A. & R. W. Greene, 1984, New Jersey Sinkholes: Distribution, Formation, Effects, Geotechnical Engineering, in Sinkholes: Their Geology, Engineering and Environmental Impact, A.A. Balkema.

Fischer, J. A., R. W. Greene, R. S. Ottoson & T. C. Graham, 1987, Planning & Design Considerations in Karst Terrain, in Proceedings of the 2nd Multidisciplinary Conference on Sinkholes and the Environmental Impacts of Karst, Orlando, Florida, A.A. Balkema (reprinted in Env. Geol. Sci., Vol. 12, No, 2).

Fischer, J. A. & R. Canace, 1989, Foundation Engineering Constraints in Karst Terrane, in Proceedings of the ASCE Geotechnical and Construction Divisions Specialty Conference, Northwestern University, Evanston, Illinois.

Palaeokarst of Yuen Long, north west New Territories, Hong Kong

DONALD V.FROST   *British Geological Survey, Newcastle-upon-Tyne, UK*
Seconded to Geotechnical Control Office, Hong Kong Government

ABSTRACT

A project in the north west New Territories of Hong Kong, undertaken by the Geological Survey of the Hong Kong Government, has provided geological maps at a scale of 1:5 000 of a region known as the Designated Area which is largely earmarked for new-town "high-rise" development.

It has proved a complexly faulted area of Carboniferous metasediments associated with Jurassic and Cretaceous granites and volcanic rocks. The rocks have been faulted and metamorphosed and are now buried beneath a superficial cover of late Tertiary and Quaternary superficial deposits to depths of 80 m.

Carboniferous limestone has been metamorphosed into marble and at some stage during the Tertiary was exposed to sub-aerial denudation resulting in extensive karstification with an epikarst zone some 10 to 30 m thick. This zone contains all sizes of solution features from widened joints to caverns 20 m high. The cavities contain clays, silts and sands, the dating of which is in progress. The epikarst surface is highly irregular with a relief of some 60 m showing hollows which are interpreted as dolines.

The maximum frequency of dolines coincides with the underlying Ma Tin Member of the Yuen Long Formation - a unit of particularly pure, white massive marble. The largest cavities also occur in this Member in close proximity to major structural lines of weakness often associated with intruded igneous rocks.

The project has provided an accurate basis for answering geotechnical enquiries in the Designated Area. The area originally thought to be underlain by cavitous marble now has been reduced by a half.

Further research into the hydrogeology of the area and geochemisty and isotope geology of the rocks is planned.

Introduction

Some ten years ago, Carboniferous marble was discovered in a borehole in the north west New Territories beneath a thick cover of superficial deposits, which mask the complex solid geology of the area. This region is now earmarked for new-town development proposing tower blocks up to 30 storeys high. Recently, however, cavities were discovered in the marble, up to 20 m in height, the presence of which could seriously affect the development of the area. As a result, a zone was delineated covering some 60 sq. km and called the Designated Area, in which all construction work was carefully monitored by the Geotechnical Control Office of the Hong Kong Government.

Funds were allocated in 1987 for a deep boring programme to locate the limits of the marble substrate and help predict the areas of potential foundation difficulty. A series of geological maps at

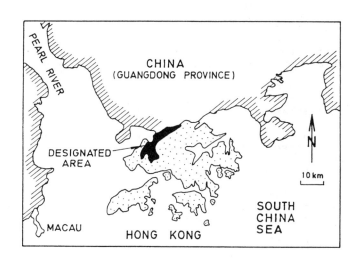

Figure 1. Location map showing limits of the Designated Area

a scale of 1:5 000 were proposed to cover the Designated Area and provide a basis for the marble project. The two year project is now past the half way stage and much of the geological mapping is complete.

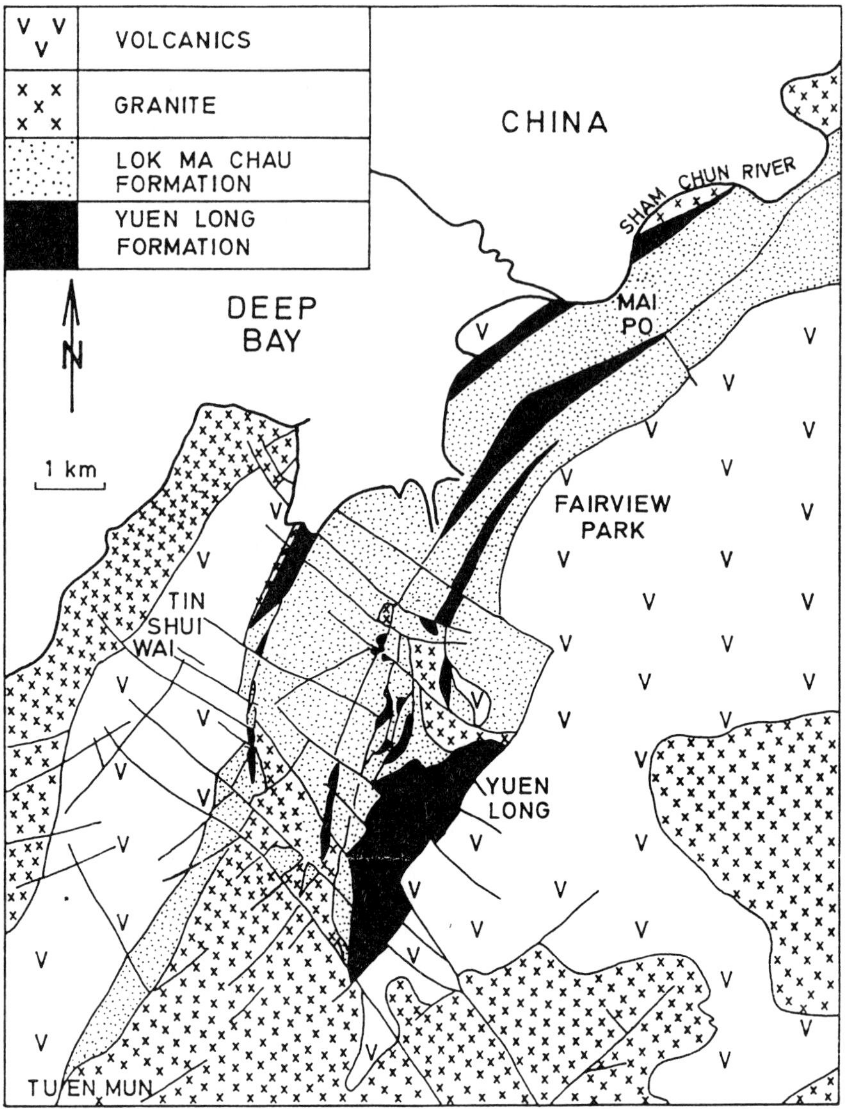

General Setting
     The Designated Area (Figure 1) occupies a low lying region most of which is only a few metres above sea level. (In Hong Kong, Principal Datum (PD) is 1.2 m below Mean Sea Level and 0.15 m above Admiralty Chart Datum.) It is some 20 km in length and 3 km wide stretching from the town of Tuen Mun in the south through Yuen Long to Mai Po and then northeastwards along the Chinese Border.

     Much of the area is agricultural, with fish and duck farming predominating in the north eastern areas bordering the Pearl River estuary and Guangdong Province.

     Tuen Mun is the largest "high rise" town of the region but Yuen Long, designated as a satellite town 10 years ago, has grown to have a population of some 90,000. Another "high rise" urban area is proposed at Tin Shui Wai whilst Fairview Park covers an area of nearly 2 sq. km but has been specified as a low density, low rise executive development.

     The combination of extremely hilly terrain, deeply weathered rock profiles and high seasonal rainfall has in the past resulted in some severe landslide problems in Hong Kong's dense urban environment.

Figure 2. Geological map of the solid rocks of the area around Yuen Long

     The Geotechnical Control Office (GCO) was established in July 1977, mainly in response to major landslide disasters. It is responsible for a wide range of getoechnical engineering activities related to the safe and economic utilization and development of land.

     A large library, the Geotechnical Information Unit, contains all the geotechnical data from these investigations. The Designated Area was originally delineated by the Geological Survey Section of GCO based on not only information from this shallow borehole database but also from nearby surface exposures around the margins of the area.

     A need for deeper boreholes was soon realised to enable detailed and accurate geological maps to be constructed to elucidate the geological structure hidden beneath the superficial deposits of the Designated Area.

Previous Research
     The first comprehensive geological map of Hong Kong was surveyed by Allen & Stephens in 1971. The Yuen Long area was shown to be underlain by alluvial and colluvial deposits, Systematic revision by the Hong Kong Geological Survey began in 1983. The north west New

240

Territories were covered by Sheets 2 & 6, Lai (1987) and Langford et al.(1989)   A general account of the geology was given by McFeat-Smith et al. in 1989.

The first detailed account of the engineering and geological aspects of Yuen Long was provided by Beggs & Tonks in 1985 with Pascall (1987) reporting on the cavernous marble with respect to specific site investigations east of Yuen Long town. Siu and Wong (1985) emphasised the hydrogeological factors relating to the area.

The past literature on the area is therefore not extensive and has largely been produced in the last 5 years.

Geological Setting

The Designated Area occupies a structurally complex, elongate, curved Hercynian basin comprising mostly Carboniferous rocks (Figure 2). They were originally limestones, mudstones, siltstones, sandstones and conglomerates but have been converted into metasediments by one or more metamorphic episodes.

The basin is surrounded, except to the north east, by granite and volcanic rocks of the Jurassic and Cretaceous periods, forming hills rising to 600 m in height. The basin extends north eastwards into the Chinese mainland where Carboniferous marble is quarried at surface near Shenzhen. To the south of Tuen Mun isolated faulted outcrops of Carboniferous rocks are also known at Mouse Island, the Brothers Islands and on Lantau Island.

The metasediments dip generally to the north west at about 60°. They were faulted during the Hercynian Orogeny with a dominant north northeast trend parallel with the basin margins (Figure 2). This trend is interrupted by a series of later north west to south east cross-faults, probably reactivated in the Tertiary, and which offset the earlier dislocations resulting in a complex Carboniferous sub-crop.

This complex pattern of solid geology is buried to depths of 20 to 80 m by superficial deposits. The general succession is shown in the following table.

GENERALISED SUCCESSION OF THE DESIGNATED AREA

Recent	Fill or made ground	
Quaternary	Pond & Estuarine muds & peat	)
	Marine clays, silts, with shelly faunas	)
		) Superficial
	Alluvial gravels, sands, silts and clays	) Deposits
		)
		)
Tertiary	Debris flow deposits	)
	Basaltic intrusions	
		)
Cretaceous	Granites and granodiorites	)
		)
Jurassic (Upper)	Volcanic ash tuffs & breccias	)
(Lower)	Metasandstones	)
		)
Carboniferous	Metaconglomerates metasandstones	) Solid Rocks
	Metasiltstones & graphite schists	)
	Marble	)
		)
Devonian	Metasandstones and conglomerates	)

Carboniferous Succession

The Carboniferous rocks of the Designated Area belong to the San Tin Group which is divided as follows:-

Lok Ma Chau Formation   (Tai Shek Mo Member
                           (Mai Po Member

Yuen Long Formation   (Ma Tin Member   (New name introduced in this paper)
                           (Long Ping Member (New name introduced in this paper)

The Yuen Long Formation comprises largely marble dark grey to black at the base (Long Ping Member) passing up into a pale grey to off-white colour at the top (Ma Tin Member). Minor intercalations of phyllite, dolomite and siliceous marble are also present.

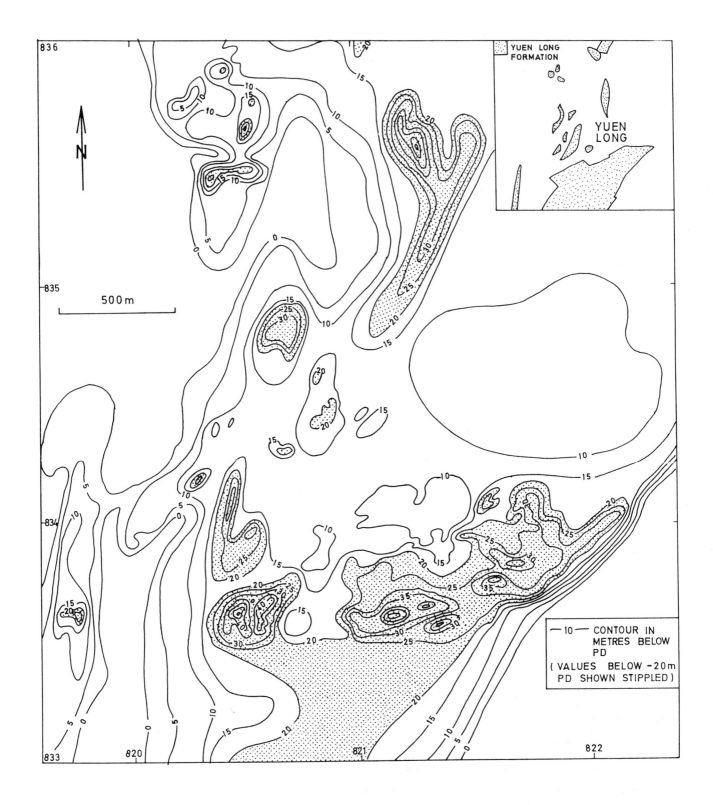

Figure 3.   Rockhead contours of the Carboniferous strata around Yuen Long.

The Long Ping Member is at least 200 m in thickness but the base has not been proved. The insoluble residue of this marble comprises fine silt and clay together with finely divided pyrites. A complex internal structure shows disruption of the original bedding, probably by a combination of both the metamorphic processes and the tectonic movements.

The Ma Tin Member is also some 200 m in thickness, usually massive fine to coarse grained recrystallised marble (0.1 to 0.5 mm) containing few impurities or variations in colour. The recrystallisation has destroyed all the original textures and faunal remains. The top few metres of the member contains in places an alternation and intercalation of metasilt-stone and metasandstone showing a passage into the overlying Mai Po Member; elsewhere the upward change is sharp. The Ma Tin Member comprises virtually pure calcium carbonate which renders it vulnerable to dissolution.

The Lok Ma Chau Formation comprises a coarsening upwards cycle of metasiltstone and metasandstone (Mai Po Member) passing into sandstone

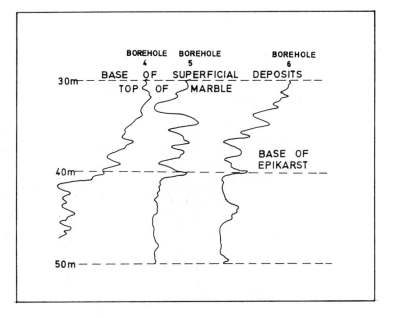

Figure 4. Gamma-log trace of three boreholes in Yuen Long showing the top 20 m of marble in the Ma Tin Member

with conglomerate (Tai Shek Mo Member). The Mai Po Member is probably 500 m thick although the total sequence has not been proved. Towards the base of the sequence graphite schist has been proved interbanded with other lithologies showing an overall thickness of about 100 m. The Tai Shek Mo Member is in the order of 300 m in thickness.

Distribution of the Yuen Long Formation

The most recent interpretation of the geological configuration of the Carboniferous sub-crop is shown in Figure 2. The location of the Yuen Long Formation is the most important for reasons of foundation safety. By far the largest area of subcrop (3 sq km) occurs in the southeast of the town and extends a few kilometres further south. It comprises largely the pure white marble of the Ma Tin Member. Other smaller areas underlain by marble are lenticular in shape, bounded by major faults and are of mostly dark grey marble of the Long Ping Member.

A large area near Tin Shui Wai was originally thought to be underlain by marble but this has been shown to comprise volcanic breccia containing marble clasts and is the subject of a paper by N.J. Darigo in this volume.

The upper part of the marble shows an epikarst zone, irregular in profile, revealing solution phenomena with widened joints, cavities and caves as common features. This epikarst zone varies between 10 m and 30 m in thickness but is underlain by further cavities to a maximum depth of about 80 m. Many of the cavities are filled or partially filled by sedimentary deposits.

The Carboniferous rockhead surface around Yuen Long is shown in Figure 3. It ranges from 0 to -60 mPD. The non-calcareous formations show a generally smooth, fairly gently sloping or undulating surface whilst that of the Yuen Long Formation is prone to many irregularities representing numerous dolines interspersed with pinnacles on the marble surface. These effects are most prevalent over the large area of subcrop of the Ma Tin Member southeast of Yuen Long. The apparent sudden end of such irregularities along the southern boundary of the town is due to lack of data beyond the town limits.

Other areas showing deep depressions and irregular surface phenomena reflect the smaller elongate fault slices of the Yuen Long Formation (Figure 3) to the north and west of the town. Again a close correlation exists between the irregularities and the presence of marble bedrock and its associated structural boundaries.

Borehole cores confirm that the top surface of the marble is moderately weathered and slightly stained (Grade III). Joint surfaces are fluted and grooved. Several boreholes have been logged by a natural gamma probe in order to assist in correlation. However, a useful sideline of this research has been the trace of the logs through the epikarst zone. Typical

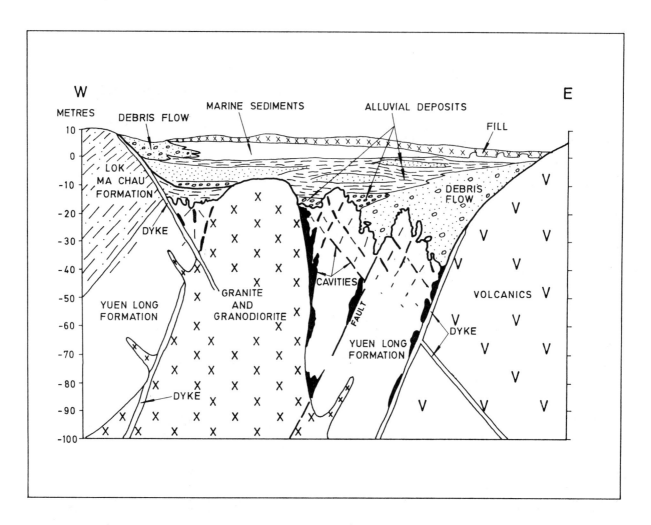

Figure 5. Generalized section of the strata in the Yuen Long area.

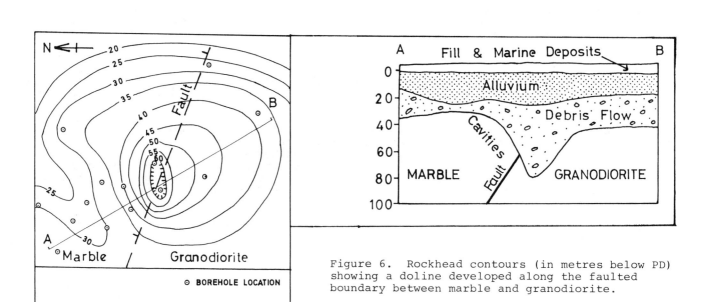

Figure 6. Rockhead contours (in metres below PD) showing a doline developed along the faulted boundary between marble and granodiorite.

logs are shown in Figure 4. A distinct gradient exists between the argillaceous-rich top to the marble and the increasing purity at greater depths. From the logs it is clear that c -41 m is the base of the epikarst - a boundary not obvious from the cores.

Figure 5 shows a typical cross-section through the karstic marble of Yuen Long based on examination of several thousand metres of borehole core and photographs.

Detailed analysis of a doline (sinkhole) north of Yuen Long (Figure 6) shows a diameter of c. 100 m with average slope to the sides of between 20 and 40°. Their maximum frequency and size can be closely correlated with the pure white marble of the Ma Tin Member and the proximity to major structural lines of weakness. In this area such structures are thrust planes and or faults usually represented by strongly fractured rock with a mylonitic texture. These zones of weakness bring the marble into juxtaposition with other lithologies including volcanic, granitic and miscellaneous dyke-rocks.

The deep weathering rock profile in tropical climates is well documented. In Yuen Long this averages some 20 m but is invariably twice that depth over major dislocations. These zones of weaker more permeable rock have provided easy avenue of access for the groundwater to circulate. The larger the dislocation the greater the size of the cavity. This may be enhanced when the adjacent rock contrasts markedly in composition or strength with the marble Tertiary igneous dykes seem particularly prone to deep weathering.

The detailed topography of a marble surface is shown in Figure 7. This small site in Yuen Long has been intensively drilled and shows a combination of karstic and structural features. The steep gradient to the marble surface in the north of the site may represent a dip slope but as the recrystallisation of the marble has destroyed all the primary structures it is difficult to assess. The steepness of the slope and the presence elsewhere along this feature of igneous intrusions indicates a tectonic origin may be more likely.

Caves and Cavity-fill

The majority of cavities in Yuen Long have been choked by Quaternary sands, silts and clays. The earliest deposits of silty and fine sand show evidence of cross-lamination but the later "fill" varies from dark grey to black sandy clay to grey-brown calcareous silty fine sand, usually with horizontal laminae. Some irregular and angular chert fragments are incorporated in the finer clays suggesting roof-fall debris as a possible origin. No dating of these deposits has yet been completed but similar caves in China have proved to range in age from Pliocene to late Pleistocene.

Engineering Aspects and Future Research

The publication of the geological maps of the Designated Area highlight the areas underlain by marble. The differentiation of the two main varieties of marble will further emphasise the location of cavitious marble hazardous to high rise construction. The correlation between geological structure and the frequency and size of cavities remains an important factor which requires further detailed investigation to enable accurate prediction of the position of the larger deeper cavities.

Now that the geology is better understood, hydrogeological research can follow. An initial project has recently been completed by the Hydrogeological Section of the British Geological Survey. Analyses of the groundwater is in hand. Early indications would suggest that very little of the resource is being used in Yuen Long. Many of the wells orginally sunk for supplies of flushing water have silted-up. The summer wet-season recharge of the Carboniferous and superficial deposits aquifer from the surrounding catchment area has, therefore, never been tested. Any overpumping of the goundwater, which is now close to 0mPD, would probably induce a rapid influx of sea-water from Deep Bay. Increases in salinity were observed in wells in close proximity to the main nullah. Rapid overpumping of groundwater producing a sudden cone of depression is, therefore, the only condition liable to induce sinkhole formation in the karst area.

A pilot scheme in Yuen Long to test the effects of such pumping in marble areas is under the direction of Dames & Moore, Hong Kong.

A better understanding of the deep geological structure can only be obtained from a complete knowledge of the Carboniferous succession in Hong Kong. Two deep (400 m) boreholes are planned in order to prove the total thickness of the marble succession and the associated clastic rocks. From this data it will be possible to redefine the Designated Area, show the limits of marble and also indicate where it occurs at depths of greater than 150 mPD i.e., at depths where it is no longer a hazard to construction.

Age-dating of the marble is presently under investigation by the British Geological Survey's Isotope Geology Centre in the United Kingdom. Recent work has successfully dated marbles from Taiwan so the discovery of at least three times the proportion of lead in the Yuen Long marbles indicate that this project should provide additional worthwhile and new data.

Conclusions

Palaeokarst has been confirmed in the Yuen Long area of Hong Kong. This has provided difficulties to the construction industry in a crowded Territory where all the easily accessible and cheaply worked sites have already been developed.

Geological mapping at a scale of 1:5,000 aided by some 30 specially sunk boreholes to depths of 150 m, has provided a basis for answering geotechnical enquiries in the Designated Area. The area, which was originally estimated to be underlain by cavitous marble has been reduced by about one half.

Two deep boreholes are proposed to try and prove the complete Carboniferous sequence and so ensure that the stratigraphy and structure of the area are fully understood.

Acknowledgement

This project was financed by the Hong Kong Government with secondment of staff from the British Geological Survey and Dames & Moore - Hong Kong.

This paper is published with the approval of the Director of the Civil Engineering Services Department, Hong Kong Government and the Director of the British Geological Survey (NERC).

My sincere thanks to all my colleagues in this project and in the Geological Survey Section who provided help guidance and backing at all times and to all members of GCO who advised and assisted in the many different facets of the research.

Figure 7. Rockhead contours of the San Tin Group of the Carboniferous in a small site in Yuen Long

References Cited

Allen, P.M. & Stephens, E.A., 1971, Report on the Geological Survey of Hong Kong 1967-1969. Hong Kong Government Press, 116 p plus 2 maps.

Beggs, C.J., & Tonks, D.M., 1985, Engineering Geology of the Yuen Long Basin, Hong Kong Engineer, p 33-40.

Geotechnical Control Office (GCO), 1988, Carboniferous solid geology of the northwest New Territories: 1:20 000 scale map, Drwg No. GS-SP/713, Civil Engineering Services Department, Hong Kong Government.

Lai, K.W., 1987. Solid and superficial geology, Sheet 2; Unpublished map at 1:10,000 scale, Hong Kong Geol. Surv. Geotechnical Control Office, Hong Kong.

Langford, R.L., Lai, K.W., and Shaw R., 1989, Solid and superficial geology of Yuen Long, 1:20,000 Sheet 6 Hong Kong Geol. Surv. Geotechnical Control Office, Hong Kong. Series HGM 20, Ed. 1-1989.

McFeat-Smith, I., Workman, D.R., Burnett, A.D., and Chau, E.P.Y., 1989, Geology of Hong Kong. Bulletin of Association of Eng. Geol., Vol. XXVI, No. 1, p 23-107.

Pascall, D., 1987. Cavernous Ground in Yuen Long Hong Kong. Geotechnical Engineering, Vol. 18, p 205-221.

Siu, K.L. and Wong, K.M., 1985, Concealed marble at Yuen Long., Proceedings of Conference on Geological Aspects of Site Investigation Bulletin No. 2., Geol. Soc. Hong Kong.

Ground subsidence movements upon the Cretaceous chalk outcrop in England – Sinkhole problems and engineering solutions

JAMES P.KIRKWOOD & CLIVE N.EDMONDS   *Applied Geology, South Lodge, Cranford, Blackdown, Royal Leamington Spa, Warwickshire, UK*

ABSTRACT

Induced subsurface ground movements in areas of pre-existing metastable ground conditions associated with natural solution features developed in Cretaceous Chalk bedrock can result in sinkholes at the surface. These ground conditions are potentially hazardous and may not, prior to surface development, be apparent at ground level. Man's activities in creating urban and infrastructure development tends to increase the incidence of subsidence. Unless the ground conditions are carefully investigated and the appropriate foundation solutions adopted severe structural damage and even a threat to public safety can result. The programme of investigatory works requires careful planning and close monitoring to ensure that all necessary data for design is obtained in a cost-effective manner and that the likely scope and nature of the problematic ground conditions are defined at an early stage. Thereafter the foundation design options have to be explored to provide a safe economic answer and any further investigations tailored to optimise the solution(s). A methodology is proposed in which the use of desk study and hazard mapping assessment is used to provide cost-effective geotechnical data to aid in the development of foundation design options to overcome these problematic ground conditions associated with chalk.

Introduction

The Cretaceous Chalk in England, covering some 250,000 square kilometres (96,000 square miles), has the largest outcrop area of any British limestone. This soluble carbonate rock is pure (98% $CaCO_3$), relatively soft, and porous, (up to 50% or so) and retains its lithological character because it is composed of low magnesium calcite (Higginbottom 1965, Hancock 1975). Although the Chalk contains very few caves it possesses numerous solution features developed by dissolution acting from the surface downwards in the form of solution pipes and swallow holes (Figures 1 & 2). Solution pipe infilling deposits are softened and disturbed following gravitational subsidence into the natural cavities concurrent with the dissolution process. Hence the infilling deposits tend to be metastable in character, which if disturbed, can reconsolidate producing a subsidence sinkhole at the surface (Figure 3). These induced sinkholes frequently afflict new and existing urban areas and infrastructure.

Subsidence triggers and mechanisms

Most ground subsidences are induced by man, the typical subsidence triggers being static or dynamic loading and more commonly, concentrated water flows associated with leaking water services. Other contributory factors include the changing of the natural groundwater flow regime by the creation of preferential drainage channels (for example backfilled trenches in which services have been placed) and the creation of areas of increased ground absorption as a result of the construction of houses, garages, roads and paved areas, concentrating rainfall in reduced areas of open ground.

There are two common forms of metastable ground conditions which give rise to induced sinkholes; these are shown schematically by Figure 4. If the cover deposit is a low cohesion permeable granular deposit, then solution subsidence can sometimes produce loose zone micro-voided ground conditions. In the case of a sand and gravel layer, the loss of basal support, below a circular area, during solution subsidence, encourages a local downward settlement of the sand grains and gravel clasts making up the layer in the affected area. Assuming the layer was compact to start with, the localised solution subsidence effectively causes a portion of the layer to take up a larger volume of space in the ground than it previously occupied. The volume increase of the layer occurs by an increase in void ratio or porosity, and hence a lowering of the bulk density. The extent to which the deposit fabric can expand to accommodate the downward movement depends on the angularity and grading of the deposit; this effect is shown by Figure 4A. These conditions commonly occur across large areas of the Chalk outcrop where permeable granular

Figure 1: Example of solution pipes

Figure 2: Example of swallow hole

conditions commonly occur across large areas of the Chalk outcrop where permeable granular deposits overlie piped chalk bedrock. Destabilisation of the loose zone micro-voided ground conditions produce ground movements which finally lead to the formation of a ground surface depression or subsidence sinkhole.

A modification of the previously described underground movements can result in the formation of a metastable cavity normally within variably cohesive granular superficial deposits overlying piped chalk bedrock. If a more cohesive, or resistant, layer is present in the superficial sequence then instead of the solution subsidence effects continuing to migrate up through the sequence, the movements halt at the underside of the resistant layer. As the solution pipe continues to deepen, the deposits below the resistant layer continue to undergo gravitational settlement creating a metastable cavity roofed over by the resistant layer. Cavities produced have curved walls with arched roofs and their volumetric size tends to be a function of the overburden pressure (confining stress), their size often decreasing with increasing overburden pressure; this effect is shown by Figure 4B.

Destabilisation of the metastable cavity by a subsidence trigger causes failure of the roofing layer by the shearing of successive arched shells of the deposits. If the destabilising agent acts only briefly then the upward migration of the cavity by roof shell failure may be arrested, but, if active for longer, will finally result in the cavity reaching the ground surface producing a subsidence sinkhole.

Ground investigation methodology

Although many sites are successfully developed on the Chalk outcrop following conventional site investigation philosophy there are a significant number of sites where this approach is inadequate and ineffectual. The standard approach rarely gives any early warning of potential sinkhole problems. The additional costs, should sinkhole problems not occur until the construction phase can be of the order of several thousands of pounds. There needs to be a better way to anticipate sinkhole problems ahead of site development in order that the risk of such costly engineering adversities can be averted or minimised.

A large natural cavity location computerised database (approximately 5000 records) has been assembled and spatially analysed to identify the geological, hydrogeological and geomorphological influential factors which control the natural cavity distribution. These factors have been ranked, numerically weighted and combined into a semi-quantitative subsidence hazard mapping model formula (Edmonds 1987, Edmonds et al.,1987a). The mapping model formula which is used to produce a subsidence hazard map was the subject of research and has now been further refined and improved during the past three years as the learning curve of experience has been extended.

Database search and hazard mapping

A greatly improved ground investigation structure and approach is now available for the Chalk outcrop. At the outset of a project a search of the computerised natural cavity database should be carried out to indicate whether the natural cavity potential at the site of interest is favourable or unfavourable. If the database search indicates favourable conditions, a second stage pre-ground investigation exercise known as Subsidence Hazard Mapping can be undertaken.

The hazard mapping techniques are applied to a site of interest in a series of steps as outlined below:

1. Undertake a Desk Study which includes a review of the historic site development as recorded on old Ordnance Survey plans and search all available published and unpublished geological records.
2. Obtain, view and interpret available aerial photographs for the site of interest. It may be possible to supplement this information with airborne multispectral scanner imagery (Edmonds et al., 1987b).
3. Carry out a site walkover survey to record, measure and evaluate the nature of the site terrain utilising engineering geomorphological mapping techniques.
4. Produce a factual database map of detected and surveyed natural and artificial cavities, geology and hydrogeology.
5. Apply predictive subsidence hazard mapping model formulae and calculate the variable subsidence hazard potential for both natural and artificial cavities.
6. Compile and produce a subsidence hazard map.

An example of a completed subsidence hazard map is given by Figure 5.

Field investigations

Following appraisal of the desk study stage information it is next required to prove whether the predicted natural cavity potential at a site exists or not. The conventional

Figure 3:   Example of subsidence sinkholes

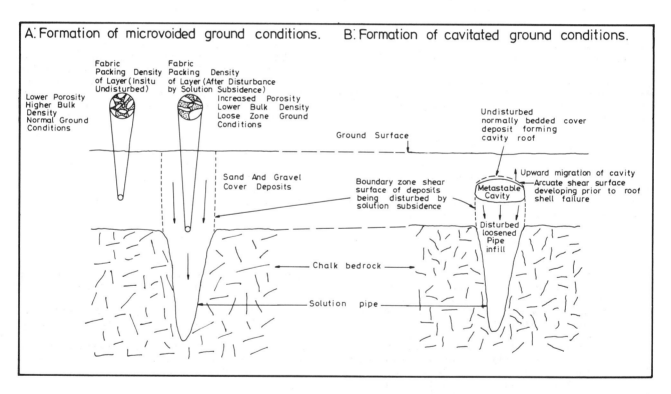

A. Formation of microvoided ground conditions.   B. Formation of cavitated ground conditions.

Figure 4:   Schematic illustration of metastable ground conditions associated with solution pipes

techniques of field investigation (Weltman and Head, 1983) are comparatively expensive techniques which provide low site coverage data per unit cost. Hence costs can be high where close centred investigations are required to locate randomly distributed natural cavities within a particular hazard zone. To minimise the cost impact the field investigation is designed as a staged approach with investigation technique options

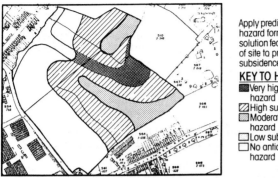

Apply predictive subsidence
hazard formula for natural
solution features to all parts
of site to produce a
subsidence hazard map.

KEY TO HAZARD ZONES

▨ Very high subsidence
hazard

▨ High subsidence hazard

▨ Moderate subsidence
hazard

☐ Low subsidence hazard

☐ No anticipated subsidence
hazard

Figure 5:   Example of a subsidence hazard map

depending on ground conditions.   The early emphasis of the field investigations is placed on proving whether a natural cavity potential exists and revealing general characteristics rather than aiming to locate every individual natural cavity present.

To achieve this picture the initial options are geophysical surveying and/or trial pitting, followed by the other techniques discussed briefly below.

Geophysical techniques using ground conductivity, resistivity and ground probing radar have been successfully utilised to identify anomalous conditions associated with the subsurface expression of natural cavities (McCann et al., 1982, 1987, Taylor and Fursman 1989) to depths of 25m.

However, it is felt that the geophysical survey should only be used for general guidance purposes of indicating the presence or absence of natural cavities across an area rather than as an absolute survey of cavities present at a site.   Starting the investigation programme with geophysics may be attractive in terms of areal coverage at reasonable cost.   Before embarking on a geophysical survey careful thought should be given to the physical contrasts being detected and the site characteristics to be certain of choosing the technique most likely to produce useful results or whether geophysics is even appropriate, since it is primarily a green field technique.

Trial pitting is used in conjunction with geophysics where appropriate, to identify soil fabric disturbance by solution subsidence activity.   This may be recognised by the presence of locally dipping strata, discordant fabric structure, mixed lithologies, brecciation, shearing, faulting, dissolution residues and precipitates.   Chalk bedrock level anomalies have also been uncovered where the cover deposits are of limited vertical extent.

Where trial pitting is not feasible or where the depth of superficial cover exceeds the depth of the pit, then in circumstances where soft to firm or loose to medium dense soils overlie the feature of interest portable continuous dynamic probing equipment such as that manufactured by Borros or Lindenmeyer (Scarff 1988) has proved efficient.   The equipment produces qualitative data by recording the penetration resistance, measured as blows/100mm of penetration.   Interpretation of the data by comparison of the blow count profile with depth can determine the ground profiles characteristic of natural cavities.

There is also an equipment variant of the above known as the Geodrive soil sampling system which can provide complementary ground truth data in the form of continuous samples 40 to 80mm in diameter.   Where conditions are favourable the probes can be driven to depths of 25m and the sampling equipment can reach 15m.

Where the overburden materials will not permit the use of any of the above techniques, particularly where dense or stiff soils overlie chalk rockhead, then recourse to conventional cable tool, rotary or rotary percussive techniques has been required.

Some element of usage of these or similar techniques is necessary at all sites to provide ground truth data with which to correlate interpretative information, and, where it is necessary, to sample at depth in order to determine design characteristics.   However, these techniques have been considered as a last resort for the bulk of the investigative work associated with the location of natural cavities.

A very important aspect of the staged investigative approach is that the proving of the general site ground conditions is only advanced until it is recognised that the natural cavity potential exists at the site and matches that predicted, or for various reasons (e.g. glacial erosion) solution features are not present and hence investigations can follow a conventional pathway. However, if the natural cavity potential is significant the emphasis of the investigative works is changed from being one of proving the general site ground conditions to one of proving stable ground conditions at the locations where the structural foundations are to be installed.

Sub-Structure Solutions

Where active solution features are suspected at a chalk site then the sub-structure options available to the engineer fall generally into one of the following categories:

(a) Conventional foundations placed on secure ground which has been proved by investigation to be free of the influence of disturbed or collapsing materials.

(b) Conventional foundations modified to withstand differential movements associated with differing engineering characteristics of the sub-surface materials due to previous solution subsidence activity.

(c) Special foundations designed to transfer structural loadings through the zones of disturbance and into underlying stable materials.

In all three cases the scope and extent of the field investigation undertaken is different and in many instances the direction of the investigation may be altered with changing emphasis on the sub-structure solution.

At the outset consideration must be given to the eventual end use of the site, the nature of the preferred sub-structure solutions likely to be utilised and the general site layout.

Figure 6 summarises as a decision flow net, key issues in the development of the investigative programme for new development upon Chalk sites using the preferred approach set out in this paper.

In most cases the optimum solution lies in determining the location of problem features and planning the development around these areas. In these circumstances the use of Hazard Mapping is ideally suited to allow the development to be planned and laid out to optimise site usage in areas remote from problematical ground conditions. Problem areas can then be utilised say, for public open space, or if a risk to personal safety is confirmed, can be treated to remove that risk but at a cost less than that required to provide a secure foundation.

If success can be achieved at this level, then conventional foundations can be utilised subject to all the normal considerations of ground strength and performance.

However, if the field investigations prove the existence of problem ground in areas where development is essential then special consideration will be required to determine the form and viability of the sub-structure system to be adopted.

Where the chalk rockhead profile has been altered by dissolution, then ideally the fieldwork should define the rockhead variation. Sub-structure solutions based upon the use of a piled foundation (Hobbs and Healy 1979), Fleming et al.,1985) carried through to natural undisturbed chalk, vibro-compaction or replacement treatment (Greenwood and Kirsch 1983) of the natural cavity infill have all been utilised.

Where natural solution features can be delineated, then the use of piled foundations, or ground beams spanning the solution feature (Bowles 1977) can provide a successful foundation system.

Under suitable conditions where predominantly granular materials overlie the solution feature, then the use of vibro-compaction techniques to break down any voids and compact the solution pipe infill with, if necessary, the inclusion of granular fill to make up volume have also been successfully used.

Although not as yet used by the authors, the technique of compaction grouting (Warner and Brown 1974) can be considered. In this system a stiff grout is injected under pressure normally utilising concrete pumping technology. The resulting grout 'football' is enlarged filling any voidage present and at the same time generating radial stress within the soil which acts to recompact disturbed ground.

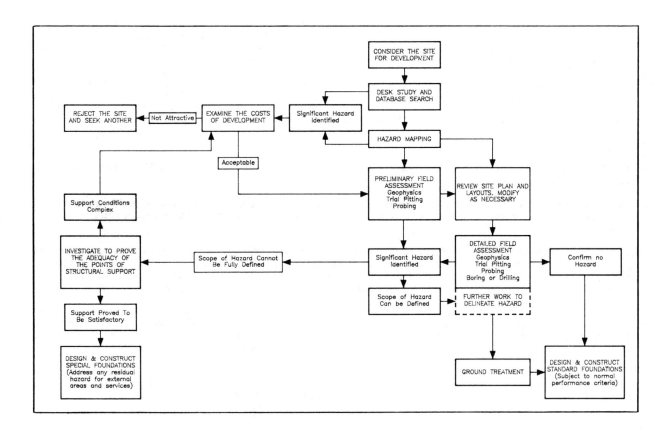

Figure 6: Decision flow net for the investigation of sites on Chalk

Providing a satisfactory ground performance is reinstated, then conventional shallow reinforced foundations are normally suitable for most residential and light-weight industrial development.

In situations where the fieldwork cannot, because of overburden cover or dense overlying layers, produce conclusive information on the nature of the problem, then the technique adopted is the use of a piled sub-structure. Here it is necessary to redirect the scope of the field investigations to prove that satisfactory ground conditions are available at the location of each support. If this can be proved then piles can be installed at each location and the structure supported from this system.

The structural solutions discussed may safeguard the structure but a risk remains for external areas of infrastructure. This risk of instability needs careful assessment to decide if this is acceptable.

Conclusions

Development upon certain areas of the Chalk outcrop can be influenced by the presence of natural solution features. If these factors are not recognised or allowed for within the initial site development planning then significant additional costs associated with ground investigation and remedial sub-structure solution can accrue.

A methodology of investigative approach for dealing with these sites is presented and it is suggested that this approach be adopted as standard for new development upon the Chalk outcrop and is also applied to existing development which may be influenced by changed or changing groundwater flow patterns.

The approach places emphasis upon the initiation of desk study and predictive hazard mapping techniques, ideally applied prior to the planning of the development. The layout of the proposed development can then be optimised to limit potential ground support problems and suitable controls built into the development to minimise the impact of site groundwater and services acting as subsidence triggers.

Having established the basic 'risk' conditions on site the use of cost effective investigative techniques to confirm, or, if necessary, modify the predictive hazard mapping model is recommended. This stage of work should be carried out in close liaison with the sub-structure designers in order that the design options can be assessed. There may come a stage in this investigative process whereby the site ground conditions will not permit the use of conventional foundation techniques and where specialised foundations may be necessary. At this stage the emphasis should be adjusted from determination of the overall picture of ground conditions to being more effectively tailored to localised proving of ground conditions within the areas of support for the structures.

References

Bowles, J.E. 1977 Foundation analysis and design, Second Edition, p 750, McGraw-Hill Kogakusha Ltd., Tokyo.

Edmonds, C.N. 1987 The Engineering Geomorphology of Karst Development and the Prediction of Subsidence Risk upon the Chalk Outcrop in England, PhD thesis London University, unpublished.

Edmonds, C.N. 1988 Induced sub-surface movements associated with the presence of natural and artificial underground openings in areas underlain by Cretaceous Chalk. In Bell, F.G., Culshaw, M.G.,Cripps,J.C. and Lovell, M.A. (eds), Engineering Geology of Underground Movements, Geological Society Engineering Geology Special Publication, No. 5,205-214.

Edmonds, C.N. and Kirkwood, J.P., 1989 Suggested approach to ground investigation and the determination of suitable sub-structure solutions for sites underlain by Chalk. International Chalk Symposium, Brighton Polytechnic, West Sussex, England (in press).

Edmonds, C.N., Green, C.P. and Higginbottom, I.E. 1987a Subsidence hazard prediction for limestone terrains, as applied to the English Cretaceous Chalk. In Culshaw, M.G., Bell, F.G., Cripps, J.C. and O'Hara, M.(eds), Planning and Engineering Geology, Geological Society Engineering Geology Special Publication, No.4, 283-293.

Edmonds, C.N., Kennie, T.J.M. and Rosenbaum, M.S. 1987b The application of airborne remote sensing to the detection of solution features in limestone. In Culshaw, M.G., Bell, F.G., Cripps, J.C. and O'Hara, M. (eds) Planning and Engineering Geology, Geological Society Engineering Geology Special Publication, No. 4, 125-131.

Fleming, W.G.K., Weltman, A.J., Randolph, M.F. and Elson, W.K. 1985 Piling Engineering, p 380, Surrey University Press, London.

Greenwood, D.A. and Kirsch, K. 1983 Piling and ground treatment for foundations, Thomas Telford, London.

Hancock, J.M. 1975 The petrology of the Chalk. Proceedings of the Geologists' Association, Volume 86, 429-535.

Higginbottom, I.E. 1965 The engineering geology of Chalk. Session A, Paper 1. a Proceedings of the Institution of Civil Engineers Symposium on Chalk in earthworks and foundations, 1-13.

Hobbs, N.B. and Healy, P.R. 1979 Piling in Chalk, p 209, DOE and CIRIA, London.

Littlejohn, G.S. 1979 Consolidation of old coal workings. Ground Engineering, Vol. 12, No. 4, 15-21.

McCann, D.M., Baria, R., Jackson, P.D., Culshaw, M.G., Green, A.S.P., Suddaby, D.L. and Hallam, J.R. 1982 The use of geophysical methods in the detection of natural cavities, mineshafts and anomalous ground conditions. Engineering Geology Unit Report No. 82/5, British Geological Survey, p.272.

McCann, D.M., Jackson, P.D. and Culshaw, M.G., 1987 The use of geophysical surveying methods in the detection of natural cavities and mineshafts. Quarterly Journal of Engineering Geology, London, Volume 20, No.1, 59-73.

Scarff, R.D. 1988 Factors governing the use of continuous dynamic probing in U.K. ground investigation. U.K. Penetration Testing Conference, Birmingham University (in press).

Taylor, D.I. and Fursman, D.C. 1989 The application of geophysical techniques to geotechnical investigations of chalk sites. International Chalk Symposium, Brighton Polytechnic, West Sussex, England (in press).

Warner, J. and Brown, D.R. 1974 Planning and performing compaction grouting. Journal of the Geotechnical Engineering Division, American Society of Civil Engineers, Paper 10606, June, 653-666.

Weltman, A.J. and Head, J.M. 1983 Site investigation manual, p. 144, CIRIA Special Publication 25/PSA Civil Engineering Technical Guide 35.

An historical review of applied science at the Institute of Karst Research, Postojna, Yugoslavia

ANDREJ KRANJC *Institute of Karst Research ZRC SAZU, Postojna, Yugoslavia*

ABSTRACT

When it was founded in 1947, the Institute of Karst Research ZRC-SAZU in Postojna had very humble goals. It was planned to be an organization exclusively oriented to basic research. However, in response to society's needs, and because of the available capacity of the Institute's staff and the need for funding, 30-40% of the Institute's activity is now in the area of applied research, which provides some 15% of its income. Approximately 80 research projects have been conducted dealing largely with water supply in karst, construction (highways and hydroelectric stations) and tourism. Other research was related to ecological problems, national defense and other special projects. The majority of these projects has been successfully completed. The biggest problem has been a lack of modern equipment and instrumentation.

The idea of establishing a Karst Institute in Postojna is relatively old. The cave "Postojnska Jama" was first opened for tourism in 1819 and its attendance has continually grown since then, attracting visitors from all over the world. To promote the borrough of Postojna (it did not become a town until 1909) and the cave into a really important international center, the cave secretary, I.A. Perko, published his plan: "Postojna must work more closely with Postojnska Jama speleological museum and speleologic institute" (Perko, 1911). In 1929, after the First World War when this territory was Italian, the "Italian Institute of Speleology" was founded with a cave register, a small museum and a biospeleological station. Following the Second World War, in 1947, the Slovene Academy of Sciences and Arts reopened the Institute under the name "Speleological Institute," which was later renamed the Institute of Karst Research. The justification for having such an institute outside the capital city was the biospeleological station, the cave museum, and the technical equipment of Postojnska Jama. These are no longer considerations: The biospeleological station and the museum no longer exist and the Institute does not now use the equipment of Postojnska Jama. However, the Institute has remained in Postojna.

The beginning of the Institute in 1947 was very humble: one person in a large, old building, badly in need of repair. However, the need for such an institute has been confirmed. The Institute has grown; it now has 21 employees and the building is almost too small. The research staff includes 12 people: four geologists, four geographers, two archaeologists, one chemical engineer and one historian.

The aspirations of the Insititute have changed considerably since its beginning. Initially the goals were solely to promote speleological investigations, to organize a cave register, and to reopen the karst museum. Over the years the need for fundamental research has become the most important. Because of a lack of both specialized scientists and funding, it has been impossible to pursue the entire spectrum of basic research on karst. In actuality, the Institute studies primarily problems in karstology (the geology of carbonate terrains, karst hydrography with special attention to underground water connections, and karst geomorphology) and speleology (speleogenesis, speleomorphology, cave sediments, and water infiltration).

In the early years, the staff of the Institute were the only professional speleologists in Slovenia and, therefore, they were soon asked for many special services connected with the karst underground. Thus, it became obvious that there was a need for specialized applied research.

During the same period, the system of financial support for the Institute slowly changed. Whereas it was once paid exclusively from the Republic's budget funds, the Institutes of the Slovene Academy of Sciences and Arts are now financed by special research funds for approved research programs, along with the possibility to enter into special contracts for applied research. This means that, in our Institute, basic research is distinguished from applied research according to the funding source: Fundamental studies are paid by research funds and applied investigations are paid directly by the client interested in the results. The ratio between fundamental and applied research varies, of course, the difference depending upon the available research capacity and the availability of outside funds. By agreement, applied research should not exceed 30-40% of the total program of the Institute.

The Institute's first applied research was begun in 1953: two projects connected with water supply. To date, the Institute's staff have conducted approximately 80 applied investigations with practical goals. Minor investigations and specialized projects are not included. It is difficult to evaluate the extent of this work, but for illustration I can say that the average report comprises 30 pages. The topics of the applied work can be grouped into a few categories: investigations of water, investigations connected with construction, studies for the development of tourism, specialized geological and speleological investigations, and investigations for national defense purposes.

Hydrologic research is the most common, demonstrating that, in our karst, water problems are the most important (more than 30% of all research). The most important of these investigations are connected directly with water supply problems: regional studies to locate possible water sources, hydrological and speleological investigations of karst springs and caves, water tracing, and determining the boundaries of drainage basins. Such studies may be on a relatively high theoretical level (such as the regional evaluation of water resources or the detailed geological mapping surrounding a karst spring) or they may be quite practical, such as pump tests or geodetic surveys. Water supply studies may cover areas of different sizes, varying from an entire region (the water supply for the Bela Krajina region) to a community, an industrial zone, or a single hotel. Most investigations include karst springs, but in some cases, water from caves is intended to be used for water supply.

Most of the applied studies have been made for local communities closely related to Postojna, such as Idrija and Cerknica. However, water supply studies have also been made for more distant communities in Slovenia (Gorica, Bela Krajina) and even for the town of Miljevina in neighboring central Bosnia). The most important work has focused in Malni, a large karst spring near Postojna, which is intended to supply water not only for the community of Postojna, but also for the tourist region along the Adriatic coast. The Institute has conducted 12 studies for the Postojna water supply including several special studies involving water protection.

Other investigations related to karst water can be true hydrological investigations (water tracing), projects related to tourism or land improvement (the experiment to change Cerknisko Polje, an intermittent lake, into a permanent one), projects related to dams and reservoir construction (for hydroelectric power supply), and projects relating to water pollution, including the effects of a water treatment plant in Postojna.

In addition to the specialized investigations concerning the problems of caverns being discovered or opened up beneath homes or factories, all the investigations connected with the construction of highways across the Dinaric karst belt between the Ljubljana Basin and the coast were entrusted to the Institute. These are of two types: investigations of the karst underground and its properites relating to the highway (caves beneath it, dolines, open fissures, collapses, subterranean streams and their protection) and the climatic conditions along the proposed route, particularly with regard to the strong northeast winter wind, the "burja" (bora), which is also characteristic of the Dinaric karst. For example, one investigation conducted a detailed examination of all the dolines along 30 kilometers of proposed highway between Vrhnika and Postojna. The slopes and the bottoms, which were infilled with sediment, were drawn on small scale maps, and each deposit of infilling sediment was drilled to determine its thickness (12 meters maximum), the thickness of individual strata, and the type of the deposit. In some cases, special analyses of loam and clay were conducted. The sediment was completely removed from the bottom of a few dolines in order to observe the underlying structure and to properly engineer the installation of the highway.

Also relatively important, but alas not financially so, were investigations connected with karst tourism. These primarily included specialized and precise surveys ("speleogeodetic") of entire caverns or one part thereof, including Postojanska Jama, Skocjanske Jama and Taborska Jama. The Institute has made the complete tourism plans for an entire cave (Celske Jama) and contributed portions of plans for other caves, such as for the railway in Postojanska Jama or the transport facilities in Skocjanske Jama. In some cases, very specialized investigations have been conducted: the geomechanical properites of the rock, the natural resources of a future tourist area, or the design of a tunnel between two caves.

Other investigations have been conducted relating to the importance of the karst as a natural defense area, or for designing cave rescue plans. Lastly, I must mention the speleological evaluation of the location of a new quarry in the karst terrain and the investigation of proposed sites for a landfill.

It is difficult to briefly evaluate the success of our applied research because some of our suggestions were ignored or "filed away." Therefore, some projects have never been realized.

Most of our work, however, has been utilized at least to some extent. There have been no errors or mistakes in the underlying theory discovered in our work. Postojna, and a large part of the surrounding community is supplied by water from Malni, and there is sufficient water of good quality for the future. The highway across the most cavernous part of the Dinaric karst was built in 1972 and to date, there have been no collapses or subsidences of the roadway, not even where the highway passes over large underground halls. Show caves, Postojna Jama included, have been opened or restored on the basis of our surveys. In determining the boundaries of a water supply catchment or the location of a local landfill, the conclusion of our research has played an important role.

Unfortunately, we have to decline many applied research projects because of the lack of suitable equipment and instrumentation which results not only in a lack of funds, but also inadequate planning for future acquisitions.

References

Čar, J.,1977: [The bases for Mining Estimation of Globoščak - Tkalca cave Break-Through].- 3 p., Annexes, Postojna

Habič, P.,1969: [Studies of Dolines in the Highway Laying out Vrhnika - Postojna].- 14 p., Annexes, Postojna

Habič, P.,1974: [Škocjanske jame. Studies of Tourist Pathes and Organization of Visit].- 61 p., Annexes, Postojna

Habič, P.,1979: [The Register of Karst Water Phenomena on Notranjska and Primorska].- 204 p., Annexes, Postojna

Habič, P.,1988: [Professional Basis for Protection of Water Sources and Water Stores of Trnovsko-Banjška Plateau].- 27 p., Annexes, Postojna

Habič, P., Kogovšek, J., Zupan, M., Kolbezen, M., Bricelj, M., Kranjc, A., 1989: [Dobličica Spring Investigations] .- 34 p., Annexes, Postojna

Habič, P., 1989: [Speleohydrological Water Tracing of Pivka near Trnje and of Stržen near Rakitnik] .- 30 p., Annexes, Postojna

Kogovšek, J.,1988: [The Quality of Water from Purifying Plant near Stara vas] .- 4 p., Annexes, Postojna

Kogovšek, J., Kranjc, A., Mihevc, A., Šebela, S., Urleb, M., Zlokolica, M., Zupan, N., 1989: [The Evaluation of Natural Conditions for Postojna Communal Dumping-Ground].- 44 p., Annexes, Postojna

Mihevc, A.,1987: [Protection against Bora and Snow-Drifts on the Highway Laying Out Razdrto - Nova Gorica and HW Razdrto - Čebulovica - Postojna] .- 25 p., Annexes, Postojna

Perko, I.A.,1911: Ein Höhlenforscher Institut in Oesterreich.- Mitt.für Höhlenkunde, 4, 1, Wien

Šebela, S., Zlokolica, M., Slabe, T., 1988: [Speleological Evaluation of Lokvica Quarry] .- 8 p., Annexes, Postojna

Zlokolica, M.,1987: [Preliminary Study of Permeability of Bottom and Slopes of Dragonja Valley] .- 10 p., Annexes, Postojna

5. Geophysical investigations of karst terrane

Subsurface investigation response to sinkhole activity at an eastern Pennsylvania site

BOHDAN L.PAZUNIAK   *NTH Consultants, Exton, Pa., USA*

ABSTRACT

A 500± acre, mixed use development in Eastern Pennsylvania experienced severe sinkhole activity during the early stages of construction. The site is in a known karst area, and the sinkholes occurred despite precautions taken from the commencement of construction. The severity of the sinkhole occurrences prompted additional geophysical and mechanical subsurface investigations, changes in proposed site plans, and development of site specific sinkhole remediation techniques.

A number of trends in sinkhole occurrences which could possibly be used to predict future occurrences, were determined. A number of subsurface investigation techniques including air surveillance, (aerial photos), geophysics and mechanical investigation techniques were tried with varying degrees of success. An evaluation of the subsurface investigation techniques is presented.

Introduction

A 500± acre, mixed-use (office, commercial, and residential) development is being built at a known karst site northwest of the city of Reading in Berks County, Pennsylvania. Although it was known that the site was sinkhole prone and precautions were taken from the inception of the project, numerous and often severe sinkholes, often in clusters, appeared during construction. The most severe outbreaks occurred at two locations which were proposed for stormwater retention or detention ponds. To forestall the occurrence of additional sinkholes and better assess the risk, additional subsurface investigations were completed, the design for stormwater detention/retention ponds was changed and site specific sinkhole remediation techniques were developed. The occurrence of additional sinkholes, which directly affect the developer's cost, has been reduced. Because we were able to observe the site for an extended period of time and use numerous subsurface investigation methods, the project became a proving site for subsurface investigation techniques. Descriptions of the site subsurface conditions, subsurface investigations, as well as the conclusions are presented in this paper. The variable nature of karst terrain from site to site is such that the techniques and conclusions developed at this site may or may not be applicable elsewhere.

Site conditions

The 500± acre site consists mostly of rolling farmland with a wooded, rocky hill located in the center. Construction activity as of this writing has been limited to approximately 35 percent of the site.

Subsurface Conditions

The study area is situated in the Reading Prong, within the New England Physiographic Province. The site overlies three carbonate formations of the Beekmantown Group ranging from the older Allentown Formation (Cambrian) to the younger Stonehenge and Rickenbach Formations (Ordovician). All three formations are primarily composed of carbonate and are characterized by a series of limestone and dolomite beds of varying thickness which are known for solution activity. Descriptions of the three formations starting with the youngest are as follows:

(1) Rickenbach Formation-Grey, very fine to coarsely crystalline, laminated dolomite with dark-gray chert in irregular beds, stringers and nodules and bands of quartz.

(2) Stonehenge Formation-Grey, finely crystalline and dark grey laminated limestone with shallow interbeds.

(3) Allentown Formation-Medium grey dolomite and impure limestone, with dark grey chert stringers and nodules.

All three formations weather to a reddish brown, clayey silt/silty clay with occasional hard, intact quartz or chert stringers. The soil/intact rock interface is extremely irregular, with both pinnacles and floating boulders.

The site is located on the top of an eroded anticline as shown in Figure 1. As part of the folding of the anticline, vertical tension cracks were formed in the bedrock. The hill in the middle of the site is a vestige of the less erodible (soluble) Rickenbach Formation.

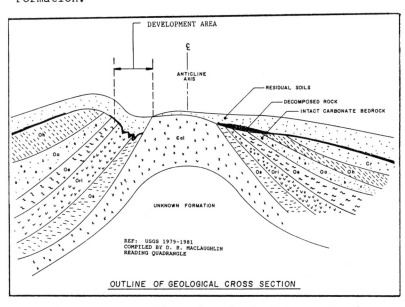

Figure 1a: Outline of Geological Cross Section

PERIOD		FORMATION	DESCRIPTION
ORDOVICIAN	Oh	HAMBURG	GRAY SHALE AND FINELY CRYSTALLINE LIMESTONE
	Oo	ONTELAUNEE	LIGHT TO DARK GRAY VERY FINE TO MEDIUM CRYSTALLINE DOLOMITE
	Oe	EPLER	MEDIUM GRAY VERY FINE CRYSTALLINE LIMESTONE INTERBEDDED WITH GRAY DOLOMITE
	Ori	RICHENBACH	GRAY VERY FINE TO COARSELY CRYSTALLINE LAMINATED DOLOMITE
	Os	STONEHENGE	MEDIUM TO LIGHT GRAY FINELY CRYSTALLINE LIMESTONE
CAMBRIAN	Єal	ALLENTOWN	MEDIUM TO DARK GRAY LIMESTONE
	Cr	RICHLAND	GRAY FINELY CRYSTALLINE DOLOMITE INTERBEDDED WITH OOLITIC LIMESTONE

REF: ENGINEERING CHARACTERISTICS OF THE ROCKS OF PENNSYLVANIA, 1982, PENNSYLVANIA GEOLOGIC SURVEY

Figure 1b: Geologic Formation Description

Site History and Conditions

The character of the karst terrain and sinkhole potential of the site was known from the onset. The corn fields at the site exhibited numerous infilled (healed) sinkhole subsidence depressions (dolines). An open file report, completed by the Pennsylvania Department of Natural Resources (PADER), identified at least 25 past sinkholes at the development site.

Although the outbreak of sinkholes during construction was not surprising, the magnitude and scope of the activity was. Site development required large amounts of both earthwork and rock blasting for utility trenches. Both operations tend to unearth and intensify zones of advanced solution activity. Initially, a number of sinkholes adjacent to blasting areas appeared, but they were both relatively small and expected. However, during a three week period which received exceptionally intense precipitation, two partially excavated basin areas at the site experienced intense sinkhole activity. At the larger site, sinkholes and depressions with tension cracks were clustered rim to rim over a one acre site. An estimated 600 to 1000 cubic yards of soil was washed away during this period. During a particularly heavy rainstorm groundwater was observed entering both basins through artisian sinkhole openings and simultaneously flowing down the throat of adjacent sinkholes. A site topographic and sinkhole distribution plan is presented on Figure 2.

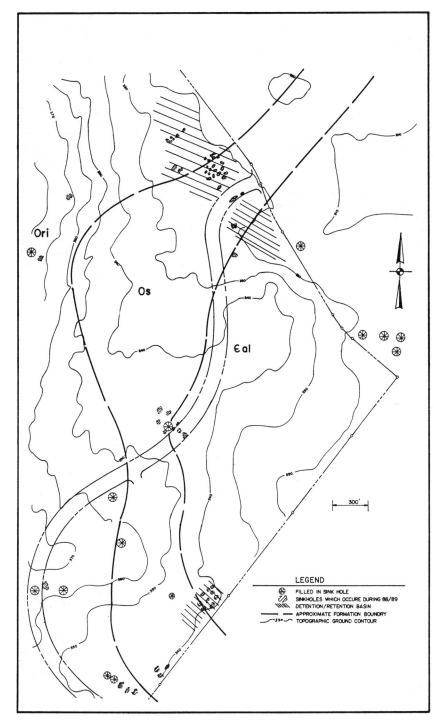

Figure 2:   Location of Major Sinkholes Occurences

During the course of the construction process, the occurrence of sinkholes appeared to follow some predictable forms, specifically

a)   along linear trends which generally run normal to the site topographic contours;

b)   along the contact between two geologic formations;

c)   in zones of relatively thin residual soil/decomposed rock adjoining shallow rock zones or pinnacles.

Subsurface Investigation

A series of overview and site specific investigations were conducted as part of the on-going development of the project.   Non-destructive testing techniques included air photo analysis, site reconnaissance, electromagnetic (EM-31 & EM-34) and resistivity surveys. Mechanical probes included air drilled rock probes, auger probes with and without SPT testing, test pits and electronic cone penetrometer (ECP) testing.   The standard procedure was to first investigate an area with non-destructive testing and then follow up with some site specific type of mechanical investigation with emphasis on previously identified anomalies.   The results of various types of investigations at this site are listed below.

Non-Destructive Testing

Air Photo & Site Reconnaissance: The simplest and most direct method for evaluating karst terrain is to examine site topography using air photos and then following up with a more detailed site reconnaissance.   Stereoscopic air photos were much more effective than individual simple air photos in locating lineaments and past sinkhole scars.

Lineaments are believed to be surface reflections of linear subsurface anomalies such as fracture patterns and faults.   Although a series of lineaments was identified from air photos of the site, there was no apparent correlation to past (healed) or recent sinkholes. Furthermore, subsequent subsurface investigations failed to locate any subsurface cavities or ravelling areas along the lineaments.

Healed sinkholes were also located using air photos followed by detailed site reconnaissance. Although healed, these conditions are a source of future problems; if a sinkhole had formed at a location, even though the site cavity is now filled in, the subsurface "plumbing" which caused the collapse is probably still in place. New sinkholes will form either in a cluster around the sinkhole or along a trend corresponding to the subsurface channels.

Electromagnetic Surveys: A relatively easy method for determining the conductivity of the subgrade is with an electromagnetic (EM) survey. The EM instrument consists of both electromagnetic transmitting and receiving coils. The transmitter sends an electromagnetic field into the ground, creating a secondary field which is measured by the receiver. The logic circuitry in the instrument then presents the inferred subsurface conductivity.

A number of instrument types and measurement modes were tried. The depth of measurement with an EM survey is dependent on a number of factors, including instrument type (EM-31 or EM-34), coil spacing (10 or 20 meters), and coil orientation (horizontal or vertical reading). After some experimentation, it was decided that the EM-34, with a 10-meter spacing and measured in both vertical and horizontal orientation was the most effective. In theory the vertical and horizontal coil orientation, assuming the same coil spacing, measure different subsurface volumes. Therefore, a conductivity change indicating a possible cavity or soil/rock interface below the ground surface can be delineated.

In reality, the measurements are clouded by a number of factors. Environmental/cultural factors, such as power lines (above and below ground), metal fences, pipes and other conductive objects influence the conductivity measurement. The averaging nature of the EM measurement also tends to hide subsurface anomalies. An example would be a zone composed of a series of relatively conductive mud filled fractures or caverns surrounded by much less conductive rock. The resultant measured conductivity may be similar to that of the overburden residual soil.

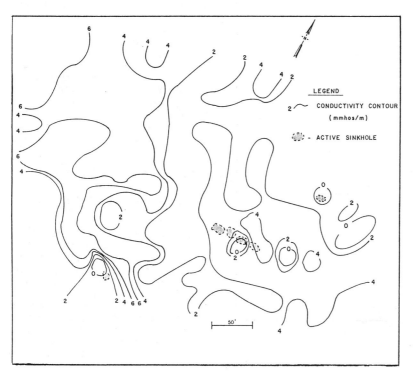

The results of the EM-34 surveys at the site were uneven. The 10-meter (shallower) horizontal (H) readings did not react to any but the shallowest intact rock outcrops (2 to 6 feet below the ground surface). The 10-meter vertical (V) readings were more responsive to the presence of shallow rock, but even then the response was limited and uneven. Limited success was achieved in identifying subsurface caverns using the vertical (V) orientation mode. At three separate survey point locations where a V measurement of effectively zero was encountered, a cluster or individual sinkholes appeared four to nine months later, as shown on figure 3. A number of interpretation/correlation differences between the vertical and horizontal measurements were tabulated in both absolute and relative terms, but a correlation to known rock depths or subsurface activities could not be established.

Figure 3: Electromagnetic conductivity plan developed with an EM-34: 10 m vertical readings at 50' centers

The EM-31 was effective in identifying man made anomalies such as pipes, but was ineffective in locating shallow rock, boulders or pinnacles.

Resistivity Survey: The resistivity (inverse of conductivity) of the subgrade can be determined with a resistivity survey.   This is a more labor-intensive method than the electromagnetic survey.   A current of known amperage is created between two electrodes; by measuring the voltage between two additional electrodes, the average resistivity of the subgrade can be determined.

There are two main advantages with this method: One is that it is somewhat less susceptible to the influence of cultural interference such as power lines and metal fences, and techniques have been devised which allow for developing resistivity cross-sections of the subgrade.

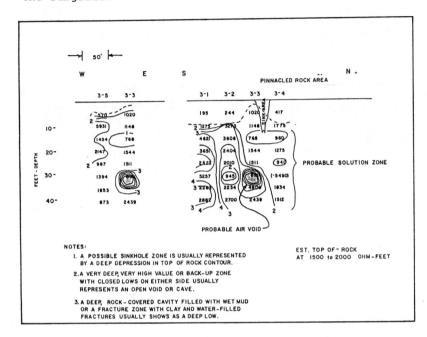

Figure 4:   Resistivity Cross Section

A resistivity survey along two intersecting lines was completed in an area which had recently experienced a sinkhole incident.   The field measurements were curve matched using a computer to develop a subgrade resistivity section (Figure 4).   In this particular cross-section, an air filled cavity is inferred, at a depth of 35 to 40 feet.   Because of the labor intensive nature of the survey and the large size of the site, further resistivity surveys of the site were not attempted. Additionally, the averaging (focused) nature of resistivity soundings limits the sensitivity of this method to only very large caverns. Therefore, the utility of resistivity surveys for sinkhole location/evaluation is limited.

During the various phases of the project a number of mechanical subsurface investigation methods were used including: auger borings, air drilling (track- and truck-mounted), test pits, electronic cone penetrometers and proofrolling.   The following results were obtained using these methods.

Electronic Cone Penetrometer: The most effective mechanical method for site evaluation was the electronic cone penetrometer (ECP).   An ECP test consists of pushing a standard cone (Apex angle of 60° and area of 10 sq. cm.) with a friction sleeve into the ground.   The friction sleeve has the same diameter as the conical tip and push rods (35.7 mm).   Both the conical tip and friction sleeve have built-in load cells which are connected to a digital recorder on the surface.   A computer

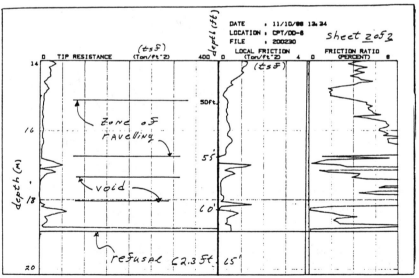

Figure 5:   Electronic Cone Penetrometer Log

generated ECP test hole log is shown on Figure 5.   A subsurface cavity can be detected on an ECP log when both the point and friction resistance drop to zero, as can be seen on Figure 5 at a depth of from 56 to 59 feet.   Indications of ravelling, as characterized by decreasing tip resistance, are also evident in the zone from 50 to 55 feet.

Over 300 ECP test holes were completed. Because they are relatively inexpensive on a linear foot basis and do not require full time field supervision, we were able to perform them at a significantly greater density (50-foot center to center spacing) than auger borings.

The ECP test holes are more prone to premature refusal due to cobbles and veins of intact rock than are auger borings. A technique to deal with this problem is to attempt an additional ECP penetration at a four-foot offset if the initial attempt encounters refusal before a predetermined depth is achieved.

The ECP survey was able to simultaneously provide subsurface cavity information, geotechnical data and the inferred top of rock at a relatively reasonable cost. The disadvantage of the technique was that rock cores could not be completed after overburden penetration. In addition, as with all point penetration techniques, the probability of encountering a subsurface void is limited.

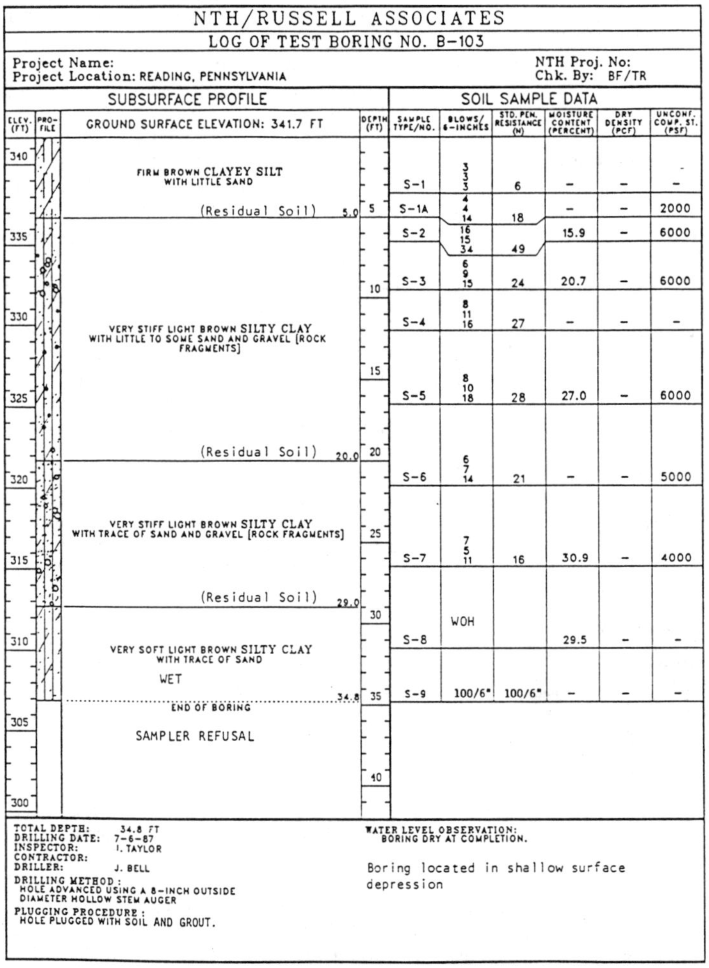

Figure 6: Auger boring log, with SPT. Possile void at 30 to 32 feet

Auger Probe: Over 70 auger borings, both with and without Standard Penetrations Tests (SPT), were completed across the site. Those performed without SPT's were effective only for locating the top of rock or floating boulders; they are not a cavity locating system per se. Auger borings with SPT were able to locate both cavities (voids) and possible ravelling zones, as shown on Figure 6.

There are two main disadvantages to this method. The cost per boring, including the need for site supervision, is significant, seriously limiting the number and area density of borings. Also, because the SPT testing is not continuous, (typically only 30 percent of the borehole is tested), smaller zones of advanced solution activity may be missed. Generally, ECP probes can be accomplished at one-sixth the cost of auger borings while providing more detailed information. The advantage of this method, other than its familiarity, is that rock cores or tri-cone penetrations may follow-up the overburden penetration.

Air Drilling: Over 50 air drilled probes were completed for exploratory purposes. While the probes were very effective in penetrating boulders and ledges, it was difficult to distinguish between ravelling decomposed rock zones and zones of residual soil. These probes were completed primarily for determining the top of rock.

Backhoe Test Pits: Over 30 test pits were completed on site. These pits normally ranged from eight to ten feet in depth. The holes were primarily for fill evaluation and for determining the top of the shallow rock. As a general method for sinkhole evaluation, the method is not particularly effective because of the shallow penetration.

<u>Proofrolling</u>: Heavy construction equipment traffic is an effective method of evaluating the subgrade for sinkhole potential, but not in the conventional way. During the course of the project there were numerous instances where sinkholes formed at locations that were subject to heavy construction traffic (loaded pans and dump trucks) only days or weeks before. A few feet of cavern roof can effectively bridge over significant transient loads, only to collapse due to gradual ravelling a short time later.

The use of proofrolling to locate healed (filled-in) sinkholes was an effective site evaluation technique. Healed sinkholes typically were filled in with washed-in (organic) topsoil. At a scarified site, the compressible organic soil was easily identifiable while under the weight of earthmoving equipment.

Conclusion

One of the most important duties that a geotechnical engineer has at the onset of a project is to warn the client of the potential for unforseen monetary and schedule implications of building in a karst area. A site with known sinkhole potential should have greater contingencies built into the budget and construction schedule than a similar project elsewhere.

Determining whether a project has such potential is not difficult and consists of a series of standard engineering procedures, including the review of site geology from the literature, and a site reconnaissance. Once such a general potential is determined, a subsurface investigation program should be set-up. The program should consist of non-destructive testing/reconnaissance to identify anomalies, followed up by a mechanical investigation of suspect areas. The program should be flexible to allow for more than one type of testing and allow the discontinuation or enhancement of any subsurface investigation method based on its effectiveness.

At this site, a number of techniques were used for subsurface karst terrain investigation with varying degrees of success. The technique of examining past site stereoscopic aerial photographs and researching site history for sinkhole activity was the most cost effective method for predicting future sinkhole activity. The EM-34 survey, with 10-meter spacing was found to be the most effective geophysical technique but the results were sometimes contradicting and difficult to interpret. The electronic cone penetrometer was found to be the most effective mechanical technique for evaluating limited target zones such as a proposed building site.

It was also found that sinkholes at this site tended to occur in a somewhat predictable manner, specifically a) along linear trends which run normal to site topography, b) along the contact of two formations and c) in zones of relatively shallow residual soil/decomposed rock adjoining shallow rock zones or pinnacles.

References

Kochanov, William E., 1988, Sinkholes and Karst Related Features of Berks Co., Pennsylvania: Pennsylvania Department of Environmental Resources, Bureau of Topographic and Geologic Survey, Open File Report 8801, 14 p.

Gyer, Alan R. and Wilshusen, J. Peter, 1982, Engineering Characteristics of the Rocks of Pennsylvania: Pennsylvania Geologic Survey, Environmental Geology Report 1, 300 p.

McNeil, J. D., 1980, Electromagnetic Terrain Conductivity Measurement at Low Induction Numbers: Technical Note TN-C, Geonics Ltd., 15 p.

Using terrain conductivity to detect subsurface voids and caves in a limestone formation

DIANE ROBINSON-POTEET   *Edwards Underground Water District, San Antonio, Tex., USA*

ABSTRACT

Two parallel rectangular grids, 9m X 32m and 18m X 32m (30ft X 105ft and 60ft X 105ft), were laid out on the ground with their longest axis trending in an east-west direction. The grids were comprised of a total of 304 stations, spaced 1.5m (5ft) apart, and marked with orange paint. The stations were surveyed and placed on the ground so that they overlaid a known, mapped cave which trended in a north-south direction to a depth of approximately 30m (100 ft) below the surface. The cave is called Bear Cave, and is located in north Bexar County, Texas within the Edwards Aquifer Recharge Zone.

Terrain conductivity readings were collected at each grid station by using instruments made by the Geonics Company, Inc., called the EM31 and the EM34. By changing the spacing between the transmitter and the receiver, as well as by changing the dipole configuration, these instruments can reach different depths of exploration. For the EM31, only the dipole configuration could be changed, since the spacing between the transmitter and the receiver were fixed. For the EM34, both horizontal dipole-coplanar and vertical dipole-coplanar readings were collected for both 10 and 20 meter spacing settings. Various computerized graphic representations of the data were analyzed, with contour mapping yielding the most useful representation. The contour maps were then compared to cross-section maps of the cave, and terrain conductivity depressions were found to correspond to the cave passageway. In addition, resistivity data was collected and compared to the electromagnetic conductivity values. Both instrumentation techniques showed the same trends. It was determined that the conductivity techniques were a viable method of subsurface void and cave detection. It was also concluded that the method could be used for pre-construction site assessments in development planning within the recharge zone of a limestone aquifer.

Introduction

During the late 1960's and early 1970's, several advanced geophysical exploration techniques and practical instrumentation systems were tested and evaluated for cavity detection (Fountain, 1981). The use of surface resistivity was among these advanced developments. Both military and civilian agencies supported the various research and development activities of these exploration techniques.

Lewis Fountain (1981) has stated that as early as 1968, during the Vietnam Conflict, the United States Army first became interested in shallow tunnel detection and mapping. At this time, the shift from seismic sources to electromagnetic detection techniques occurred. By the middle of the 1970's, cavity detection also became an interest to the Federal Highway Administration. Problems in the southeastern states with sinkholes forming under existing highways and along proposed roads led to further exploration of resistivity methods. Today, the usefulness of terrain resistivity and conductivity for geological investigations is firmly established.

Through educational conferences, the Edwards Underground Water District (EUWD or the District) staff learned about the applications of electromagnetic investigative techniques and decided to apply them to their own geologic problems. The District monitors the area where the limestones of the Edwards Group and associated limestone formations outcrop. This area is known as the Edwards Aquifer Recharge Zone (ERZ). Because the Edwards Aquifer is a sole source water supply for over a million people, urban development within this zone is one of many activities which concern the EUWD. This concern stems from the District's dual responsibility of preventing polluted waters from entering sensitive recharge features, such as caves and sinkholes, and at the same time protecting and preserving significant sources of recharge.

The District's current practice of pre-construction site inspections, however, does not fully reveal caves and subsurface voids. In addition, contractors report the subsequent discovery of such features only infrequently. Full knowledge of such features is important in planning and directing development and for the prevention of pollution from such sources as private septic systems or leaky sewer lines and hydrocarbon storage facilities. For this reason, the EUWD has been willing to invest time and money in understanding the applications of this new technology with regard to the geology of this sensitive region.

The focus of the following report is to use an established method of electromagnetic measurement of terrain conductivity in order to confirm the ability of the technique to detect a known cave system in a limestone formation. The cave which is studied is called Bear Cave and is located in north Bexar County, within the ERZ. This study does not try to re-invent the electromagnetic technique but rather is an applied study in which the goal is to build confidence in the data typical to a specific limestone terrain. In this report, the conductivity values are presented in contour-map form and then compared to a compass and tape map of the cave. The results will also be analyzed with regard to their applications to geologic assessments concerning development within the ERZ.

Site Geology

Bear Cave, the site where the conductivity readings were recorded, lies within the general outcrop area of the Kainer Formation, a member of the Edwards Group. The Edwards Group and the Glen Rose Formation are regional stratigraphic units of the Cretaceous System which outcrop in the area and which are covered sporadically and thinly with Quarternary Fluvial Deposits. These limestone beds have a very slight regional dip to the southeast. Regionally, this area is known as the Edwards Aquifer Recharge Zone. This zone consists of cavernous limestones which are heavily faulted and fractured. Thus, when surface waters traverse this zone, large quantities of water may enter the Edwards Aquifer through the various karst features. Several flood control dams constructed within the recharge zone add significantly to the quantities of water entering the subsurface by slowing the runoff process, allowing the waters to infiltrate to the subsurface.

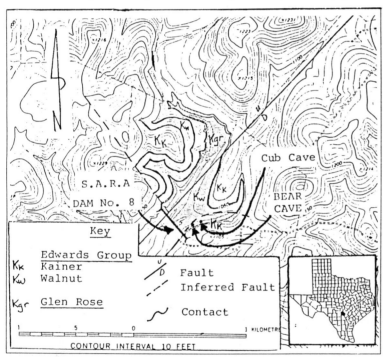

FIGURE 1.— Map of Texas showing location of Bexar County and a geologic map of the study site and surrounding area.

In Figure No. 1, a geologic map of the study site and surrounding area has been prepared by use of aerial photographs and ground surveys. The base map used is the United States Geologic Quadrangle Sheet - Bulverde, Texas (1973). Bear Cave lies in the area immediately upstream of San Antonio River Authority (S.A.R.A) Dam No. 8. It is to be noted that another large cave, Cub Cave, lies approximately 100 m (328 ft) to the northeast. The dashed line is a possible fault running on the southern side of Mud Creek and adjacent to the cave locations.

The mere existence of the two caves in the vicinity of the proposed fault allows us to surmise that increased dissolution along the fault caused the caves to form. Moreover, since the construction of the dam, the impoundment of water behind the dam has probably increased the dissolution of limestone, increasing the size of the caves (as described in "Bear Cave"). This area, where streams often form along faults and where streams are important recharge features themselves, is representative of the Edwards Aquifer Recharge Zone.

Topographic Setting

The elevation of the area ranges from approximately 900 to 1200 ft (270 to 400 m) above sea level. The area comprises the flood plain known as Mud Creek, which is flanked with rugged slopes of limestone covered with oaks, cacti, and bunch grasses. Snakes, mosquitos, spiders, deer, and armadillos are all known to reside in the vicinity of the cave. The surrounding area has been developed into the Champions Equestrian Center, comprised of a stable area and natural riding paths. For the development in the area, a large sewer line has been installed which follows the north bank of the creek, cuts through the dam and heads southward towards the City of San Antonio Sewer Treatment Plant. In addition, on the slopes of the north bank, Champions Residential Subdivision has been under development, with one of its units completed with utilities, sewers, and streets. During the construction of the sewer lines for this subdivision, some subsurface voids were discovered and reported by the contractor to the authorities.

Bear Cave

Bear Cave is located approximately 305 m (1000 ft) northeast of the San Antonio River Authority (S.A.R.A.) Dam No. 8, on the southeast side of the flood plain of Mud Creek. This area is situated in north Bexar County within the Stone Oak Subdivision, of which Champions is a part. The S.A.R.A Dam was built in the Spring of 1973. The cave, however, predates the dam, and has been known for many years as a major site of recharge in the county for the Edwards Aquifer.

From the entrance, the cave trends basically in a slight northwest - southeast direction, with its deepest and largest portion extending to the north-northwest out under the flood plain. George Veni (1988), in his book <u>Caves of Bexar County</u>, describes the extent of the cave, its biology and its history. The cave received its name from bones that were found in it which were believed to be from a bear. The cave was surveyed on August 1, 1965 by R. Cuzzi, J. Jasek, R. Summar, and R. Szalwinski. For this report, the cave was resurveyed on December 31, 1988 by D.R. Poteet, D. Poteet, and D. Robinson, and several changes were made to the original drawings. Between 1965 and 1988, many people explored this cave and the local speleologists have often used it for training. In fact, a minor rescue occurred in the Fall of 1984 when a corpulent person became stuck in a small entrance which existed at that time. As a result, and fearful that someone would get hurt, the owners of the property filled the two small cave entrances with sand, gravel, and large boulders. The fill was then graded, leaving no trace of the cave entrances. Flood waters in the Spring of 1985, however, washed out the fill, leaving a single but much enlarged entrance. Biology surveys performed in the cave have revealed that salamanders, crickets, spiders, mosquitos, and even bats have at one time or another inhabited the cave.

Principles of Operation

To measure the earth's conductivity or resistivity, electromagnetic inductive techniques (EM) can be used. In this case, the ease with which an electromagnetic field can be transmitted through a material is quantified by conductivity. Millimhos per meter (MKS units) are the units of measurement for conductivity and may be abbreviated as mmho/m. Resistivity is the reciprocal of conductivity and has the units of ohm-meters. (To convert from conductivity to resistivity, divide 1000 by the conductivity.) Conductivity is preferred in inductive techniques since the response is generally proportional to conductivity and inversely proportionally to resistivity. EM is based on the simple principle that magnetic fields can be generated from electrical currents and vice versa.

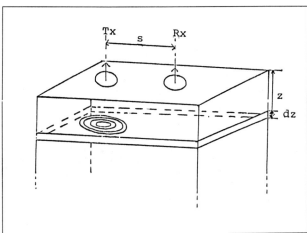

Figure 2.Vertical Coplanar

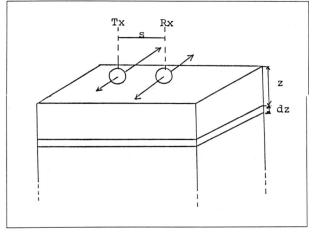

Figure 4.Horizontal Coplanar Dipoles

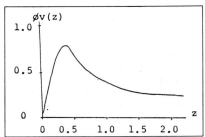

Figure 3. Relative Response Vs. Depth for Vertical Dipoles

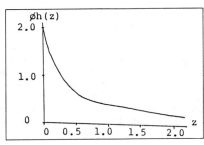

Figure 5. Relative Response Vs. Depth for Horizontal Dipoles

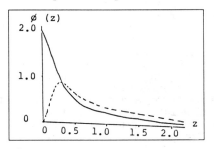

Figure 6. Comparison of Relative Response for Vertical and Horiz. Dipoles.

273

For any EM technique, a time-varying electric current flows through a transmitter coil to create a magnetic field. This magnetic field penetrates into the earth. The magnetic field causes current loops to flow in the ground material. (Some rock types conduct better than others. For example, air voids will have lower conductivity than limestones, sands, and gravels; while clays, saline soils, and metals will have relatively much higher conductivity values.) A secondary magnetic field is created by these induced flowing currents in the ground. This field then generates current flow in the receiver coil. The receiver indicates a weighted average of the conductivity of the underground in the vicinity of the instrument.

In Mc Neill's (1980) "Technical Note TN-6," he has determined the relationship between depth and the apparent conductivity contribution measured for a thin layer at depth z. Thus, to explain the instrument response (apparent conductivity) as a function of depth, Figures 2 through 6 must be noted. Tx and Rx in the Figures Nos. 2 and 4 represent the transmitter and the receiver magnetic fields at a given spacing s for either the EM31 or the EM34. The secondary magnetic field in the receiver arising from all current flow within a thin layer, dx, at a depth z, can be calculated in the following manner.

Let the function $Øv(z)$ describe the relative contribution to the secondary magnetic field arising from a thin layer at any depth z. The graph in Figure No. 3, shows that for a coil configuration of "dipoles vertical and coplanar" a material located at a depth of approximately 0.4 s gives a maximum contribution to the secondary field and material at a depth of 1.5 s still contributes significantly. Also note that at or near surface level or zero depth, the ground material contributes very little to the secondary magnetic field, and thus, shows that this coil configuration (dipoles vertical and coplanar) is insensitive to changes near the surface.

Figure Nos. 4 and 5, show the depth response to the function $Øh(z)$ for coil configuration "dipoles horizontal and coplanar." (This configuration is often used more with the EM34 than the EM31.) It can be observed that the relative contribution from material near-surface is large and the response falls off monotonically with depth. (McNeill, 1980.)

To emphasize the differences in response with depth with regards to the coil configuration, the graph in Figure No. 6, has been presented. It is to be noted on this figure that for regions greater than one intercoil spacing in depth the vertical transmitter/receiver dipole gives approximately twice the relative contribution of the horizontal transmitter/receiver dipole. (McNeill, 1980.)

Instrumentation
Geonics EM31-D
The operating controls, which include a mode switch, conductivity range switch, phasing potentiometer, a coarse inphase compensation and a fine inphase compensation, are on a small box (dimensions = 24 x 20 x 18 cm or 9.45 x 7.87 x 7.09 in, and weight = 11 kg or 2.43 lbs) from which a boom is attached on either side. The total length of each boom is approximately 1.4 m (4.59 ft). The power source of the instrument is 8 disposable alkaline C-cell batteries which are stored on the underneath side of the operating control box. On one boom is the receiver coil and on the other end is the transmitter coil. (Geonics Limited Catalog, 1988)

This instrument provides a measurement of terrain conductivity without ground electrodes or contact. The EM31 can be strapped over one's shoulder and easily carried. The operator simply walks from station to station, taking readings at each station. The intercoil spacing is 3.7 m (12.14 ft) and yields an effective depth of exploration of about 6 m (19.68 ft). The instrument can also be operated on its side (horizontal dipoles), in which case the effective depth of exploration is reduced to approximately 3 m (9.84 ft).

Geonics EM34-3
This instrument operates on the same principles as the EM31 but is designed to achieve a greater depth of exploration. For this instrument, the measured value of terrain conductivity is determined solely by the instrument geometry: intercoil spacing and coil orientation (see Figure No. 1). The EM34-3 can be operated at three fixed spacings of 10, 20, or 40 m (32.81, 65.62, or 131.24 ft) and in the verical coplanar or horizontal coplanar mode. The instrument systems includes two coils (both of a diameter = 63 cm or 24.8 in, and weight = 5.6 kg or 12.35 lbs for the receiver coil and 8.8 kg or 19.4 lbs for the transmitter coil), a self-contained dipole transmitter (dimensions = 15 x 8 x 26 cm or 5.91 x 3.15 x 10.24 in and weight = 3 kg or 6.61 lbs), a self-contained dipole receiver (dimensions = 19.5 x 13.5 x 26 cm or 7.68 x 5.32.x 10.24 in and weight = 3.1 kg or 6.81 lbs), and a lightweight 2-wire shielded cable for each spacing. The power source is 16 batteries: 8 D-cells for the transmitter, and 8 C-cells for the receiver. (Geonics Limited Catalog, 1988)

The EM34 requires two people to operate, one for the receiver coil and operating box, and one for the transmitter. The intercoil spacing is measured electronically so that the correct spacing of 10, 20, or 40 meters can be easily achieved. These changes in spacing directly vary the effective depth of exploration as shown in table 1 below:

Table 1.	Exploration Depth In Meters (Feet)	
Intercoil Spacing In Meters (Feet)	Horizontal Dipole	Vertical Dipole
10 (32.81)	7.5 (24.6)	15 (49.2)
20 (65.62)	15 (49.2)	30 (98.4)
40 (131.24)	30 (98.4)	60 (196.86)

(From Mc Neill's (1980) "Technical Notes TN-6".)

Field Methods and Data Collection

The grid sizes were first determined. Due to a 3.05m (10 ft) surface topographic change near the middle of the cave study area, two separate but adjacent and equal in length grids were laid out. Both grids extended beyond the cave boundaries. The topographic change divided the study area into the "cave entrance area" and the "flood plain area." Grid spacings were adapted to match the station-spacing for each instrument.

Alteration of the station-spacing was necessary because of the different distances between the receiver and the transmitter on the EM31 and the EM34. For all the EM34 settings, the stations were kept at 15 ft spacings regardless of the change in spacing between the transmitter and the receiver (i.e. 10 or 20 m spacing) due to the fact that the values were collected at the same time. This was accomplished by changing the setting on the instrument and also by using the 20 m spacing cord for all readings. On the instrument, a gage would alert the operator when the correct spacing had been reached. Thus, both EM34 types of readings could be collected by moving the transmitter and receiver further apart for the 20 m readings and decreasing the distance between them for the 10 m readings.

The station spacings were measured by tape and Brunton Compass surveys. Each station was marked on the ground with orange paint and sometimes also with a flag. The north-south rows of the grids trended in the same direction as the longitudinal direction of the cave, which is approximately 11 degrees west of north.

The EM31 values for the flood plain area were collected with a data logger. EM31 values for the cave entrance area, as well as all the EM34 values, were manually read and recorded. The EM31 was simply carried from station to station, with the instrument centered over the station in an east-west alignment. For the EM34, the receiver coil was placed at each station, with the transmitter coil moved to the proper distance (i.e. 10 or 20 m). Both the receiver and the transmitter coils (for either vertical or horizontal dipole arrangements) were in an east-west alignment.

No readings could be accepted for collection near or on the topographic change because of what is called "geologic noise." This noise caused the readings to be much higher due to the interference of the elevated rock. Also, an occasional reading could not be collected because of trees, cacti, or the missing ground of the cave entrance. An explanation for these missing values will be given in the next section regarding data calculations.

Data Calculations

The data collected in the field program was analyzed by computer. Two-dimensional plots of the EM31 and EM34 conductivity data were produced by GRAPH-WRITER. Each of these plots corresponds to one grid row in the east-west direction. With LOTUS 123-PRINTGRAPH, 2-dimensional graphs for each grid row in the north-south direction were created. In LOTUS 123, missing data was obtained by interpolating between adjacent data points when possible. These averages were checked against the averages in the east-west direction as well. The values for the EM31 which would have been collected over the cave entrance were given "0's" on the grids. This was done because the depth of exploration was equal to the void space of the cave entrance. The data was stored on GRID, then TOPO, and finally SURFER.

Results

It was found that the contour maps displayed the data in the clearest fashion for interpretation and communication while the 2-dimensional graphs were found to be adequate but

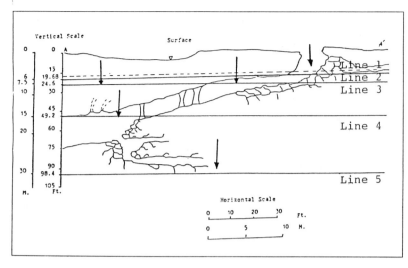

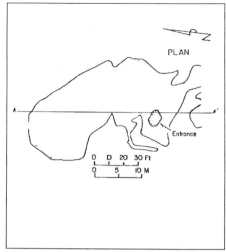

Figure: 7. Profile of cave with depth of
Exploration Lines

Figure: 8. Bear Cave (Plan View)

limiting. The SURFER 3-dimensional graphs were determined to be too difficult to interpret. The 2-dimensional graphs were later used to verify trends along the profiles which are described below.

To best understand the results, each contour plot of the conductivity measurements that were made with TOPO were overlaid with a plan-view of the cave outline (Figure 8). In Figure 7, a cross-section of the cave is shown. On the cross-section, lines have been drawn to show the depth of exploration of the instrument for each contour map. It is to be noted that the instrument indicates a weighted average of the conductivity of the underground (area between the depth of exploration and the surface) in the vicinity of the instrument. The cross-sectional area is taken on a north-south transect, longitudinally across the cave (see Figure 8). Each map is explained below.

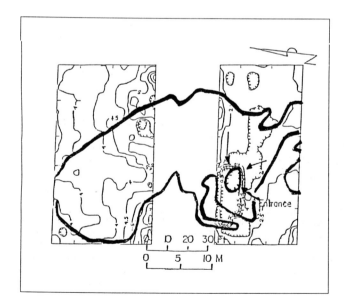

If the instrument was placed on the surface then Line 2 (Figure 7) in the cross-section view would represent the depth of exploration. However, the instrument was held at waist height, approximately 0.81 m (2.7 ft) above the surface, and Line 1 (Figure 7) represents this depth of exploration.

The contour map for the EM31 represents the lows and highs for the conductivity values collected. The small depression (circles with hack marks pointing toward the centers) on the "cave entrance area" grid, around the mouth of the cave, represents the lowest area of conductivity. This corresponds with the area between the surface and the solid line which cross the cave passageway on the cross-section map. An arrow points to these areas of low conductivity on both maps.

Figure 9. EM31 - Vertical & Coplanar
Dipoles

276

Line 3 (Figure 7) represents the depth of exploration for this instrument and its setting. (It is to be noted that this instrument is laid on the surface to collect the values.) Almost in the center of the "cave entrance area" grid, a conductivity depression exists which corresponds to the area between the solid line and the surface on the cross-section view which intersect the cave passageway. Arrows point to the low conductivity areas on both Figures 7 and 10. There are also three low spots within the "flood plain area" grid. These possibly represent dissolution areas which were unmappable by physical exploration. In the lowest room in the cave, shafts could be seen extending upward from the ceiling of this cave. These shafts appear to be in the same vicinity as these northern conductivity depressions.

Figure 10. EM34 - 10 Meters Spacing With Horizontal & Coplanar Dipoles

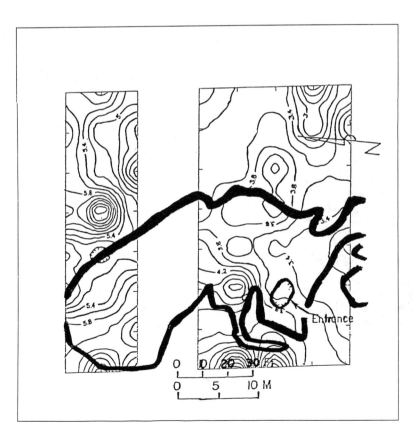

For the depth of exploration shown in the cross-section view for this instrument configuration (Line 4, Figure 7), the contour map shows no relevant conductivity depressions which would correspond to the cave passageway. The effectiveness of the configuraton of the dipoles was explained in the subsection entitled "Principles of Operation." Reviewing Figure No. 6, the vertical dipole configuration is more effective at deeper depths of exploration than the horizontal dipole arrangement: the response of the vertical dipole peaked at a deeper depth of exploration than the horizontal dipole response. As can be seen below in the "20 meters spacing/horizontal & coplanar dipoles" arrangement, at the same exploration depth of 15 m (50 ft), relevant conductivity depressions were recorded.

Figure 11. EM34 - 10 Meters Spacing With Vertical & Coplanar Dipoles

Line 4 (Figure 7), marking the depth of exploraton, passes through the cave in the far north half of the cross-sectional view. This corresponds with the "flood plain area" contour map which has a conductivity depression in the northwest corner. Another conductivity depression exists on the "cave entrance area" contour map which may represent blocked passageways extending in a northeast direction from the deepest room.

It is also to be noted that the "cave entrance area" contour map has generally much lower readings than the "flood plain area" map — more so than any of the other EM34 configurations. In speculation, this difference is probably caused by the "cave entrance area" having more subsurface areas of dissolution and collaspe (which were not mappable through a compass and tape survey) than the "flood plain area" between the surface and a depth of exploration of 50 feet.

Figure 12. EM34 - 20 Meters
Spacing With
Horizontal &
Coplanar Dipoles

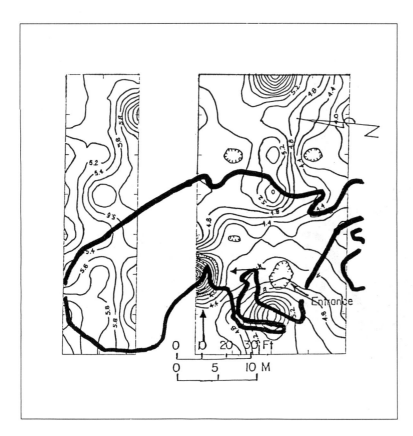

In the cross-sectional view (Figure 7), on the vertical scale, between 22.87 m (75 ft) and 32.0 m (105 ft) and near Line 5, passageways in the cave could not be explored further due to the smallness of the passageways or because they were blocked by debris. However, the contour map shows a corresponding conductivity depression where these passageways continued.

Figure 13. EM34 - 20 Meters
Spacing With
Vertical & Coplanar
Dipoles

To verify the accuracy of the EM data that was collected, some of the trends displayed in the 2-dimensional graphs were compared to a resistivity survey that Dr. Blaney (Engineering Professor at the University of Texas at San Antonio) performed in March of 1989 over some of the same profiles. In addition, EM34 readings were re-taken by the District in order to verify the consistency of the data. It was found that generally the trends were the same for the resistivity measurements and for both the new and old EM34 measurements. VLF waves were also used to confer the trends. The results of the VLF (WADI) also confirmed the same trends, as well as the fault which was found to parallel the change in topography.

Conclusion

In general, the EM31 and the EM34 has proven to be reliable and fairly inexpensive for detecting subsurface voids. However, several short-comings have been noted:
- When these instruments are used near sharp changes in topography, the readings may be much higher than the ground's apparent conductivity.
- When comparing the Bear Cave Site to more developed areas, problems may arise associated with "cultural interferences". These interferences, such as power lines, metal fences, underground cables, etc., may also cause the EM readings to become very high and not reflect the ground's apparent conductivity.
- Depth of exploration is limited to 200 feet even when the EM34 with a 40 meter spacing is used.

These short-comings appear to not interfere with this study. Even though the sharp change in topography made it impossible to collect data in some parts of the study area, sufficient amounts of apparent conductivity values were collected which could be compared to the compass and tape map of the cave. Moreover, the study site was in a relatively remote place, and the depth of exploration to be studied was not deeper than 200 feet.

In view of the data presented in this paper, the viabiltiy of finding subsurface voids in the limestones of the Edwards Group with the electromagnetic inductive techniques is indicated. Applications in geologic assessments for developments within the ERZ appear to be many. The usefulness of detecting subsurface voids before blasting drainfields for septic tanks, sewer lines, or below ground hydrocarbon storage faciltes is obvious. For these types of development structures, which are installed in the first thirty feet below the surface, the EM31 and EM34 are concluded to be practical and inexpensive for use in performing geologic assessments. Thus, these instruments would aid in the planning of the placement of potentially harmful facilities within the ERZ, which the District believes is one of the important goals behind master plans and geologic assessments. Further development of EM and other geophysical instruments with regard to ERZ construction activities would be needed in order to gain expertise and public acceptance of these techniques.

References

Blaney, Geoff, Ph. D., 1989. Telephone and Personal Visits. Professor of Engineering, University of Texas at San Antonio.

Fountain, Lewis, 1981. "A Chronology of Developments in Tunnel Detection." Southwest Research Institute. San Antonio, Texas. Presented at the Symposium on "Tunnel Dection" sponsored by U.S. Army Mobility Equipment Research and Development Command and the Colorado School of Mines, Golden Colorado, July 21-23, 1981.

Geonics Limited, 1988. Geophysical Instrumentation Catalog. Geonics Limited. Ontario, Canada.

McNeill, J.D., 1980. "Technical Note TN-6: Electromagnetic Terrain Conductivity Measurments at Low Induction Numbers." In Manual of Operation for the EM34-3. Geonics Limited. Ontario, Canada.

Raba-Kistner Consultants, 1980. "Preliminary Engineering Geology Investigation For Proposed 4475 Acre Development, Bexar County, Report # F80-13". In the 1986 "Water Pollution Abatement Plan for Champions Equestrian Center at Stone Oak, San Antonio, Texas" prepared by Simmons, Aykroyd, & McKnight.

Veni, George, 1988. The Caves of Bear County, Second Ed. Texas Memorial Museum, University of Texas at Austin.

Seismic expression of solution collapse features from the Florida Platform

STEPHEN W. SNYDER   *Department of Marine, Earth and Atmospheric Sciences, North Carolina State University, Raleigh, N.C., USA*

MARK W. EVANS, ALBERT C. HINE & JOHN S. COMPTON   *Department of Marine Science, University of South Florida, St. Petersburg, Fla., USA*

ABSTRACT

A variety of subterranean solution collapse features have been identified from very high-resolution seismic reflection profiles collected in Crooked Lake (Central Florida Highlands) and the St. Johns River (northeast Florida). The seismic data have been correlated to adjacent borehole sites for stratigraphic control. Collectively, these data demonstrate that dissolution within several different lithostratigraphic units of the relict Florida carbonate platform has produced at least 3 distinct subbottom solution features: (1) deep, narrow, cylindrical shafts, tens to a hundred meters wide and tens of meters deep; (2) broad solution valleys showing vertical displacements of greater than 10 meters of relief stretched across a horizontal distance of 1 to 5 km; and (3) large-scale, cavern-collapse features several hundred meters wide and of undetermined depth. The type of solution feature constructed appears to be largely determined by which of the many aquifers within the Florida Platform has actively undergone dissolution.

INTRODUCTION

Peninsular Florida was once part of an enormous shallow-water Mesozoic carbonate platform originally extending along the entire continental margin of the eastern North America from the Bahamas northward to Nova Scotia (Paulus, 1972; Sheridan et al., 1981; Jansa, 1981; Schlager and Ginsburg, 1989; Ladd and Sheridan, 1987). This megaplatform produced over 5 km of mostly shallow-water, bank-top, carbonates (Meyerhoff and Hatten, 1974). Typical rimmed bank-margin environments consist of low-magnesium calcite skeletal grainstones, while interior platform environments are usually rich in aragonitic mud. Both of these carbonate mineralogies are metastable, and are easily dissolved via percolation of meteoric waters. Thus, they are susceptible to subterranean dissolution.

The Mesozoic megaplatform of eastern North America began to falter by middle Cretaceous time, and it eventually drowned in the Late Cretaceous as clastic sheets spread seaward across the platform environments. Only a few isolated carbonate-producing, shallow-water banks survived. They included the Bahama Banks and the Florida Platform. The latter was isolated from the deleterious influence of Late Cretaceous siliciclastic deposition due to the presence of a seaway known as the Suwannee Channel (Puri and Vernon, 1964; Chen, 1965). This seaway flowed northeastward across the northern end of peninsular Florida (Pinet and Popenoe, 1985). It served as an effective barrier to fine-grained sediment transport from the Southern Appalachian alluvial and deltaic plains. Consequently, the Florida Platform slowly aggraded vertically during the early Tertiary via the production of Eocene and Oligocene shallow-water carbonate sequences. Again, lithologies which are vulnerable to dissolution processes. This Paleogene platform finally began to flounder when siliciclastic sediments invaded via progradation of late Paleogene through early Neogene sediments across a remanent of the former Suwannee Channel known as the Gulf Trough (McKinney, 1984; Popenoe et al., 1987).

During the Neogene, a mixed siliciclastic-carbonate depositional regime occupied much of the Florida peninsula (Puri and Vernon, 1964; Schmidt, 1984; Scott, 1988). This was followed by a invasion of quartz-rich siliciclastic sediments during the Quaternary. The latter is responsible for most of the present day geomorphic features of Florida such as the many relict shorelines and beachridge plains described by White (1970) and illustrated in Figure 1.

The depositional evolution of the Florida Platform has thus provided a suite of Mesozoic through mid-Tertiary, shallow-water, carbonate sequences, which are overlain by siliciclastic-to-mixed carbonate/siliciclastic sequences. These stratigraphic sections are finally capped by a blanket of surficial, quartz-rich sands. Collectively, the lithologic fabric presented by these stratigraphic sections forms the first-order framework for groundwater transport and associated dissolution processes in the Florida Platform. That

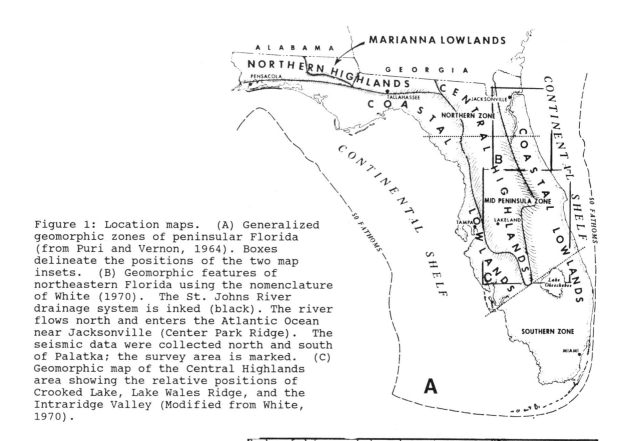

Figure 1: Location maps. (A) Generalized geomorphic zones of peninsular Florida (from Puri and Vernon, 1964). Boxes delineate the positions of the two map insets. (B) Geomorphic features of northeastern Florida using the nomenclature of White (1970). The St. Johns River drainage system is inked (black). The river flows north and enters the Atlantic Ocean near Jacksonville (Center Park Ridge). The seismic data were collected north and south of Palatka; the survey area is marked. (C) Geomorphic map of the Central Highlands area showing the relative positions of Crooked Lake, Lake Wales Ridge, and the Intraridge Valley (Modified from White, 1970).

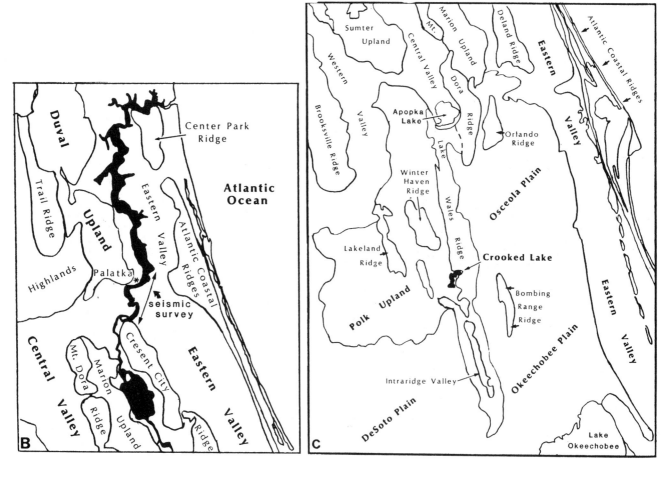

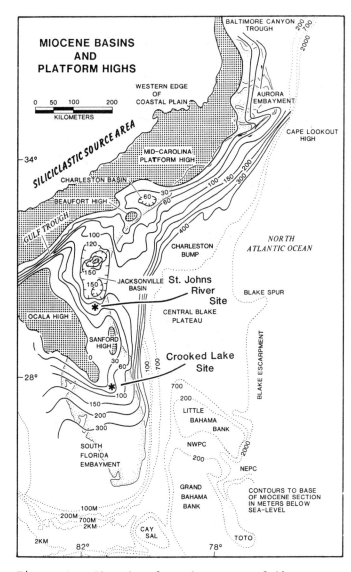

Figure 2: Structural contour map of the Base of the Miocene section depicting the shape of the Paleogene carbonate platform of peninsular Florida and its geographic relationship to the two study sites.

is, the older, shallow-water, carbonate sections comprise the regional Floridan Aquifer system, while the Neogene siliciclastic-rich sections act as the regional aquitard.

Recent very high-resolution seismic surveys from Crooked Lake (Central Florida Highlands; Fig. 1c) and the St. Johns River (Northeastern Florida (Fig. 1b), have revealed several different types of deformation features originating from dissolution via subterranean fluid transport. These data show that the type of solution-tectonic feature developed is intrinsically-related to the local lithostratigraphic framework. They also indicate that solution-collapse events originate chiefly by dissolution via lateral groundwater transport within selective aquifers, and not by percolation of groundwater from surface aquifers or waterways down to and through the buried Tertiary limestones of the Florida platform.

METHODS

Subbottom Profiles

Seismic reflection data were collected using a mix of components from the EG&G UNIBOOM and ORE GEOPULSE profiling systems. They were deployed from a 6 meter skiff powered by an outboard engine. The frequency spectrum of these profiling systems range from 300 Hz to 14 kHz. All data are single-channel, analog profiles. Power levels on the seismic source varied, but generally ranged between 100 to 300 joules. The electromechanical transducer was consistently fired every 0.5 seconds. The incoming signal was collected on a multi-element hydrophone and filtered between 500 Hz and 5 kHz. These data were graphically displayed using an EPC 1901 recorder. This seismic profiling system is capable of resolving reflecting strata less than 1 meter apart (vertical resolution < 1m); however, it is generally limited to less than 100 m of total penetration.

Bathymetric Surveys

A Raytheon RTT 1000 7.0 kHz seismic profiler was deployed in the Crooked Lake site in order to generate a bathymetric map of the lake floor. The transducer was mounted to a surfboard and towed behind the outboard skiff. The graphic chart recorder was calibrated to account for the depth of the transducer in the water. In addition to providing a data base for construction of the bathymetric map, these data were also utilized to evaluate whether the subbottom collapse features identified in the boomer profiles exhibited any surface expression in the lake floor.

Stratigraphic Control

The Crooked Lake seismic data set was correlated to a well-log provided by the South West Florida Water Management District (T. Gilboy, personal communication 1986). Published well/borehole logs and archived samples provided by the Florida Geological Survey were used for the stratigraphic ground-truthing of the St. Johns River profiles.

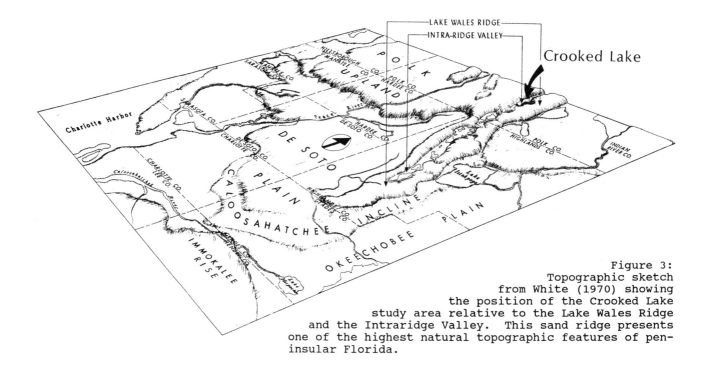

Figure 3:
Topographic sketch
from White (1970) showing
the position of the Crooked Lake
study area relative to the Lake Wales Ridge
and the Intraridge Valley. This sand ridge presents
one of the highest natural topographic features of peninsular Florida.

Two vibracores were also collected in Crooked Lake. These cores were acquired using a portable piston-vibracorer similar to the one described in Lanskey et al. (1978). The cores were split, megascopically logged, and subsampled primarily for grainsize analyses.

CROOKED LAKE DATA

Setting
Crooked Lake is located in central Florida and overlies the southern margin of a relict Eocene carbonate platform which floundered as late Paleogene siliciclastic sediments encroached from the north and prograded across the Suwannee Channel (Chen, 1965; McKinney, 1984; Pinet and Popenoe, 1985) and later the Gulf Trough (Popenoe et al., 1987). The combined outline of the Ocala and Sanford Highs marks the general position of the rim of this former carbonate platform. Figures 1 and 2 illustrate the geographic relationship of these features to the Crooked Lake site.

Crooked Lake also overlies the Lake Wales Ridge (LWR), a more recent (Pliocene/Pleistocene) geomorphic feature which trends N-S across central Florida (Fig. 1c). It is the most prominent geomorphic feature of peninsular Florida, locally reaching over 90 M above present sea-level. The Lake Wales Ridge is believed to be a linear sand massif deposited in association with several late Cenozoic shorelines (White, 1970). This sand ridge is divided by an intraridge valley, and the valley is characterized by a linear series of small lakes (Fig. 3). Crooked Lake lies just north of this intraridge valley. Its southwestern half lies proximal to the intraridge valley, while the northeastern section overlies the eastern flank of the Lake Wales Ridge (Figs. 1c, 3, and 4).

Crooked Lake actually consists of multiple coalescing ponds labeled ponds A through G in Figure 4. The individual circular ponds range from 0.3 to 2.6 km in diameter. Historical data show that lake level has dropped over 6 meters in the past 3 decades, and the lake shoreline has receded as much as 1 km based on aerial photogammetry surveys between 1972-1982 (see Figure 4). A series of relict shorelines associated with the receding lake level are easily identified in aerial photos (Fig. 5).

Bathymetry
A bathymetric map constructed from a network of approximately 50 km of 7.0 kHz profiles depicts the floor of ponds located near the intraridge valley as shallow (< 1.5 meters), flat, and featureless, while the ponds overlying the eastern flank of the LWR exhibit many ridges, topographic bumps, and local depressions -- some exceeding 3 meters of relief (Figs. 4 and 6).

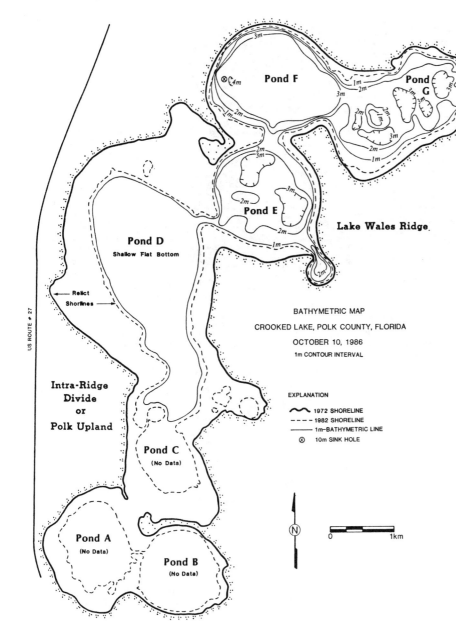

Figure 4: Bathymetric map of Crooked Lake constructed from a network of 7.0 kHz seismic reflection profiles. No data was collected in Ponds A, B, or C because these areas were too shallow to profile.

BATHYMETRIC MAP

CROOKED LAKE, POLK COUNTY, FLORIDA

OCTOBER 10, 1986

1m CONTOUR INTERVAL

EXPLANATION

〰 1972 SHORELINE
--- 1982 SHORELINE
— 1m-BATHYMETRIC LINE
⊗ 10m SINK HOLE

The flat and featureless bathymetric portrait of pond F in Figure 4 is actually an artifact poor data control. Bottom profile data was obscured by a thick, organic-flocculent layer settling above the hard lake bottom. This layer attenuated the seismic signal before it reached the lake bottom (Figs. 6 and 7). Thus, the bathymetry for pond F as shown in Figure 4 only represents the depth to the top of the flocculent layer and not to the lake floor.

Solution Collapse Features

Subbottom UNIBOOM and GEOPULSE seismic profiles collected from the shallow flat ponds within the intraridge valley of LWR (mostly pond D) reveal numerous, sediment-filled, solution shafts 8-12 meters deep (Figs. 8, 9, and 10). They exhibit little surface expression on the lake floor (< 2 M), but their subbottom expression penetrates an additional 20-40 meters through the Miocene siliciclastic aquitard of the Hawthorn Group. The solution shafts or "pipes" all appear to terminate at the top of a Miocene carbonate aquifer. Vibracores were targeted to penetrate a few of these sediment-filled solution features (Fig. 11). The core data show the shafts were primarily infilled with reworked LWR sands; however, the sands are interbedded with thin (<20 cm), organic-rich, clay horizons. These finer-grained horizons may act as impermeable plugs within the shafts, retarding transmission of fluids between the Miocene carbonate aquifer (45 meters below the lake floor) and the surficial sand aquifer overlying the less permeable clays of the Miocene siliciclastic sequence (above 30-35 meters). Several of these sediment-filled shafts are visible in aerial photo mosaics as small, dark, circular depressions within the lake. They also appear as circular patches of wetland vegetation in the adjacent land areas (Fig. 5).

Similar sediment-filled solution shafts penetrating down to the Miocene carbonates of the Hawthorn Group were found in all the eastern ponds (Fig. 12). A couple of additional solution features were also found in these ponds. One modern solution shaft was identified in pond F (see Figure 4 for location). It is represented by a hole approximately 50 meters in diameter and a minimum of 11 meters in depth (Fig. 13). Even though this feature was traversed several times, the total depth of the shaft could not be confidently defined due

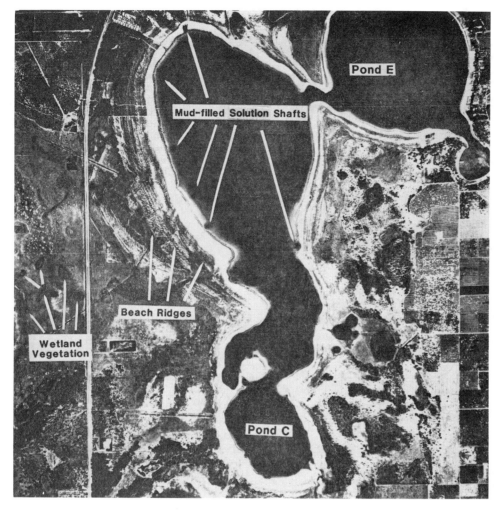

Figure 5: Low-altitude aerial photograph of Ponds C, D, and E illustrating the relict beach ridges associated with the recent recession of the lake level, as well as mud-filled solution shafts along the western shoreline of Pond D. These dark depressions are ubiquitous, but only show along the shoreline due to the eutrophic nature of the water column. Note also the patches of wetland vegetation to the west. These features appear to be former solution shafts, and the saturated water conditions suggest that there remains some groundwater communication between the surface and Floridan Aquifers.

to a combination of interference patterns generated by side-wall diffractions and attenuation by the organic flocculent partially filling the shaft.

Several dolines consisting of central sink surrounded by 2 to 4 adjacent solution pipes were identified in ponds E and G. Figure 14 is an example. The central shaft penetrates to approximately 12 - 15 meters below the lake floor. It shows less than 2 meters of surface expression and is almost totally infilled with sediment -- some of which is gas charged as demonstrated by the attenuated subbottom seismic signal. Its subsurface expression penetrates deep into the Miocene carbonate section 40-50 meters below the lake floor. The surrounding smaller shafts show no lake floor expression; they have less than 5 m of total subbottom relief, and barely penetrate to the Miocene carbonate aquifer. These multiple-shaft dolines form some of the broadest solution subsidence features found in Crooked Lake. Some appear to be genetically related to the shallow depressions depicted in the bathymetry of pond E. These depressions range up to approximately 400 m in diameter, but only have 2-3 m of lake-bottom relief.

Large-scale solution-collapse events were also found in pond G. They are marked by the abrupt lateral termination of continuous, high-amplitude, Miocene reflectors, as well as the sudden appearance of multiple diffraction patterns originating as side-wall echoes (Figs.15 and 16). Subbottom stratal gaps exist where the highly-reflective Miocene section has apparently disappeared (fallen?). These gaps range between 150 to 600 m in breadth. In one of the profiles, large blocks interpreted to be part of the Miocene section were found near the side wall of a collapse feature (Fig. 15). The vertical offset indicates the collapse of underlying caverns resulted in at least 20 meters of displacement. However, this was found in only one site. In most profiles traversing these collapse features, segments of the fallen Miocene section were not resolved within the down-dropped area of the collapse feature. This suggests that the Miocene stratigraphic section had fallen below the depth of penetration of our seismic tools (i.e., more that 70 meters of vertical displacement).

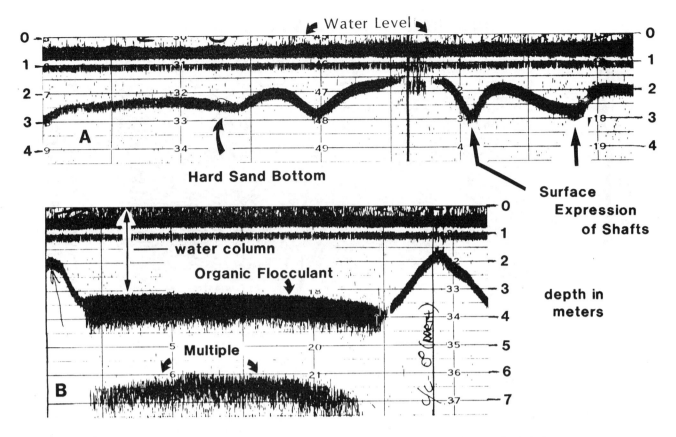

Figure 6: (A) 7.0 kHz seismic reflection profile from Pond E of Crooked Lake showing the topography of the sandy bottom. Note the surface expression of two partially-filled solution shafts. (B) 7.0 kHz seismic reflection profile from Pond F illustrating the top of the organic flocculent layer covering the lake bottom.

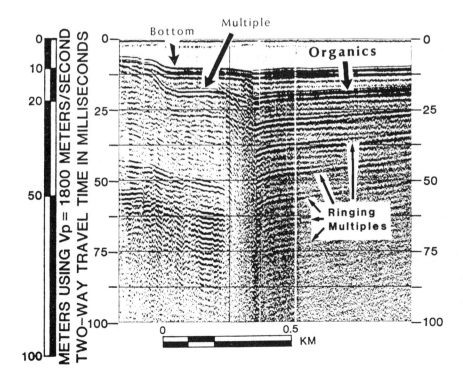

Figure 7: UNIBOOM seismic reflection profile showing the transition from a hard sandy bottom near the shoreline (left side) to the organic flocculent layer in Pond F. Note that the seismic signal is totally attenuated by the organics, and only the bottom multiple is rendered (i.e., ringing multiples).

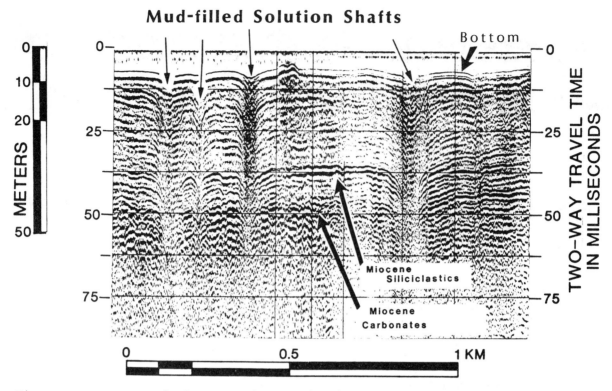

Figure 8: UNIBOOM seismic reflection profile illustrating several mud-filled solution shafts from Pond D. Note that all of these features penetrate the Miocene siliciclastic aquitard and terminate in the Miocene carbonate aquifer.

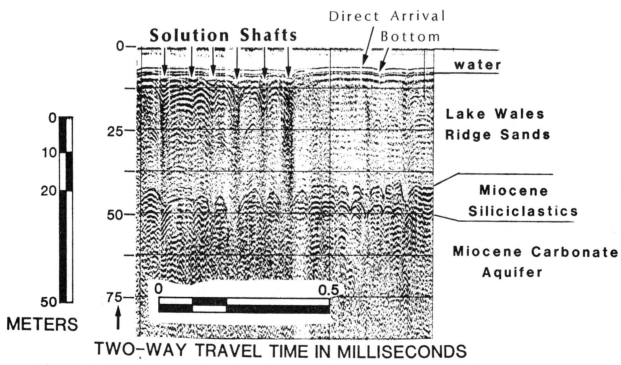

Figure 9: UNIBOOM seismic reflection profile from Pond D depicting a series of sediment-filled solution shafts all penetrating through the thin Miocene aquitard and terminating in the Miocene aquifer.

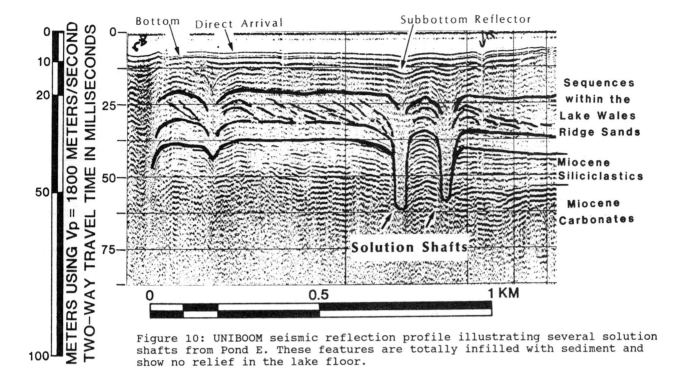

Figure 10: UNIBOOM seismic reflection profile illustrating several solution shafts from Pond E. These features are totally infilled with sediment and show no relief in the lake floor.

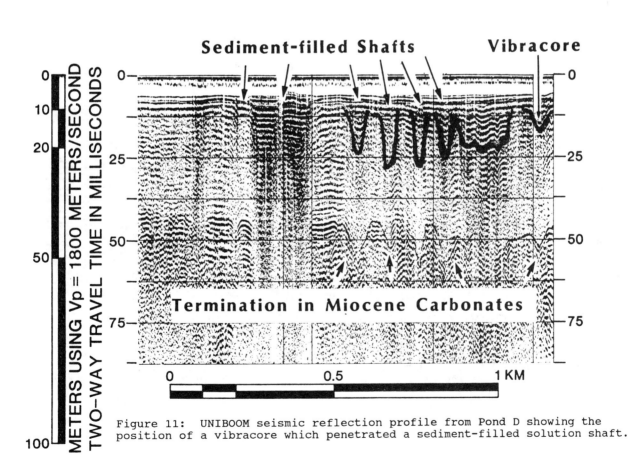

Figure 11: UNIBOOM seismic reflection profile from Pond D showing the position of a vibracore which penetrated a sediment-filled solution shaft.

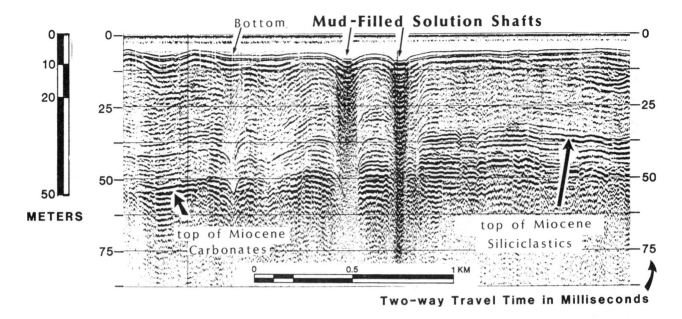

Figure 12: UNIBOOM seismic reflection profile illustrating several deep penetrating solution shafts from Pond G. The attenuated signature indicates the present infill is dominated by organic-rich sediments. The surface relief of these features is less than 2 meters.

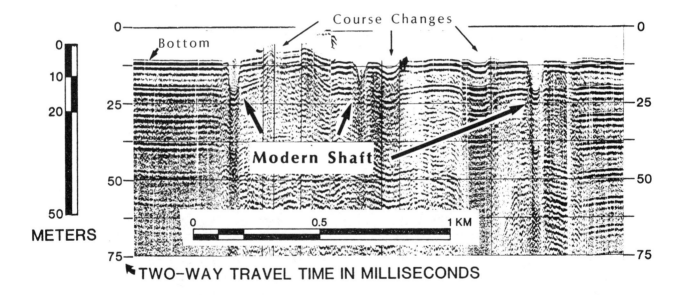

Figure 13: UNIBOOM seismic reflection profile illustrating a solution shaft approximately 40 to 60 meters wide and 11 to 13 meters deep. This feature was found in Pond F and is marked as a 10 m sink in Figure 4. The profile depicts 3 traverses of the same shaft.

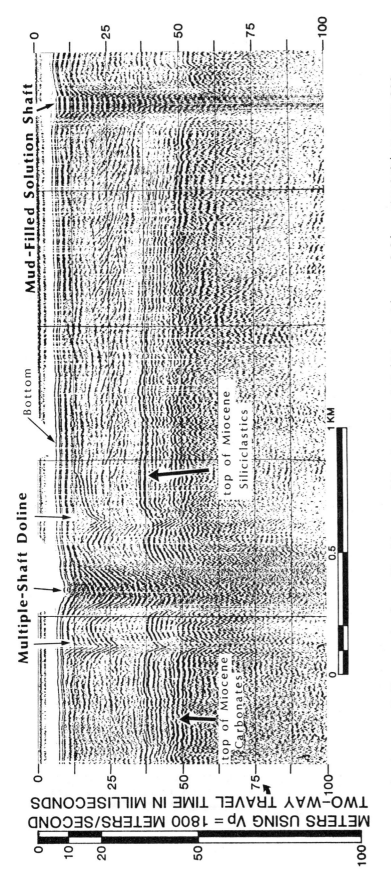

Figure 14: Uniboom seismic reflection profile from Pond E illustrating a doline stretching over 800 meters wide and consisting of multiple shafts. The central shaft penetrates well into the Miocene carbonate section and shows some relief in the lake floor. The adjacent shafts are smaller and barely penetrate the Miocene carbonate aquifer. Note the progradational sequence within the Lake Wales Ridge sand section is severely deformed near the solution features.

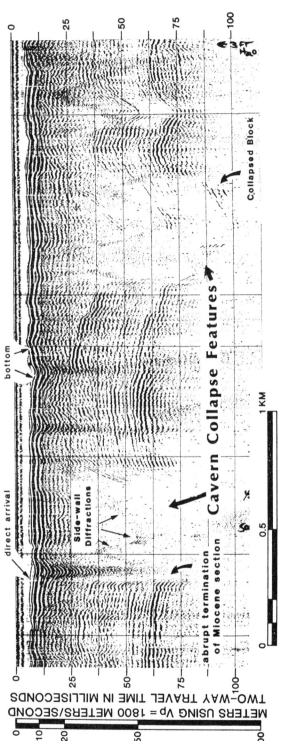

Figure 15: UNIBOOM seismic reflection profile depicting two, large, cavern-collapse features from Pond G. Note that the highly reflective Miocene section is missing where side-wall echoes appear.

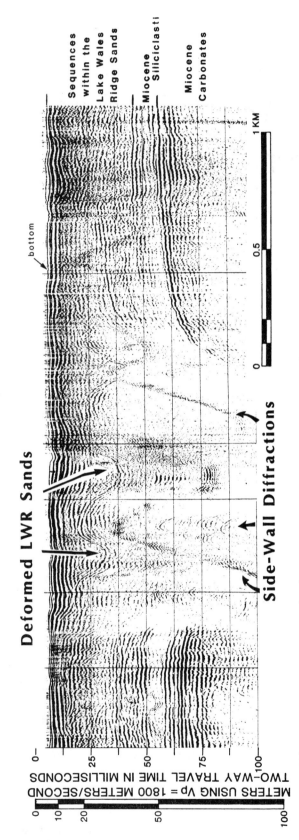

Figure 16: UNIBOOM seismic reflection profile illustrating an enormous cavern-collapse feature from Pond G. Note that the Lake Wales Ridge (LWR) sands are severely deformed.

292

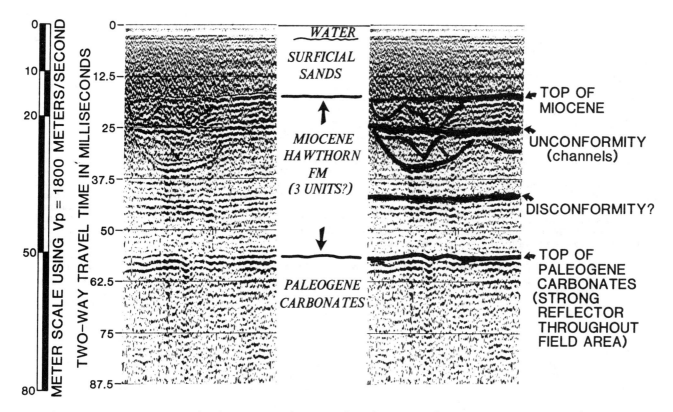

Figure 17: UN⊥BOOM seismic reflection profile illustrating the general stratigraphy within the St. Johns River near Palatka. Left panel is raw data; right panel depicts the interpretations.

Reflectors within the LWR sands overlying these large-scale collapse features all exhibited multiple flexures generated by subsidence originating from the subbottom collapse. Their morphology indicates that the subsidence was sudden, and that the resulting depressions were slowly-filled with reworked LWR sands and organic-rich muds.

Every collapse site showed the Miocene section of the Hawthorn Group was severely disturbed, if not completely missing. This indicates that the caverns formed below the Miocene section rather than within it. The relatively deep origins and the apparent rapid evolutionary nature indicate that these large-scale collapse features are a consequence of wholesale dissolution deep within the underlying Paleogene carbonates of the Floridian Aquifer. The cavernous Eocene carbonates of the Ocala Group, Avon Park, Lake City, and Oldsmar Limestones as described by Puri and Winston (1974) all represent potential sources for these features.

ST. JOHNS RIVER DATA

Setting

The St. Johns River flows north within the Eastern Valley, a geomorphic lowland area bound to the east by the Atlantic Coastal Ridge and to the west by the Duval and Marion Uplands (Fig. 1b). From Palatka to its mouth east of Jacksonville, the St. Johns River is in an estuarine condition. Its water is brackish (> 8 %), and the river experiences semidiurnal tides.

The estuarine section of the St. Johns River also overlies the southern rim of the Jacksonville Basin (Fig. 2). This large Miocene depositional basin was infilled with a mix of siliciclastic, carbonate, and authigenic lithologies of the Hawthorn Group (Riggs, 1979; Scott, 1983; 1986; 1988). Figure 17 shows the Miocene section overlies a sequence of Paleogene carbonates. This Paleogene section was previously shown to exhibit multiple karstic surfaces and solution-related subsidence features (Popenoe et al., 1984).

<u>Solution Features</u>
     Seismic profiles were collected in the Palatka area (Fig. 1b), and several solution
valleys were identified. They are characterized by broad flexures having approximately 12-
20 meters of relief stretched over several kilometers (Fig. 18). These flexures are
expressed through the entire Miocene section, and even through the upper part of the
Paleogene carbonate section as indicated by the presence of the faint, contorted, deep
reflector in Figure 18. This type of deformation within the Miocene section has previously
been interpreted as a consequence of coastal plain faults (Fig. 19; Scott, 1983). Those
interpretations were based on first-order correlations between boreholes only. The
continuous nature of the seismic reflection data shows however that the subsidence is a
consequence of solution within the underlying Paleogene carbonates.

     Note that the surficial sands and sandy clays overlying the subbottom flexure are
dissected by a series of channels (Fig.18). These channels probably represent the thalwegs
of former river valleys occupied during Pleistocene lowstands of sea-level. Reoccupation
of the same site appears to be a consequence of the topographic low originating from the
solution flexure. This suggests that the present St. Johns River as well as its Pleistocene
ancestors did not cut the valley via fluvial excavation processes. Instead the solution-
subsidence flexure has presented a persistent trough or low area in which the surficial
fluvial processes have been concentrated.

     Termination of the solution-valley subbottom flexures in the underlying Paleogene
stratigraphic section could not be observed due to the penetration limitations of the
seismic tool used. However, these features are interpreted to be the product of
subterranean flow within a more or less linear conduit, and that subterranean path was
apparently formed by dissolution of lesser resistant carbonates within the Floridan Aquifer.
The difference between these solution valleys and the major cavern collapse features of
Crooked lake appears to be the rate of settlement. In the solution valleys, subsidence
seems to take place gradually allowing overburden settlement to keep pace with the formation
of solution vugs. Although no large-scale cavern-collapse features have yet been identified
in the St. Johns River area, we expect to encounter them as our seismic surveys continue
since both the solution valleys and the wholesale cavern-collapse dolines appear to
originate from within the Floridan Aquifer.

     Also, no solution-tectonic features related to dissolution within the Miocene section
were identified from the St. Johns River data set. This may be a product of less carbonate-
rich Miocene facies in the Jacksonville Basin when compared to that of the Central Florida
Platform. The Jacksonville Basin is proximal to the source of the Neogene siliciclastics
(Fig. 2), and non-carbonate lithofacies seem to have dominated the basin's infill history
(Riggs, 1979; Scott, 1988).

<u>DISCUSSION</u>
     The depositional evolution of the Florida carbonate platform has rendered a series of
stacked aquifers separated by siliciclastic and dolomitic aquitards. These lithologies
represent the first-order framework which dictates groundwater flow and dissolution
processes in peninsular Florida. During the Paleogene, a series of carbonate bank margin
and interior platform sequences were deposited (Chen, 1965). This carbonate depositional
regime produced the bulk of the lithologies comprising the Floridan Aquifer. The platform
persisted through most of the Paleogene, floundering only when siliciclastics from the
southern Appalachians began to invade the Florida platform via progradation across the Gulf
Trough (McKinney, 1984; Popenoe et al., 1987). This Neogene siliciclastic invasion rendered
most of the non-carbonate lithologies of the Miocene Hawthorn Group, and they represent the
chief aquitard of the Floridan Aquifer.

     Subterranean flow within the Floridan Aquifer (Paleogene carbonates) originates
primarily from the watershed areas of the Sanford and Ocala Highs where Eocene and Oligocene
limestones are exposed or lie in the shallow subsurface. As meteoric waters flow downdip
away from these upland interior platform highs, they selectively dissolve the least
resistant carbonates in the Floridan Aquifer system. Microscale vugs to megascale caverns
and chambers result. Some of these voids begin to collapse. When settlement keeps pace
with the dissolution processes, solution valleys may be produced. They exhibit 10 to 25
meters of relief via a broad flexure stretching across a horizontal distance of 3 to 30 km
(e.g., St. Johns River). Both modern and ancestral surficial fluvial processes are
concentrated within these solution valleys. It is important to note however that the
surface waterways did not excavate the valleys themselves, but rather their flow appears to
be confined to the valley lowlands created by dissolution and related subsidence.

     Where large caverns develop in the Floridan Aquifer, the eventual collapse of the roof-
rock results in large-scale collapse dolines. These features are typically hundreds of
meters in breadth and tens (to hundreds?) of meters deep. They extend through the Paleogene

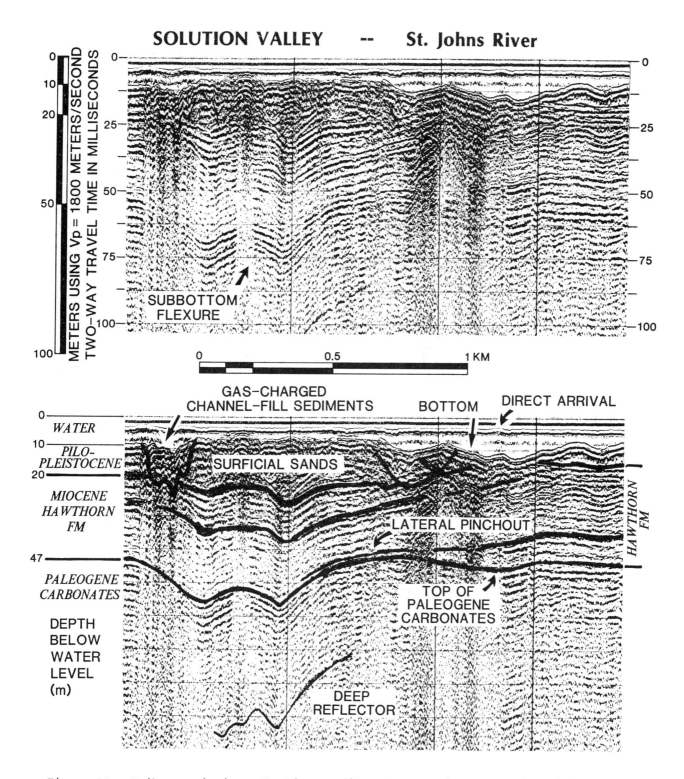

Figure 18: Uniboom seismic reflection profile of a solution valley identified via a north-south transect across the St. Johns River. Upper panel presents the raw data, and the lower panel illustrates the interpretations. Total relief of the solution valley is approximately 16 to 21 meters. These features originate in the Paleogene carbonate section.

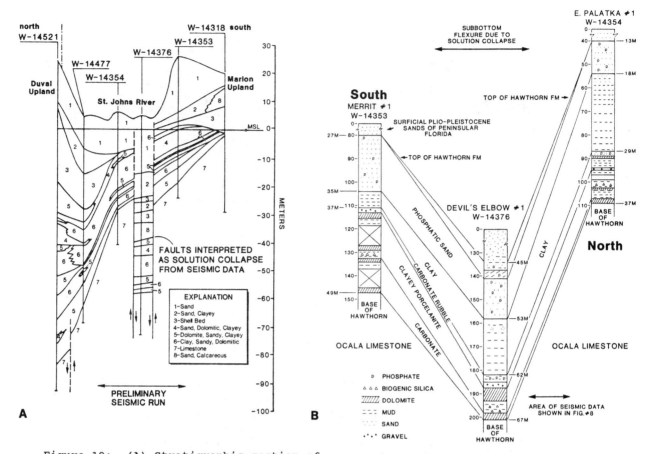

Figure 19: (A) Stratigraphic section of
the Palatka area based on correlations between boreholes. Note that the vertical
displacement between boreholes W-14354, W-14376, and W-14353 has been interpreted as
a product of faulting (modified from Scott, 1983). (B) Detailed lithostratigraphic
correlations across the St. Johns River solution valley based on the analysis of
archived samples.

carbonate section up to the surficial Quaternary sands, and they severely deform the Miocene
section. The destruction or collapse of the Miocene siliciclastic aquitard creates zones of
relatively high transmissivity. This provides potential pathways for groundwater exchange
between the surficial aquifer and Floridan Aquifer. Examples were found in pond G of
Crooked Lake.

The Miocene Hawthorn Group contains a number of carbonate sequences, especially in the
lower part of the section (Early Miocene). These carbonate sections dissolve through
subterranean flow as well, and they produce small-scale solution chambers. The solution
cavities collapse to produce deep, narrow, cylindrical shafts, tens to a hundred meters wide
and tens of meters deep. They are expressed in the surface as small circular depressions (<
2 m of relief). In the subbottom record, they penetrate through the surficial Pliocene and
Pleistocene sands, the Miocene siliciclastic aquitard, and finally terminate in the
underlying Miocene carbonate section. They are filled with reworked surficial sands, muds,
and organic detritus. These solution shafts were the most common type of sinkhole observed
in the Crooked Lake data set.

CONCLUSIONS
Three types of sinkholes were observed from very-high resolution seismic reflection
profiles. None appear to be related to the percolation of surface runoff or surficial
groundwater down through the overlying sediment column and into the underlying carbonate
aquifers. Instead, all of the solution features observed appear to be the product of
dissolution via subterranean flow within different aquifers comprising the ancestral Florida
carbonate platform. The subterranean fluids originate in upland recharge areas where
Eocene, Oligocene, and Lower Miocene carbonate sections lie near the surface. There,
groundwaters flow laterally downdip through the permeable Tertiary aquifers. Dissolution
within the Miocene carbonate aquifers results in steep, narrow, cylindrical shafts, while

large-scale collapse dolines originate via wholesale dissolution and catastrophic collapse of caverns formed deep within the Floridan Aquifer. Broad solution valleys also originate from dissolution processes within the Paleogene carbonates of the Floridan Aquifer; however, subsidence and settlement of the overlying Neogene and Quaternary stratigraphic sections is non-catastrophic in the solution valleys and seems to keep pace with the rate of subterranean dissolution.

ACKNOWLEDGEMENTS

Collection of the Crooked Lake data set was funded by the Southwest Florida Water Management District via a contract with the University of South Florida (A.C. Hine, Principal Investigator). The St. Johns River data was acquired via a Faculty Research Grant awarded to J.S. Compton from the University of South Florida. MEAS @ NCSU provided the resources for the production of this manuscript.

REFERENCES

Chen, C. S., 1965, The Regional Lithostratigraphic Analysis of Paleocene and Eocene Rocks of Florida: Florida Geological Survey Bull., No. 45, 105p.

Jansa, L. F.,1981, Mesozoic Carbonate Platforms and Banks of the Eastern North American Margin: Marine Geology, V. 44, p. 97-117.

Ladd, J. W. and R. E. Sheridan, 1987, Seismic Stratigraphy of the Bahamas: American Assoc. Petroleum Geologists Bulletin, V. 71, p. 719-736.

Lanskey, D. E., B. W. Logan, R.G. Brown and A. C. Hine, 1979, A new approach to Portable Vibracoring Underwater and on Land: Journal of Sedimentary Petrology, V. 49, p. 654-657.

Meyerhoff A. A. and C. W. Hatten, 1974, Bahamas Salient of North America: in C. A. Burk and C. L. Drake (eds.), The Geology of Continental Margins: Springer-Verlag, p. 429-446.

McKinney, M. L.,1984, Suwannee Channel of the Paleogene Coastal Plain: Support for the "Carbonate Supression" Model of Basin Formation: Geology, V. 12, p. 343-345.

Paulus, F. J., 1972, The Geology of Site 98 and the Bahama Platform: in C. D. Hollister, J. I. Ewing et al., (eds.), Initial Reports of the Deep Sea Drilling Project, US Government Printing Office, Wash., D.C., V. 11, p. 877-897.

Pinet P. R. and P. Popenoe, 1985, A Scenario for Mesozoic-Cenozoic Ocean Circulation over the Blake Plateau: Geological Society of America Bulletin, V. 96, p. 618-626.

Popenoe, P., V. J. Henry and F. M. Idris, 1987, Gulf Trough -- the Atlantic Connection: Geology, V. 15, 327-332.

Popenoe P., F. A. Kohout and F. T. Manheim, 1984, Seismic-Reflection Studies of Sinkholes and Limestone Dissolution Features on the Northeastern Florida Shelf: in Beck, B. F. (ed.), Sinkholes: Their Geology, Engineering, and Environmental Impact, A. A. Balkema Press, p. 43-57.

Puri, H. S. and R. O. Vernon, 1964, Summary of the Geology of Florida and a Guidebook to the Classic Exposures: Florida Geological Survey Spec. Pub. No. 5, 312 p.

Puri, H. S. and G. O. Winston, 1974, Geologic Framework of the High Transmissivity Zones in South Florida: Florida Geological Survey Bull., No. 20, 101 p.

Riggs, S. R., 1979, Phosphorite Sedimentation in Florida -- A Model Phophogenic System: Economic Geology, V. 74, p. 285-314.

Schmidt, W., 1984, Neogene Stratigraphy and Geologic History of the Apalachicola Embayment, Florida: Florida Geological Survey Bull., No. 58, 146p.

Scott, T. M., 1983, Evaluation of the Miocene Hawthorn Formation Carbonates of Northeastern Florida: Florida Bureau of Geology Report of Investigation No. 94.

Scott, T. M., 1986,A revision of the Miocene Lithostratigraphic Nomenclature, Southwestern Florida: Transactions -- Gulf Coast Geological Societies: V. 36, p. 553-560.

Scott, T. M., 1988, The Lithostratigraphy of the Hawthorn Group (Miocene) of Florida: Florida Geological Survey Bull., No. 59, 148 p.

Schlager, W. and R. N. Ginsburg, 1981, Bahama Carbonate Platforms-- the Deep and the Past: Marine Geology, V. 44, p. 1-24.

Sheridan R. E., J. T. Crosby, G. M. Bryan, and P. L. Stoffa, 1981, Stratigraphy and Structure of Southern Blake Plateau, Northern Florida Straits, and Northern Bahama Platform from Multichannel Seismic Reflection Data: American Assoc. Petroleum Geologists Bulletin, V. 65, p. 2571-2593.

White, W. A. , 1970, The Geomorphology of the Florida Peninsula:Florida Geological Survey Bull., No. 51, 161 p.

6. Case studies of engineering in karst terrane

Radon in homes, soils and caves of north central Tennessee and implications for the home construction industry

PAUL D.COLLAR & ALBERT E.OGDEN   *Center for the Management, Utilization and Protection of Water Resources, Tennessee Technological University, Cookeville, Tenn., USA*

ABSTRACT

A survey of radon concentrations in 59 basements and crawl spaces within houses built on karst in the Highland Rim and Central Basin provinces of central Tennessee had a measurement range of 0.1 to 37.6 pCi/L with twenty-two percent of the measurements exceeding EPA's 4 pCi/L recommended maximum. Excepting the highest measurement, the population had an arithmetic mean of 2.6, a median of 2.2 and a standard deviation of 1.95. Although the Chattanooga Shale and lower rocks were poorly represented within the survey compared to Fort Payne and Warsaw strata, the ranking of the means according to rock type was as follows: Bigby-Cannon (4.4 pCi/L), Fort Payne (3.4 pCi/L), Chattanooga (3.3 pCi/L), and Warsaw (1.6 pCi/L). The similarities between the means suggests that no single lithology other than perhaps the Warsaw was found to have a significantly different radon occurrence potential.

The degree of home weatherization proved to be a poor indicator of the magnitude of likely indoor radon levels but was significant in its effect on the distribution of radon progeny within 11 individual homes. Highly weatherized homes demonstrated a lower average reduction factor in radon progeny from basements to first floors than did average and poorly insulated homes.

Homes with concrete foundations and basements had a slightly higher average radon concentration than did homes with crawl spaces. The depth of home excavation correlated positively with radon concentration at a 99.9% confidence level.

A case study of a house overlying a small near-surface cave system indicated the likelihood that radon-bearing cave air was entering the soil via enlarged fractures visible in outcrop. Anomalous indoor concentrations, ranging from 3.5 to 37.6 pCi/L, were thought to result from the advective transport of radon through the foundation via small fractures in the concrete foundation and loose utility fittings.

A soil radon progeny traverse perpendicular to a near surface cave passage of known location and radon content surprisingly revealed the lowest progeny concentration immediately overlying the cave passage of interest. Anomalously high radon progeny concentrations were unexpectedly found 220 feet normal to the passage within a soil of unusually high porosity and permeability.

Radon's abundance and transport are controlled by a large number of different factors, and radon occurrence is not accurately predictable based on geological and home construction characteristics and should be measured using a time-integrated device to accurately assess the radon content of an existing home. Where development is planned overlying karst or other terrains implicated as radon sources, soil radon surveys are recommended to evaluate the potential need for preventive construction techniques.

Introduction and Objectives

During the 1970s, karst scientists became aware that high concentrations of radon gas were naturally present in some limestone caves. In response to the potential health threat to cave guides and employees, the National Park Service instituted a radon progeny monitoring program for its caves which has been in effect nationwide since 1976 (Carson, 1989). In central Tennessee, The Chattanooga Shale and phosphatic horizons within the karstic, Ordovician carbonates were both implicated as rocks with high radon occurrence potential within a statewide survey of indoor radon concentration (TDHE, 1987). The results of the TDHE report prompted the research for the present paper.

Radon-222 is a colorless, odorless, inert, radioactive gas naturally present in the environment as an intermediate product of the decay sequence from 238-U to 206-Pb; Rn-222 has a half-life of 3.8 days. Health studies (NAS, 1988; Cross, 1989) have shown that radon's short-lived daughter products can stimulate lung cancer within laboratory animals. Based on epidemiological studies of uranium miners who developed lung cancer, the EPA has recommended a maximum indoor airborne concentration of 4 pCi/L radon, at which Nero (1988) projects there is a 0.8% probability of developing lung cancer as a result of radon progeny exposure over a lifetime.

Despite the fact that considerable controversy currently exists over the actual health hazard posed by low levels of radon, the real estate industry will be increasingly required to respond to the problem as perceived by consumers, regardless of the eventual scientific consensus on low level risk assessment. Even without regulation of radon levels within building codes, home buyers can be expected to be increasingly aware of the radon problem and to expect contractual provisions assuring them of a "radon free" home.

Because preventive construction techniques can be more effective in controlling radon impact than radon mitigation within an existing structure, the conscientious home builder should be aware of the radon issue and take steps to ensure that his product will not be susceptible to radon entry. Specifically, tracts of land slated for development within an area known to have occasionally high radon levels should be systematically analyzed for soil radon content using nuclear track-etch monitors. Boyle (1988) proposes that radon content be measured and published within soil survey publications, an action which would greatly facilitate preliminary site evaluations. The variability of radon concentration can be significant even very locally as a result of geological and soil heterogeneities, and it is always best to make as many measurements as possible to avoid suboptimal coverage.

The measured radon distribution within the soil permits the developer to: 1) design the lay out so as to minimize construction over high radon soils, and 2) evaluate the potential need for radon preventive construction techniques, summarized by EPA (1987, 1986-a).

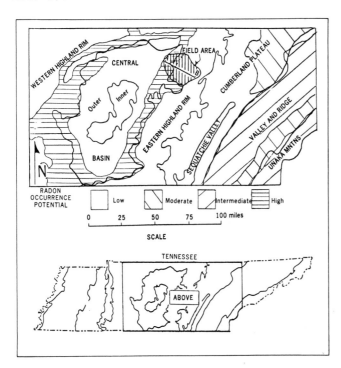

Figure 1: Physiographic provinces of central Tennessee showing the location of the field area and TDHE's (1987) classification of statewide radon occurrence potential.

The present study draws upon measurements of radon and radon progeny in homes, soils and caves in north central Tennessee to determine the effects of local geology and home construction characteristics on indoor radon concentrations. Two case studies are included to illustrate the applicability of soil radon testing to home design and construction within karstic terranes.

Study Area

The study area includes portions of the Eastern Highland Rim and eastern outer Central Basin provinces in north central Tennessee (Figure 1). As the generalized cross-section in Figure 2 indicates, the field area encompasses rocks ranging in age from mid-Ordovician (Hersey & Maher, 1985) to mid-Misissippian (Milici et. al., 1979).

The mid to late Ordovician section is karstic and contains phosphate-rich horizons within the Hermitage, Bigby-Cannon and Catheys-Leipers formations (Hershey & Maher, 1985). The Fort Payne and Warsaw formations are highly siliceous, and overlying Mississippian carbonates are relatively pure. The St. Louis formation and portions of the Monteagle and Warsaw comprise a sinkhole plain which flanks the Cumberland escarpment, trailing its erosional re-

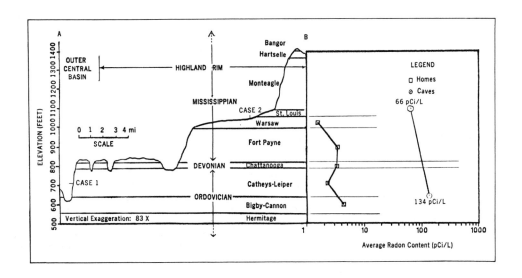

Figure 2: Cross-section of traverse A-B (Figure 1) showing corresponding average cave and home radon concentrations.

cession (Crawford, 1980). The Fort Payne contains abundant seams of chert and forms small "pocket" caves. Caves are locally abundant in the upper Warsaw. Other Mississippian and Ordovician formations contain greater numbers of caves due to the higher solubility of the relatively more pure carbonates.

The Devonian-aged, uranium-rich Chattanooga Shale (Conant & Swanson, 1961) is exposed within the field area along the Highland Rim Escarpment. It possesses an average thickness of 30 feet in the field area and is a valley-forming lithology; consequently, the Chattanooga is exposed almost exclusively within steep terrain, and houses are infrequently constructed upon the shale itself. Based on the highly uraniferous nature of the Chattanooga (Glover, 1959), as well as the findings of their own study, the Tennessee Division of Health and Environment (1987) classified its exposure and downslope terrains statewide with the state's highest radon occurrence potential ranking (Figure 1; Chattanooga outcrop corresponds approximately to the boundary between the Highland Rim and Central Basin provinces). Despite the fact that the Highland Rim Province as a whole was characterized within the same report (TDHE, 1987) with the lowest state-wide radon occurrence potential ranking, (Figure 1) measurements made by the Civil Defense in Cookeville and environs found that out of 94 measurements using an integrating radon gas monitor (precision = +/- 25%; Powers, 1989), 44% exceeded the 4 pCi/L recommended maximum level (average nationally: 7%; Nero, 1989). Locally high measurements (up to 80.7 pCi/L) were found in houses in close proximity to known caves (Smith, 1989).

Average home radon concentrations were calculated for houses built upon Bigby-Cannon (n=2; mean=4.35), Cathye-Leipers (n=7; mean=2.88; median=3.0), Chattanooga (n=3; mean=3.33; median=3.1), Fort Payne (n=24; mean=3.43; median=2.4) and Warsaw (n=17; mean=1.56; median=1.5) strata. Although the Bigby-Cannon and Chattanooga are highly under-represented, the mean concentration of each lithology was plotted against the stratigraphic profile in Figure 2, indicating that soils overlying the Fort Payne and possibly the Chattanooga and Bigby-Cannon may have somewhat higher radon content on the average than do soils associated with the Warsaw and Catheys-Leipers.

Yarborough et. al. (1976) report a range in equivalent radon concentration (assumes 50% equilibrium between radon and daughters) from caves around the U.S. of <1 pCi/L to 404 pCi/L, with a mean of weighted averages equal to 85 pCi/L. Webster & Crawford (1987) found that three caves in Mississippian strata of southern Kentucky ranged from 570 pCi/L to 1080 pCi/L equivalent radon. Radon concentrations were measured in 23 caves in central Tennessee by Neff (1987) and within the present investigation. Seventeen Mississippian caves ranged from 17.1 to 285 pCi/L and averaged 66 pCi/L whereas 6 caves within Ordovician limestones ranged from 40 to 240, averaging 134 pCi/L (Figure 2). The higher radon content of Ordovician cave air probably reflects a background uranium concentration within the Ordovician rocks higher than that found within Mississippian rocks. Uranium's ability to enter the lattice of apatite minerals through ionic substitution for Ca++ has resulted in the enrichment of uranium within phosphorite lithologies in Florida (Humphreys, 1987)

Tennessee, and elsewhere. AEC (1970) reports that the uranium content of phosphatic strata within the Hermitage and Catheys were 4 ppm and 2 ppm respectively. Phosphates are more abundant within Ordovician strata than Mississippian strata, and it is likely that uranium content may be resultantly higher in the older carbonates.

Methods

Radon concentration in indoor air was measured using carbon canisters. Described in greater detail by Cohen and Nason (1986), charcoal canisters permit the passage of radon-containing air into the canister while excluding radon progeny. Decay products resulting from entering radon are adsorbed onto activated carbon. Following a seven day exposure period, the carbon is analyzed with either a scintillation or gamma detector and concentration is determined based on calibration using the same procedure within an atmosphere of known radon concentration. Canisters used in the present study were purchased from and analyzed by The Radon Project, in association with the University of Pittsburgh.

Radon progeny concentration in indoor air and soil air was measured using a variable flow pump and a Ludlum 2200 scaler ratemeter with a scintillation detector. Radon progeny were captured from soil gas and indoor air by pumping air at a flowrate of either 3 or 5 Lpm through a 0.8 micron filter for 5 minutes. Soil samples were taken in augered holes 4 inches in diameter and 3 feet in depth; packing was used in the borehole around the collection hose to minimize atmospheric dilution. Airborne particulates, including the ions 218-Po, 214-Pb, 214-Bi, and 214-Po, collectively known as radon progeny, were assumed to have been completely captured by the filter. The filter was subsequently removed and transferred to a scintillation detector within a window of time 40 minutes to 90 minutes subsequent to the sample's collection. Counts per minute were averaged from either a 5 or a 10 minute reading. Data reduction to working levels (total airborne alpha activity) was accomplished using the Kusnetz model, described by EPA (1986-b).

The efficiency of the counter was calculated using Cs-12, an alpha-emitter with a disintegration rate of 8455 disintegrations per minute. The check source was the same diameter and approximate thickness as the filters to be analyzed and was monitored on a monthly basis. During 6 months of operation the efficiency varied from .42 to .58.

In order to avoid the awkwardness of comparing radon progeny (working levels) with radon (pCi/L), an "equivalent" radon concentration can be calculated based on progeny concentration. The precise conversion requires knowledge of the degree of equilibrium between radon and its daughters. Nero (1989) recommends an assumption of 50% equilibrium between radon and radon progeny when the exact equilibria are not known. At 50% equilibrium, 1 WL = 200 pCi/L equivalent radon. For consistency all measurements in working levels have been converted using a factor of 200 to pCi/L. The progeny concentrations can be used only as a relative measurement and should not be considered absolute due in part to the imprecision resulting from the assumption of 50% equilibrium. The diurnal and seasonal variation of radon concentration and the difficulties of representative sampling adds to the misrepresentation that a single grab sample can provide when compared to time-averaged track-etch or carbon canister measurements. Due to these factors, the radon progeny grab measurements were used to evaluate the relative distribution of radon progeny within a given home only when all measurements were made within an hour or two of one another. Based on poor overall correlation between carbon canister and grab measurements within the present study, grab progeny measurements are not deemed sufficient as a radon screening measurement unless periodic measurements under highly controlled conditions are performed. Consequently, carbon canisters and track-etch monitors are the best choices for home screening purposes.

Indoor Air Radon Survey

Radon concentration was measured with carbon canisters in the basements or crawl spaces of 60 homes throughout the study area. The measurements ranged from 0.1 pCi/L to 37.6 pCi/L. Excepting the high measurement, the range was from 0.1 to 10 pCi/L; mean =2.58, median = 2.2, and standard deviation = 1.95. Twenty-two percent of all measurements were above 4 pCi/L. Compared to the national average of 7% above the recommended maximum and arithmetic average indoor concentration of 1.5 pCi/L (Nero, 1988), the study area could be characterized as a "high radon area". However, excepting the one measurement of 38 pCi/L (a home in close proximity to a cave), all measurements were less than or equal to 10 pCi/L. Measurements were taken in either crawl spaces or basements, and it is likely that living area concentrations are considerably lower than the initial screening measurements. Radon progeny reduction between basements and first floors within 11 homes in the present study averaged a factor of 1.8. Consequently, an average home would have to have a concentration greater than 7 pCi/L in the basement or crawlspace before living area concentrations would be above 4 pCi/L. Within the present set of data (n= 59), only 3 measurements (5%) were above 7 pCi/L. Therefore, although a greater percentage of homes greater than 4 pCi/L is typical of the study area when compared to the national average, the magnitude of the offensive measurements is so low that on the whole the potential health impact is suspected to also be low.

The effectivity of home design and construction was evaluated on the basis of carbon canister measurements of indoor air as well as grab radon progeny measurements in soil air, basement air, and first floor bathroom air. Factors thought to play a role in radon entry which were characterized for analysis included: 1) the presence or absence of a foundation, 2) the depth of excavation, and 3) the degree of home weatherization.

Basements Vs Crawl Spaces

All homes tested had either 1) basements with concrete slabs, or 2) crawl spaces with exposed earth. Excluding the 37.6 pCi/L measurement, homes with slabs and basements had an average of 2.83 pCi/L (Median = 2.6, n = 19), whereas homes with crawl spaces had an average of 2.63 pCi/L (median of 2.2; n = 37). Twenty-six percent of the homes with basements exceeded 4 pCi/L, whereas 19% of homes with crawl spaces exceeded the limit.

In fifteen homes, basement and crawl space radon progeny measurements were compared with soil progeny measurements. The two were expressed as a ratio of basement/soil in an effort to evaluate how effective home substructures were in excluding naturally present soil radon from indoor air. The ratios of basement/soil for homes with crawl spaces had a range of 0 - 2.16, and a mean of 0.46. Homes with basements had an indoor/soil ratio ranging from 0.19 to 2.41 with a mean of 0.99. The results suggest that within the houses studied, crawl spaces exclude radon from the indoor environment more effectively than do basements with concrete foundations. Two homes with basements and one with a crawl space demonstrated basement concentrations higher than soil concentrations, suggesting that these houses may play an active role in the development of pressurized airflow. All three of these houses were coincidentally highly weatherized; consequently, radon introduced from the soil was less able to migrate to the exterior of the home than in poorly insulated homes.

The results suggest that the presence of a concrete slab does not inhibit radon transport in general and may in fact promote higher radon entry than does the presence of a crawl space. This observation is thought to reflect the possible confluence of three separate factors: 1) the presence of cracks or loosely fitting utility lines within the slab and basement walls, 2) the fact that basements present a larger surface area to soil radon flux, and 3) the possibility that greater ventilation rates within crawl spaces promote a proportionally greater amount of radon transport to the exterior.

Degree of Weatherization

Based on qualitative characterization of homes as 1) poorly insulated, 2) well insulated and 3) average, degree of home weatherization proved to be a poor indicator of radon levels. Average houses possessed the highest average equivalent radon concentration (3.03 pCi/L; median = 2.2 pCi/L), followed by well-insulated (mean = 2.48 pCi/L; median = 2.2 pCi/L) and poorly insulated (mean = 2.21 pCi/L; median = 1.7 pCi/L). Although the qualitative characterization contributes to imprecision in evaluating the effects of home weatherization, more significant is the fact that where soil concentrations are low to begin with, the difference in concentration between a highly insulated house and a poorly insulated house is so low that no clear definition of trends was possible.

The degree of home weatherization was determined to have a significant effect on the distribution of radon progeny within a given house, however. The ratio of first floor progeny to basement progeny was calculated for 11 houses. Those that were well insulated or superinsulated had an average ratio of .80, whereas poorly insulated and average houses demonstrated average ratios of .53 and .51, respectively. The results imply that within average and poorly insulated houses, high concentrations of radon in the basement are more likely to be dispersed by air exchange to the exterior whereas within highly insulated homes there is a lower degree of concentration reduction between basements and living areas. Therefore, if radon is able to migrate into a house substructure, it is less likely to pose a health hazard within the living areas of poorly insulated houses than it is within well insulated houses.

Depth of Excavation

A correlation between depth of home excavation and indoor radon concentrations yielded a Pearson coefficient of .55 at a .0001 confidence level, indicating the strong probability of a direct relationship between excavation depth and radon content. The mediocre coefficient, however, suggests that the relationship is not confidently predictable with these two variables alone. A linear regression yielded equation (1).

$$\text{Radon (pCi/L)} = 0.6 + 0.7 * (\text{Depth [ft]}) \quad (1)$$

The scatter of the data resulted in a low confidence (46%) in the intercept (0.6) but a very high confidence (.01%) for the slope (0.7). The function predicts that the 4 pCi/L indoor limit is exceeded at excavation depths of 5 feet or greater. The wide scatter in the data are thought to reflect the importance of other factors including basement ventilation rate and the relative importance of advective transport.

One subterranean home and another highly excavated, partially subterranean home had basement concentrations of 10 pCi/L and 38 pCi/L respectively, the two highest measurements within the survey.

The likelihood of a house possessing high radon concentrations therefore appears to be a complex function of geological factors and home construction characteristics. If radon is not naturally present in high concentration in the soil, home construction characteristics are irrelevant in predicting its occurrence. Where radon concentrations are high in the soil, however, home construction techniques determine whether or not radon is able to gain entry into a structure as well as whether radon that has gained entry is readily transported out-of-doors or is concentrated in the indoor air.

Although all homes are subject to some degree to the flux of radon from the soil by molecular diffusion, a variety of natural and anthropogenic factors can stimulate pressure driven airflow through the soil matrix. Where substructures are neither tight nor highly ventilated, such advective transport of radon can introduce potentially large levels of radon to the indoor environment, as the following case study suggests.

Case 1: Radon Distribution Within a House and Adjacent Soils

The home containing the highest radon concentration measured within the initial screening survey (38 pCi/L) was revisited, and the distribution of radon and radon progeny was measured within different rooms of the basement and first floor as well as within the soil in two places adjacent to the home. Although the orifice of a nearby cave tapers within 15 feet of the mouth to a prohibitive squeeze, the home owner reported that a very small caver had been able to follow the cave back several hundred feet in the approximate direction of the house, where the passage was reported to have opened sufficiently for him to walk upright; he reported having heard hammering from the garage, demonstrating the proximity of the cave passage to the house.

The approximate location of the house within the stratigraphic section is depicted in Figure 2. Figure 3-b shows the house vicinity in cross-section and plan, indicating that the depth of excavation varies from 0 feet in the front to 25 feet in the rear of the house. The hill-slope is located 20 feet from the back of the house, and several fractures are evident within the exposed Ordovician limestone bedrock. The largest fracture had a trend normal to the approximate trend of the cave system and was thought to intersect the cave passage in the subsurface. A soil air sample was collected from a hole augered between the house and the hillslope along the strike of the fracture (soil 1, Figure 3-b). An equivalent radon concentration of 10.25 pCi/L was measured. Approximately 100 feet perpendicular to the trend of this fracture a second soil sample (soil 2) yielded an equivalent radon concentration of 0.86 pCi/L. The soil penetrated in both samples was very similar in character; the very large difference in radon concentration was thought to reflect the influence of the fracture system in promoting advective transport of radon through the soil.

Figure 3-a shows a plan view of the basement and first floor levels of the house in question, with measurements noted within the rooms in which they were taken. The initial screening measurement made in room 4 in December was 37.6 pCi/L. Follow-up measurement of room 4 yielded the highest follow-up radon concentration of any of the rooms measured, 29 pCi/L. The difference in concentration is thought to reflect the seasonal variability of radon concentrations (higher in winter/lower in summer) that suggest that thermally driven advection of radon and not diffusion may be the predominant transport mechanism. During the winter, greater thermally driven pressure gradients exist between indoor/outdoor air and indoor/cave air, provided that an interconnected fracture system exists that allows the transport of gases between cave and soil and vice-versa. Thus, radon-laden cave air drawn into the soil due to a thermally induced pressure gradient is likely to supplement the convective air flow into the basement resulting from the thermal gradient existing between inside and outside airmasses, as postulated by Nazaroff et al. (1988).

In the summer, however, temperature gradients between indoor air and cave air are probably small due to the insulating effect of deep basement excavation. Therefore, pressure driven cave air flow into the basement is likely to be lower. The upper level of the house is opened during the summer, leading to essentially isothermal conditions. The basement, however, stays cool even in the heat of summer. The thermal gradient between indoor air and outdoor air is such that convective air flow within the soil would be expected to be minimal on the whole, and in a direction reversed from the winter, i.e. out of the house, further minimizing the advective entry of cave air.

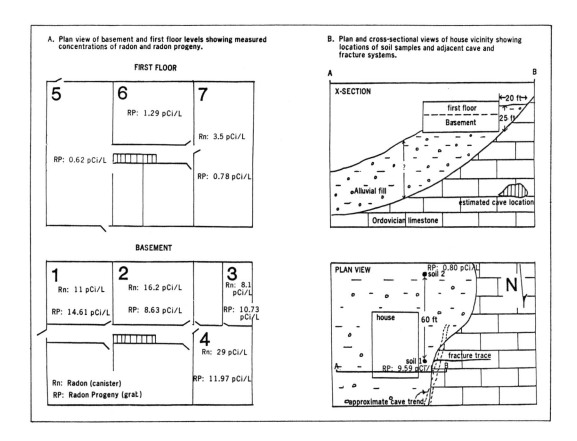

A. Plan view of basement and first floor levels showing measured concentrations of radon and radon progeny.

FIRST FLOOR

| 5 | 6 RP: 1.29 pCi/L | 7 |

Rn: 3.5 pCi/L

RP: 0.62 pCi/L

RP: 0.78 pCi/L

BASEMENT

| 1 Rn: 11 pCi/L RP: 14.61 pCi/L | 2 Rn: 16.2 pCi/L RP: 8.63 pCi/L | 3 Rn: 8.1 pCi/L RP: 10.73 pCi/L |

4 Rn: 29 pCi/L

RP: 11.97 pCi/L

Rn: Radon (canister)
RP: Radon Progeny (grat.)

B. Plan and cross-sectional views of house vicinity showing locations of soil samples and adjacent cave and fracture systems.

X-SECTION

first floor
Basement

20 ft
25 ft

Alluvial fill

estimated cave location

Ordovician limestone

PLAN VIEW

RP: 0.80 pCi/L
soil 2

N

house

60 ft

soil 1
RP: 9.59 pCi/L

fracture trace

approximate cave trend

Figure 3: Case Study One.

While the average radon progeny reduction factor between basements and first floors within 11 houses was found to be 1.8, this house exhibited a 13-fold diminution in average radon progeny concentration and a 5-fold reduction between the averaged basement radon measurements and the single first floor radon measurement. The observation is thought to reflect the unusually high basement ceiling (12 feet) and more importantly the low air exchange in the basement as compared to the higher apparent air exchange on the first floor.

A soil radon survey conducted prior to construction of the house would have revealed that the soil contained locally anomalous concentrations of radon. Siting the house 50 feet away from the existing site might have obviated the radon contamination the homeowner must now contend with. Even at the same site, however, an awareness of a potential problem with radon would have enabled the implementation of preventive design and construction methods.

Case 2: Radon in Soils Overlying a Near-Surface Cave

Indian Cave is developed within the upper Warsaw formation and is located approximately one mile to the southwest of the point indicated on the cross-section in Figure 2. The portion of the cave mapped for the present study is shown in Figure 4-a. Radon canisters were exposed within the cave at the approximate locations shown in Figure 4-a and provided measurements of 15.8 and 24.7 pCi/L; the high measurement was made during winter and the low measurement during late spring.

A surface resistivity traverse was conducted normal to the trend of a cave segment as shown in Figure 4-a; the resistivity profile is shown in Figure 4-b. Since air-filled caves and zones of increased porosity are typically characterized by higher resistivity than adjacent solid rock, the resistivity traverse was deemed applicable in more accurately defining the cave's location.

Five holes (Figure 4-a) were augered along the same traverse for radon progeny

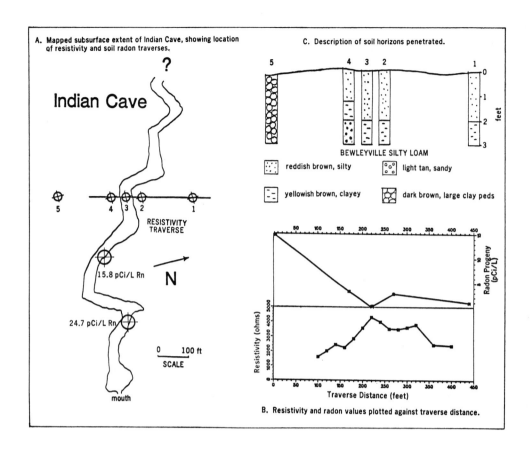

A. Mapped subsurface extent of Indian Cave, showing location of resistivity and soil radon traverses.

C. Description of soil horizons penetrated.

BEWLEYVILLE SILTY LOAM

reddish brown, silty   light tan, sandy
yellowish brown, clayey   dark brown, large clay peds

Indian Cave

RESISTIVITY TRAVERSE

15.8 pCi/L Rn

24.7 pCi/L Rn

N

0   100 ft
SCALE

mouth

B. Resistivity and radon values plotted against traverse distance.

Figure 4:   Case Study Two.

measurement.   One sample was taken in soil coincident with the resistivity peak under the presumption that it corresponded to the cave trend.   Two equidistant sets of holes were augered on either side of the first hole.   The distribution of measurements is shown in Figure 4-b.

The results demonstrate that soil immediately overlying the resistivity peak had the lowest measured progeny concentration (0 pCi/L).   The highest value (15 pCi/L) was coincidentally the highest of twenty other soil measurements made regionally and was measured at site 5 in a soil 220 feet south southwest of the cave passage.

All 5 boreholes penetrated soils mapped by the Soil Conservation Service (1963) as Bewleyville silty loam.   Soils at sites 1, 2, and 3 were essentially identical (Figure 4-C), consisting of a surficial two feet of reddish brown loam with a high silt content underlain by a yellowish brown soil of higher clay content.   The two soil horizons are of reduced thickness at site 4, where a third soil type was found to consist of a very light tan, sandy loam.   The soil at site 5 was highly porous and permeable in nature and wholly different from the other soils, consisting entirely of walnut-sized peds of clay. Whereas five minutes of hard work with a gasoline powered auger were required for the other boreholes, the machine cut through site 5 soil in less than 1 minute.   The high porosity and permeability at site 5 may provide a path of least resistance for pressurized cave air flow within the soil. It is also possible that a separate fracture system may be present in the subsurface of site 5, contributing to the concentrations measured independently of, or only peripherally affected by, the cave passage of interest.

Again, the implications in terms of a planned development upon the tract of land in which this study was conducted are significant.   The construction of a house upon the soil typical of site 5 without regard to radon-preventive construction practices could result in an indoor radon problem of greater magnitude than the case study presented earlier.   Although rational geological expectation would predict that radon transported from a cave system would affect immediately overlying soils, the fact that high measurements

308

were detected at an unexpected site reinforces the practical necessity of soil monitoring to satisfactorily characterize the radon distribution within a soil or soils.

Conclusions

The classification of houses in terms of the lithostratigraphy they were built upon revealed that the Warsaw Formation had the lowest average radon measurements. Lithologies possessing higher radon concentration were the Bigby-Cannon, Chattanooga and Fort Payne. Most of the measurements are thought to represent situations in which molecular diffusion is the predominant mechanism of radon transport. Proximity to a cave system was shown, however, to be potentially capable of stimulating pressure-driven cave and soil air flow into house substructures. In homes subject to advective radon entry, the potential for hazardous indoor radon concentrations appears to be significantly higher. Because caves are developed within most of the lithologies exposed in the field area, the possibility of localized soil radon "hot spots" is thought to exist throughout the Highland Rim and Central Basin provinces.

Houses with basements and concrete slabs had radon concentrations slightly higher on the average than concentrations measured within earthen crawl spaces. Similarly, basements were found to have higher basement/soil progeny ratios than crawl spaces.

The degree of weatherization had apparently no predictive significance in terms of radon content but did affect progeny distribution patterns within houses, with more insulated homes having higher first floor/basement progeny ratios than average or poorly insulated homes.

Deeply excavated homes were statistically more likely to have higher radon concentrations than were homes characterized by shallow excavation.

Regardless of the actual threat posed by radon progeny within the indoor environment, it is clear that the perception of a threat by the consuming public will increase as information and national discussion continues to focus on the radon issue. Although many people are not interested in measuring the radon content of their own existing homes, an increasing number of buyers require this information when purchasing a house. As a result, the real estate industry in certain parts of the country has established a growing relationship with the radon measurement and mitigation wing of the environmental consulting industry.

Radon concentration within indoor air is a complex function of many variables, and radon concentration cannot be accurately predicted based on known geological and home construction factors. Consequently, radon screening of existing homes and soils where development is planned is recommended for individuals who have a vested interest in indoor radon concentrations. The measurement technology is simple and affordable, leaving this option within the reach of most individuals who want to know. However, within areas known to contain occasionally high radon concentrations, soil screening should always be conducted prior to development because the expense and effort beforehand is small in comparison to the future expenses that a radon-laden house could engender. In an uncertain near-future regulatory climate with regard to radon concentrations in homes and a market that is increasingly more acquainted with the radon problem, the value of a developed property may be significantly affected by indoor radon concentrations. Since preventive construction does not require a significant capital outlay, merely minor design modifications in many cases, the economic incentive for incorporating wide-spread preventive construction techniques is already in place in light of possible lawsuits and property value reduction. Consequently, it appears that radon testing of soils could actually offer a business edge to developers by providing them with increased information upon which to base decisions as well as a potential public relations boost and advertising device.

References

AEC, 1970: Uranium in the Southern United States; U.S. Atomic Energy Commission, Div. Raw Matls., Document WASH-1128 TID UC-51, Contract #AT (49-6)-3003, Atlanta, GA, pp. 91-96.

Boyle, M., 1988: Radon testing of soils; Environmental Science & Technology, v. 22, #12, pp. 1397-1399/

Carson, B., 1989: personal communication; National Park Service.

Cohen, B.L., and Nason, R., 1986: A diffusion barrier charcoal collector for measuring Rn concentrations in indoor air; Health Physics, v. 50, #4, pp. 457-463.

Conant, L.C. and Swanson, V.E., 1961: Chattanooga Shale and Related Rocks of Central Tennessee; U.S.G.S. Professional Paper 357.

Crawford, N.C., 1980: The Karst Hydrogeology of the Cumberland Plateau Escarpment; Part IV; Cave and Karst Studies Series, Center for Cave and Karst Studies, Dept. of Geog. & Geol., Western Kentucky University.

Cross, F.T., 1988: Evidence of lung cancer from animal studies; IN: Radon and its Decay Products in Indoor Air, Nazaroff W.W. and Nero, A.V. eds., Wiley-Interscience; 518 pages.

EPA, 1986-a: Radon Reduction Techniques for Detached Houses: Technical Guidance; EPA/5-86/019.

---, 1986-b: Interim Indoor Radon and Radon Decay Product Measurement Protocols; EPA 530/1-86-04, 50 pages.

---, 1987: Radon Reduction in New Construction: an Interim Guide; EPA/OPA-87-009, 50 pages.

Glover, L., 1959: Stratigraphy and uranium content of the Chattanooga Shale in north-eastern Alabama, northwestern Georgia, and eastern Tennessee: U.S.G.S. Bulletin 1087-E, pp. 133-168.

Hershey, R.E., and Maher, S.W., 1985: Limestone and Dolomite Resources of Tennessee, 2nd Ed.; Tenn. Dept. of Cons., Div. of Geol., Bulletin 65, 252 pages.

Humphreys, C.L., 1987: Factors controlling uranium and radium isotopic distributions in ground waters of the west central Florida phosphate district; IN: Radon in Ground Water, B. Graves, ed., Lewis Publishers, 546 pages.

Milici, R.C., Briggs, G., Knox, L.M., Sitterly, P.D., and Statler, A.T., 1979: The Mississippian and Pennsylvanian Systems in the United States--Tennessee: U.S.G.S. Professional Paper 110-G.

NAS Committee on Biological Effects of Ionizing Radiations, 1988; Health Risks of Radon and Other Internally Deposited Alpha Emitters---BEIR IV. Natl. Acad. Press, Washington D.C.

Nazaroff, W.W., Moed, B.A., and Sextro, R.G., 1988: Soil as a source of indoor radon: generation, migration and entry; IN: Radon and its Decay Products in Indoor Air, Nazaroff W.W. and Nero, A.V. eds., Wiley-Interscience; 518 pages.

Neff, B., 1988: personal communication.

Nero, A.V., 1988: Radon and its decay products in indoor air: an overview; IN: Radon and its Decay Products in Indoor Air, Nazaroff W.W. and Nero, A.V. eds., Wiley-Interscience; 518 pages.

Powers, T., 1989: personal communication; Sun Nuclear Corporation.

Smith, S., 1989: personal communication; Putnam County Civil Defense.

Soil Conservation Service, 1963: Soil Survey; Putnam County, Tennessee; U.S. Dept. Agri., Series 1960, #7, 114 pages.

Tennessee Division of Health and Environment, 1987: Summary Report of the Tennessee Radon Survey, 34 pages.

Webster, J.W. and Crawford, N.C., 1987: Preliminary results of investigation of radon levels in the homes and caves of Bowling Green, Warren County, Kentucky; IN: Radon Workshop Manual; Center for Cave and karst studies, Dept. of Geog. and Geol., West. Kent. Univ., Bowling Green, Kentucky.

Yarborough, K.A., Ahlstrand, G.M., and Fletcher, M.R., 1976: Radiation study done in National Park Service Caves; Natl. Spel. Soc. News, v. 34, #8, 146.

Building construction over an erosive sinkhole site: A case study in Lancaster County, Pennsylvania, USA

HENRY GERTJE   *Dunn Geoscience Corporation, Albany, N.Y., USA*
ANTHONY E.JEREMIAS   *Dunn Geoscience Corporation, Mechanicsburg, Pa., USA*

ABSTRACT

Observations made during the growth of a 3000 cubic meter sinkhole in Lancaster County, Pennsylvania, U. S. A. provided insights into developing an innovative technique for sinkhole remediation. The successful remediation allowed for subsequent building construction over the sinkhole prone site. A building addition, with first floor slab on grade, covering 15,000 square meters was constructed over the repaired sinkhole with a triple layer geogrid safety net system. Building columns are supported by deep caissons drilled to competent bedrock. The sinkhole was caused by roof drain runoff from a 37,000 square meter manufacturing plant and with runoff created by upgradient development. In general, most sinkholes in Lancaster County are caused by concentrated surface water flowing into the subgrade resulting in erosion of the overburden soil. Successful remediation of sinkholes depends on controlling surface water. Sequencing the earthwork to minimize exposure of the subgrade was essential in order to prevent other sinkholes from developing during construction. In this case, conventional earthmoving and compaction equipment were used to backfill the sinkhole by placing cobble stone, crushed gravel and native soil into geotextile wrapped plugs.

INTRODUCTION

On May 5, 1988, after an extended drought, 1.5 centimeters of rain fell on a proposed building addition site at a large manufacturing plant in Lancaster County, Pennsylvania, U.S.A. Near the center of the site, a sinkhole was discovered at the confluence of the plant's roof drain outlet and drainage ditch. Water rushing from the 2.6 meter wide drainage ditch and 0.6 meter diameter roof drain penetrated the undermined grass mat which lined the ditch. The widening hole in the ditch revealed several near-surface, steeply dipping ridges of dolomitic limestone. As the rainfall continued, its runoff began to erode the clayey overburden soil into the solutioned subsurface and downstream along the ditchline.

The next day, a second rain of similar magnitude (1.8 cm) fell on the area. On that day, the uncontrolled drainage entered into the sinkhole, promoting continued erosion of the soil upgradient along the ditchline. Heavy rainfalls continued over the next two weeks, overwhelming several attempts to backfill the hole with gravel and concrete.

As the sinkhole grew, more solution channels into the subsurface were observed. By May 26, 1988, 21 days after its discovery, the sinkhole covered an area of 630 square meters (6900 square feet); it had grown to within 5 meters of the manufacturing plant and its depth ranged from 2.5 meters to greater than 8 meters (8 feet to 25 feet). The sinkhole had completely undermined a 10 meter section of the facility's perimeter truck road. On the opposite end of the sinkhole, the building's fire water line, pressurized to over 1400 kilopascals (200 pounds per square inch), was also undermined. At least five separate chimneys formed beneath the floor of the sinkhole. The total eroded volume was estimated at greater than 3000 cubic meters.

This report describes the conditions which contributed to the development of the sinkhole. The observed conditions led to the engineering designs and construction practices which successfully remediated the sinkhole and allowed construction of the new building addition over the damaged site.

Physiography of Lancaster County Pennsylvania

The part of the Lancaster County surrounding the site is referred to as the Northern Piedmont Physiographic Region. It is characterized by an east - northeast trending landscape of smooth valleys and moderately sloping low hills. In essence, the hills are the lowest part of the Appalachian Mountain Range. The broad flat valleys are typically underlain by limestones, dolomites and shales. The region lies south of the southernmost advance of Pleistocene Glaciation.

For many generations Lancaster County has fostered the production of agriculture. Only light manufacturing and retailing have prevailed in the small rural towns. The land is owned mostly by traditional agrarian Pennsylvania Dutch and Amish people, who have limited the modern development of the land. Large buildings which have recently been developed in the county frequently develop significant geotechnical problems with karst because of increased rainfall runoff caused by roofs and paved surfaces.

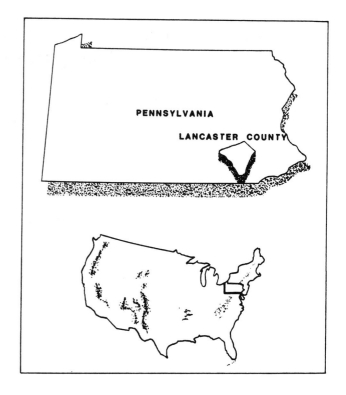

Figure 1 - Site location.

Sinkhole History at the Site

Sinkholes have been a common occurrence in many locations at the facility. They have formed along the sides of the perimeter truck road and near the edges of the parking lots. The sinkholes have usually been less than 3 meters in diameter at the surface and have appeared as funnels in the soil or as open wedges beneath the flanks of near-surface rock pinnacles.

Occasionally, sinkholes have caused damage to the building and disrupted the operations at the facility, especially at areas of the building which house vibrating equipment. The original two story building is supported by shallow spread footings and has a ground floor slab on-grade. On several occasions, the interior floor slabs have settled, cracked, and required jack grouting. By one account, an operator was drilling through a cracked floor slab when the jackhammer broke through the concrete and fell from his hands. It fell 9 meters into a sinkhole.

The most significant sinkhole at the site occurred beneath the southwest corner of the building adjacent to the proposed building site. Several building columns settled there and required underpinning.

SURFACE AND SUBSURFACE CONDITIONS AT THE SITE
....

AD = Soils

The site soil is mapped as the Hagerstown silt loam. The soil is predominantly an orange-brown to brown silty clay with minor components of sand and gravel. The soil has the potential to shrink and swell with changes in moisture content. Atterberg limit tests indicated an average plasticity index of 30.3% and an average liquid limit of 48.5%.

This residual soil is derived from the weathering of the original bedrock. Some of the soil exhibits a relict bedrock structure characteristic of saprolitic-type soils. Typically, the soil directly above the bedrock is granular in nature, containing angular gravel-size fragments of highly weathered limestone and dolomite. The thickness of overburden soil at the site is highly variable, ranging from 0.3 to greater than 14.5 meters.

Bedrock
The carbonate bedrock underlying the proposed building addition is mapped as part of the Lebanon Valley Sequence. Observations at the site suggest that the Epler Formation is present beneath the sinkhole and building site. The Epler Formation is a thick-bedded, medium to medium-dark-gray, finely crystalline limestone, interbedded with medium dark-gray, finely crystalline dolomite.

The Allegheny Orogeny strongly affected the structure of the Epler Formation along with the remainder of the Paleozoic rocks of Pennsylvania. Locally, the stratigraphy is tilted, folded and faulted such that the beds beneath the site strike to the east-northeast and dip approximately 45 degrees to the southeast. Differential solutioning of the carbonate beds has resulted in an irregular pinnacled "ridge and slot" bedrock surface. The spaces between the pinnacles are filled with residual or colluvial soils. Voids are also common. Voids within the loose soils between the dipping pinnacles (particularly beneath the overhanging edges of the pinnacles) provide primary drainage conduits into the subsurface. Severly fractured but intact slickenside zones in the bedrock pinnacles are believed to provide a secondary flow path for water.

Other Related Conditions at the Site
In early May of 1988, the facility's main building covered approximately 37,200 square meters (9.2 acres) of land surface. The entire roof surface was pitched to the west so that all rainfall runoff was collected along the west wall into a single precast concrete drain pipe. The pipe consisted of 2.4 meter (8 foot) sections which were bell and spigot connected end-to-end and sealed with flexible plastic O-ring gaskets. The drain pipe led away from the building to the southwest, several feet below the surface. The pipe daylighted at a drainage ditch, approximately 30 meters from the building, near the center of the proposed building addition.

Each section of the pipe weighed approximately 500 kilograms. The sections could not be mechanically fastened end to end and thus were not capable of cantilevering the pipe weight. To make matters worse, the concrete pipe was aligned almost parallel to the strike of the bedrock pinnacles in the area. These conditions caused the pipe to erode along the length of a soil trough and to undermine itself, breaking off section-by-section towards the main building. A lighter, mechanically connected and ductile drain pipe may have remained relatively intact and could have limited the size of the sinkhole.

The grass lined drainage ditch was parallel to the west wall of the building. It was cut approximately 2 meters into the original ground surface of the site. The bottom of the ditch was approximately 2.5 meters wide. The banks were sloped at approximately 2 horizontal to 1 vertical. The ditch collected and transported runoff from a 27 hectare drainage basin upgradient of the roof drain.

A majority of the upper drainage basin is covered with pasture and grain fields and slopes between 3 and 8 percent. Another building is located within the drainage basin, upgradient and across the highway from the site. The roof area of this second building was approximately equivalent to the size of the first. In previous years, both facilities had increased the size of their parking lots within the lower reaches of the drainage basin. Both facilities had detention basins designed to delay the release of runoff. The detention basins on both sites have been plagued with sinkholes throughout their history.

According to the climatological data recorded at a nearby station, the rainfall conditions from March through June 1988 departed significantly from precipitation normals. The early spring was very dry and probably resulted in shrinking of the clayey soils. The reduced soil volume may have allowed channels to open into the subsurface. In addition , the groundwater levels were probably much lower than normal, due to well irrigation of the nearby farmland. The rainfall which finally did fall on the site in early May, was reported to be short and heavy.

Table 1

Rainfall Records at the Landesville, Pennsylvania Station

	March	Apr.	May	June
Precipitation Normals (cm)	8.1	8.8	9.1	10.4
Recorded Precipitation (cm)	5.4	4.5	15.2	2.9
Percent Departure	-33%	-49%	+67%	-72%

An estimate of the peak flow into the sinkhole during its growth was made using the maximum recorded daily rainfall of 3.9 cm on May 19, 1988. The roof drain carried no less than 0.4 cubic meters of water per second and the ditch provided an additional 1.7 cubic meters per second, giving a combined peak flow into the sinkhole of 2.1 cubic meters per second. The reported intensity of this rainfall at the site suggests that the peak discharge may have been somewhat higher than is estimated here. Considering the peak discharges capable of flowing into the sinkhole, the substantial rainfall deficit recorded in June was most fortunate for the efforts to remediate the sinkhole.

Observations and Analysis of the Sinkhole

Unlike the large collapsing type of sinkholes which are more commonly known, the sinkhole at the manufacturing plant was produced by surface water eroding the fine-grained soils in solutioned bedrock. At least five separate chimneys were observed in the sinkhole. The largest and deepest chimney was located beneath the intersection of the roof drain and drainage ditch. It was a narrow fissure bounded by two dolomitic pinnacles. The remainder of the chimneys appeared as funnels in the soil. Excavation of these chimneys revealed that they were located above soil troughs between steeply dipping dolomitic pinnacles and they were similar in nature to the narrow central fissure.

Initial attempts to repair the sinkhole were unsuccessful because the erosion caused by runoff advanced upgradient away from the original sinkhole location along the ditch and roof drain. Hastily poured concrete plugs were bounded by soft native soils. Subsequent erosive surface water easily reestablished the throats through the soil along the flanks of the plugs. The gravel, loosely poured into the hole, did not remain separated from the wet,

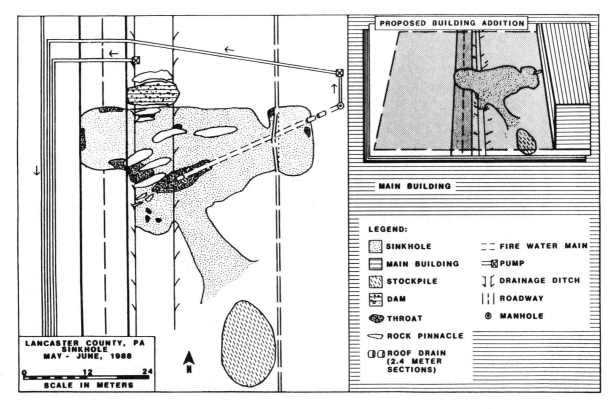

Figure 2 - Features of the May 1988 Sinkhole during remediation. Inset (upper right) shows layout of the sinkhole relative to the main building and proposed building addition.

cohesive soils between the pinnacles. The high permeability of the gravels allowed the throats to reestablish themselves by piping the silts and clays through the gravel into the subsurface.

At the sinkhole's greatest extent, there was no evidence that the rock mass had collapsed. This indicated that it would be safe for heavy equipment to enter into the base level of the sinkhole to perform the remedial work.

REMEDIATION OF THE SINKHOLE

Remediation of the sinkhole was accomplished between May 26 and June 10, 1988 under the field guidance of two engineers from Dunn Geoscience Corporation. The undermined roof drain and fire water main presented an emergency situation, so that the remediation team worked long hours while the weather remained dry.

In order to limit runoff into the sinkhole, the roof drain was sealed off inside a manhole next to the building. Two small dams were also constructed upgradient of the sinkhole in the drainage ditch. Large pumps were stationed at each of the three locations, standing-by for use 24 hours a day. Two additional pumps were available for backup capacity. All water was pumped via pipes to the site detention basin.

The first task accomplished within the sinkhole was to stabilize the soft soil beneath the fire water main near the building. A steel joist was used to temporarily straighten and support the pipe until the pressure in the undermined section was relieved.

An excavator equipped with a 3/4 cubic meter bucket and 7.2 meter reach was used to slope-back the soils along the building edge of the sinkhole. Since no distinct bedrock was encountered in that area, a 10-ton drum roller entered the sinkhole and compacted the drying soil.

At the same time, a track-mounted front end loader equipped with a 2 cubic meter bucket carved an entrance ramp into the central portion of the sinkhole. The ramp allowed access for a concrete truck to descend to the central fissure beneath the juncture of the roof pipe and drainage ditch. The fissure was partially filled with cobble stones, then a sand mix concrete was poured over and through the cobbles to seal the opening.

After the concrete work was completed, the loader and excavator were used to shape the sinkhole bottom **and to move the slumping soft soils to a nearby stockpile location.** Large detached boulders were removed from the excavation and not reused in backfilling the sinkhole. As the excavation proceeded from within the base level of the sinkhole, the pin-nacled nature of the subsurface was revealed. Soil troughs between the ridge-like pinnacles hid numerous small throats and soft compressible areas, usually underlain by detached boulders. In most cases, when the boulders were removed, the soil remaining in the troughs would compact tightly. The compacted, overexcavated troughs were lined with geotextile and backfilled with a layer of 10 cm rounded cobble stone. The stone was overlain with a com-pacted lift of well-graded, crusher run gravel. The gravel was capped with a compacted lift of native clayey soil. Enough extra geotextile remained in the trough so that the stone and soil package could be completely enveloped. These geotextile plugs were generally limited to troughs less than 2.5 meters wide below the base level of the sinkhole.

As the soils dried in the sun, broad areas of the sinkhole bottom were densified by numerous passes of the front-end loader and the heavy drum roller. Beginning from the build-ing end, sloping gradually downward toward the center of the sinkhole, broad areas of geotextile were laid over the sinkhole floor. A lift of crushed gravel averaging about 0.3 meters thick was then compacted over the geotextile, followed by two compacted 0.3 meter lifts of the stockpiled native soils. Enough geotextile was laid out so that the compacted layers of gravel and soil could be enveloped. Three to five laborers were required to place the geotextile.

The selected geotextile was a spun bonded, polypropylene heavy support fabric in roll widths of 4.7 meters (15.5 feet). The burst strength was 1850 kilopascals (265 psi). The ef-fective opening size was 140 to 170 microns, making the material ideal for retaining silt particles. The geotextile appeared to conform well with the backfill materials and subsur-face structure during placement and compaction. It occasionally punctured when impacted by cobble stones or when the gravel was overly compacted above it. A more elastic geotextile may have been a better selection in terms of resisting puncture. However, it is more impor-tant that the geotextile provide separation of the cobble and gravel materials from the plastic clayey soils below. The geotextile plugs were also flexible enough to follow any deformation of the surrounding soils.

By wedging a tight interlocking mass of dense granular material above the native clayey soil troughs, the tendency for the clays to shrink and swell is reduced. Using coarse back-fill materials at the bottom of the sinkhole provides bridging for the soil fill components placed above. Capping the sinkhole area with well compacted native soil serves to reduce the permeability of the soil and also helps to limit the shrink/swell potential.

Construction of the Building Addition

The foundation for the proposed building addition was constructed between June 10 and November 22, 1988 immediately following remediation of the sinkhole. The addition now covers an area of approximately 140 meters (460 feet by 300 feet) by 90 meters.

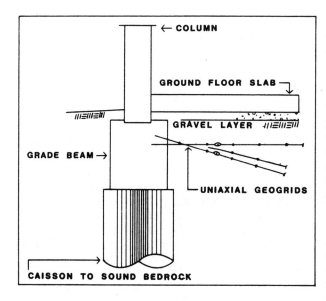

Figure 3 - Detail illustrating connection of the geogrid safety net system into the gradebeams.

To reliably support the structure, drilled caissons were specified. Approximately 100 caissons were drilled into rock with end-bearing contact pressures of 20 tons per square foot. The caissons ranged in depth from 1.5 to 18 meters below the surface. The caisson bottoms were test drilled from the surface using a pneumatic air-track percussion drill. Thin pinnacles, boulders and highly altered rock zones were frequently drilled through to achieve a sound bearing rock for the caissons. Drilling water was not allowed during placement of the caissons.

Preparation of the subgrade for the ground level floor slab was performed concurrently with the caisson installation. Excavation of the subgrade was limited to two column bay widths, parallel to the length of the building addition in order to minimize the surface area of the exposed uncompacted subgrade. The subgrade was overexcavated by one meter to accommodate a basal geotextile layer plus three layers of biaxial polyethylene geogrids which were placed between 0.3 meter lifts of compacted backfill select soil. The upper and lower geogrid layers were tied into the foundation grade beams along the building perimeter.

The function of the geogrid layers is to provide a "safety net" in the event sinkholes ravel upward through the subsurface and undermine the floor slab. It is not the intent of the geogrids to prevent the floor slabs from cracking. Rather, they are to prevent a catastrophic failure of the slab and to reduce the possibility of injury and damage to expensive equipment housed within the building. In the event of a developing sinkhole, the geogrids develop a "pull out" resistance in the soils surrounding the undermined area. The available resistance transmitted by the grids is proportional to the pressure overlying the intact soil area and is inversely proportional to the weight lying on the undermined area. If insufficient embedment length is available to support an undermined weight near the building perimeter, the grade beams are designed to carry the lateral loads transmitted by the grids.

The sequence of geogrid placement worked well in the field. The most difficult task was to provide adequate compaction along the edges of the excavated strips, especially in the vicinity of the exterior grade beams where the geogrids required a change in grade for the tie-in. Gradual slopes of 5 horizontal to 1 vertical were specified for changing grade, in order to minimize slack in the geogrids.

Conscientious site work by the excavation contractor helped to prevent any serious sinkholes from developing during the construction operation. Very few soft areas or throats were observed during excavation. Observations made during construction suggested that the remediated sinkhole area was the most stable part of the subgrade beneath the building addition.

CONCLUSION

Sinkholes appear frequently in Lancaster County U.S.A., particularly as a result of modern development of the land surface. The sinkholes are usually caused by uncontrolled surface water eroding the overburden soils into the karstic subsurface. Sudden large collapses of the ground surface into caverns are not a common cause of sinkholes because of the steeply dipping interbeds of dolomite. While the near-surface limestone beds have generally dissolved or altered into residual soils, the dolomitic beds have persisted in the soil matrix forming steeply inclined tabular-shaped pinnacles and "floating" boulders.

Voids within the soil zones between rock pinnacles provide the primary drainage conduits into the subsurface. The conduits apparently begin developing beneath the overhanging surfaces of the pinnacles. For this reason, ground areas with pinnacles near the surface are likely to be susceptible to sinkhole development. A more serious situation is presented when surface water flows parallel to the dip direction of the pinnacles.

The remediation method for this sinkhole focused on conforming to the pinnacled structure of the subsurface. Crusher-run quarry stone, native cohesive soils and geotextile were the primary backfill materials used in repairing the sinkhole. The geotextile is intended to separate the well-graded compacted gravel from the cohesive native soils below. The geotextile also provides filtration and retention of fine grained soil particle, and is used to form flexible plugs which have an appreciable tensile stress/strain capability. The compacted clayey native soil, provides an excellent low permeability cap over the backfilled sinkhole. Concrete is considered an appropriate material only to seal open conduits which are bounded by bedrock.

With the possibility that sinkholes could develop beneath the new building addition, the ground floor slab was underlain with a geogrid safety net system. The safety net included three layers of geogrid installed in the fill over a basal geotextile layer. In the event that a sinkhole should develop beneath the floor slab, the geogrids would mobilize a frictional pullout resistance around the perimeter the ravelled sinkhole throat to prevent a catastrophic failure of the floor slab. The geogrids were tied into the grade beams along the perimeter of the building structure. Sequencing the subgrade stripping, grading, compaction and geogrid placement was essential to minimize the area of exposed subgrade, thereby reducing the risk of sinkhole development. Controlling surface water during remediation and construction was of vital importance.

REFERENCES

Geyer, A. I., Pennsylvania Geological Survey, 1980. Geologic Map of Pennsylvania: Commonwealth of Pennsylvania, Department of Environmental Resources.

Koerner, Robert M., Ph.D., P. E. 1986. Designing with Geosynthetics, Prentice Hall 424 p.

McGlade, W. G., Geyer, A. R., Wilshusen, J. P., 1972. Engineering Characteristics of the Rocks of Pennslyvania: 200 p. Commonwealth of Pennsylvania Department of Environmental Resources, Bureau of Geological Survey, Harrisburg, PA.

Newton, J. G., 1986. Development of Sinkholes Resulting from Man's Activities in the Eastern United States: 40 p. U. S. G. S. Circular 968.

United State Department of Agriculture, Soil Conservation Service, 1985. Soil Survey of Lancaster County Pennsylvania: 152 p. plus Soil maps. (In cooperation with Pennsylvania State University College of Agriculture and the Pennsylvania Department of Environmental Resources State Conservation Commission)

U. S. D. A. Soil Conservation Service Engineering Division, 1975. Urban Hydrology for Small Watersheds Technical Release No. 55: 7 chapters plus appendices.

Mann road sinkhole stabilization: A case history

R.L.GOEHRING & S.M.SAYED   *Jammal & Associates, Inc., Winter Park, Fla., USA*

ABSTRACT

On September 21, 1988, a sinkhole suddenly developed beneath a 54 inch diameter effluent transfer line. The sinkhole occurred along Mann Road in Orange County, Florida, approximately 0.8 kilometer (0.5 mile) south of the intersection of S.R. 545 and Tilden Road. The sinkhole was initially about 9 meters (30 feet) in diameter and 2.4 meters (8 feet) deep at the center, and was beginning to undermine the pipe. During corrective deep grouting, the sinkhole suddenly opened to a diameter of 15 meters (50 feet). While emergency measures were being considered, the prestressed concrete pipeline spanned 15 meters (50 feet) of unsupported length. The vertical extent of the ravelled zone was approximately 38 meters (125 feet) below ground surface.

Given the implications of a possible structural failure of the pipe at the unrestrained joints, it was necessary to react immediately and provide judgement based on field observations. With this in mind, the sinkhole was filled with sandy soil immediately. This filling operation maintained support for the pipeline until remedial measures could be taken.

After partial stabilization of the pipeline was obtained through filling it with sand, a deep cement-grout stabilization program was implemented to densify ravelled soil conditions and retard future internal soil erosion. The stabilization program consisted of pumping sand-cement grout at depths 38 to 6 meters (125 to 20 feet) below ground surface. Pre- and post-grouting Standard Penetration Test (SPT) and auger borings, as well as Cone Penetration Test (CPT) soundings were made to assess the merit of the grouting program.

Post-grouting borings showed that the deep conditions at the site greatly improved due to the cement grouting program. This was evidenced by the increase in the SPT blow count. Because upper sands could not be pressure grouted without heaving the pipeline, additional measures were also proposed to preserve the long-term integrity of the pipeline.

The cement grouting program proved to be effective in stabilizing the sinkhole. However, additional remedial measures were also proposed to safeguard the integrity of the pipeline on a long-term basis. Alternatives included chemical grouting, shallow support beam, rerouting of the pipeline, and do-nothing. From these options, the chemical grouting of the near surface soil was adopted as the long-term remedial measure.

Background

On September 21, 1988, a sinkhole occurred at Mann Road beneath the transmission/distribution pipeline of Conserv II in Orange County, Florida. The effluent transfer line is a prestressed concrete cylindrical pipe (PCCP), 1.37 meters (54 inches) in diameter with unrestrained joints. The pipeline is buried a few feet beneath the centerline of a clayey sand surfaced road. The sinkhole, approximately 0.80 kilometer

(0.5 mile) south of the intersection of S.R. 545 and Tilden Road, was initially 9 meters (29.5 feet) in diameter and about 2.4 meters (8 feet) deep near the center and later, on October 13, 1989 during grout stabilization, the sinkhole suddenly enlarged to 15 meters (50 feet) in diameter and 6.0 meters (20 feet deep). A location plan is given in Figure 1.

The City of Orlando retained Jammal & Associates, Inc. within a few hours of the original sinkhole development to provide observation, consultation and to make recommendations regarding remedial measures.

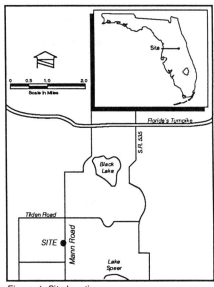

Figure 1. Site location

Site Conditions

The original site conditions described herein are based on the records of Standard Penetration Test (SPT) borings, TB-1 and TB-2, as well as Cone Penetration Test (CPT) soundings, CPT-1 and CPT-2. The locations of SPT borings/CPT soundings are shown in Figure 2 and the test results are given in Figure 3. These borings provide limited information regarding the approximate boundaries between different soil strata due to the high variability of soil conditions in the collapse feature.

Based on these records, the stratigraphy generally consists of very loose to loose fine sand to slightly silty fine sand (fill material) with thickness varying from 0.3 to 0.9 meters (1 to 3 feet). Beneath this fill material, there is a very loose to medium dense fine sand layer with thickness of 12 to 19 meters (40 to 62 feet), becoming silty and/or clayey fine sand at a depth of 2.4 meters (8 feet) below the existing grade. The underlying stratum is sandy silt with thickness varying from 4.5 to 6 meters (15 to 20 feet), containing phosphates in the lower 1.8 meters (6 feet). Consolidated clay/silt layers were encountered at depths of 21 to 23 meters (70 to 75 feet) below the existing grade. Below a depth of 25 meters (84 feet), TB-1 indicated the presence of 1.5 meters (5 feet) of thick hard clay underlain by a nearly incompressible sandy silt to silt layer to the termination depth of 30 meters (100 feet).

Groundwater table was measured at the time of the field investigation on September 24 through September 26, 1988 at depth of 3.3 to 7.6 meters (11 to 25 feet) below existing grade.

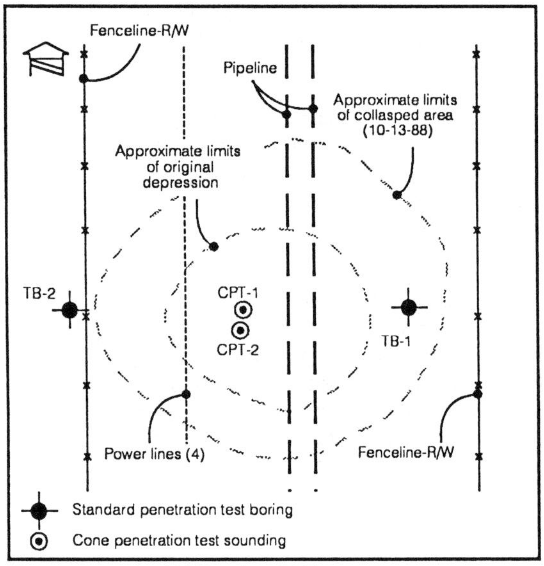

Remedial Measures-Deep Subsoils

To prevent the undermining of the pipeline, it was necessary to take immediate actions. Firstly, the sinkhole was filled with sandy soil immediately. The main purpose of this filling operation was to maintain support for the pipe. Secondly, in order to densify ravelled soil conditions and retard future internal soil erosion at this location, a program of deep subsoil stabilization was recommended. The program consisted of pumping cementaceous grout at various locations in the collapsed area.

A total of ten (10) injections of cement-grout were made in the depth interval of 38 to 6 meters (125 to 20 feet) below the ground surface. A layout of the cement-grout injections is shown in Figure 4. A summary of the grout quantity at various locations along with observations made during the grouting operations is given in Table 1.

Figure 2. Layout of SPT borings/CPT soundings

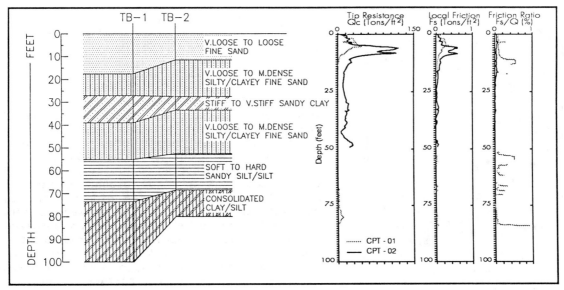

Figure 3. Test results (SPT/CPT data)

TABLE 1					
GROUT INTAKE AT MANN ROAD SINKHOLE					
Date Start/Finish	Injection No.	Max.Depth (Ft.)	Type Of Grout	Total Pumped (yd3)	Remarks
10/13	Original Center(C1)	32.6 m (107 ft)	Low-Slump	18.3 m³ (20.0yd³)	Sinkhole Opened to 15.2 m(50 ft.)
10/17	P1	16.8 m (55 ft)	Slurry	73.2 m³ (80.0 yd³)	
10/18-10/21	P2	36.3 m (119 ft)	Slurry	121.4 m³ (132.8 yd³)	No Pressure During Pumping
10/21	P3	17.9 m (59 ft)	Slurry	72 m³ (78.yd³)	
0/24-10/25	New Center (C2)	38.4 m (126 ft)	Low-Slump	87.3 m³ (95.5 yd³)	
10/25-10/26	P4	19.8 m (65 ft)	Slurry	42.5 m³ (46.5 yd³)	
0/26-10/27	P5	22.9 m (75 ft)	Slurry	38.7 m³ (42.3 yd³)	
0/27-10/28	P6	21.3 m (70 ft)	Slurry	40.1 m³ (43.9 yd³)	Pipe Settled 1.3 cm (0.5 in.)
0/28-10/29	P7	22.9 m (75 ft)	Slurry	37.8 m³ (41.3 yd³)	Pipe Came Back Up 0.48 cm (3/16 in.)
11/1	P8	32.3 m (106 ft)	Slurry	88.2 m³ (96.5 yd³)	
TOTAL 619.5 m³ (677.5 yd³)					

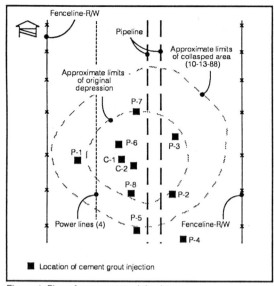

Figure 4. Plan of cement grout injections

Typically, the grout intake in the bottom of the injections was about 19 cubic meters (25 cubic yards). This initial mass of grout was pumped in entirely unless refusal to pumping was encountered as indicated by high pumping pressures on the order of 2000 kPa (300 psi). Minor ground surface settlement was observed during P1 injection. The pumping pressures during P2 injection were predominately "zero" until about 9 meters (30 feet) deep. After this depth, the pressures were about 690 kPa (100 psi). At the onset of the grouting operations at P3 injection, there was a minor ground surface settlement. However, the pipeline didn't move. The remainder of the grouting work was finished at reasonable pressures. Little or no settlement of the ground was observed until pumping P6 injection. After about 13.7 cubic meters (15 cubic yards) of grout was pumped into the bottom of P6 injection, the ground surface began to settle at a noticeable rate and pumping was stopped. After approximately 1 hour, the settlement ceased and pumping was resumed at a much slower rate. During injection P8, the pumping pressures dropped to 72 kPa (25 psi) using a wet slurry grout. After approximately 3 loads of slurry grout caused upward ground surface movement, pumping was discontinued at 11 meters (36 feet) deep. This required premature termination of the injection to prevent possible uplift of the pipeline.

Following completion of the cement-grout program, three (3) Standard Penetration Test (SPT) borings and two (2) auger borings were drilled. The SPT borings TB-3, TB-4 and TB-5 were drilled in the central portion of the collapsed area, adjacent to the pipeline. Borings TB-3 and TB-4 were drilled to depths of 40 and 38 meters (130 and 125 feet), respectively, and TB-5 was drilled to a depth of 10.6 meters (35 feet) below the subsidence surface.

Two (2) auger borings were performed near the perimeter of the collapsed area to a depth of 15 meters (50 feet). The purpose of these auger borings was to determine post-grouting water levels and to install 5 centimeters (2-inch) diameter monitoring wells. These wells will be read in the future during hydraulic loading of the nearby percolation ponds. The location of these borings is shown in Figure 5. The post-grouting water table was measured at a depth of 6 to 7.5 meters (20 to 24 feet) below the existing grade.

An evaluation was made of the Standard Penetration Test (SPT) borings TB-3, TB-4, and TB-5 to assess the effectiveness of the grouting program (Goehring and Jammal 1988; Sayed et al., 1989). The deep conditions at this site improved significantly due to cement-grouting. This was

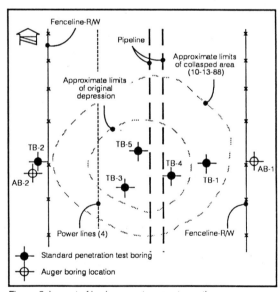

Figure 5. Layout of borings post-cement grouting program

evidenced by the increase in the SPT values throughout the boring depth (Goehring and Jammal 1988; Sayed et al., 1989). However, some loss of drilling fluid circulation was observed after grouting. These conditions did not indicate a serious deterioration of overlying soils because the circulation losses generally occurred in the deep strata at the interface into the underlying rock. For these reasons, no additional deep grouting was recommended. However, alternate methods of increasing the integrity of the shallow subsoils were examined.

Remedial Measures-Shallow Subsoils

As mentioned in the previous section, cement-grouting was used for stabilizing the deep subsoil conditions at the site. However, the pressure grouting operations had to be discontinued at about 6 meters (20 feet) depth to avoid heaving the pipeline. It was thought that if no further remedial measures were taken, the near surface sand fill placed by loose dumping would consolidate with time. This consolidation and any deeper long-term relaxations due to future ravelling could remove support from beneath the pipeline.

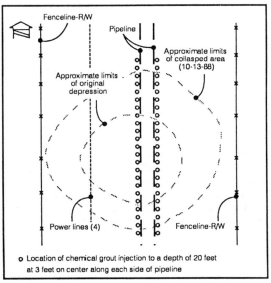

Figure 6. Option I: Chemical grouting

The long-term stability of the pipeline is primarily governed by two factors; namely, stabilization of the main sink area and construction of an adequate pipe support system. The deep cement grouting done during the period October 13 to November 1, 1988 stabilized the collapsed area to a large extent. Improving the shallow soil conditions to provide adequate support for the pipeline could be achieved by additional shallow grouting and/or implementation of an innovative foundation system. Proposed long-term measures are briefly discussed in the following paragraphs.

Option I: Chemical Grouting

One option for stabilizing the upper fill material would be to chemically grout the soil into a strong sandstone-like matrix. Using a low-viscosity chemical grout, this strong matrix could be formed without exposing the pipeline to heave. Once formed, the strengthened soil mass would tend to bridge over any future minor relaxations. As an added benefit, the chemical grout would minimize surface water infiltration into the collapsed area.

A total of 50 injections, spaced at approximately 1.0 meter (3 feet) on center along each side of pipe, were proposed to stabilize the soils beneath the pipeline. Injections would

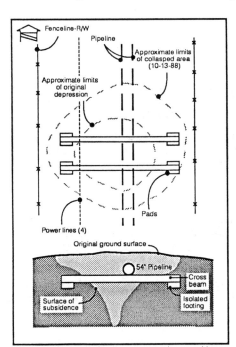

Figure 7. Option II: Shallow support beam with isolated footings

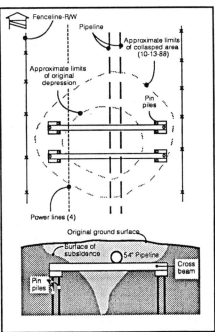

Figure 8. Option II: Shallow support beam with pin piles

extend to 6 meters (20 foot) depth and receive at least 3785 cm³ (15 gallons) of SIROC chemical grout per 0.3 meters (1 foot) vertically. Figure 6 shows a conceptual layout of chemical grout injections.

Option II: Shallow Support Beam

The pipeline in the collapsed area could be supported on two or more beams extending across the collapsed area. Deep or shallow foundations embedded in stable, natural soil at the fringe of the largest perimeter of the collapsed area could be used to support the beams. In principle, the beams should support the pipeline near the center of the depression. In the event of future soil relaxation/ settlement beneath the pipeline, the cross beams would prevent total loss of support and reduce the unsupported length of pipe. In addition, the beams would provide a base to reference future settlements. This would allow an assessment of the long-term behavior of the pipeline.

A conceptual scheme for supporting the pipeline with two beams is presented in Figures 7 and 8.

Option III: Rerouting Pipeline

Another feasible solution proposed was to reroute the pipeline alongside the perimeter of the collapsed area. Limited space between the boundaries of the collapsed area and right-of-way on both sides of the pipeline would require buying new right-of-way and encroaching into the area near an existing percolation pond. Based on the failure zone observed at the time of the collapse, it was recommended the pipe be rerouted at least 50 feet from the collapse limits, starting at least 50 feet north or south of the failure limits. This scheme is depicted in Figure 9.

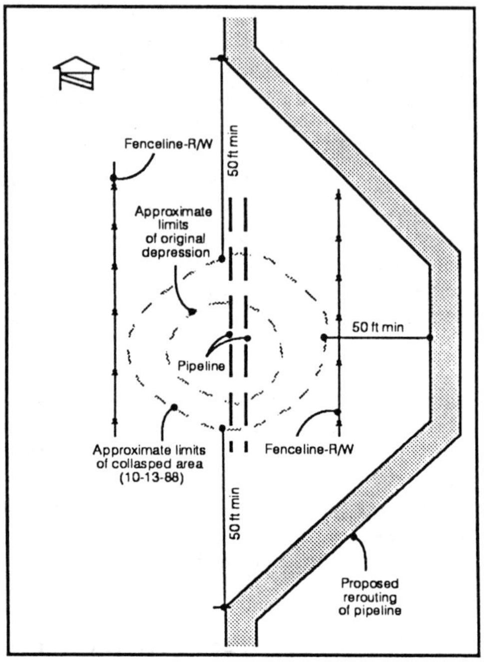

Figure 9. Option III: Rerouting pipeline

Option IV: Do-Nothing Option

The do-nothing solution implies "let us wait and see". The effectiveness of the grouting completed November 1, 1988, in stabilizing the collapsed area may support the adoption of such an option. Because the magnitude and timing of surface relaxations cannot be accurately predicted, a do-nothing approach has a higher risk than the other options. As a minimum, it was recommended that Mann Road be permanently closed to through traffic if the do-nothing option was selected.

It was also recommended that adequate surface drainage be provided away from the collapsed area and vicinity since it is an essential precaution in this sinkhole-prone area. This positive drainage is also applicable to the other options proposed in this paper.

Conclusions

The cement grouting program was successful in stabilizing the deep subsoils within the collapsed area. However, the additional remedial measures proposed herein would improve the long-term integrity of the pipeline with the exception of do-nothing option.

Based on relative cost, security/peace of mind, as well as feasibility of implementation, the shallow subsoil remedial solutions were ranked in the following order of preference:

1.   Shallow Depth Chemical Grouting
2.   Shallow Support Beam
3.   Do-Nothing Option
4.   Rerouting Pipeline

After discussion with the project owners, the shallow depth chemical grouting option was adopted and was being implemented at the time of finalizing this paper. An assessment of this option will be reported in the future.

References

Goehring, R. L., and Fulford, L. Daniel, "Recommendations for Remedial Stabilization of Subsidence at Mann Road, Conserv II Project, Orange County, Florida, Report No. 1, Jammal & Associates, Inc., Winter Park, Florida, October 1988.

Goehring, R. L., and Jammal, S. E., "Remedial Stabilization of Subsidence at Mann Road, Conserv II Project, Orange County, Florida," Report No. 2, Jammal & Associates, Inc., Winter Park, Florida, November 1988.

Sayed, S. M., Goehring, R. L., and Jammal, S. E., "Assessment of Subsidence Stabilization at Mann Road, Conserv II Project, Orange County, Florida," Report No. 3, Jammal & Associates, Inc., Winter Park, Florida, January 1989.

Ground modification techniques applied to sinkhole remediation

JAMES F.HENRY   *GKN Hayward Baker Inc., Tampa, Fla., USA*

Abstract

The use of ground modification techniques to treat sinkholes has advanced from remedial grouting after the problem has occurred to also encompass pre-construction site improvement. A range of techniques is available, including dynamic deep compaction, vibro-compaction, vibro-replacement, slurry grouting and compaction grouting. Each has merit for particular applications and has been closely monitored in the field. A description of each technique as applied to sinkhole remediation is presented and their relative advantages and disadvantages evaluated. Case histories covering each technique illustrate the growing awareness of the ability to pre-treat sites and thus greatly reduce the potential for future sinkhole subsidence. The range of options available permits the engineer to select the most appropriate solution for the site in question so that project objectives may be achieved in an efficient and cost-effective manner.

INTRODUCTION

The term ground modification was coined to describe the full range of techniques now available to densify or otherwise improve the ground as an integral part of the construction system. Ground modification techniques alter the in-situ soil and rock conditions and these changes can usually be confirmed by in-situ geotechnical tests. In the remediation of sinkholes, ground modification is an excellent concept to build upon.

Sinkholes are the result of very unfavorable soil and rock conditions. Most frequently, they are found in Karst formations where solution activity has created:

1.) voids in the underlying rock formation,

2.) eroded, voided and loosened overburden soils,

3.) differential groundwater heads.

In the evaluation and remediation of sinkholes, a number of important steps are involved:

1.) A thorough geotechnical study should be performed by geotechnical engineers and geologists who are familiar with the specific sinkhole profiles of a given area. Tools that are available for their studies include: standard penetration testing, cone soundings, ground penetrating radar, seismic refraction, electrical resistivity, geologic studies and maps, groundwater monitoring wells, etc. The more that is known, the better and more reliable the remediation.

2.) As a result of the geotechnical understanding gained from the above mentioned studies, specific objectives to be met by the remediation effort should be listed.

3.) A remediation plan should be developed to meet the specific objectives.

4.) The plan should be properly executed in the field.

5.) Comprehensive monitoring of the remediation program and evaluation of test results should be conducted to confirm that objectives have been met.

The purpose of this paper is to establish a range of techniques that can be applied to sinkhole remediation projects. Thus, the engineer may select an approach appropriate to site conditions and project requirements.

GROUND MODIFICATION OPTIONS

The following ground modification techniques have been used to remediate sinkholes. It is not intended that all possibilities should be included. However, these techniques have been used most recently, the results monitored, and advantages and disadvantages evaluated.

1.) <u>Dynamic Deep Compaction</u> (DDC) - DDC is a system generally used to densify granular soils to depths of about 3m (10 ft.) to 12m (40 ft.). The technique is a method of dropping heavy weights, 4.5 metric tons (5 short tons) to 27 metric tons (30 short tons), on the ground surface from a pre-determined height, (Figure 1). The impact sends energy waves deep into the ground which densifies the soils at depth. If limestone cavities are not much deeper than 6m (20 ft.) it is possible to cause collapse of voids in the rock and soil. However this technique has rarely, in our experience, caused sinkhole collapse to occur. The advantages of DDC are that it does economically densify loose overburden granular soils and, if the limestone cavities are shallow enough, it can collapse cavities. It does not significantly affect soils or rock much deeper than 9 m (30 ft), nor seal the limestone from further solutioning. It probably does not significantly change the soil/rock permeability and, as the water table fluctuates, soils could still migrate into cavities and subsidence could continue to be a problem. This technique should not be used close to existing buildings because of the potential for adverse effects caused by vibrations.

2.) <u>Vibro-Compaction</u> and <u>Vibro-Replacement</u> - These techniques use large diameter depth vibrators to densify granular soils or to create stone columns in mixed or layered fine grained soils. High pressure water jetting is used to help the vibrator penetrate soils and to help feed in the replacement stone or sand around the vibrator. Vibro-compaction and vibro-replacement can densify and strengthen overburden soils, and the jetting action and vibrations of the vibrator can create sinkhole subsidence. Vibro techniques have been used to a depth of 33.5 m (110 ft) (Figure 2). The vibrator is capable of penetrating through thin limestone caps to expose voids. Either sand or stone can then be fed to voids in the limestone.

Stone or sand columns are also potential drainage paths which may be a disadvantage when clay caps are penetrated in zones of differential ground water head. This technique should not be used for sinkhole remediation close to existing structures because it has created sinkhole subsidence at distances of up to 30 m (100 ft) from the probe locations.

Figure 1: Dynamic Deep Compaction (DDC)

Figure 2: Vibro-Compaction and Vibro-Replacement

3.) <u>Slurry Grouting</u> - This technique involves the injection of various mixtures of very fluid grouts into the ground. Slurry grouting will fill cavities at virtually any depth that can be drilled (Figure 3). It can run along the plane of weakness of the limestone and overburden forming very effective seals. However, little to no densification of overburden soil takes place, and grout quantities can be excessive. To totally seal the limestone and fill all voids may be too expensive a solution for most applications and since it does not densify the loose overburden, a potential remains for building settlement problems. Slurry grouting may also aggravate an active sinkhole and cause additional ravelling and settlement. This technique should be used cautiously around and under existing structures.

4.) <u>Compaction Grouting</u> - This technique involves the injection of, typically, less than 50 mm (2 in) slump cement based grout mix into the ground under high pressure (Figure 4). It can be used to partially seal and fill limestone voids, and to densify overburden soils. It is the safest and most acceptable technique for fixing sinkholes that occur under or near existing structures. Quantities of grout can be controlled thus generally making the process cheaper than slurry grouting, but typically more expensive than DDC and vibro techniques. If the engineer desires a better seal he may choose to use a combination of slurry and compaction grout. Compaction grout would be used to initially fill the cavities and loose zones and densify overburden. Slurry would follow to provide additional sealing of the formation.

Figure 3: Slurry Grouting

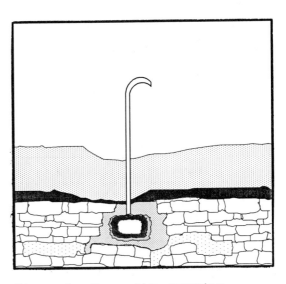

Figure 4: Compaction Grouting

The following case histories illustrate how the above principles are being used to treat sinkhole prone sites prior to construction.

Pasco County Resource Recovery Facility
Spring Hill, Florida

Project specifications for the Pasco facility dictated a maximum total settlement of 25 mm (1 in.) and differential settlement of 20 mm (0.75 in). However, geotechnical evaluation determined that, given the large building and machinery loads, and soil contact pressures of up to 192 kN/m^2 (4 ksf), settlements would substantially exceed specifications unless the upper soils were improved.

The soil profile consisted of 3.3 m (11 ft) to 10.3 m (34 ft) of loose clean sands underlain by a 0.75 m (2.5 ft) to 6.5 m (21.5 ft) thick zone of medium dense silty sand. This, in turn, was underlain by a 0.75 m (2.5 ft) to 6.5 m (21.5 ft) thick stratum of clay resting on the limestone formation. Although the limestone [surface at a depth of 4.5 m (15 ft) to 18m (60 ft)] had numerous solution cavities, the absence of a significant hydraulic driving force led the geotechnical engineer to conclude that the potential for ground loss due to sinkhole activity was quite low. This factor, together with the proven success of dynamic deep compaction in the type of sands present, resulted in this technique being recommended for the project.

An 18 metric ton (20 short tons) weight dropped from 18 m (60 ft) to 24 m (80 ft) was selected to densify the upper 6m (20 ft) to 7.5 m (25 ft) of loose sands. This 6 m (20 ft) thick mat of dense sand is thought to slow down the rate of subsurface ravelling, thus mitigating some sinkhole related subsidence risks.

The densification program lowered the entire treatment area by approximately 0.6 m (2 ft), with post treatment soil borings confirming densification of the upper 7.6 m (25 ft) of soils. No sinkhole dropouts were created during the DDC process.

University of South Florida Eye Institute (USF)
Tampa, Florida

The four story USF Eye Institute Building plans called for continuous footings with a design bearing pressure of 192 kN/m^2 (4 ksf). Site soils consisted of a three-layered stratigraphy. A sand stratum 6 m (20 ft) to 9 m (30 ft) deep overlaid a mixture of sandy silt and clays which extended to the highly irregular limestone surface at depths ranging from 7.6 m (25 ft) to 18 m (60 ft). The top 3m (10 ft) to 4.5 m (15 ft) of the noncohesive layer was very loose, with SPT values averaging between 4 to 8 blows per foot. Below 4.5 m (15 ft) in depth, sands became more silty and clayey. The cohesive strata on top of limestone was very soft, with SPT values less than 2 blows per foot. Several significant voids were encountered in the borings. As a result of this soil profile, the building design and the history of sinkholes in the USF campus area, ground modification was selected to collapse and fill voids, and densify and strengthen the overburden. Vibro-replacement was selected to achieve these objectives.

The vibro probe was advanced to about 12 m (40 ft) with full flushing and stone was fed down the side of the vibrator until voids were filled with stone. Compacted stone columns were then constructed to the ground surface. The vibro-replacement triggered numerous sinks and large quantities of stone were charged into cavities. As the vibro-replacement program proceeded across the site no new sinks developed in areas treated. Post treatment testing verified that the modified ground had achieved the engineers' objective.

Southeast Paper Manufacturing
Dublin, Georgia

Subsurface exploration for expansion of the paper mill showed that overburden soils were competent. Numerous large voids were present in the underlying limestone formation at a depth of 15m (50 ft). Topographic maps and aerial photographs also revealed circular depressions and ponds, indicators of previous sinkhole activity in the vicinity.

The overburden soils had not been damaged or loosened by the limestone voids. Static and dynamic loads on the new paper machine foundation would be substantial and, since the foundation mat covered such a large area, additional load would be transferred to the limestone, and could cause collapse of voids.

The geotechnical consultants selected a slurry grouting program, with the objective of substantially filling the voids in the limestone under the machine mat foundation.

Grout holes were drilled on a 6 m (20 ft) by 6m (20 ft) grid pattern and 200 mm (8 in) slump slurry grout, consisting of sand, cement and bentonite was injected into voids ranging between 1 m (3 ft) and 4.5 m (15 ft) in depth.

When large grout takes were observed in adjacent holes, secondary locations were drilled and grouted to ensure grout coverage. In a 150 m (500 ft) x 30 m (100 ft) area, 4,880 m (16,000 ft) 'of drilling was performed and approximately 2,294 cu. meters (3,000 c.y.) of grout was injected.

Alachua County Landfill
Archer, Florida

During the construction of additional lined cells to increase the Alachua Landfill capacity, sinkhole activity was observed. This created a concern that sinkholes could cause a deformation in the liner and allow leachate to be introduced into the limestone aquifer.

The geotechnical engineer used ground penetrating radar to survey the entire landfill bottom and 21 potential sinkhole areas were detected. SPT borings were taken at each location and the number of potential small sinkhole areas was then reduced to sixteen. Compaction grouting, providing the most control over grout volumes, was selected to treat the isolated areas. Based on the geotechnical data provided, anticipated total grout quantities and injection spacings were established. Generally, a 4-hole pattern was used for each area, with treatment extending to depths of 9m (30 ft) to 12 m (40 ft).

The primary objective of the grouting program was to seal the top of the limestone and densify the loosened overburden. SPT tests taken after the compaction grouting confirmed that the remedial design objectives were met.

Plaza International
Orlando, Florida

A prime commercial 64,700 sq. meters (16 acres) site in Orlando had an existing subsidence feature in the corner of the property. Soil borings taken to explore the possibility of sinkhole activity confirmed that this depression was in fact an old sinkhole.

The geotechnical consultant recommended that ground modification techniques would successfully reduce future sinkhole potential, and would allow multi-story commercial construction on this site.

Three ground modification techniques were used for the project. Compaction grouting was initially performed to fill cavities and densify deep, loose overburden soils (up to 48 m (160 ft) deep). This was followed by a secondary slurry grouting program to further insure an impermeable barrier at the limestone surface and minimize future internal soil erosion. The upper 12 m (40 ft) of soils were treated by vibro-replacement which was determined to be the most economical method of densifying this zone.

Test borings indicated that the sinkhole area had been improved to an equivalent or better condition than the remainder of the site. A significant weak zone in the old sinkhole consisting of clayey fine sand with an average blow count of 3 was modified to an average blow count of 9. Selected zones in this soil were changed from weight of rod material to 12 blows per foot.

CONCLUSIONS

The ground modification techniques of dynamic deep compaction, vibro-compaction, vibro-replacement, slurry grouting, and compaction grouting can be used effectively to remediate sinkholes. Although grouting techniques have long been used to correct sinkhole problems under existing facilities, there is a growing awareness of the ability to treat sites prior to construction, thus greatly reducing the potential of future sinkhole subsidences.

References

Guyot, C. A. "Collapse and Compaction of Sinkholes by Dynamic Compaction" Proceedings, First Multidisciplinary Conference on Sinkholes, Orlando, Florida, October 1984.

Henry, J. F. 1986. "Low Slump Compaction grouting for Correction of Central Florida Sinkholes," Proceedings, National Water Well Association Conference, Bowling Green, Kentucky, October 1986.

Henry, J. F. 1987. "The Application of Compaction Grouting to Karstic Foundation Problems" Proceedings, Second Multidisciplinary Conference on Sinkholes, Orlando, Florida, February 1987

Welsh, J. P. 1988. "Sinkhole Rectification by Compaction Grouting". Proceedings, Geotechnical Aspects of Karst Terrains, Nashville, Tennessee, May 1988

Highway engineering aspects of karst terrane near Alpha, New Jersey

JAMES S.MELLETT   *New York University, and Subsurface Consulting, New Fairfield, Conn., USA*
BERNARD J.MACCARILLO   *New Jersey Department of Transportation, Trenton, N.J., USA*

ABSTRACT

Carbonate rocks of early Ordovician (Beekmantown) Age extend into northwestern New Jersey, where caverns and sinkholes occur in a number of areas. Construction of Interstate Highway 78 near Alpha, NJ, encountered problems associated with karst develop- ment in and around the Epler Formation, a fine-grained limestone that is not noted elsewhere in the state for cavern or sinkhole development. Engineering problems included gabion collapse, cracked concrete structures, undermined pavements, and tension cracks and collapse features at the surface. Problems were ameliorated by excavating to bedrock, and pouring or pumping grout into the cavities. Geotextiles were used to prevent washing out of fines in the grout. Ground-penetrating radar (GPR) was used to monitor the engineering solutions and to attempt to locate any sinkholes that were missed in the remed- iation process.

Introduction:

       Karst terranes and their associated engineering problems are generally thought of as being exclusively a Floridian problem, but karst areas extend as far poleward as carbonate rocks exist. Although New Jersey does not leap to mind when talking about karst related problems, hundreds of caverns and sinkholes have been discovered in the state (Dalton, 1976). This paper will discuss sinkhole and subsidence problems encountered during construction of a portion of Interstate Highway 78 (I-78) near Alpha, New Jersey. Alpha lies about 5 km (3 mi) southeast of Easton, Pennsylvania, near the Delaware River (Fig. 1).

    The area lies within the southeastern margin of the Appalachian fold belt, and is about 25 km (15 mi) from the edge of the Allegheny Plateau. The Wisconsinan terminal moraine lies about 8 km (5 mi) to the north. This is a significant factor, because glacial deposits frequently plug caverns and sinkholes in glaciated regions; most problems associated with karst topography in the U.S. are found south of the terminal moraine (Newton, 1986).

Bedrock Geology:

    Early Paleozoic carbonates dominate the bedrock in this area (Fig. 2). The stratigraphic section of 30 years ago lumped all of these units as the "Kittatiny limestone", but that unit has since been subdivided into a number of more distinct formations (Dalton and Markewicz, 1972). Note in Fig. 2 the apparent reversal of stratigraphy. The Alpha Antiform is an inverted syncline, where the youngest unit, the Martinsburg Formation (Om) lies <u>below</u> the older Cambrian age Allentown Dolomite. This configuration is the result of massive Paleozoic thrust faulting, which overturned the beds.

    The basal unit around Alpha is the Cambrian Age Allentown Dolomite (€a, in Figs. 1-2, which has a thickness of 520 m (1700 ft), and contains one of the largest cave systems in the state. The Rickenbach Dolomite (Or) overlies the Allentown, and reaches a thickness of 200 m (650 ft). Sinkholes of a variety of sizes are reported in this unit, although it typically occurs in low, swampy areas. The overlying Epler formation (Oe) has the most outcrop exposure around Alpha. Its thickness is about 250 m (800 ft), and because of its fine-grained and compact nature, sinkholes tend to be rare in this formation in the rest of New Jersey. A major unconformity (equivalent to the Knox) is developed on the upper surface of the Epler, marking a long period of early Ordovician subaerial erosion, during which a karst surface was established. The uppermost unit is the Jacksonburg Limestone (Ojr), which is about 90-180 m (300-600 ft) thick around Alpha.

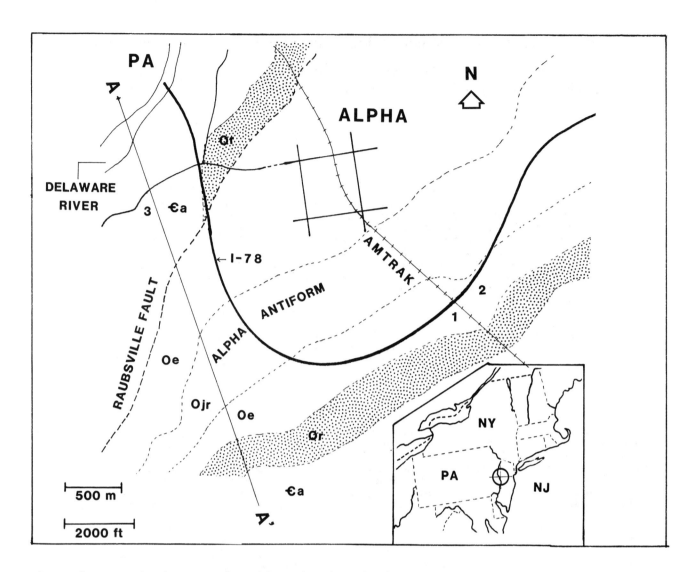

Figure 1: Map showing stratigraphic units in relation to I-78, near Alpha, NJ.
Legend: €a, Allentown Dolomite; Or, Rickenbach Dolomite; Oe, Epler Formation; Ojr,
Jacksonburg Limestone cement rock facies. (1) Location of Amtrak railroad overpass footing;
(2) Location of small new sink on highway; (3) Sink near Carpentersville Road.

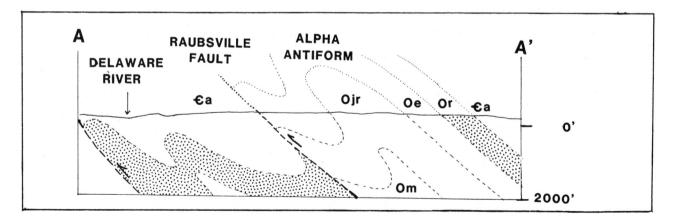

Figure 2: Structure section A-A' showing relationships of units around Alpha, N. J.

No caves are known from the cement-rock facies of this formation, which is exposed around Alpha.

Unlike the rest of the state of New Jersey, where the fine-grained Epler Formation yields few Karst features, in the vicinity of Alpha, virtually all of the sinkhole activity is within the Epler, or at the Epler-Jacksonburg contact. This abrupt change may be simply the result of being south of the terminal moraine, or of some structural condition (e.g., faulting) around Alpha. Fractured rock along fault zones may act as conduits to bring surface water into potential karst-producing formations.

Highway Construction:
Planning for this 6.5 km (4 mi) segment of the Interstate Highway system began in the 1980, and exploration crews encountered immense caverns during drilling operations. Local people in the area know about sinkholes in the region, and as recently as a few years ago, an entire pub in neighboring Phillipsburg disappeared down a sinkhole in a three day period. This section of I-78 is one of the last to be completed in New Jersey, and the location of the route was "locked in" because of completed Interstate segments east and west of this area, as well as restrictions on highway construction through parklands. Although from a highway engineering standpoint, the large loop made by I-78 as it skirted the town of Alpha was obviously an appropriate choice, it also brought the highway in to parallel the Jacksonburg-Epler contact, and to remain in a karst-rich environment until it finally crossed the Delaware river and entered Pennsylvania.

Construction of the highway began in the summer of 1984, and was completed in 1989. Many problems involving ground subsidence were encountered in that 5 year period, particularly after heavy rains. Construction is known as a major activator of sinkholes, including "...those caused by erection of structures, by impoundment or diversion of surface water, and by modifications of the land surface" (Newton, 1986).

Specific problems that were encountered included offsets in concrete slabs, sinkholes on the roadway and embankments, gabion distortion, and wing wall cracks on overpasses.

Remediation:
Techniques used in solving sinkhole problems were modifications of methods learned during the short course "Engineering and Geology of Karst Terranes", which was offered 1-5 Feb 89 by the Florida Sinkhole Research Institute.

Where sinkholes were encountered prior to construction, the surrounding area was excavated to sound rock. The sinkhole outlets were located and then plugged. Holes were lined with geotextile filter fabric, and the opening was backfilled with dense graded aggregate. The fabric was used to prevent the fines from washing out of the filling, and possibly going down secondary outlets. It was important to excavate these holes down to bedrock; if one did not get the hole filled the first time, subsequent heavy rains would reopen the sink.

Because early survey work indicated that sinks were going to be a problem along this section of I-78, grout was injected into the ground before construction of the bridge and trestle footings. For these structures, holes were drilled on 2 m (6 ft) centers twice the width of the footing, and a grout mix (2 parts sand, 1 part cement, and enough water to produce a product that would flow through a 2.5 cm (1 in) diameter pipe. After pumping, the spread of the grout was checked by drilling holes at random over the affected area.

When problems appeared after construction (e.g., gabion sag), grout was poured or pumped under the structure to stabilize it.

Ground-Penetrating Radar Scans:
Following the final phases of construction, Subsurface Consulting, Ltd. was called in to run scans along the eastbound shoulder of the road to check for any sinks that might still exist along the route, and to verify the grout take in areas around large overpass footings.

Equipment used in the radar runs was a SIR-3 system manufactured by Geophysical Survey Systems, Inc., of Hudson, NH. Data were collected and displayed on a PR-8304 graphic recorder. Scans were made with a 300 megahertz (MHz) antenna (GSSI Model 3105 AP) run in a

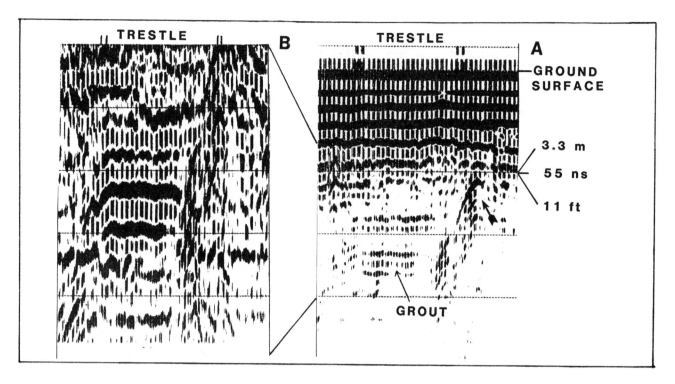

Figure 3: (A) GPR scan of area near Amtrak railroad trestle footing. Note position and spread of grout at depth, and small void signature (arrow) on west (right) side of printout. 90/300 MHz bistatic, at 200 ns. (B) Same area, at 100 ns. Width of section about 30 m (100 ft).

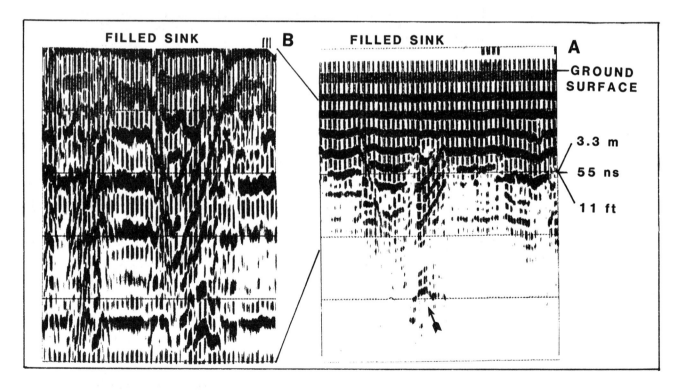

Figure 4: (A) GPR scan of area near newly developed surface sink. Void signature at depth (arrow) is probably related to new sink. Note large collapse feature, which has already been filled with grout. 90/300 MHz bistatic, at 200 ns. (B) same area, at 100 ns. Width of section about 37 m (120 ft). Vertical ticks near surface sink.

bistatic mode with a 90 MHz unit (GSSI Model 3207A-T/R). Although the latter unit has a nominal frequency of 250 MHz, it actually operates around 90 MHz in the field. In bistatic operation, the 90 MHz unit transmitted the radar pulses, and the 300 MHz received them. Even though the antennas operate at different frequencies, the transducer frequency is the center of a broad band width, and there is sufficient frequency overlap to allow antennas to be used in tandem. The advantage of this particular arrangement is that one transmits with a long wavelength, which allows great penetration depth, and receives with a shorter wavelength unit, which provides good signal resolution.

Penetration depths attained were on the order of 135 nanoseconds (ns), which corresponds to a depth of about 8.2 m (27 ft), assuming a dielectric constant of 6.25 for the subsoil and bedrock.

Radar scans were made on the shoulder of the highway; the roadbed could not be used because rebars in the concrete would have reflected all the transmitted radar energy back to the receiving antenna. In addition, the concrete had not fully cured and was highly conductive, reducing any further possible penetration.

Areas of major concern were the Amtrak railroad overpass footings (1, in Fig. 1) and any possible new sinks that might lie below the surface, but were not as yet active.

Results:

Radar scans made parallel to the railroad overpass footings detected the grout that had been pumped down prior to construction. (Fig. 3); the grout reached a thickness of 20 ns (thickness of 1.2 m (4 ft), at a depth of 78 ns (4.5 m, or 15 ft) below grade. Only one small void echo was seen on the west side of the trestle; because of its small size, it is unlikely to be a problem.

Fortuitously, on 9 April 89, the day before scans were made, a small (1 m, or 3 ft) diameter sink developed on an embankment 3 m (10 ft) north of the shoulder, and about 100 m (300 ft) east of the trestle. The sink extended to a depth of at least 5 m (15 ft) below grade. Scans made south of the sink (Fig. 4) indicated a large collapse feature to the east that had been stabilized with grout, and a small target at 51 ns (3 m, or 10 ft) below ground, which probably communicates with the newly developed sink. The opening has since been filled with grout.

Summary and Conclusions:

Sinkhole problems associated with construction of a segment of I-78 were remediated by pouring or injecting grout into the cavities. Results of the remediation were checked with ground-penetrating radar. Grout takes around footings were verified, and only a few small voids were detected along the route.

References:

Dalton, R.F. (1976) Caves of New Jersey. New Jersey Geological Survey. Bull, vol 70, pp. 1-51.

Dalton, R.F. and Markewicz, F.J. (1972) Stratigraphy of and characteristics of cavern development in the carbonate rocks of New Jersey. Nat. Speleol. Soc. Bull., vol 34, 4, pp. 115-128.

Drake, A.A. (1967) Geologic map of the Easton Quadrangle, New Jersey-Pennsylvania. U.S. Geological Survey, Map GQ-594.

Newton, J.G. (1986) Development of sinkholes resulting from man's activities in the eastern United States. U.S. Geological Survey Circ. 968, pp. 1-41.

7. Planning, governmental and legal implications of karst terrane

Land use regulations in the Lehigh Valley: Zoning and subdivision ordinances in an environmentally sensitive karst region

PERCY H.DOUGHERTY   *Geography Department, Kutztown University, Kutztown, Pa., USA*

ABSTRACT

Karst subsidence is a big problem in the Lehigh Valley, the portion of the Great Valley of the Appalachians that cuts diagonally from northeast to southwest across Pennsylvania. More than $1,000,000 damage occurs each year with several single episodes exceeding $500,000. In addition, groundwater pollution has endangered the productive aquifers of the area. This paper is based on the premise that the best way to keep subsidence and groundwater problems from occurring is to stop them before they occur by placing karst concepts and safeguards in local zoning ordinances, comprehensive plans, and subdivision ordinances. Since few municipalities have zoning ordinances with karst precautions, a model ordinance was created for Lower Macungie Township in the Lehigh Valley in order to serve as a guide for other municipalities. The Lower Macungie Township Zoning Ordinance of 1989 is an example of what can be done under Act 247, the Pennsylvania Municipalities Planning Code, in protecting an area from unwise building practices on karst features. The ordinance defines certain karst features, delineates major karst features on an official map, and sets up zones where construction may not occur unless geotechnical expertise can show that there will be no ill-effect. The onus of legality is thus moved from the government body to the developer. This ordinance can be used as a model for new and or improved ordinances in other municipalities in Pennsylvania and other states.

Introduction

Structural damage to buildings, highway subsidence, and disruption of utility lines results in over $1,000,000 damage a year in the Lehigh Valley of Eastern Pennsylvania. There is a long history of karst related problems in the Lehigh Valley with several deaths related to karst collapse at the turn of the century (Wittman, 1988). The Allentown-Reading metropolitan area contains over one million people in a confined carbonate valley; therefore, there is a large population at risk. In the United States, the Lehigh Valley ranks second only to the Florida karst in terms of total damage (Dougherty and Perlow, 1988). If this were a rural area the problem would undoubtedly be overlooked. With a dense population, even the very smallest sinkholes are reported. The area has a long history of karst subsidence and it is assumed that the same processes will continue to work in the future.

If it were known that a problem is going to occur, it makes sense to do something about it today. None of us would build a home on top of a known sinkhole, for an area experiencing subsidence in the past will probably do so again. Why should we allow someone else to build on the same sinkhole? This is especially relevant if it is going to be our tax money which pays the expensive repair bill. Locally, the Vera Cruz Sink cost in excess of $700,000 to repair and the Macungie Sinkhole cost over $600,000; both using tax money (Dougherty and Perlow, 1988). It is the purpose of this paper to show that it is wise to investigate legal codes as an avenue of decreasing the amount of damage from karst subsidence. The best way to do this is through the adoption of a zoning ordinance or a subdivision and land development ordinance that will keep development from occurring on subsidence prone areas. The model for this study is Lower Macungie Township in Lehigh County, Pennsylvania.

The Study Area

Lower Macungie Township is located in Pennsylvania's Lehigh Valley, the local name for the Great Valley of the Appalachians. It is located southwest of Allentown and is one of the most rapidly growing suburbs of the Allentown-Bethlehem-Easton metropolitan district. It is in the southwestern portion of Lehigh County adjacent to Berks County and straddles the contact between the Reading Prong and Great Valley physiographic regions (Figure 1). Contrary to the bad publicity in the song "Allentown" by Billy Joel, expansion in tertiary activities has more than made up for the decline in manufacturing jobs. A low unemployment rate, access to the New York and Philadelphia metropolitan regions, the low cost of land, and low taxes have resulted in a migration of both people and business. The combined Allentown-Reading metropolitan area already contains in excess of 1,000,000 people and continues to grow. Lower

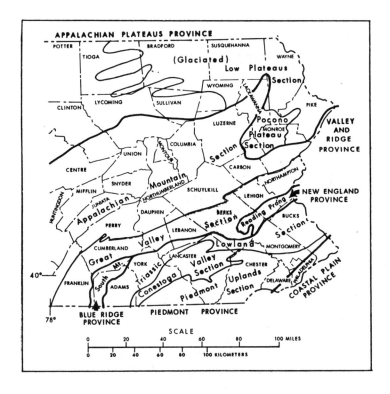

Figure 1. Physiographic map and location of the study area (after Wood, 1978).

Macungie is well situated to participate in this growth because of its location at the intersection of route 222 and the new Interstate 78; and, it is transected by routes 309, 100, and the Pennsylvania Turnpike Northeast Extension.

As long as the Township remained rural, there were few problems with subsidence, but with urbanization the problems were exacerbated by denudation of the land (Myer and Perlow, 1986), drawdown of the water table (Wood et al., 1972), and filling of karst features (Kochanov, 1988). Between 1950 and 1984 the Township grew from 2,997 to 14,081 people, with the estimated 1988 population at over 15,000 (Lower Macungie, 1988). There are over 1,000 housing units already approved, and nearly 2,000 units are pending before the Planning Commission or are in litigation this year alone. By the year 2000, the projected population may exceed 25,000.

Being located in Pennsylvania creates extra problems since planning, zoning, and subdivision activities are vested in the excess of 1,500 minor civil divisions by Act 247, the Pennsylvania Municipalities Planning Code (Commonwealth of Pennsylvania, 1989). Minor civil divisions in Pennsylvania are cities, boroughs, and townships. County planning commissions serve only an advisory role and are not the powerful decision making bodies as in other states. This results in fragmentation and numerous legal challenges to local code enforcement. The positive aspect of this is the relative ease of instituting karst protection measures on the local level. This is more easily facilitated than by changing county or state codes.

Geology and Geomorphology

The Lehigh Valley is physiographically located between the Reading Prong/Blue Ridge Province, known locally as South Mountain, composed of Pre-Cambrian and Cambro-Ordovician rocks including granitic gneisses, quartzites, and sandstones; and Blue Mountain, the first ridge of the Appalachians, composed of Silurian sandstones and quartzite. The resistant ridges stand in marked contrast to the 10-12 km (16-20 mile) wide structural valley resulting from the thrust faulting of the older South Mountain materials over the younger Ordovician and Cambro-Ordovician limestones and shales forming the relatively flat valley bottom (Figure 1).

Figure 2 shows the geologic map of the Allentown West 7.5 minute quadrangle on which most of Lower Macungie Township is located. Notice that the limestones of the valley bottom occur in a northeast to southwest strike with five major formations evidenced at the surface. The Township is located almost entirely within the limestone members and nearly surrounds the independent boroughs of Alburtis and Macungie. Of the five formations there is a great difference in the karst processes. The area near the base of South Mountain is underlain by the Leithsville Formation which has been covered by colluvium in excess of 33 m (100'). This area is prone to the formation of suffosion dolines as evidenced by the Macungie Sink which resulted in a collapse measuring 24 m (81)' across and 12.5 m (41') deep in 1986 (Dougherty and Perlow, 1988). On the other hand, the Richenbach and Allentown Formations do not have sinkholes as large but the number and density is much greater, 24 and 22/ sq. km (9.3 and 8.6/square mile) (Myer and Perlow, 1986). The subdivision activity in the Township has been the greatest on the outcrops of the latter two formations. The Little Lehigh Creek through this area has also been shown to be influent, thus pointing to a possible aggravating influence on karst subsidence (Myer and Perlow, 1986).

A perusal of local newspapers shows a constant barrage of sinkhole collapses in the carbonate formations of Lower Macungie Township and surrounding areas; including, the Macungie Sinkhole causing an estimated $600,000 damage (Dougherty and Perlow, 1989), the infamous 46 m

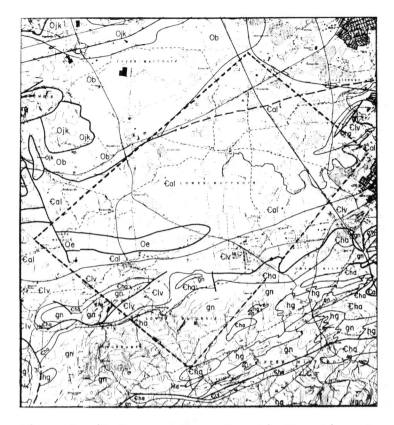

Figure 2. Geology of Lower Macungie Township and vicinity: gn, granitic gneiss; hg, hornblende gneiss; Cha, Hardyston Fm; Clv, Leithsville Fm.; Cal, Allentown Fm.; Oe, Epler Fm.; Ob, Beekmantown Fm.; Ojk, Jacksonburg Fm. (after Berg, 1981).

(150)' by 17 m (55') deep Vera Cruz Sinkhole which required $700,000 to repair (Bonaparte, 1987), the closure of Washington Street in Allentown (Sanchez, 1988), the closure of Hillcrest Drive in Lower Macungie (Morning Call, 1989), the sinking of a dormitory at Muhlenberg College (Youngwood, 1988), the collapse of a church into a sinkhole in Allentown (Clark and Reaman, 1988), an investigation of sinkholes under the Allentown Airport runways (Morning Call, July 27, 1988), the closure of Interstate 78 in Easton, and the unsubstantiated story about a high school football coach being engulfed by a sinkhole along the sidelines at a Friday night game in front of several thousand people. Another sinkhole, the Voortman Site, has been classified as a superfund site by the US EPA because of the dumping of old battery casings (Morning Call, June 8, 1988). There is definitely a problem with karst subsidence and potential for human and groundwater problems.

Methodology & Analysis

Using the old adage that an "ounce of prevention is worth a pound of cure," existing ordinances in the Lehigh Valley were investigated to find what provisions were used to minimize the danger of karst subsidence. It was the intent to take the best of existing ordinances and create a new zoning ordinance for Lower Macungie Township incorporating the best of the rest. In addition, a karst overlay district similar to the flood plain overlay districts common in most zoning ordinances was planned. Several existing data bases were used including a detailed fracture trace analysis and sinkhole location study done for the Delaware River Basin Commission (DRBC, 1985), a study done for Pennsylvania Power and Electric by VFC, Inc. (Perlow, on-going), and interviews with township officials. To this was added sites known by personal knowledge and by having a Joint Planning Commission intern identify sinkholes from USDA aerial photographs. This resulted in a data base of existing karst features that can be updated periodically so there can be a historic record of problem areas. Although not envisioned early in the project, a comparison of zoning and subdivision ordinances was made in order to determine the best place for karst hazards enforcement. It must be realized that the material mentioned in this paper reflects the vagaries of Act 247, the Pennsylvania Municipalities Planning Code, and cannot be used directly in other states. The concepts presented can be modified to fit the planning codes of other states on either the municipal or county level.

Only two ordinances were found in the Lehigh Valley that contained karst language, and both were subdivision ordinances. One subdivision Ordinance from Upper Saucon Township was overly restrictive and was thought to be construed as anti-development and unenforceable in court by the Lower Macungie Township Planning Commission Solicitor (Donald Miles, Personal Communication, 1989). In addition, it made extensive use of fracture traces and lineaments, generally accepted karst features but a legally indefensible concept in court (Upper Saucon Township, 1988). If any three karst experts were asked to draw their interpretation of fracture traces from the same aerial photograph, three entirely different spatial patterns would result. Some of the lines would even be cultural manifestations such as tree lines, utility line scars, cropping patterns, excavations, etc. It was decided to delete the use of fracture traces from the present ordinance. The other subdivision ordinance is one developed by the Joint Planning Commission of Lehigh and Northampton counties (Taremae, 1988). It is not detailed enough to restrict development near karst features and does not allow for geotechnical investigations which allow a developer a way out without resorting to a challenge of the local zoning ordinance via the Seventh Day Adventist "taking" procedure which was upheld by the United States Supreme Court. No individual may be restricted from developing a property unless just cause can be shown.

There is a debate over where the karst mitigation factors should go, in the subdivision or zoning ordinance. It was decided upon the zoning ordinance for that is where similar environmental problems are addressed such as the floodplains, historic places, wetlands, hydric soils, and steep slopes. Since most of these are environmental constraints on the property and pertain to public safety, this appeared to be the logical place for karst controls. In addition, language in the zoning ordinance is less easily changed than a subdivision ordinance. Under Pennsylvania law, a local planning commission can waive any requirements under the subdivision ordinance. Provisions in the zoning ordinance must first be reviewed by the local planning commission and if the developer fails to comply with the zoning ordinance, he or she must apply to the Zoning Hearing Board for a variance before returning to the planning commission for approval. A zoning hearing board is a legal entity where policy decisions are made that are legally binding; whereas, a planning commission is an advisory board which only makes recommendations to the governing body of the municipality. In other words, it is much more difficult for a developer to overcome the controls in a zoning ordinance.

The heart of the proposed zoning ordinance is the Karst Hazard Overlay District and the Karst Hazard Overlay Map (Figure 3). Developers seeking a building permit, conditional use, or special exception must submit a map at a scale of at least 1"=100' showing the karst features listed in the Ordinance. The Zoning Officer has the responsibility of informing applicants that they have karst features on their property; and, if necessary, may perform a site visit to delineate such features, or procure the necessary expertise to delineate such features. The applicant is also required to have an engineer visit the site and assess the presence of karst features. If further testing is necessary, the results should be submitted to the Township Engineer who will report to the Zoning Officer and Planning Commission on the adequacy of the report. In special cases the Township Engineer may request further testing by the applicant's engineer.

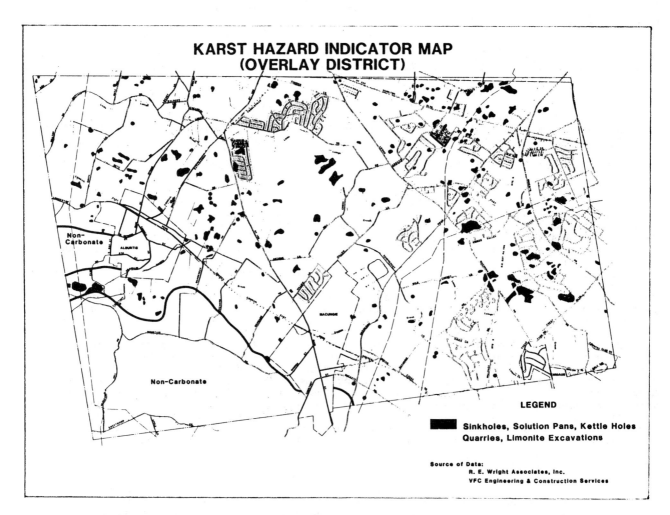

Figure 3. Karst Hazard Indicator Map showing karst features in Lower Macungi Township, PA (Lower Macungie Twp., Zoning Ordinance, pending).

A set of performance standards is also adopted in the Ordinance. These include the provision of having no stormwater detention facility within 30 m (100') of a karst feature. This is necessary, for many developers try to site detention basins in dolines with disastrous consequences. The eye of the sink can reopen due to excess water and the added pressure. In addition, no stormwater swale or sewer pipe may be constructed within 30 m (100') of a karst feature unless special precautions are taken including installation of liners or impermeable beds. No throughflow is allowed along utility trenches and impervious dikes are required at 10 m (30') intervals. No buildings or accessory structures are allowed within 15 m (50') of a karst feature unless detailed geotechnical work shows there will be no negative impact. No septic tanks, swimming pools, solid waste disposal area, transfer facility, salt storage, blasting, well enhancement activities, gasoline and other chemical storage is allowed with 30 m (100') of the listed karst features. To guard against the stripping of land cover and the associated opening of dolines, a soil conservation plan must be submitted with the County Soil Conservation Service to safeguard against sinkhole formation. When there is a disagreement over the delineation of a karst feature, the applicant shall bear the burden of showing that such conditions do not exist on the property in question. This may require extensive field surveys and geotechnical work, the expense to be borne by the applicant. An appeal procedure is set up in another section of the Zoning Ordinance for use by applicants.

Development within 100 of karst features such as sinkholes, sinking streams, ghost lakes, cave entrances, closed depressions, lineaments, faults, and any other recognizable karst feature is expressly forbidden unless expert geotechnical work shows that it can be done safely. In addition, the same restrictions apply to limonite excavations, the remnant of the mining legacy of this area. Many of the iron pits were undoubtedly in old sinkholes where the ore became concentrated, or are areas where the interior drainage leads to a subterranean drain with potential for subsidence or collapse. Seasonal high water indicators were also added to the definitions for it is found that many sinkholes are often ponded in the early spring for the shaft of the doline is often clogged by fine sediments. This is a good indicator of shallow dolines which do not appear on topographic maps or are not normally visible on air photos. Contacts between geologic formations are also shown on the overlay map. Especially at the contact between carbonate and non-carbonate rocks, there is a great potential for subsidence due to the highly aggressive water coming from the non-karstic rock. Even between carbonate formations there is often a difference in the solubility of the limestones. Another feature identified is clumps of trees in plowed fields. When farmers leave an area unplowed, it often indicates a depression, an old iron mine, or a rock outcrop. By requiring developers to investigate each clump of trees, many extra karst features can be identified that were not identified on the overlay map provided in the Ordinance.

As an extra backup to the zoning procedure, a subdivision application must also come before the Township Planning Commission for several reviews: sketch plan, preliminary plan, and final plan. At each step there is ample opportunity to discuss the karst hazards in the zone. In fact, at the preliminary plan stage the Township Engineer is expressly directed to flag any such zones on the applicant's property and/or tell the applicant that there is a potential for karst features on the land in question. The Planning Commission can deny the application for subdivision or conditional use if the applicant fails to go before the Zoning Hearing Board for a zoning variance. Extra safeguards can also be written into the Comprehensive Plan, a document required by all municipalities in Pennsylvania. Under the new Act 170, revisions to the Pennsylvania Municipalities Planning Code, zoning and subdivision ordinances must be compatible with the Comprehensive Plan.

Summary and Conclusions
Although it is stated that the zoning ordinance is a good way to avoid problems associated with karst subsidence, there are many short-comings with this approach. Most karst researchers will probably feel that the wording in the Ordinance is not strong enough. Any debatable language or legal uncertainty, like the identification of fracture traces has to be left out of the Ordinance. Only language that can be defended by the Township Solicitor in court and only readily identifiable features can be used in the Ordinance. Although not shown in this paper, all karst terms are defined in the beginning of the Zoning Ordinance. This points up another problem--the status of karst feature research is still in its infancy and detection methods are primitive at the best. Until we get better methods of delineating karst features and identifying incipient sinkholes, we cannot make the language or requirements in the Ordinance any tougher. There is also a problem with the public and government officials not realizing the threat of karst subsidence to life and property. Education must also accompany the push for implementation of codes or else the effort is bound to fail. Developers can present a convincing argument against the implementation of codes because of the extra expense they have to pay. Probably the biggest problem with a karst hazards code, be it subdivision or zoning, is the lack of enforcement. Many municipalities have no way to follow up on developers to ensure that the provisions agreed upon are followed. Many of these problems only come to light if there is a bad problem which is protested by someone in the community.

There are several assumptions in the current ordinance that may also cause some problem. It is assumed that karst subsidence is not a random event and indeed follows some pattern.

This means that areas of karst subsidence are areas where future subsidence is also likely to occur. This assumes a cyclical development of sinkholes which will be debated by some. In addition, known karst features are assumed to have been caused by processes which are still operating at or near that feature and may continue to occur into the future, a point which may be open for debate. After a sinkhole forms, it is easy for researchers to tell why it formed there, but seldom are such occurrences forecast before they occur. This chance occurrence demeans the karst scientist in the same way that a chance thunderstorm destroys the credibility of the weatherman.

In conclusion, it is the position of this paper that implementation of zoning and subdivision ordinances is the best way to mitigate the danger of karst subsidence. The zoning ordinance has been selected as the best way to enforce karst hazards enforcement for it is tougher to change and the appeal procedure is longer and contains more checks and balances than a subdivision ordinance. In addition, the karst hazard zone fits in with other environmentally sensitive parameters which are usually addressed by zoning, i.e., the floodplain. Although the identification of problem areas has not reached a high degree of reliability, we must institute such ordinances so we can monitor and archive karst features. With increases in technology and advances in karst research, our ability to detect problem areas will continue to improve. With a zoning ordinance in place, it is an easy procedure to update and modify it with advances in karst knowledge.

APPENDIX I

Lower Macungie Township Karst Ordinance

1205. <u>Karst Hazard (Overlay District)</u>
  1. <u>Purpose</u>
     The purpose of this district is to recognize the potential for damage to public and private improvements, human injury or death, and the disruption of vital public services which may arise by the potential for sinkholes and/or subsidence within areas of carbonate geology. A further purpose of this district is to minimize the potential for such sinkhole and/or subsidence occurrence and to protect the ground water resource.
  2. <u>Application</u>
     The Karst Hazard District operates as an overlay district to the districts otherwise found in the Township Zoning Ordinance. Should the regulations of this overlay district and other applicable regulations conflict, the most stringent regulations shall apply.
  3. <u>Disclaimer of Liability</u>
     Whereas the exact occurrence of sinkholes and/or subsidence is not predictable, the administration of these regulations shall create no liability on behalf of the Township, the Township Engineer, Township employees, or Township agencies as to damages which may be associated with the formation of sinkholes or subsidence. That is, compliance with these regulations represents no warranty, finding, guarantee, or assurance that a sinkhole and/or subsidence will not occur on an approved property. The municipality, its agencies, consultants and employees assume no liability for any financial or other damages which may result from sinkhole activity.
  4. <u>Delineation of Area Affected</u>
     The Karst Hazard Overlay District is portrayed on the Karst Hazard Indicator Map which can be found at the end of this Ordinance. The areas affected by the Karst Hazard Overlay District are the "carbonate" areas shown on the map. The sinkholes, solution pans, kettle holes, quarries and limonite excavations delineated on the Karst Hazard Indicator Map were taken from two sources: (1) Sinkhole Occurrence and Geologic Maps prepared by VFC Engineering & Construction Services as part of the Lehigh-Northampton Sinkhole Study; (2) the Little Lehigh Creek Carbonate Prototype Area Closed Depression Map prepared by R.E. Wright Associates, Inc. Should dispute arise as to the boundary of the district or the location of any of the karst features shown on the map, the applicant may present information to the Zoning Officer in support of his or her position. This information shall be prepared by a recognized professional with competence in the field. The Zoning Officer shall make a decision on the proper extent of the karst hazard based on the information presented, with the assistance of any technical review deemed appropriate.
  5. <u>Procedures</u>
     Whenever an application for a building permit, conditional use or special exception is made, the Zoning Officer shall determine from the Karst Hazard Indicator Map whether or not karst features are likely to be present and shall so notify the applicant. The applicant must provide the Township with a map at a scale of 1"=100' that shows the karst features listed in 5.2 of this section.
     5.1 Whenever notified by the Zoning Officer that karst features are likely to be present, or when the applicant knows these features are present, the applicant

shall engage a qualified engineer to review the existing aerial photos, soils, geological and related data available to him as it may pertain to the subject property and to make a site inspection of the property.

5.2 A site inspection by the applicant's engineer, using all available data and with such assistance as is needed, shall determine the presence or absence of karst surface features on the site, and locate the same if present on a site plan at a scale no smaller than 1"=100'. In particular, the following features shall be located, if present, on the site:

5.2.1 Closed depressions

5.2.2 Open sinkholes

5.2.3 Seasonal high water table indicators

5.2.4 Unplowed areas in plowed fields

5.2.5 Surface drainage into ground

5.2.6 "Ghost lakes" after rainfall

5.2.7 Lineaments and faults

5.2.8 Limonite excavations and quarries

5.2.9 Contacts between geologic formations

5.2.10 Any karst feature shown on the Karst Features Indicator Map

5.3 Based upon the site inspection, the applicant's engineer shall determine what further testing should be done by the applicant to ensure compliance with the performance standards set forth in Section 6. Testing methodology shall be reasonable under the circumstances, including (1) the scale of the proposed development; and (2) the hazards revealed by examination of available data and site inspection.

5.4 The applicant shall cause the additional testing, if any, to be effected and shall submit test results to the Township Engineer.

5.5 The Township Engineer shall report to the Zoning Officer and Planning Commission, with a copy to the applicant, his opinion concerning the adequacy of the report submitted based upon the scale of the development and the hazards revealed by the report, and shall make recommendations to the Planning Commission based upon the report submitted concerning the layout of utility lines, and building location. The Township Engineer may require the applicant to perform such additional testing as may be appropriate.

6. **Performance Standard**

6.1 All applicants for building permits, subdivisions, land developments, conditional uses and special exception uses shall comply with the requirements of the Karst Hazard Overlay District.

6.2 No stormwater detention facility shall be placed within one hundred (100) feet of the features listed in Section 1205.5.2.

6.3 No stormwater swale in excess of ten (10) cubic feet per second for the ten (10) year flood may be constructed within one hundred (100) feet of the features listed in Section 1205.5.2.

6.4 No storm sewer pipe shall be constructed within one hundred (100) feet of the features listed in Section 1205.5.2 unless it is concrete pipe utilizing O-ring joints.

6.5 No principal or accessory building, no structure, and no impervious surface shall be located closer than fifty (50) feet from the edge of the features listed in Section 1205.5.2 unless a detailed geotechnical solution to the subsidence, pollution, and safety problems of the karst feature has been presented by a competent professional in carbonate terrain.

6.6 No septic system or tile field, no swimming pool, no solid waste disposal area, transfer area or facility, no oil, gasoline, salt or chemical storage area, and no blasting for quarrying or well enhancement activities shall occur within one hundred (100) feet of the features listed in Section 1205.5.2 unless a detailed geotechnical solution to the subsidence, pollution, and safety problems of the karst feature has been presented by a competent professional in carbonate terrain.

6.7 Soil conservation plans filed with the County Soil Conservation Service shall detail safeguards to protect karst features from runoff changes.

6.8 All underground utility lines located in the Karst Hazard Overlay District shall be so constructed as to not permit the flow of water along the utility line trench, and shall be imperviously diked at thirty (30) foot intervals.

1206. **Appeals to Zoning Hearing Board**

1. **Appeal of Zoning Officer's Determination of Environmental Protection District**
Where the Zoning Officer's determination of the extent to which a lot or lots lie within the Environmental Protection District is appealed to the Zoning Hearing Board, as provided in Section 2103.5.1 of this Ordinance, the appellant shall bear the burden of establishing, through actual field surveys or otherwise, that such conditions do not exist on the land in question.

References

Berg, Thomas M. and Christine M. Dodge (eds.), 1981.   Atlas of preliminary geologic quadrangle maps of Pennsylvania. Harrisburg, PA, Topographic and Geologic   Survey.

Bonaparte, R. and R. R. Berg, 1987.   The use of geosynthetics to support roadways over sinkhole prone areas.   In Barry F. Beck and William L. Wilson (eds.).   Karst hydrogeology: engineering and environmental applications. Rotterdam: A. A. Balkema.

City firm awarded contract to find sinkholes at ABE (Allentown Airport), July 27, 1988. Allentown, PA: The Morning Call.

Clark, John and Denise Reaman, February 18, 1988.   City church collapses into sinkhole: Allentown, PA: The Morning Call.

Commonwealth of Pennsylvania, January, 1989.   Act 247 of 1968, as amended by Act 170 of 1988: Analysis of revisions to the Pennsylvania Municipalities Planning Code.   Harrisburg, Pennsylvania, Local Government Commission, General Assembly, Commonwealth of PA.

Darrah, Tim, January 10, 1987.   Upper Saucon horror: return of the sinkhole. Allentown, PA: The Morning Call.

Delaware River Basin Commission, 1985.   Special groundwater study of the middle Delaware River Basin and open file map.   Middletown, PA: R.E. Wright Associates, Inc.

Dougherty, Percy H. and Michael Perlow, 1989.   The Macungie Sinkhole, Lehigh Valley, PA: cause and repair.   Environmental Geology and Water Science, Volume 12, Number 2.

Keep watching Voortman site, June 8, 1988. Allentown, PA: The Morning Call.

Kochanov, William, 1987.   Sinkholes and karst related features of Lehigh County, Pennsylvania.   Harrisburg, PA, Open File Report, Topographic and Geologic Survey, 1987.

Lower Macungie Township, 1988.   Comprehensive plan.   Macungie, PA, Lower Macungie Township Planning Commission, pp. 138.

Lower Macungie Township, pending. Zoning ordinance.   Macungie, PA, Lower Macungie Township Board of Supervisors.

Miller, B. L., 1934.   Lehigh County, Pennsylvania: Geology and Geography. Harrisburg, PA, Pennsylvania Topographic and Geologic Survey.

Myers, P. D. and M. Perlow, 1986.   Development, occurrence, and triggering mechanisms in the carbonate rocks of the Lehigh Valley, eastern Pennsylvania.   In Barry F. Beck (ed.). Sinkholes: their geology, engineering, and environmental impact. Rotterdam: Balkema.

Perlow, Michael, on-going.   Maps of sinkholes in the Lehigh Valley.   Open File Report. Allentown, PA: Valley Foundation Consultants.

Pritchard, Sandra F. and Percy H. Dougherty, 1988.   Physiographic transect of the Great Valley of Pennsylvania.   In C. Gil Griswall and Charles H. Fletcher, III (eds.).   Guide to fieldtrips. 1988 Eastern Section Meeting, Nat Assoc Geo Teachers: West Chester, PA.

Sanchez, Leonel, February 20, 1988.   City crews to repair washout below street. Allentown, PA: The Morning Call.

Street collapses in Lower Macungie neighborhood, April 16, 1989. Allentown, PA: The Morning Call.

Taremae, Olev, 1988.   Minimizing sinkhole occurrences: an initial inquiry into regulatory approaches. Allentown, PA, Joint Planning Commission, Lehigh-Northampton Counties.

Upper Saucon Township, 1988. Zoning Odrinance. Upper Saucon Board of Supervisors.

Wittman, Bob, April 10, 1988.   Deep history of sinkholes in Allentown. Allentown, PA, Morning Call.

Wood, Charles R. and David B. MacLachlan, 1978.   Geology and Groundwater Resources of Berks County, PA.   Harrisburg, PA: Topographic & Geologic Survey.

Youngwood, Susan, February 19, 1988.   Muhlenberg sinkhole nearly repaired. Allentown, PA, Morning Call.

A national study of natural underground cavities in Great Britain – Data collection, spatial analysis, planning and engineering significance

CLIVE N.EDMONDS & GEORGE MARTIN   *Applied Geology, South Lodge, Cranford, Blackdown, Royal Leamington Spa, Warwickshire, UK*

BRIAN R.MARKER   *Minerals Division, Department of Environment, London, UK*

ABSTRACT

A national survey of natural underground cavities in Great Britain has been commissioned by the Department of Environment. A large natural cavity database (in excess of 10,000 records) is being collected and entered into a database held on a PC-based computer system which is linked to a CAD package to enable the information to be presented in plan format. The natural cavity types include not only those developed on soluble carbonate and non-carbonate rocks but also certain non-soluble rocks, as produced by the processes of dissolution, piping, erosion and cambering. Particular attention has been paid to cavities occurring in urban areas and those affecting infrastructure. The study has revealed general patterns of subsidence incidence, mechanisms, triggers and effects on existing and proposed surface development. The interaction of natural cavities with man-made cavities has also been explored. Spatial analysis of the database demonstrates factors influencing the broad geographic distribution of cavities. Areas which are more prone to subsidence over natural cavities pose an increased hazard to existing and new urban and infrastructure development. Such areas require special sub-structure and drainage measure designs or ground treatment. In Great Britain the engineering significance of these problematic ground conditions needs promotion to improve awareness amongst engineers and planners. The national study provides a basic information resource to guide planners and engineers on the circumstances in which consideration of potential land instability is appropriate which should help to reduce future problems and links to government policy guidance. This information forms the basis of a geographic information system.

Introduction

In Great Britain, during the past five years the Department of Environment has been funding a series of national and regional studies of land instability. Work on quarry face stability, mine shafts and landslides has been completed. Studies of natural underground cavities, mined ground, general foundation conditions, and seismicity are in progress. Work on natural "contaminants" such as methane, radon and heavy metals will commence soon and studies of flooding, erosion and deposition are being considered. This paper describes the current national study of problems associated with natural underground cavities. The principal aim of the study is to assess the scale, nature and extent of engineering and ground stability problems arising from natural underground cavities. The objectives of the study are:

(a) to establish the geological and geographical extent of significant occurrences of natural underground cavities;
(b) to review the types and extent of surface instability and other effects of such cavities and the consequent risks to life, property and infrastructure;
(c) to review the techniques for investigating the presence and condition of cavities and for assessing the stability of ground overlying them;
(d) to review remedial measures for mitigating the risks to life and property;
(e) to identify gaps in existing knowledge and to set out priorities for investigation of topics and of localities.

The study is essentially concentrating on case history experience in Great Britain, but will also incorporate an international review particularly as reported within published literature.

It is intended to raise the threshold of awareness of natural cavity problems of engineers, land use planners and developers and is linked to government policy guidance.

<u>Natural cavity data collection</u>

The ultimate success of the study is dependent on the large-scale natural cavity data collection exercise. It is not intended to collect every possible natural cavity record, but to assemble a sufficiently large and representative database of the diverse natural cavity population. The database target size comprises approximately 10,000 individual natural cavity records but consideration is being given to whether this should be further expanded.

The natural cavity types included within the study are listed below:

(a) Vadose and phreatic karstic cavities in soluble carbonate rocks.
(b) Vadose and phreatic pseudo-karstic cavities in soluble non-carbonate rocks.
(c) Cavities present within non-soluble rocks.
(d) Cavities resulting from decalcification of calcareous matrix from rocks.
(e) Cavities associated with gulling and fissuring during cambering.
(f) Cavities associated with walls of irregular fault planes.
(g) Cavities produced by sea cliff erosion.
(h) Cavities resulting from piping in non-cohesive material.

The terminology used to describe karstic cavities present in limestones is the internationally recognised terminology as discussed by Walsh et al (1973) and Beck (1984).

In order to satisfy the aims of the programme of work and particularly the benefits to land use planning, it was determined that the natural cavity data would be collected according to a priority based system. Wherever possible, this system has been adhered to, the assembled database being divided into three classes of information.

<u>Class 1</u>
1A Zones where natural cavities potentially exist and specific data locations are readily available within urban centres.
1B Zones where natural cavities potentially exist and specific locations are readily available in the vicinity of urban centres.
1C Zones where natural cavities potentially exist and the available specific data locations fall along a corridor occupied by a road, railway or canal.
1D Zones where natural cavities potentially exist and the available specific data locations fall within publicly accessible open space away from urban centres or infrastructure.

<u>Class 2</u>
2A Zones where natural cavities potentially exist, but only generalised data locations are available e.g. non-specific larger mapped geological units or geographical areas where the presence of natural cavities is generally recorded.

2B Zones where natural cavities potentially exist but where no data locations are known.

<u>Class 3</u>
3 Zones where no apparent natural cavities potentially exist and where no data locations are known.

The natural cavity data has been collected from a large and diverse number of data sources as set out in the following table.

<u>Table 1</u>   <u>Natural Cavity Data Sources</u>

1.Civil, Structural and Geotechnical Consultants and Contractors	10.National House Building Council
2.Academic and Research Institutes	11.Department of Transport
3.British Geological Survey	12.County,Borough,District and City Councils
4.National Parks Authorities	13.Water Authorities
5.Nature Conservancy Council	14.Welsh and Scottish Offices
6.British Rail	15.Waste Disposal Authorities
7.British Coal	16.Caving organisations
8.British Waterways Board	17.Mineral extraction companies
9.Inland Revenue Mineral Valuers	18.Published literature

The natural cavity database collection has been undertaken utilising a set of specially prepared data collection pro forma sheets which record the essential details of the natural cavity location in a series of logical steps. The sheets have been prepared to streamline the entry of recorded information by using a system of coded answers which are selected and entered into the appropriate box. Completed data collection pro forma sheets are then entered onto a dedicated computer facility. The computerised database is stored and can be accessed through an IBM compatible PC computer using dBase III + software which is linked to Autocad software. Hence it is possible to generate scaled plans (paper or film) of cavity information as required. Using a standard digitiser the geological boundaries of cavity prone strata (Class 2/3 boundary) across Great Britain have been entered onto computer. Currently only 1:625,000 scale digital topographic map data is available from Ordnance Survey, the organisation responsible for producing topographic maps of Great Britain, but 1:250,000 scale data is due to become available in 1990, too late for incorporation within this project. Therefore it is presently intended to produce film overlays of the natural cavity data to be viewed against paper map underlays showing topographic and administrative details.

Spatial distribution of natural cavities

Great Britain, despite its small size, has a very varied geology and tectonic structure. Many different soluble carbonate rocks are present including hard crystalline low permeability limestones, soft high porosity limestones, dolomitic limestones and metamorphosed limestones. There are some 20 British limestones of varying geological age spanning from Pre Cambrian to Tertiary. The geographical spread of soluble carbonate rocks across England, Wales and Scotland is shown by Figure 1.

Karst landforms abound in these limestone areas and natural underground cavities are widely developed. They have formed and evolved in response to many and varied weathering cycles , climatic environments, tectonic and groundwater table level changes resulting from eustatic sea level changes through geological time. These landforms currently present themselves as denuded upland, interstratal and covered karst regions containing active and fossil vadose and phreatic elements.

Natural cavity types include swallow holes; sinkholes; solution widened joints, bedding planes and fault planes; vadose and phreatic caves. These features are generally best developed within the crystalline limestones especially Carboniferous Limestone. In contrast the softer porous limestones of Jurassic and Cretaceous age tend to contain relatively few caves, the effect of dissolution being most in evidence at the upper surface of the limestone. Dissolution acting from the surface downwards has preferentially formed point features such as swallow holes, sinkholes and solution pipes. These natural cavity types are particularly well developed upon the Cretaceous Chalk outcrop in England.

Apart from limestones there are also soluble non-carbonate rocks present which include salt-bearing and gypsum-bearing strata of Permian and Triassic age (Figure 2). Within the main saliferous basins the subsidence effects of natural dissolution have been largely masked by the long history of salt extraction utilising techniques such as wild brine pumping, rock salt mining and controlled solution mining. The areas of former brine pumping which suffered major catastrophic subsidence have now generally stabilised as the natural groundwater flow regime has been re-established. There are in evidence some geologically old subsidence hollows which are presumed to be naturally formed, and current monitoring suggests that in certain areas natural brine runs have reactivated. However there is no positive evidence to suggest a new cycle of natural cavity generation posing a renewed subsidence threat.

Significant numbers of gulls and fissures (Figure 3) have been recognised across the country reflecting the widespread phenomenon of cambering. A large proportion of these natural cavities have been found to be associated with Permian, Jurassic and Cretaceous strata in particular (Figure 2). The lithostratigraphy of these geological units is especially favourable since widespread conditions exist whereby upland areas composed of competent sandstone and limestone strata overlie thick clay sequences. The topography is characterised by steep hill and valley sideslopes with gulls and fissures developed sub parallel to the slope directions (Waltham 1989; Hawkins & Privett 1981).

Other natural cavities have developed as piping cavities within aeolian cover deposits in certain areas e.g. Merseyside. Sea caves are known to be extensively distributed around the coasts of Great Britain and are developed in most non-cohesive geological units.

The geographical spread of natural cavities throughout Great Britain generally reflects the controlling influences of geological, hydrogeological and geomorphological factors. The degree of influence and interaction between combinations of factors has not yet been appraised for the individual rock types, though an example of how this exercise can be undertaken has been presented for the Cretaceous Chalk in England (Edmonds et al, 1987). The research work demonstrates how a natural cavity prone rock can be subdivided into areas of greater and lesser potential for the presence of natural cavities. For other

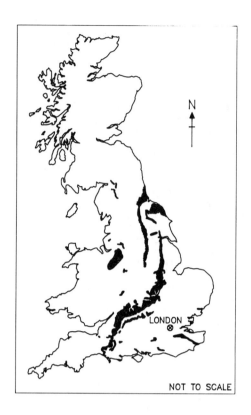

Figure 1:   Distribution of Cavity prone soluble carbonate rocks

Figure 2:   Distribution of cavity prone evaporitic rocks and other rock types affected by gulls and fissures

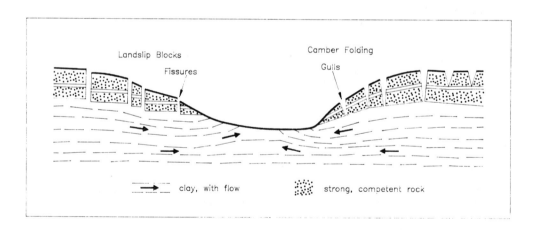

Figure 3:   Example of the development of gulls and fissures (After Waltham 1989)

rock types it should be possible to develop similar models by research.

Natural cavities as problems and resources

Urban development has suffered structural damage due to the subsidence effects and collapse of natural underground cavities within underlying soluble carbonate rocks.

Infrastructure such as roads, railways, canals and services (water, gas, electricity and telephone) have similarly suffered damage. Most of the problems arise due to the formation of subsidence sinkholes triggered by heavy rain, leaking services or soakaways. In cavity prone areas, overcoming subsidence problems requires special precautions such as reinforced cruciform ground beams, raft or piled foundations; impervious lining of service trenches with remote water collectors; flexible leak-proof pipework and couplings; ground surface and road pavement reinforcement utilising high tensile strength geotextiles. Incorporation of these measures whether from the outset of a project or retrospectively greatly increases the order of costs incurred. Within existing urban areas the problems are compounded by a risk to the safety of an unsuspecting public. The repair of properties damaged by subsidence often presents an engineering challenge, though in a number of instances the costs of repair outstrip the insured value of the property and hence the property becomes an insurance write-off. The unfortunate house-owner may only receive the insured value of the property which could be significantly below that of the market value of the property, with the consequential effect of financial ruin or hardship for that individual. In Great Britain it is the responsibility of the private individual to safeguard against subsidence risk due to natural underground cavities by appropriate insurance.

Quarrying operations can be disrupted by the presence of natural cavities which can cause a variety of difficulties. These include excess flying debris during blasting due to cavitated rock being intersected by blast drillholes; quarry face slope instability; contamination of the rock being quarried where sediment infilled cavities are revealed in the quarry face; loss of haul roads due to subsidence or collapse into underlying cavities. Recognition of these potential hazards requires careful appraisal and increased expenditure on exploration programmes when extending or developing quarry reserves.

Natural cavities can present advantages and disadvantages to water resource development. Groundwater flow along solution widened fissures for example can enhance the volume of water available for abstraction. By contrast these favourable conditions for water abstraction can be hazardous in terms of the potential for pollution of the aquifer. The relatively high transmissivity rates mean that should pollutants enter the bedrock at some point such as a swallow hole, then the contaminant will spread quickly and be difficult to control and clean up afterwards. Another problem encountered concerns artificial groundwater lowering by pumping where hydrostatic support for cavities in the process of upward migration to the surface is removed. The changed circumstances allow downward percolation of groundwater through the cavity increasing the seepage forces leading to degradation and failure of the cavity roof which is triggered to migrate upwards eventually producing subsidence at the surface. Water storage facilities can also suffer should natural cavities be present, particularly as a linked system, allowing significant through-flow of water. The water losses by seepage can even increase with time by the effects of washing through of materials formerly infilling cavities thereby effectively increasing the permeability of the bedrock and flow rates. If these conditions persist for a long time and artificially induced dissolution takes place, this could be particularly critical in areas of gypsiferous and saliferous beds (James & Kirkpatrick 1980).

When waste disposal is carried out by the backfilling of old quarries in cavity prone areas, then in certain circumstances hazards may arise. If insufficient precautions have been taken to seal off and contain a waste disposal site leachates and methane could migrate away from the site causing problems some distance away. Another possibility is that during the active life of the waste disposal site a natural cavity below the floor of the clay lined containment area could collapse providing a pathway for leachates and methane to escape. To date, however, there are no known case histories proven where these types of problems have developed in Great Britain.

Finally, there are a number of instances where natural cavities intersecting with man-made tunnels and mines have produced some spectacular subsidence problems. The subsidences have normally taken place where the tunnel or mine gallery passes through the basal portion of a large deep infilled natural cavity and the infill is washed down into the open tunnel or mine gallery by downward percolation of concentrated groundwater flows. Some dramatic surface collapses have occurred where buildings have collapsed and lives have been lost when the downward movement of the natural cavity infill has been triggered by a water main burst, or other leaking drains and services. The concentrated water flows saturate and soften the infilling deposits until they are able to flow downwards into and laterally along the tunnels or mine galleries (Edmonds 1988).

A knowledge of natural underground cavities is also important because many are of scientific value and interest (e.g. provide habitats for bats), while others form valuable amenities visited by many tourists and cavers each year.

Engineering and land-use planning significance

Prior to this study natural cavity problems have tended to be treated in isolation. Consequently the engineering and land use significance of natural cavity prone areas has not been appreciated. There has not been, with the exception of the research carried out for the Cretaceous Chalk in England, any previous attempt to collect natural cavity data to form a national archive which can be analysed to provide a balanced overview. There are a number of areas in Britain where many planners and engineers have surprisingly failed to realise the recorded level of subsidence incidence. This is because, as a geological hazard, subsidence over natural cavities has tended to be viewed as a random event. The general awareness of natural cavity land instability problems is either poor or so generalised as to be of no practical use. A common problem encountered within local and county administrative authorities is that very few authorities formally record subsidence events in their administrative area. The events may be recounted by an officer who has lived and worked in the area for a long period of time and has taken an interest in the phenomenon. The unfortunate effect of this circumstance is that when the officer leaves or retires from that authority, the fund of knowledge is lost and the successor is unaware of the problems that have occurred. Similarly, an engineer unfamiliar with a natural cavity prone area is liable to misinterpret or entirely miss evidence of natural cavities being present at a site. Therefore, against this background, the natural cavity database has an important role to play in terms of engineering and planning significance and guidance in the future.

It is hoped that by widely publicising the results of this national study, the general awareness of natural cavities and subsidence will be greatly enhanced. The study will provide a much needed overview of the engineering and land-use planning significance by the existence of a large computerised natural cavity database forming a national archive which can be searched. Additionally, natural cavity location plans will become available at 1:625000 and 1:250000 scales. The study will also provide an up to date state-of-the-art review of natural cavity detection and investigation techniques, ground treatment and foundation solutions.

Future developments of a geographic information system

The natural cavity distribution could be spatially analysed to determine the influential controlling factors upon the distribution as a basis for creating subsidence hazard mapping models for various rock types. The qualitative factors can then be ranked, numerically weighted and combined into a quantitative subsidence hazard mapping model formula. This methodology has been successfully utilised by Edmonds (1987) to create a subsidence hazard mapping model for the Cretaceous Chalk outcrop in England. A further enhancement would be to carry out statistical spatial analysis looking at the relationship between the influential geological, hydrogeological and geomorphological factors in three dimensions. This would require a sophisticated computerised geographic information system, where each factor is modelled by a three-dimensional layering of information. This ultimate development of subsidence hazard mapping would form a very powerful tool for use by engineers and planners to assess natural cavity significance.

Conclusions

When the national study of natural underground cavities in Great Britain is complete the awareness of associated problems should be greatly improved. Accessing the natural cavity database to find whether cavities are recorded within or near areas of interest should become a standard procedure for engineers at the desk study stage of projects. Similarly, the availability of natural cavity location maps for planners should transform the current limited appreciation of subsidence problems when considering areas suitable for particular types of development in the future. This will be aided by a document (Department of Environment, 1989) which sets out guidance for planners and developers on the various types of land stability. The document outlines the responsibilities and liabilities associated with or incurred in dealing with stability problems, and the appropriate methods of considering such issues in planning of land use and development control. It also reinforces the important concept that land instability can be a material planning consideration which needs to be taken into account, where appropriate, when planning applications are made. Additionally, the study will define the best practice for detecting and overcoming natural cavity problems in particular circumstances and the most appropriate remedial measures for particular ground conditions.

Acknowledgements

Thanks are due to the Department of Environment for their permission to refer to the natural cavity study commissioned by them, and currently being carried out by Applied Geology. The views expressed in the paper are, however, those of the authors alone and do not necessarily represent those of the Department of Environment.

References

Beck, B.F. 1984. Sinkhole terminology In: Beck B.F. (ed) Sinkholes: their Geology, Engineering and Environmental Impact. Proceedings of the 1st Multidisciplinary Conference on Sinkholes, Florida. Balkema Press, ix-x.

Department of Environment, 1989. Planning Policy Guidance: Development on Unstable Land (in press).

Edmonds, C.N. 1987. The Engineering Geomorphology of Karst Development and the Prediction of Subsidence Risk upon the Chalk Outcrop in England, PhD thesis London University, unpublished.

Edmonds, C.N. 1988. Induced sub-surface movements associated with the presence of natural and artificial underground openings in areas underlain by Cretaceous Chalk. In: Bell, F.G., Culshaw, M.G.,Cripps, J.C. and Lovell, M.A. (eds), Engineering Geology of Underground Movements, Geological Society Engineering Geology Special Publication, No. 5,205-214.

Edmonds, C.N., Green, C.P. and Higginbottom, I.E., 1987. Subsidence hazard prediction for limestone terrains, as applied to the English Cretaceous Chalk. In: Culshaw, M.G., Bell, F.G., Cripps, J.C. and O'Hara M. (eds), Planning and Engineering Geology, Geological Society Engineering Geology Special Publication, No.4, 283-293.

Hawkins, A.B & Privett, K.D., 1981. A building site on cambered ground, Radstock, Avon. Quarterly Journal of Engineering Geology, Volume 14, 151-167.

James, A.N. & Kirkpatrick, I.M., 1980. Design of foundations of dams containing soluble rocks and soils. Quarterly Journal of Engineering Geology, Volume 13, No. 3, 189-198.

Walsh, P.T., Edler, G.A., Edwards, B.R., Urbani, D.M., Valentine, K. & Soyer, J., 1973. Large scale surveys of solution subsidence deposits in the Carboniferous and Cretaceous limestones of Great Britain and Belgium and their contribution to an understanding of the mechanisms of karstic subsidence. In: Sinkholes and Subsidence-Engineering geological problems associated with soluble rocks, Proceedings of the International Association of Engineering Geology Symposium, Hanover, T2.A1-T2.A10.

Waltham, A.C., 1989. Ground Subsidence. Published by Blackie, Glasgow & London, 202p.

A karst ordinance – Clinton Township, New Jersey

JOSEPH A. FISCHER   *Geoscience Services, Bernardsville, N.J., USA*
HERMIA LECHNER   *Mayor of Clinton Township, N.J., USA*

ABSTRACT

The Township of Clinton, New Jersey is underlain by solution-prone Cambro-Ordovician aged carbonate rocks. As the locale has been identified by both State agencies and regional planners as a prime population growth area, the township is the target of development pressures. As a result of the environmental awareness of the Township leadership, a lay, legal, and technical group prepared a geotechnically oriented Ordinance which mandates consideration of the problem of karst in a multi-disciplined, multi-phased investigation. The results of a developer's investigation are reviewed by the Township Planning Board and an experienced geotechnical consultant. To date, the Ordinance appears to function as intended, has been well received by the Township officials and generally accepted by the developers.

Introduction

The location of Clinton Township in relation to New Jersey Cambro-Ordovician aged carbonates is shown on Figure 1. As may be seen, the Township is not unique. Further to the west, in Pennsylvania, portions of Bucks County and most of the burgeoning urban/suburban Lehigh Valley are underlain by similar rocks.

Voids in the overlying soils and in the rocks themselves in these areas, provide an unobstructed pathway for undiluted slugs of pollution that could reach local water supplies. Wells supply all of Clinton Township's water. In addition, enhancement, and/or formation of soil voids atop the cavity-prone rocks, can lead to sinkholes and safety related concerns for surface-supported structures. Obviously, thin or weathered rock ceilings over large cavities can lead to foundation failures or sinkhole formation as a result of construction loads above these weak zones.

Sinkholes and caves are found in the Township. It is likely that some of the earliest observers of doline formation in the area were Clinton Township farmers. They plowed around these annoying holes and backfilled them with rocks, soil, cornstalks and tree trunks. To our knowledge, the first documented enumeration of the possible environmental concerns in the areas of the Township that are underlain by carbonate rocks, is a 1987 letter from the State Geologist, Haig F. Kasabach. Dr. Kasabach noted a number of potential concerns, with specific references to the carbonates underlying a high growth area of the Township.

As the carbonate formations below the Township are some 400 million years old, the obvious question is why did it take so long to recognize the problem? Actually, the question should concern itself with: why did the knowledge of karst-related problems take so long to get to the public and decision makers? There is an 1862 study that indicates a recognition of the solutioning of carbonates in the formation of river valleys in Ireland. The Natural Bridge of Virginia is formed in carbonate rock. George Washington was the surveyor, Thomas Jefferson was the purchaser, (in 1774). Many local residents have certainly visited or are familiar with the caves of Virginia and Pennsylvania. These caves were formed in the same rocks that underlie Clinton Township.

Despite the relatively long history of scientific recognition of the problem, and the secular observations of the results of solutioning in carbonate rocks, it is only recently that this information was disseminated in other than the occasional article in an obscure technical journal or National Geographic. Man's ventures into areas previously considered less suitable for development, as well as the actual processes of development, which can exacerbate the formation of dolines, have fueled the transfer of information. Perhaps, also the media quest for "news of disaster" has helped the transfer of knowledge from scientists to the public.

In 1976, government-sponsored, regional studies were underway in Appalachia to catagorize and manage subsidence-prone lands. Laws have not yet been passed, but the Pennsylvan-

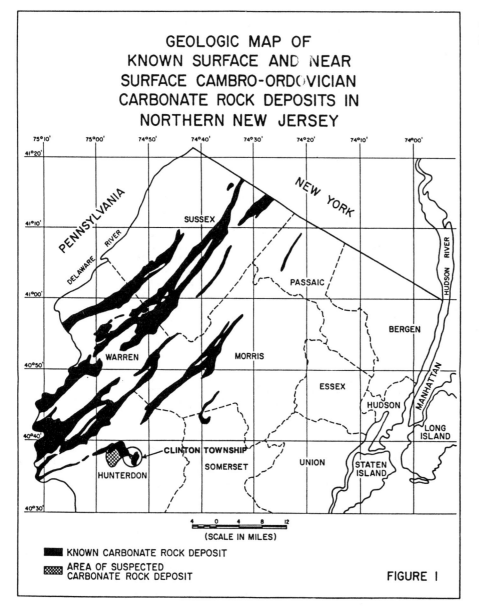

GEOLOGIC MAP OF
KNOWN SURFACE AND NEAR
SURFACE CAMBRO-ORDOVICIAN
CARBONATE ROCK DEPOSITS IN
NORTHERN NEW JERSEY

(SCALE IN MILES)

KNOWN CARBONATE ROCK DEPOSIT
AREA OF SUSPECTED
CARBONATE ROCK DEPOSIT

FIGURE I

ia, Virginia and West Virginia Geologic Surveys have initiated long-term, comprehensive studies of the carbonate rocks of these states. Bucks County, Pennsylvania commissioned a "limestone" study (completed in 1978) in relation to sewage system suitability. A major step was taken in 1987, when the Water Resources Agency for New Castle County, Delaware, proposed stringent land management practices in areas underlain by carbonate rocks.

At about the same time as Clinton Township officials became aware of the possible problems below their Township, several nearby Pennsylvania communities in the fast-growing Lehigh Valley, also began to respond to development pressures by designing and passing restrictive ordinances. Clinton Township also recognized the need to respond to the planning and economic realities that have identified what was previously a rural area, as the nucleus of both planned and unplanned growth in northwestern New Jersey. In an attempt to control (but not ban) new construction in a realistic manner, the Township Council convened a broad-based committee of lay and technical people. Included were representatives from; the local Water Shed Association, the

Township Engineers office, the Township Sanitary Engineers office, the New Jersey Geological Survey, the New Jersey Department of Environmental Protection, the County Health Department, the Town Council, and a geologic engineer with experience in investigation and construction in karst terrane. An attorney, experienced in State land laws, reviewed the final committee drafts of an Ordinance, converting primarily a technical presentation into a legal document. Amazingly, the legal structuring of the technical concepts did not significantly alter the wording or intent of the committee drafting the Ordinance.

In May of 1988, the Ordinance was passed by a majority of the council. The Ordinance is shown as Figure 2.

Operations
    The Ordinance is designed to force an applicant to follow a study process. The investigation is phased in a manner that has proven effective on a number of projects, both in Clinton Township and in other areas. An applicant has the opportunity to cancel a project if the problems seem insurmountable at an early stage without large economic penalties. The phased program provides the opportunity to utilize both planning and engineering solutions. Throughout the various phases of investigation and design, the Ordinance either provides specific requirements or suggested methods of investigation, as well as indicating preferred and alternate procedures. Thus, both direction and assistance are built into the Ordinance checklists, without, what is believed to be inordinate restrictions on either a major devel-

oper or a small home builder.

The Ordinance has built-in quality control aspects. It requires Township peer review of the investigation program and the engineering solutions proposed by an applicant if cavity-prone rocks and soil voids are found during the investigation. The Ordinance also requires both the applicant and the Township to inspect construction operations to assure that not only are design requirements met, but that any unexpected subsurface conditions are addressed by appropriate field changes.

A construction moratorium had been in existence for some seven months prior to the passing of the Ordinance, hence, a backlog of unhappy applicants was immediately on the books. Since that time, all developers have apparently accepted; the Ordinance as written, the review process, and the recommendations of the geotechnical consultant and the Planning Board. The Ordinance may be tested in court as one of the moratorium developers has indicated he may challenge its' constitutionality as part of a much larger action against the Township.

The process is relatively simple (see Figure 2) and has been accepted with few problems. As the Ordinance requires that an experienced consultant perform the investigation, one of the engineering firms initially doing several studies in the Township has secured many of the subsequent projects, as a result of their earlier success in the review process. The early (and present) procedures, where they have been successful, is to make a preliminary identification of the Township's technical concerns and attempt to obtain all the available information prior to submitting their phased investigation (Checklists I and II of Figure 2). In addition, they have generally used the same team, resulting in a uniform quality of product. Where less experienced individuals have been incorporated into a project team, the difference is noticable.

The other organization's successfully performing work in the Township have followed the same procedures. Those who have problems have not done the homework required by the Ordinance.

Conclusions

The authors are undoubtedly biased in this assessment, but we believe the Ordinance has been administered successfully. There has been a quantum leap in awareness in those attempting to build within the Township and surrounding region. Other towns are considering similar Ordinances, or are imposing its' concepts in a de facto manner. Several developments have been either cancelled or significantly altered by the need to: protect ground water flows in carbonates, and protect the safety and investment of future inhabitants. The Township's largest proposed development (some 850 acres) is proceeding in complete accordance with the Ordinance requirement of a phased, practical, state-of-the-art investigation, with complete cooperation between the developer's consultant and Township representatives. The office facility constructed at this same site (some 10 - 12 years ago) resulted in millions of dollars of over-runs in investigation and construction.

We believe that the Ordinance appears to be working successfully and has improved the development process within Clinton Township.

TOWNSHIP OF CLINTON
ORDINANCE #387-88
COUNTY OF HUNTERDON
STATE OF NEW JERSEY

AN ORDINANCE TO PROTECT THE WATER RESOURCES OF CLINTON TOWNSHIP LOCATED IN THE CARBONATE ROCK FORMATIONS OF THE TOWNSHIP

Be it ordained by the Township of Clinton, State of New Jersey, County of Hunterdon as follows:

Pursuant to the provisions of N.J.S.A. 40:55D-32, et seq., the zoning map of Clinton Township is amended to provide an overlay zone which is designated The Critical Geologic Formation Zone and shall be superimposed upon maps of those areas of the Township where carbonate rock formations are found. Within this zone all Environmental Impact Statements required by §72-68 of the Code of the Township of Clinton as amended herein shall include the additional information required by this Ordinance. The Critical Geologic Formation Zone shall consist of the Critical Geologic Formation Area (C.G.F.A) and the adjacent Critical Formation Watershed Protection Area (C.F.W.P.A.). The Critical Geologic Formation Zone shall be shown and delineated on the Official Map of Clinton Township and shall be available for inspection and will be on file, after adoption, in the office of the Township Clerk pursuant to §72-77 of the Code of the Township of Clinton.

(To be inserted on Page 7307 as Article XI, Section 72-65.1)

General Requirements in the Critical Geologic Formation Zone: For all development proposals in the Critical Geologic Formation Area whether residential or non-residential, a geologic investigation program to determine the potential for development shall be prepared and conducted in accordance with the requirements of subsection 72-68C. The geologic investigation program shall be designed to produce information and provide recommendations for site planning, engineering design and construction techniques which shall meet or exceed the standards of the geology study provisions of the Geologic Segment of the EIS (§72-68). The geologic segment of the EIS may be completed and filed prior to the completion of the other portions of the EIS at the applicant's option.

(To be inserted on Page 7330 as Section 72-68)

Section A is amended to read as follows:

(A) When required: An EIS is required as part of any application for development involving new buildings or any land disturbance where Planning Board approval is required. Additionally, any proposal for development within the Critical Geologic Formation Area as shown on the official map of the Township requires an analysis in accordance with the geologic segment of the EIS. The Planning Board may grant an exemption from the requirement of part or all of an EIS under subsection (G) below. An EIS is also required for all public and quasi-public projects unless they are exempt from the requirements of local law or by supervening county, state or federal law.

The following is added to subsection 72-68 (C) on page 7331 after the words, "...available from the township." The geologic segment of the EIS shall in addition to the information referred to herein shall include a discussion of the probable effects of the proposed development upon Township water resources as related to existing geologic conditions and investigation results, a presentation of proposed engineering solutions (specifically as to design and construction aspects, including alternate solutions where appropriate), provisions for inspection and monitoring procedures during construction, and any long-term monitoring/inspections which may be recommended.

(To be inserted on Page 7336 after 72-68 C (3))

Section 72-68 C (3) is hereby amended to include subsection (n) as follows:

(n) Geologic Conditions Report. For all tracts located in the Critical Geologic Formation Area, the EIS shall contain data obtained during an appropriate site investigation. A comprehensive site investigation program shall be conducted by the applicant to provide the Planning Board with sufficient data to define the nature of all existing geologic conditions that may limit construction and land use activities on the site. Specifically, the investigations shall yield information, which shall demonstrate that the proposed development will identify any existing geologic conditions for which appropriate engineering solutions may be necessary to minimize any adverse environmental impact caused by the proposal.

Section 72-68C(7), is hereby amended to add Critical Geologic Formation Area Investigation Program:

(a) An investigation program shall be commenced by completing checklist #I. Said checklist #I shall be submitted to the Planning Board Secretary and shall be reviewed by the Township Geotechnical Consultant (GTC) and a report rendered to the Planning Board within 30 days of submission. The GTC in his report shall recommend to the Planning Board that checklist #II be prepared and submitted, or in the alternative, that portions or all of the requirements required by checklist #II be waived. If checklist #II is required by the Planning Board, the applicant shall then prepare and submit checklist #II to the Planning Board Secretary. Checklist #II shall be reviewed by the GTC and a report shall promptly be made to the Planning Board advising whether the checklists are complete. The Planning Board shall rule on the completness of the checklist within 45 days of the date of submission. The report shall also advise the applicant as to whether any proposed testing methodology is prohibited because of the potential danger the methodology may pose to the integrity of the site or the health, safety, and welfare of the community. The geotechnical consultant may also recommend waiver of some or all of the required investigations in appropriate cases pursuant to subsection (G). At the applicant's option, both checklist I and checklist II may be submitted simultaneously in which event the Planning board shall deem the checklists completed or incomplete within 45 days of submission.

After checklists I and II have been deemed complete by the Planning Board and the GTC has advised the Planning Board that the testing methodology poses no danger to the integrity of the site or the health, safety and welfare of the community, a permit shall be issued to the applicant authorizing the commencement of the testing procedure.

(1) Any on-site investigations and tests shall not begin until the applicant has received approval of his investigative plan and a permit has been issued. Additionally, actual notification at least 15 days in advance of the testing to commence in writing by certified mail, return receipt requested or personal service of said notice on the Township Clerk shall be given.

(2) The applicant shall arrange to have the proposed development site open for on-site inspection by the GTC or designated township inspectors at all times while the field investigation program is in progress, and testing data and results shall be made available to township officials and township inspectors on demand, but at no less frequent intervals than bi-monthly.

(3) At the completion of the field investigation a formal written report including the following shall be submitted: a description of the project, a general plan, to scale, of the entire project, showing the location of the project with respect to surface water, existing wells (within ½ mile) and adjacent property owners, logs of all borings, test pits and probes including evidence of incipient cavity formations, loss of circulation during drilling, voids encountered, type of drilling or excavation techniques employed, drawings of monitoring or observation wells as installed, time and dates of explorations and tests, reports of chemical analyses of on-site surface and ground water, names of individuals conducting tests if other than the P.E. referred to in the checklist, analytical methods used on soils, water samples, and rock samples, a 1" to 100' scale topographic map of the site (at a contour interval of two feet) locating all test pits, borings, wells, seismic, or electromagnetic, conductivity or other geographical surveys, an analysis of the ground water regime with rate and direction of flow; a geologic interpretation of the observed subsurface conditions, including soil and rock type, jointing (size and spacing), faulting, voids, fracturing, grain size, and sinkhole formation.

(4) The site investigation report should define the extent of geotechnical concerns at the site in relation to the planned development or land use. The proposed engineering solutions to minimize environmental impact as a result of the project, both during construction and in the foreseeable future, must be clearly detailed, together with the bases for the conclusions reached. Special consideration should be given to innovative control of surface water flows, as well as protection and replenishment of ground water.

(5) All samples taken shall be preserved and shall be available for examination by the Township upon request until final action is taken by the Planning Board on the application.

To be inserted on Page 7338, Section 72-68 D).

Section 72-68D, Environmental Impact Statement requirements, is amended by adding to EIS ITEM (§72-68 C) (3) (n) Critical Geologic Formation for all uses.

Section 72-68E, Planning Board Review, is hereinafter designated as Section 72-68E(1).

(A) The following shall be added as Section 72-68E(2):

At the applicant's option result of the investigation program required for the geologic segment of the EIS may be submitted to the Planning Board prior to the completion of other segments of the EIS, and if so submitted, shall be reviewed by the GTC and a report of his findings, conclusions, and recommendations shall be submitted to the Planning Board within 45 days of the submission of the data. The GTC shall confer with the Township Environmental Commission and request their input and non-binding recommendations within said 45 day time period. If the geotechnical consultant recommends the disapproval of the program, the recommendation shall include suggestions on alternate methodology which the GTC suggests would provide the requisite data.

(B) During his review of the geologic segment of the EIS for proposed development in the C.F.W.P.A. the GTC shall consider the data, formal reports, maps, drawings and related submission materials and shall advise the planning board whether or not the applicant has provided the Township with:

1. Sufficient design, construction and operational information to insure that the proposed development of the tract will not adversely impact on the health, safety, and welfare of the community.

2. The proposed method of development of the tract, will: minimize any deleterious effects on the quality of surface or subsurface water, and will not alter the character of surface and subsurface water flow in a manner deleterious to known conditions on-tract or off-tract;

3. Specific details insuring that design concepts and construction and operational procedures intended to protect surface and subsurface waters in critical zones will be properly implemented;

4. The submission provides specific details on inspection procedures to be followed during construction.

(4) The Planning Board shall at the request of the applicant within 45 days of the receipt of the report from the geotechnical consultant approve or disapprove the proposed geotechnical aspects of the development plan and associated construction techniques. In the event the Planning Board disapproves of the proposed development plan and associated construction procedures the Board shall state in the resolution its reasons for disapproval.

Section 72-13 is amended to add subsection I.

I. Geologic Segment Review. For any application requiring an EIS geologic segment submission, there shall be an application fee of $500.00 plus $100.00 for each acre of the site included within the Critical Geologic Formation Area. Additionally, there shall be posted with the Township a review escrow of $1,000.00 plus $500.00 per acre for each acre within the Critical Geologic Formation Areas, the escrow to be administered in accordance with the review escrow provisions of the Clinton Township Code. With regard to applications requiring review where the site is located all or in part of the C.F.W.P.A., the escrow fee shall be $500.00 plus $200.00 per acre.

FIGURE 2A

Owner's name and address:
Developer's name and address:
Location of proposed development site:
Tax Block, Tax Lot(s)
Type of proposed development:
Residential
Commercial
Industrial
single-family
multi-family
Proposed density (units per acre or lot coverage)
Any other data that applicant wishes the Planning
Board to consider:

CHECKLIST #1
INSTRUCTIONS

A. In compliance with the Land Use and Development Provisions of the Clinton Township Code, all development proposals for development in C.g.f.a. and/or C.F.W.P.A. shall submit completed checklists to the Township Planning Board as step one of the required geologic investigation required by the geologic segment of the E.I.S.

B. Procedure for submission of documents:

1. The applicant shall submit the completed checklist I to the Planning Board Secretary for distribution to the Township Geotechnical Consultant (GTC). Applications shall also submit the required application fee and escrow for geologic segment review (§72-13). ·

2. Checklists I and II may be completed and filed prior to the completion of the other portions of the E.I.S. at the applicant's option.

3. The applicant and the Planning Board will be advised within 45 days of submission of checklist I whether a waiver of completion of checklist II is being recommended by the GTC. The GTC may recommend a waiver of some or all of the required investigations as provided in subsection (G) of the general E.I.S. requirements.

C. Checklist I is intended to ensure that the information to be submitted by the applicant demonstrates that the applicant has sufficient information available on geotechnical issues to enable the applicant to prepare a plan for investigation of the proposed development site.

D. Any applicant with questions regarding whether applicant is entitled to a waiver of some or all segments of the geologic investigation is encouraged to contact the GTC prior to the commencement of the preparation of the geologic segment of the E.I.S.

CHECKLIST I

Clinton Township C.G.F.A./C.F.W.P.A. investigations program preliminary requirements:

Review of C.G.F.Z. map of Clinton Township

Review of New Jersey geological survey maps

Review of USDA publications and maps on Hunterdon County, Natural Resource Inventory prepared by the Southbranch Watershed Association (Vol. 1)

Review of Special Report #24, Geology and Groundwater Resources of Hunterdon County, New Jersey, Division of Water Policy and Supply, Department of Conservation and Economic Development, N.J. State Geologic Survey, 1966.

Submit with this checklist a site plan map at a scale of 1:24,000 identifying proposed development site and boundaries of site that are within the C.G.F.A. and/or C.F.W.P.A. as designated on the C.G.F.Z. Map.

Aerial photographs for the proposed site and surrounding area (at a minimum scale of 1" - 1,000', obtained during periods of little or no foliage cover.).

Summary of all known water production well logs and previous known subsurface investigations in the immediate area.

Submission of a site map at a scale of 1"-100' with a contour interval of two feet identifying existing surface water bodies, topography of the site, location of any existing water production wells.

Submit any other published geologic information available to applicant which applicant deems pertinent. Please specify.

Owner's name and address:
Developer's name and address:
Location of proposed development site:
Tax Block, Tax Lot(s)

Type of proposed development:
Residential
Commercial
Industrial
single-family
multi-family
Proposed density (units per acre or lot coverage)
Any other data that applicant wishes the Planning
Board to consider:

CHECKLIST II
INSTRUCTIONS

A. In compliance with the Land Use and Development Provisions of the Clinton Township Code, development proposals for development in C.G.F.A. shall submit completed checklist II to the Township Planning Board if required to do so by the Planning Board after the review of checklist I.

B. Procedure for submission of documents:

1. The applicant shall submit the completed checklist II to the Planning Board Secretary for distribution to the Township Geotechnical Consultant (GTC). Applicants shall also submit the required application fee and escrow.

2. The applicant and the Planning Board will be advised within 45 days of submission as to completeness of the submission. The GTC may recommend a waiver of some or all of the required investigations as provided in subsection (G) of the general E.I.S. requirements.

. 3. A permit shall be issued to the applicant authorizing commencement of field investigations when the Planning Board deems the geologic segment checklists to be complete.

4. Checklists I and II may be completed and filed prior to the completion of the other portions of the E.I.S. at the applicant's option.

C. Checklist II is a detailed outline of the proposed Investigation program, including reference to site specific investigation techniques, equipment and program objectives.

CHECKLIST II

Proposed investigation program to be conducted in C.G.F.A. in Clinton Township.

A. General Requirements:

1. Test borings and test pits are to be used as the primary means of identifying potential geologic hazards. Percussion probes geophysical techniques (e.g. seismic refraction and reflection, ground penetrating radar, magnetic, gravity and conductivity) can be used to provide data between test borings and pits.

2. Proposed exploration techniques which are not outlined in this checklist may be submitted to the GTC for review and possible inclusion in the approved investigation program. Alterations to the planned program can be made during the progress of the field investigation by request to the GTC if so required by the nature of the encountered subsurface conditions.

B. The intention of the site investigation program is to define the nature and limits of possible design, construction and operating concerns that could result from the existence of carbonate soil and/or rock formations underlying the proposed development site.

C. List name and address of New Jersey licensed engineer:
List name and address of New Jersey licensed well driller:

1. DIRECT TESTING PROCEDURES
TEST BORINGS

(a) number proposed

(b) depth anticipated

(Note: If rock encountered is within 40' of ground surface, a minimum of 10' of rock is to be cored. Rock cores shall be a minimum of 2" in diameter, to be obtained by double tube, split barrel coring device)

(c) boring techniques to be utilized:

(Note: unless written approval is authorized, all test borings will be drilled using rotary wash boring procedures without use of drilling muds. Water losses in borings are to be monitored as to depth and quantities.

(d) proposed bore hole grouting techniques

(e) soil and rock sampling to be performed in accordance with ASTM Standards D 420, D 1586, D 1587, and D 2113

(g) logging of all test borings or test pits in accordance with the Unified Soil Classification System and in relation to the geologic origin of the constituents of the encountered materials, e.g. light yellow brown silty clay (CH), with occasional angular

dolomite fragments, moderately stiff, residual soils, some stained paleo jointing.
TEST PITS

(a) number and depth of proposed pits

(Note: to be acceptable, minimum bottom area of pits shall be 10 square feet and shall encounter rock surface over 50% of the pit area).

(b) method of backfill to be employed

(Note: Test bit backfill shall be composed of excavated material, placed in layers and compacted to pre-excavation density, unless authorized otherwise by GTC)
PIEZOMETERS, LYSIMETERS AND WATER TABLE DATA

(a) number, locations and types of to be used

(b) other methods to be used

Note: These shall be installed and monitored in sufficient locations to identify depth to seasonally high water table and rate and direction of ground water flow.

GEOCHEMICAL TESTING OF PROPERTIES OF SOILS, ROCK AND WATER:

A. Methods proposed:

2. INDIRECT TESTING PROCEDURES
Percussion Probes

a. number proposed

b. depths anticipated

c. measuring techniques to be utilized (air loss, drilling speed and rod drops must also be monitored)
Geophysical Studies

(a) seismic refraction and reflection; location and number of runs anticipated; equipment to be used

(b) ground penetrating radar specify procedures and location of traverses

(c) magnetic, gravity or conductivity techniques -specify procedures and location of surveys
Geologic Reconnaissance

Factors to be examined - soil types, rock types, vegetative changes, observable seeps, or ground water discharge, circular depressions, swales.

Additional field investigation techniques proposed
MAPS, DRAWINGS AND OTHER DOCUMENTATION

(a) location of site on 1:24,000 scale USGS topo map (See checklist I)

(b) general site plan showing location of all field testing procedures in relation to planned development at a scale of 1"-100'.

(c) Timetable of proposed field investigation, laboratory testing, test data receipt and final report to the Township.

(e) Proposed technical inspection procedures during investigation (continuous technical supervision of field investigations is strongly recommended)

(f) Has an EIS without a geologic segment report already been submitted for this site? Date:

(g) Submission of application fees (§72-13)
Amount:
Date:
Future payments anticipated:

(h) Special factors or conditions applicant wishes to bring to the attention of the GTC:
TOWNSHIP GTC REVIEW
Approval of checklists I and II
Date of completion of checklist I:
Date of completion of checklist II:
Conditions to be imposed on approval
Date investigation to commence:
Denial of Checklists I and II
Items needed for completion
Waiver(s) (if any) deemed appropriate by GTC
PLANNING BOARD REVIEW
Date receipt of initial submission
Dates checklists I and II deemed complete by GTC:
Permit to be issued
Date:
Date of accepted start and completion of program (Subsection C, item 4 Timetables)
Permit denied

I hereby certify that the above ordinance was introduced on March 28, 1988. Public hearing was held on April 20, 1988 continued to April 25, 1988 and is further continued to May 9, 1988 7:30 P.M.

CAROL A KORBOBO, RMC
Township Clerk

FIGURE 2B

Karst mapping and applications to regional land management practices in the Commonwealth of Pennsylvania

WILLIAM E.KOCHANOV   *Department of Environmental Resources, Bureau of Topographic and Geologic Survey, Harrisburg, Penn., USA*

ABSTRACT

The Pennsylvania Geological Survey is currently conducting a program detailing karst regions throughout the state. This program is a two-phase approach to map karst features and to analyze the data for practical use in regional land management activities.

The mapping program incorporates the use of aerial photography, field surveys, and contact with the public to define environmentally hazardous areas. These methods comprise the majority of the data and are used in conjunction with carbonate geologic basemaps. The data forms a base from which inferences regarding karst development and distribution can be used to evaluate the components of urbanization and infrastructure design.

The availability of sinkhole data expands the role of the geologist, engineer, and planner. Decisions regarding the placement and design of communities, utility services, and foundations can be achieved at the beginning phases of a project rather than after subsidence problems have already begun.

Introduction

The Pennsylvania Geological Survey is actively investigating problems associated with sinkhole occurrence throughout the carbonate regions of Pennsylvania. Much of the Survey's work regarding sinkholes is to provide information and aid to local governments, state and federal agencies, geo-technical consulting groups, and individuals who have had or are currently having problems with sinkhole subsidence. Questions dealing with development, prevention, and repair of sinkholes, as well as land management planning in karst areas, has led to the development of a state-wide karst mapping program (Kochanov, 1988). The program, which has been in effect since 1985, is structured to:

1. Delineate the major karst regions within Pennsylvania.
2. Outline those areas with existing sinkhole problems and those areas most prone to sinkhole development.
3. Define the parameters, geologic or otherwise, which have an impact on subsidence development.
4. Relate those parameters to land management practices as a means to provide the necessary background data to help solve karst-related problems.

The major karst regions of Pennsylvania occur within two physiographic provinces, the Ridge and Valley and the Piedmont (Figure 1). The Ridge and Valley province is characterized by linear, northeast-trending valleys underlain by shale and carbonate bedrock and flanking ridges composed of sandstone and shale. The Piedmont is characterized by carbonate and shale valleys with quartzitic and metamorphic rock ridges. The karst regions occur primarily in structurally deformed lower Paleozoic carbonates within the valley sections.

Historically, population and economic growth has been more desirable within the carbonate valleys than the adjacent ridges as sites for homes, farms, industry, and transportation routes. As population increases in these regions, the rural areas are targeted for urban expansion. Figure 2 shows the location of population centers within, or in close proximity to, the major carbonate regions of Pennsylvania. With expansion of the urban centers, the potential for development in karstic areas and subsequent subsidence problems increases (Wilshusen and Kochanov, in prep.).

Karst Mapping Program

The karst program in Pennsylvania is divided into two phases. The first phase of the project is a detailed inventory of karst features on a county-wide scale. The

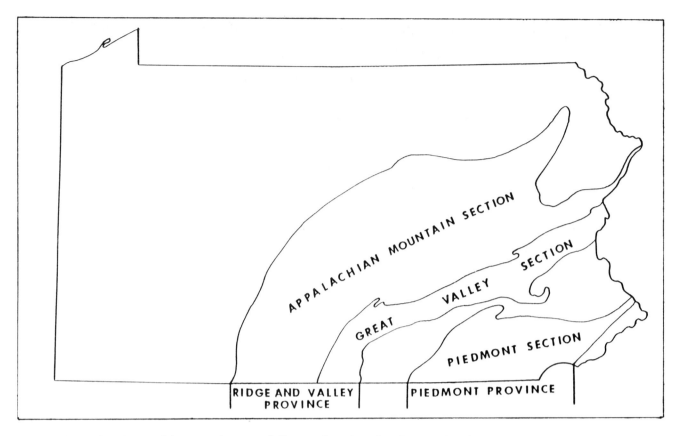

Figure 1: Physiographic provinces of Pennsylvania showing major karst regions.

county reports consist of a series of 7½ minute topographic maps (scale 1:24,000) delineating the carbonate geology, sinkholes, closed depressions, caves, bedding and joint
orientation, and past and present surface mines. To date, seven out of thirty-one counties
have been completed.

As part of the inventory, a questionnaire is mailed out to municipalities to gather
data on existing sinkholes within their districts. The questionnaires request information
on sinkhole location, physical characteristics, causative factors, and repair methods.
All of the karst features compiled in the inventory phase are transferred to geologic
base maps (Figure 3). Data pertaining to individual sinkholes received from the mailing
survey is compiled in the computer database.

The second phase is designed to interpret the data compiled in phase one and to
be published as part of the Survey's Environmental Geology series. In general, the
report will present a broad perspective of karst development in Pennsylvania with key
reference maps for the two physiographic provinces. The text portion of the second
phase will cover karst development, karst landforms, karst-related problems, suggestions
for regional planning, local zoning and land development guidelines, and methods of
sinkhole repair. The reference maps will be at a 1:100,000 scale and will cover specific
information pertaining to a particular province. This will include geology, hydrology,
and segments on how lithology, regional and local structure, paleokarst, and glaciation
have affected recent karst development in each province.

Methods
With completion of the inventory phase, the mapped karst features can be used
to establish relationships between the carbonate units, structure, and sinkhole distribution. This primary analysis of the data allows one to determine where existing sinkholes
are located and which carbonate units are most prone to sinkhole development.

The majority of the data is derived from aerial photographic interpretation. The
use of air photos however, is often the most subjective of the data collection methods.
As with the recognition of fracture traces, the locating of sinkholes and closed depres-

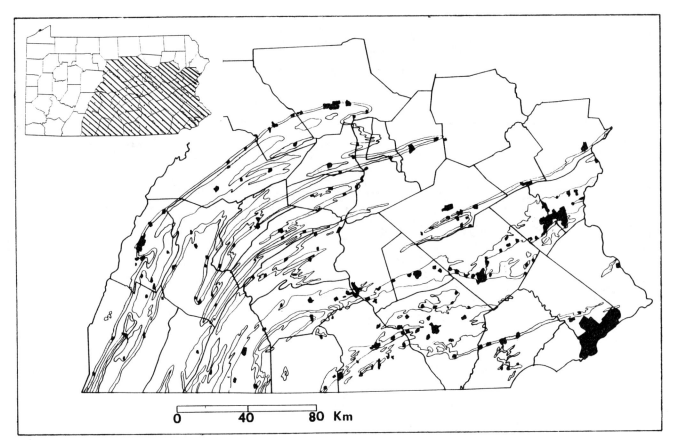

Figure 2: Major population centers (black), carbonate bedrock (gray).

sion features can vary with the reviewer. It is important that field surveys compliment
each review of aerial photographs. This helps to attach a degree of confidence to the
interpretation as well as helps define the relationships between topography and tonal
variations observed when reviewing the photographs. It is also important to select
those photographs which reflect optimum moisture conditions in the soil. U.S. Department
of Agriculture, black and white, stereo-paired aerial photographs (scale 1:20,000)
were found to be the most useful in locating karst features due to their consistently
good quality and coverage of large areas. They are taken during different times of
the year and at approximately ten year intervals. This allows one to view similar areas
under a variety of climatic conditions and at different stages of land development.

Through aerial photographic review and verification of karst surface features
by field surveys, a number of karst and bedrock features have been observed. They include:
mottled areas, fracture traces and bedding attitude, sinkholes, and surface mines.

1. Mottled areas: The most common surface feature observed in carbonate terrain, they
   normally appear as circular to irregularly shaped patches in a modified lattice
   or boxwork type of pattern (Figure 4). The amplitude of closed depressions varies
   with the depth and type of regolith. Some of the depressed areas have relief less
   than three meters, making it difficult to note from a stereoscopic projection. Since
   the closed depressions retain moisture, they can be identified by differences in
   neutral color tones. The depressions will range in tone from black to various shades
   of gray, contrasting noticeably with the surrounding terrain. In instances of prolonged
   rainfall, a heavy precipitation event, or the time of year (after winter thaw),
   the depressions may impound water and will appear white or light gray due to the
   reflection of the sky (Figure 4). The depressions may also appear white in instances
   where the impounded water has drained into the subsurface and has prevented vegetation
   from growing in the once wet soil.

2. Fracture traces and bedding attitude: Solution channels coinciding with fractures
   may be apparent as straight linear features that often connect with depressions
   or sinkholes. Similar linear features can be attributed to bedding orientation within

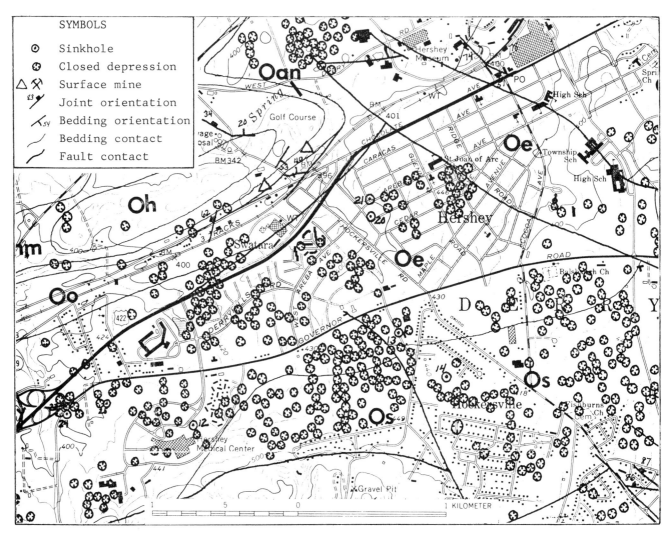

Figure 3: Geologic base map showing sinkholes, closed depressions, surface mines, and bedding and joint orientations, Hershey 7½ minute quadrangle, Dauphin County.

a particular carbonate unit. With a high moisture content in the soil, the linear features will appear darker than the surrounding terrain (Figure 4).

3. Sinkholes: Because of the scale of most aerial photographs (1:20,000; 1:24,000) and the average size of the majority of sinkholes in Pennsylvania (usually less than six meters), most sinkholes cannot be easily located. Larger sinkholes can be as much as ninety meters in diameter however, and can be easily recognized. There are surface features which can be used to help locate some sinkholes. The most common feature is isolated clumps of trees or other non-cultivated vegetation located in fields. These islands of vegetation stand out in marked contrast to crops normally grown in cultivated fields. Small, forested areas, adjacent to known karst areas, can also be located and checked during the field survey. Low altitude aerial photographs (scale 1:4800) have been used in locating smaller sinkholes with excellent results.

4. Surface mines: Surface mines present a problem in that they appear similar to large sinkholes. They can be distinguished from sinkholes in aerial photographs by having a raised rim around the perimeter of the hole, evidence of roads leading down into the hole, or not being entirely closed (one end of the hole will be breached and is flush with the adjacent land surface). Some iron deposits in Pennsylvania formed in sinkholes and were mined during the middle to late 1800's. Review of the literature dating back to those time periods has helped to identify and locate these features.

Figure 4: Aerial photograph showing mottled areas (M), water filled closed depressions (CD), and fracture traces (FT). USDA photograph, Allentown West 7½ minute quadrangle, Lehigh County. 1000 feet = 304.87 meters.

The aerial photographic review and the field surveys do leave gaps in the record. Surface expression of sinkhole development in an area changes over time because of weathering, land development, and individual sinkhole repairs. As a result, sinkholes identified in early aerial photographs may no longer appear in the field and, conversely, new ones have developed that were not apparent on the photographs. To supplement the data, review of existing literature and mailed questionnaires are used. Municipalities that commonly deal with sinkhole problems are one of the best sources of information. State and federal agencies, consulting groups, and the general public can also provide input into the sinkhole database.

One important aspect of the inventory is to announce their availability upon completion. It is felt that public awareness, of the karst landscape and the potential problems related to karst, is essential to sound land management practices. Not only should the geologic community be notified but the general public should also be included. Announcements for the sinkhole reports in Pennsylvania are made through the mail (legislative representatives and municipalities), local newspapers, and through the Pennsylvania Geological Survey's bimonthly publication Pennsylvania Geology.

The sinkhole project in Pennsylvania is prepared in an open-file format. This has the advantage over published reports in that the open-file system allows updating of the data without having to go through formal revisions and editing. For any particular area, the carbonate geology, karst features, and surface mines can be located on the base maps. Specific sinkhole information can be printed out from the computer database for any municipality.

Applications
Requests for information on sinkholes and related karst features from state and federal agencies, consultants, and planning agencies usually entails location of sinkholes and sinkhole-prone areas and the carbonate geology. The information is used in locating and designing waste disposal facilities, permit review, and incorporating the data into local zoning ordinances.

Bureaus within the state Department of Environmental Resources have used the information as part of the permit review process. Areas underlain by carbonate bedrock and sinkhole-prone areas are excluded from the siting of sewage disposal systems and landfill and land treatment facilities. The proximity of sinkholes and closed depressions to active surface mines is carefully considered in permit review due to past incidences of subsidence events related to groundwater withdrawal. Storm water cannot be discharged into sinkholes in Pennsylvania. Storm water is considered a potential polluting substance and is prohibited from being disposed of into the subsurface. In regard to these examples, the use of the open-file reports can allow quick reference to excluded areas, indicate potential problem sites, and serve as a source of information for permit applicants, base studies, and engineering projects.

Data from the open-file reports has also been used as background data in establishing environmentally sensitive zones (Upper Saucon Township, Northampton County), modeling zoning ordinances (Lehigh-Northampton County Planning Commission, Lancaster County Planning Commission), and the siting of on-lot sewage disposal systems (Oley Township, Berks County).

Individual homeowners normally wish to know how the sinkholes form, where existing sinkholes are located, their location in relation to their home, and repair methods. With established background data from the sinkhole inventory, the answers to some of these questions can be visually apparent. This is particularly useful in handling requests for assistance in the field.

Summary

The sinkhole project in Pennsylvania has been established to provide a source of background information for land development and foundation design to help avoid subsidence problems related to sinkholes. The maps should be used to develop information leading to specific site investigations for land use and development but do not take the place of a thorough subsurface investigation.

With increased use of the system, additional applications can be developed. One of the foreseeable applications is a completely computerized inventory system. Information from the maps can be entered into existing mapping software. Once entered, the data can be plotted for any geographic location and at any scale. An example of such an application was completed in the summer of 1988 and was used in a model GIS (Geographic Information System) for Lehigh County in eastern Pennsylvania. In addition to the computer generated maps, the sinkhole database can be used to provide a complete history of sinkhole occurrences for a given area. Add a telecommunications network and a statewide information service can be provided. This has broad reaching applications not only with regard to the sinkhole program, but also to related geographic, environmental, and engineering applications.

References

Kochanov, W.E. (1988), The inventory of sinkholes and related karst features in the Commonwealth of Pennsylvania, USA, in Daoxian, Y., ed., Karst hydrogeology and karst environment protection, Proceedings of the IAH 21st Congress, Guilin, China, 10-15 October, 1988, v. XXI, part 2, p. 1163-1168.

Wilshusen, J.P. and Kochanov, W.E. (in prep.), Land subsidence, carbonate terrane in Shultz, C.H., ed., The Geology of Pennsylvania, Pittsburgh Geological Society and Pennsylvania Geological Survey, part IX, Environmental and Engineering Applications, ch. 49 A.

Late paper:
Subsurface reservoirs and karst geomorphology

SONG LIN HUA   *Institute of Geography, Academia Sinica, Beijing, People's Republic of China*

Abstract

The well developed karst commonly causes the limestone terrains lack of surface water and very dry, while the aboundant underground water is not adequately used. Recent decades, a lot of subsurface reservoirs and hydropower stations have been built in South China. The underground reservoirs are constructed predominantly in the locations of karst geomorphology: (1) the connecting between fengcong depression and fenglin basin. It is the monoconduit pattern of reservoir and characterizedd by the small capacity and big regime; (2) on the monoconduit in the block between two basins with the hydraulic connecting as the Wangou reservoir with small regime and high capacity for water storing in the net work of karstic fissures and conduits under the basin in the up reaches; (3) at the topwindow of the underground river in fengcong region. The capacity of storage is rather small but with high gradient that is good for generating like Beiyan underground reservoir and hydropower station; and (4) when the emergence of underground river on the limestone cliff, the dam may be built just up the resurgence to form the reservoir generally for irrigation and power station. Usually, the dam of the reservoir is built on the underground river up the knick point of the resurgence.

Since 1958, especially 1970, the serious drought and flood disasters have been forced the hydrogeologists and local people to explore and exploit the karst water resources and harness the flood disasters in karst terrains in South China. First a lot of surface reservoirs had been built in 1958 and 1959, but most of them just stored a little water or nothing as the terrible leakage problem that makes hydrogeologists headache to accumulate water in karst areas. The very serious drought and short of food in 1972 gave great surpress to the local governments and the people. In 1974, a small underground reservoir was succesefully built in South Dushan, Guizhou Province. Since then a number of underground reservoirs have been built up in Guizhou, Hunan, and Guangxi provinces. Many hydrogeologists are very experienced to build this kind of reservoirs in limestone areas, but how to determine the right dam sites on the underground river and why we can succesefully build them now are still a big question.

This paper will be stressed on the characteristics of karst hydrogeology in limestone regions in South China and the relationship between the evolution and karst geomorphology and underground reservoirs.

Physical Setting

The carbonate rocks distribute about 450,000 Km^2 in South China including Guizhou, Guangxi, Hunan, Hubei, Sichuan and Yunnan Provinces, especially it occupied about 70% of the total area (176,128 Km) of Guizhou Province. The total thickness of carbonate rocks deposited from the pre-Cambrian to Triassic Period in South China varies in the ranges of 5,000-6,000 m. The rocks were folded and fractured during the tectonic movements such as Caledonian, Hercynian, Yenshan and Himalayas movements.

It belongs to the tropical and subtropical climatic zones since Teriary. At present the annual average temperature is about 15-20°C, and the precipitation 700-2,000 mm, about 80% of rainfall is in the Summer and Autumn with the high temperature and lower in the Winter and Spring. So that it is favourable to the development of karst geomorphology. With the intermittent uplifting of Qinghai-Tibet Plateau created by the Himalayas Movement, 3 regional geomorphologic terraces have been formed, in the Qinghai-Tibet Plateau with the mean altitude of 4,000 m above sea level as the first terrace, the Yunnan Plateau with the elevation of 1,500-2,000 m a.s.l. and the Guizhou Plateau at the altitude of 900-1,300 m a,s,l, are the second and third terraces (traditionally they are second grade), and the middle Guangxi and Hunan as the lower plain of peneplain. The areas between two levels of terraces constitute the geomorphological slope. Thus when the warm and humid currents are moving from the southeast towards northwest along the landform slope, the precipitation will be higher on the front of upper levels of geomorphology than that in the back part of the lower level (Table 1).

In the favour conditions of climate, geomorphology and geology, the karst geomorphology has well developed. Generally, the karst geomorphology is characterized by the order of the karst plane or peneplane, fenglin plane, fenglin basin, fengcong depression, fenglin basin, fenglin peneplane from the lower level terrace to the higher level terrace through the transission zone (Fig. 1).

As the intermittent uplifting of the Neo-tectonic Movement, the drainage base level descends down and increase the hydraulic gradient of the surface and subsurface drainage systems, which makes the water course deep cutting by the physical and chemical erosion of karst water. Therefore, the deep canyon of surface rivers have been developed in the slope zones. The main course and its tributaries cut the karst blocks as to tear them to the pieces. The lower drainage base level and stronger vertical karstification will complicate the geomorphological patterns formed in the previous periods and renew the sequences of karst development.

Table 1 Precipitation and elevations from Nanning to Ziyi

Site	Elevation (m)	Precipitation (mm)
Nanning		1318
Duan		1739
Nandan		1452
Dusha	1064	1346
Duyun	1200	1446
Guiyang	1070	1175
Zinyi	900	1098

In the Neogene, it was hot and humid climate (Paleogeography, 1984) and the earth crust was at stable and planation period, the residual red soil was gradually formed and filled in the solutional fissurs and covered on the ground surface. In the early and middle Pleistocene, the climate was still warm and humid, the red soil contains about 60% of Gaolin (Paleogeography, 1984), the vegetation was vigious. Therefore, the karstification was very strong at that time. In the late Pleistocene, South China was effected by the glacio-climate or snow mountains, it became drier and cooler. In the Holocene, it became warm and humid again and with the strong uplifting of the earth crust, the vertical climatic characteriustics is getting more and more visiable. Under the conditions of the complicated climate,geologic and geographical environment, the karstification appears the regional differentiation.

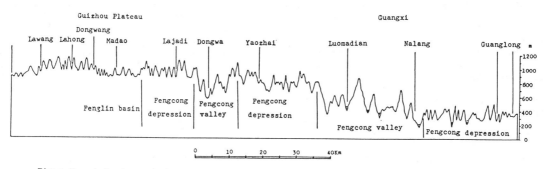

Fig.1 Karst landscapes in the transition zone from Guangxi upto Guizhou plateau

Evolution Of Karst Geomorphology And Aquifer

Since Davis set up the theory of karst geomorphology evolution,many authors have made very strongly arguement about whether existence of karst developing sequences. Some people recognized that Davis's theory was not to emphasize the karstification process and the dynamics, but othors support Davis's theory though it there are some shortages. Recently, the discussion have been carried on in china,A lot of geomorphologists and hydrogeologists have explained the characteristics of karst geomorphology development and karst aquifer evolution. Mr. Xong (1985) explained the geomorphology evolution by means of the quantitative analysis, the Fengcong should develop toward to fenglin landscape if the time longer enough. Zhang and his colleque set up the evolution model based on studying karst in south Guizhou (1987). But zhu et al (1988) described that the fenglin and fengcong are two different landscapes which may develop along their own evolution directions at the same time and no relations between their development.

In Guizhou, Hunan and Guangxi provinces,the characteristics of karst aquifer is commonly controlled by the evolution process of karst geomorphology.In order to discuss the process of karst geomorphology evolution, we have to assume some the theoretical conditions as to simplify the discussion: (1) the limestone terrain is consisted of the uniform, pure and thick limestone and has undergone a strong tectonic movements; (2) the earth crust is stable nor uplifting neither subsidence; (3) no climatical variation, that means the annual precipitation and temperature will be constant for a long time enough for the karst geomorphological evolution; (4) the carbonate rocks should be gentle and spreaded a far distance; (5) the ground water table deeply below the ground surface; (6) the earth crust uplifts or the drainage base lows down after the karst geomorphology evolution for several stages.

Under the assumptive conditions above, the karst geomorphology will evolve in four stages as Fig.2. At the first stage, the carbonate rocks may be situated in the wide syncline part or anticline part, so carbonate block will be homogeneously karstified along the openings. If they are too tight, the syncline would form a water collecting trough and just the rock in the central of trough will soak in the water and most rock is out of the water, that will cause the karst geomorphology unhomogeneously developing. The tight anticline structure will make the water fed by the rainfall flow down to syncline part along the bedding or the fractures. Hence the karst morphology will develop very slow and no space for the typical fenglin and fengcong landscape developing. If the syncline and anticline areas wide enough for karst landforms evolution, it may develop as the models in Fig. 2 (Song, 1985).

If in the evolution process of karst geomorphology, the earth crust is uplifting by the Neotectonic movement or the drainage base level is deciinding, the evolution of karst geomorphology will be turned back in the general point of view, that is to simulate the rejuvenation of karstification. The degree and rate of the rejuvenation are greatly different from place to place. If it is happened in the region from the drainage base level to the watershed, the intension and rate of karst rejuvenation are gradually decreasing with the distance increasing away from the base level. As the rate of drainage base lowing down is rather higher than that of the branches and underground drainage systems, the hydraulic profiles of underground drainage systems or tributary streams take

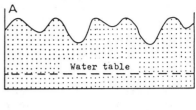

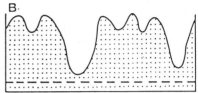

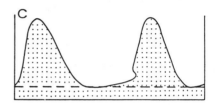

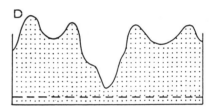

Fig.2 Stages of Karst Geomorphology
in South Dushan, Guizhou

A. Funnel and depression stage; B. Fengcong-depression stage;
C. Fenglin-basin stage; D. Spring-returned (Fengcong-depression
or canyon) stage.

the shape of upward convex (Fig.3) (Song, 1981). As Fig. 2 shows that in the upper reaches or near the watershed area, the karst hydraulic gradient is very low as several to 10 parts per thousands in Huanghe underground drainage system, but in the lower reaches it is higher. In fact, the hydraulic profile is in the convex shape, however the convex profil is not a smooth line rather a multi-terraces shape. There commonly are a lot of knick points of hydraulic gradient for the undergropund drainage system in the geomorphological transiting zone. With the headward erosion continuously moving upwards to the watershed, the knick points will be drawn back upto the

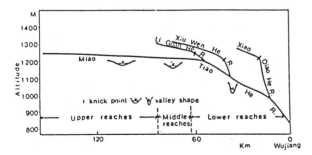

Fig.3 Longitudinal profile branch and major valleys,
Maotiaohe River.

watershed too. If the Neotectonic movement is intermittent elevating, the number of uplifting will make the same quantity of the knick points. If the earth crust is stable for a long time, the knickpoints finially will be disappeared and the equalibrium of hydraulic profile will be reached.

Based on the evolution model of karst geomorphology as Fig.2, Liuzhou plain and Rongshui karst fenglin plain belong to the evolution stage 3 and the karst geomorphology in the west parts of Guangxi and Hunan,Guizhou and Yunnan now is at the stage 4. The karst geomorphology is strongly rejuvenating in the transition zone from the higher level of karst geomorphology surface to the lower level.

The tectonic fractures including the faults, joints and cleavages, pores and bedding constitute the water bearing space and the passages for the water movement. The porosity of limestone usually is 5-15% and the connecting is not very well.In fact, its transmition is very low generally less than 10 m/d Therefore the main water bearing structure is like the conglate frame building, bedding surface is the floor or ceiling, the solutional fissures and conduits and the fractures are as the cement collume, and the pores are the bricks.

In the different stage of karst geomorphology evolution, there are different properties of karst underground hydrologty. In the original stage, the depth of fractures including joints, faults and cleavages are very different, the faults may be several tens meters to several hundreds meters upto thousands meters deep, the depth ofjoints are only several ten cm to several m and the cleavages only several mm to several tens cm. According to the opening dimensions, they may be divided into open system and closed system . During the deformation of the formations, the open systems commonly appear near the ground surface and the closed systems occur quite deep. Based on the survey of karst fissures in Niukouyu area, Beijing and Xizhuo in Yunnan province, the depth of opening joints are only 7-10 m below the ground surface (Table 2). In this case, the opening fractures will form a natural water storage as the subcutaneous aquifer after rain. When the storage space has been filled up or the recharge is greater than the seapage rate, the extra water will produce the surface runoff. Therefore, the surface drainage systems may be formed as White (1988) and Tang (1988) described.

In the karst funnel and depression stage, the opening fractures are dissolved and enlarged by the rainwater, later the residual materials would be left in the solutional fissures. The soil contains more water for longer time and the moisture environment makes the microbe vigorously acts in the soil and they increase the CO_2 content and organic acid in the soil, that will enhence the solutional ability of the soil water against limestone. Because the depressions are collecting the water, the fractures in the bottoms will be fastly enlarged than anyother places and constitute the master pass of water moving down, also other fractures in the depression absorb and store some rainwater. Therefore, in this stage, there are two water-bearing units, the subcutaneous and satuated zone (Mangin, 1975-78; Williams, 1983).

Table 2 Depth of opening fissures in Niukouyu area, Beijing

Below Ground surface (m)	Width of Fissures (mm)
0--1	20--100
1--2	20-- 30
2--3	3-- 12
3--5	1-- 3
> 5	closed

In the karst fengcong stage, the depression is a large collecting basin with the area of several hundred m² to 1 Km² . In the depression, the sinkholes, shafts and larger fractures well developed and absorb the collecting water rapidly. But the sinks or shafts solely work after the surface runoff produced. The soil water covered on the bottom of depression will recharge the fracture flow for several days even whole year, that we discovered the fracture water droped down for a year from the cave ceiling in the upper Lion cave with 20 m thick of limestone block covered by 0.5-1 m thick of soil.

Due to the ground water level being several tens to hundreds meters below the ground surface, the shafts are several to hundreds meters deep to connect the groundwater level, conduits or cave rivers (Table 3).

The hydraulic gradient of the underground rivers or conduits commonly are high as 10-100 o/oo or more, velocity of the flow generally is greater than 0.5 m/s. As the fluctuation of water level is very strong, the conduits are often filled with water and the regime of water level may rise upto several m to more than 100 m after heavey rain and the discharge may be increased 100 times as the flow in the lowest period but several days, later the water table and discharge greatly drop down. The discharge of conduit only is several to hundreds l/s, The drainage capacity of the conduit is limited to drain off the flood water after the heavey rain, therefore the flood logging usually appear in the karst depressions, for example, there are 375 depressions or more with 18500 mu (1233 Hec.) farmlands about 15.5% of the total including the slope lands, pot lands and terrace lands on the slope in the limestone areas flooded for 7 days to 9 months every year.The hydrogeological properties are very unhomogeneous.

Table 3 Depth of shafts and sinkholes in Duan, Guangxi and Dushan, Guizhou

Site	Fluct. of water Table (m)	Depth of shafts and Sinkholes (m)
No.1 Zhuanxing	62.08	62.08
No.2 Shanglin	27.99	27.99
No.15 Guocha	35.11	35.11
No.16 Guocha	29.83	29.83
No.17 Guocha	33.71	33.71
No.20 Guocha	35.44	35.44
Nuongtuandong		160
360 Steps	60	60
Muzhudong	91	90

In the stage of fenglin basin and peneplain, the thick soil with abundant organic materials covered on the basin and plain, so the subsoil corrosion is rather homogeneous against limestone fissures. The ground water table is near the ground surface and the hydraulic gradient is very gentle, about 1-10 o/oo. There are some surface streams or surface section of underground drainage system in the basin. The seapage water with high contents of CO_2 and organic acid is mixed with karst water to strongly and uniformly corrode and enlarge the limestone fissures as to form the network of fissures, the main fissures will develop into the master water courses as the local drainage base level to discharge the karst water. Along the flow direction, the monoconduit develops in the limestone block between 2 karst basins or large depressions and poljes, The water level in the basin or plain little fluctuate in a year, generally in the range of 1-10 m. The water course of underground drainage system is characterized by the interconvension of subsurface sections with the surface sections.

In the karst rejuvenation stage, the lowing of drainage base level is resulting the headward erosion from the base level up towards the watershed and it is changed to vertical circulation of groundwater from the horizontal. Thus this vertical karstification gradually prevails over the lateral function and the homogeneous property of karst hydrogeology will be in stead of the unhomogeneous. Therefore the binary structure of underground karst system, upper uniform fissure system and lower part being the monoconduit system, appear in the rejuvenation zone of the fenglin plain. As the lowing rate of the drainage base level is higher than that of its branches, the hanging spring or stream valley appear from several tens to hundreds meters above the drainage base level. But the effect of the rejuvenation distance is rather limited, for example, it is only 10 Km or little more far from Wujiang, the main branch of Changjiang River. Out of the interferencial zone of visible rejuvenation, the hydrogeological properties before the rejuvenation are still maintained.

Subsurface Reservoirs

In the limestone terrains, the problem of water supply for agriculture, industry and daily life is getting more and more serious. Though a great amount of money and materials had been spent to built water tanks and other accumulations each year, the results are very different in different places. Take instance, the people of Duan County, Guangxi Province, made the cement boats with pumps to be put in the top windows of the Disu underground river system, it may work whole year no matter the water table is higher or not. They also setlle the pump on the trucks into the declining window, sinkhokle and interval cave spring, it should be moved on the trucks according to the groundwater table, if the groundwater table is low, the pump will be put down further, and in the period of high water level in the rainy season, it will be moved up. In some cases, the pump is settled in the closed house under the water table. The water may be diversed from the spring to irrigate the farmland.

In Guizhou province, the first closed underground reservoir with the storage capacity of 100,000 m³ water, named Fengfa reservoir, was built in 1974 2 years later of the serious drought in 1972. As the dam in the cave was too close to resurgence of the underground stream by the fault, there was a little leakage problem causing the efficiency being not so idea. In the spring time, the water is not enough to irrigate the rice fields, so later, a new closed dam was rebuilt on the complete conduit 100 m inside far from the first dam avoiding the fault zone and the water is well sealed in the cave reservoir. By the end of 1977, 28 subsurface reservoirs with the range of 100,000-500,000 m³ of the accumulation capacity had been set up in South Dushan. It is the visible comparison of the underground reservoirs with the surface accumulations as follows:

1. The Leakage Problem

In 1958, there were 3 surface reservoirs built in the depressions and basins, but no one succeeded. Mongla reservoir with the designed capacity of 200,000 m³ was built in C2hn limestone terrain in the Shangzhai karst basin in South Dushan in 1958-1959. In the first year, it could be full of the water and got excellent efficiency. But unfortunately, the serious leakage appeared in 1960 and it completely dried out in the third year, so the people called it stored the sun in the daytime and the moon at the night. Another example is Rerong reservoir on the surface section of Huanghe underground drainage system. Since 1958, 3 dams had been built for 3 times about 300 m, 200 m and 10 m far from the Rerong cave, respectively. First and second dams were distroyed by the serious leakage and flood water, and the new one was set to close to Rerong cave, however the strong leakage still happens from the dam base rock and shoulder fissures, that caused the negative results of the reservoir.

Based on the statistics of 1252 surface reservoirs in limestone regions of Guangxi province, 644 reservoirs of which are seriously leaking known as the sick reservoirs. They were set up by the dams or blocking the sinkholes in the karst depressions, basins and poljes. Now it is very difficult to prevent the reservoirs from the leakage.

Up to now, almost the underground reservoirs built on the conduits or cave rivers there are no any leakage problem, except very few set up unsuitable sites or the technical problems that make the leakage. Therefore no leakage problem is the great advantage of the subsurface reservoirs.

2. No Loss of Submergence

Some good quality farmland, villiges and buildings will be submerged by the surface reservoirs, so that a lot of local people have to move out their living and production base from the reservoir basin that will spend much money, while because the subsurface reservoirs are built on the underground water courses, all water will be stored in the conduits, caverns, fissures and connecting pores, it will not occupied the farmland and be neccessary to ask the people moving out.

3. Low Cost

To build the surface reservoirs it should be very carefully done the hydrogeological and engineering geological surveys including a great amount of drilling, hydrogeological tests, geophysical work, mechanic experiments of base rocks and constructional materials. a huge amount of constructional material should be carried, transported. Therefore the cost to set up a reservoir and treat the leakage problem if exists is very high. For example, Wulichong reservoir with 70,000,000 m³ of the designing capacity will take 180 million yuan RMB to build the dam and to prevent it from the serious leakage problem.

4. It is easy to manage the underground reservoirs and hydropower stations.

5. The storage of the subsurface reservoirs is small, generally less than 1,000,000 m³ of water.

To build a underground reservoir the geological survey is much easyer than to build the surface reservoir. It just need to survey the cave geology and hydrogeology and a small amount of construction materials and generally no neccessary to pay attention to the leakage problem if the dam site is right.

According to the storage spatiality, it may be divided into the real underground reservoir and combination reservoir (part surface and part subsurface), the former is the principle pattern like as Tiechangping reservoir and the Wangou underground reservoir is the later example (Fig.4. 5) If it is possible, the combination reservoir will accumulate more water than the conduit reservoir. But it should be paid great attention to the draining flood.

Based on the degree of accumulation, it might be classified to the closed system and unclosed (or opened) system. The closed system is completely to seal the conduit or cave river as to form bthe underground reservoir, and the opened system that is the dam does not seal the underground water course. During the raining seasons, the flood water may flow over the dam. To determine whether to build the closed system or unclosed system, it is depended on: (1) the flood discharge. If the flood discharge is not very high, the

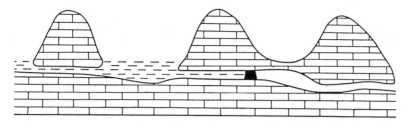

Fig.4 Wangou combination reservoir, Dushan County

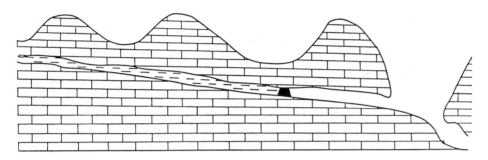

Fig.5 Beiyan subsurface reservoir, Dushan County

capacity of the re-servoir may keep all the water, then it may be considered to construct the closed system. If the flood water is too much for the reservoir, it must be designed to build the opened system to avoid the economic loss from the flood disaster; (2) degree of karst development, if the conduit is complete and there is no any hydraulic relations between the multi-levels of caves, it is possible to build a perfect closed re-servoir.If there are multigrades of caves originated by the intermittent rising of earth crust or lowing down of the drainage base level. when the conduit is closed, the reservoir will rise up and the water flows out from the upper level of caves or lateral passage, that will result the leakage of the reservoirs.In this case, it is better to build the unclosed reservoir.

It is very important to select the dam site:
1. Based on the evolution model of karst geomorphology, the conduit system with high hydraulic gradient will be well developed in the fengcong-depression stage and the net-conduits and -solutional fissures system are developed in the fenglin stage. In the different karst area, there is a time difference in the evolution history of karst geomorphology, or the geological factors such as faults cause the difference of geomorphological evolution in a region, so the different karst geomorphological and hydrogeological units will be closed together. If the fenglin basin is neiboured with the fengcong depression, the dam might be build on the monoconduit at the upper side of connecting between the fenglin basin and fengcong-depression. For instance, the Muzhudong underground reservoir in Puding County, Guizhou, is a good example of this kind of reservoir. The dam was built about 50 m far from the cave entrance. But it must be very careful that near the connecting zone, the solutional fissures usually are developed well, that will lead to the leakage problem of the reservoir. Now when the Muzhudong reservoir is full of the water, the 1-2 l/s of water leakes through the dam base rock.

2. When fengcong large depression or basin is developed, the monoconduit between the depressions or basins and or poljes along the conduit flow direction is generally formed. If there is a elevation difference between two large depressions or basins, the dam may be built on the conduit near the resurgence at the upside of lower level of depression and basin. It will be failed if the dam built near the entrance of the conduit. Because, the water from the basin or depression contains high content of CO_2, organic acid and other corrosive materials, it is very strong to dissolve the limestone and enlarge the fissures near the entrance to produce a solutional net-fissures which make the water leakage. This is why the depression reservoirs damed at the entrance of conduit or underground river take place the serious leakage problem.

3. If there are some knick points of conduit flow, the dam should be constructed at the upperside of knick point. There are 3 factors: (1) it is possible to seaze the large accumulation as possible; (2) the high and effect water head may get for the power generation; (3) lowest cost to build the reservoir, for example, it is the Beiyen underground hydro-power station that uses the closed Beiyen reservoir water to generate the electrocity. The elevation difference at the knick point is about 10 m high and the capacity of generation set is 100 Kwh, also the closed reservoir system is used to work out 100 Kwh electrocity of Tiechangping station.

4. If the karst conduit is ganging up above the local drainage base, it is possible to build a dam to accumulate water for generator. For instance, in Dejiang County, Guizhou Province, the Niaoshuiyen underground reservoir with 280 m elevation difference and the dam about 20 m high now has been built. The hydropower station produce 1,600 Kwh electrocity.

Conclusion

To build the subsurface reservoirs is a good way to exploite the karst water resources in the fenglin and fengcong regions, especially in the karst rejuvenation areas. There are number of advantages for the underground reservoirs comparing with the surface reservoirs: (1) it is fit to solve the water supply for irrigation and local people life as the small pieces of farmland spreading in the karst depressions and basins; (2) it is lowest cost to build the subsurface reservoirs; (3) it takes less construction materials to make the dam or block the conduit; (4) it is very simple to manage and run the subsurface reservoirs; (5) as the water is accumulated in the conduits and cavers, it will not submerge the farmlands, villiges and buildings.

It is very important to chosen the correct dam site according to the karst geomorphological features. Commonly the dam will be set up at the site as follows: (1) when the fenglin basin is neiboured with the fengcong depression, the dam might be build on the monoconduit at the upper side of the connecting belt between the fenglin basin and fengcong depression; (2) if the monoconduit develps in the limestone block between two karst basins, or poljes and or depressions, the dam may be build on the monoconduit near the resurgence at the upside of the lower level of the basin and depression; (3) if there is a knick point on the monoconduit, the conduit may be damed upside the knickpoint; (4) if the karst conduit is ganging up above the local draining base level, it is possible to build a dam to accumulate water for the hydropower station and irrigation.

References

Editorial Commission of <PHYSICAL GEOGRAPHY OF CHINA>, Academia Sinica: Paleogeography. Science Press, 1986.

Gao Daode, Zhang Shichong et al, KARST IN SOUTH GEIZHOU, CHINA. Guizhou People's Publishing House, 1986

Song Lin Hua, 1981, Some characteristics of karst hydrology in Guizhou plateau, China. Proc. 8th Intern. Congr. of Speleol. p139-142.

Song Lin Hua, 1985, Evolution of karst geomorphology and hydrology in South Dushan, Guizhou Province, China. Annales de la Societe Geologique de Belgique, T 108, p227-231.

Song Lin Hua, 1986, Principle characteristics of karst hydrology in China. <Communication, 9 Congreso Internacional de Espeleologia Espana>, p55-59.

Tan Ming, 1988, The mathematical model of the structure of karst drainage ages in Guizhou, China. M.S thesis.

W. B. White, GEOMORPHOLOGY AND HYDROLOGY OF KARST TERRAINS. Oxford University Press, 1988, 464pp.

P. Williams, 1983, The role of the subcutaneous zone in karst hydrology. J. Hydrology, 61, p45-67.

Zhu Xuewen et al, STUDY ON KARST GEOMORPHOLOGY AND CAVES IN GUILIN. Geological Publishing House, 1988.

Late paper:
Cockpit karst and geological structures in South Yunnan, China

SONG LIN HUA & LIU HONG *Yunnan Institute of Geography, Kunming, People's Republic of China*

Abstract

The study area, Xichou County of Yunnan Province, at the mean elevation of 1400 m a.s.l. is located 23°05'-23°37' N and 104° 22'-104° 58' E. and the Tropic of Cancer is through. Topographically, it is higher in the northwest and lower in the southeast. The annual average temperature is 15.9°C and precipitation about 1300 mm a year. The limestone, dolomitic limestone and dolomite with total thickness of 4000 m occupy 72.3% of the total area. Under the function of Yenshan Movement, the carbonate formations have been revolved, folded and fractured. They distribute as the shape of an "ear" striking from NW to NE and EW. The funnels, sinkholes, depressions, stone teeth.fengcong extensively develop in the area. As the clastic rocks distribute besides the carbonate rocks and the surface rivers develop in the southwest and northeast, the middle region of carbonate terrain forms the watershed. The karst geomorphology is characterized by (1) the large depressions distribute in the waterdivid region and the small one in the two ends of the "ear"; (2)the surface rivers deeply cut down about 200-300 m, the depressions by the river valleys are more than 100 m deep in the northeast, but the rivers in the southwest are shallow, the depressions about 50-100 m deep; (3)the long axises of the depressions and fengcong in the central zone of the carbonate terrain are coincide with the geological structural line as the shape of the "ear", but in two zones of sides that vertical to the carbonate terrain boundary; (4) the dimention of depressions in pure thick limestone is larger than that in dolomitic limestone; (5) in the watershed area, the uplifting of earth crust causes the rejuvenative fengcong-depressions, while the relic fenglin-plain still remains in the watershed area.

The developing intension, evolution precesses, compound characteristics and distribution of karst geomorphology are controlled more or less by the geological structures. This paper will be stressed on the relationship between the karst geomorphology and geological structures.

Physical Geographic Setting

The studied area is situated in the southeastern Yunnan, 23°05'-23°37'N and 104° 22'-104°28'E, and the tropic Cancer passes through the area from the east to the west. Due to the area in the transition zone from the Yunnan Plateau to the lower mountain and hills in Vietnam, topographically, it is higher in the northwest and lower in the southeast, the highest is Baoyada Shan at the altitude of 1962.9 m above the sea level (a.s.l.) near Jijie and the lowest is Jiabu stream with the elevation of 667.9 m a.s.l. near Pingzhai in the southeastern part of the studied area. The total height difference reaches to 1295 m. It belongs to the south subtropical climatic zone with the character of neither cold in the Winter nor hot in the Summer. The annual average temperature is 15.9° C and the precipitation 1260 mm, the range of 1075.7-1615.3 mm in the different sites. The rainfall in the rainy season (May-October) may reach 1095.8 mm that occupies 86% of the total and the precipitation in the dry season from November to April next year only occupies about 14% of the total. The Devonian-upper Triassic limestone, dolomitic limestone, dolomite and mudy limestone universaly distribute in the area. The carbonate rocks with about 4,000 m of the total thickness occupy 72.3% of the total studied area. The middle Triassic sandstone and shale of Triassic Feixianguan Group distributes to the south of the carbonate area and the Silurian sandstone and shale and Cambrium phyllite distribute to the north and west of the carbonate terrain (Fig.1). Under the function of the rotational geoforece during the Yeshan Movement, the rotational structural belt was formed around the centre Babu. The carbonates had been strongly folded and faulted. The characteristics of the geologic structures will be discussed in detail next section.

There are three surface drainage systems: Bajia river in the north, Jijie river and Hongha river in the northeast and southeast, Chouyang river and Panlong river in the south and southwest. both Jijie river and Hongha river are the tributories of Babu river originate from the sandstone and shale regions, however Chouyang river with the interval of the surface and sub-surface sections in the upper reaches comes from the limestone terrains. The Panlong river flows on the southwest boundary of the researched area. All these rivers become the draining base level of the area.

Under warm and humid climate since Neogene, the purple-red clay, grey dark subsoil and brown yellow clay commonly deposited. The depressions and poljes in which deposited the thick sediments are very important farm lands. The forest only covers about 18.2% of the total area.

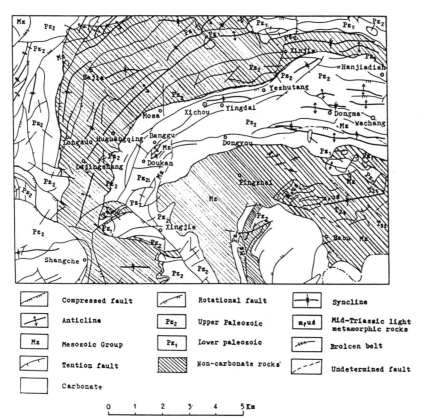

Fig.1 Sketch of geological structure in Xichou

Geological Situations
During the late Paleozoic and Mesozoic Eras, the studied area was in the environment of shallow marine and near 4,000 m of carbonates deposited.Table 1 shows the chemical positions of the carbonates. The contents of CaO and MgO are 31.25-53.27% and 0.21-18.37% of the carbonates, repectively, the insoluble materials only 0.12-6.12%. The ratio of CaO/MgO varies from 1.7 to 251.3, most in the range of 11.3-251.3, that means most of the carbonates are limestone that provide a good condition for karst development.

As the formations in the studied area are situated in the interior belt of Babu rotational tectonics, all the carbonates were distributed as the arc. The striking of carbonate rocks gradually changes from NE to EW and SW, also the striking faults and shear faults and conjugate joints with SE, NE, NEE and NWW developed very well (Fig.1).

1 Main Folds
(1) Daqingshan-Dongguajing anticlinorium
The Cambrium dolomite and dolomitic limestone and muddy band limestone distribute in the axial part of the Daqingshan-dongguajing anticlinorium and the late Paleozoic rocks discordantly covered on them in the two limbs. It extends from NNE to EW about 80 km long. The studied area is in the southern limb of Daqingshan-Dongguajing anticlinorium with a lot of sub-foldings such as: (a) Xichou- Yezhutang syncline with the striking EW and the upper carboniferous limestone in its axial part; (b) Jijie SEE syncline, its axial zone is occupied by the middle devonian dolomitic limestone, limestone and dolomite; (C) Laolong EEN syncline constituting of lower Permian limestone as its axial belt; and (d) Wangjiatang-Dongma EW syncline with lower Permian limestone as its central part.

(2) Pingzhai synclinorium
The axial region of the Pingzhai synclinorium about 40 km long is occupied by the middle Triassic Falang light metamorphic clastic rocks with EW striking. Several subfolds developed on the north limb such as Longguao anticline with 5 km long and 4 km wide, of. which, the axial part is occupied by the middle Carboniferous limestone and the upper Carboniferous limestone and the lower Triassic argillaceous limestone distribute on both limbs.

Table 1 Chemical compounds of carbonates in Xichou, Yunnan

| Number | Form-ation | Site | Lithology | Chemical composition % | | | | | CaO/MgO |
				Lost	SiO$_2$	R$_2$O$_3$	CaO	MgO	
1	D$_2$d	Changqing chongzi	Limestone	48.50	0.17	0.09	51.10	0.48	106.46
2	D$_2$d	Changqing chongzi	limestone	47.95	0.25	0.11	52.15	0.32	162.97
3	D$_2$d	Ganlongtang	limestone	44.69	4.76	2.05	47.23	0.90	52.48
4	D$_2$g	Xiaomichong	Dolomitic Limestone	50.07	0.44	0.31	31.25	18.37	1.70
5	D$_2$g	Longzheng	limestone	47.80	1.84	0.21	49.77	0.48	103.69
6	D$_2$g	Mosa	limestone	48.55	0.37	0.11	49.55	1.41	35.14
7	D$_2$g	Xisa	limestone	48.32	0.11	0.10	49.10	3.53	13.91
8	D$_2$g	Laolong chong	Mudy Limestone	45.18	6.01	0.11	47.71	0.56	85.20
9	D$_3$	Laoxisa	limestone	48.49	0.18	0.09	51.11	0.39	131.05
10	D$_3$	Mojiao	limestone	47.98	0.28	0.14	50.98	0.36	141.61
11	D$_3$	Shuitou	limestone	47.60	0.07	0.056	50.93	0.30	169.77
12	D$_3$	Shuiyandi	limestone	47.97	0.08	0.05	51.81	0.27	191.89
13	C$_1$d	Laozhai	limestone	47.66	0.13	0.07	52.77	0.21	251.29
14	C$_1$d	Banggu	limestone	48.00	0.06	0.05	51.69	0.23	224.74
15	C$_2$w	Dadichong	limestone	47.18	0.10	0.032	53.27	0.28	190.25
16	C$_2$w	Zhaitou	limestone	47.88	0.07	0.05	52.58	0.26	202.23
17	C$_3$m	Dongyou	Mudy limestone	47.14	1.56	0.70	50.64	0.76	66.63
18	T$_1$	Shangnanqiu	limestone	47.54	0.37	0.16	52.28	0.46	113.65
19	T$_1$y	Erhaizi	limestone	44.68	4.78	2.05	47.23	0.90	52.48
20	T$_1$y	Banggu Haizi	limestone	47.26	1.55	0.59	46.39	4.11	11.29
21	T$_2$g^2	ShuiHaizi	limestone	47.94	0.46	0.25	50.62	0.35	144.62

2. Faults

There are 4 regional faults in the study area: (1) Huguangqing-Hanjiadian fault with 60 km long and dipping to south developed on the north limb of Xichou-Yezhutang syncline. Near Huguangqing, it extends to NNE, then changes its trending to NE in the region of Xichou-Xinjie and EW and SEE in the area to the south of Xinjie. (2) Piaopiaochong-Wachang fault is very similar to Huguangqing-Hanjiadian fault in the trending and dipping, but it is characterized with the fault breccias which may reach upto 100 m wide in some places. The calcite veins had been filled in the fault fracture. (3) Yingdai-Yezhutang fault striking to EW and dipping to southward is developed on the south side of Xichou-Yezhutang syncline and constitutes the branch of Piaopiaochong-Wachang regional fault. (4) Xiaochang-Yaopuzi compressed fault is developed on the Pingzhai syncline. It trends to EW and dips to southwards.

Fig.1 shows that the folds and faults are characterized with the southwestwards comvergence and northeast-or east-wards spreading. Under the function of the clockwise rotating geoforce outside of the rotational structure and unticlockwise turning force inside of the tectonic belt, therefore the strikings of the structural lines have greatly changed. There are a lot of sheared faults distroying the distribution and completation of the geological structures and formations.

3 Joints

The tectonic joints are very developed in the studied area and their trends extensively change from place to place based on their positions in the geological units. There are NNE, NNW and SN joints in the southern and northern zones, NWW, NEE and EW joints in the central zone in the eastern area, while there are NE and Nw joints in the central zone, and NEE and NW joints in the northwest side and NNE and SN joints in the southwest side of the southwest region studied.

4 Neotectonic movements

Since Quaternary, the earth crust has strongly uplifted in the different ranges in different regions, greater in the north region and smaller in the southwest. The Jijie river has cut down about 300-400 m and formed the conyon. Its hydraulic gradient reaches upto 10o/oo near Chongzhai. The deep drainage base has increased the thickness of the vadose zone, about 100-200m in the east limestone region. For example, there is a shaft over 200 m deep near Dongma. The karst rejuvenation has highly hanged the fossil karst landforms over the river valley and makes the new closed depressions about 100-150 m deep at present develop in the relic basins or peneplain.The conduit flows are dominant and pour into Jijie river at the higher hydrogradient.

In the southwestern region of the study area, the Chouyang river is with the characteristics of the plateau rivers in low hydraulic gradient and the channel meanders. Its water level is only 3-5 m below the groundsurface in the Xinjie valley and the hydraulic gradient is sole 1%. It seems the neotectonic uplifting is not so strong to impact the development of Jijie river. The fengcong-depression develops well in the area. The chickness of vadose zone varies in the range of 50-100 m and the groundwater level is about several to tens m below the ground surface of the depressions. The subsurface drainage system well develop and drain into Chouyang river in the 32 o/oo of hydrogradient in the distance of 7.35 km from Bawei to Nanqiu spring. The block near Xisa constitutes the watershed of the surface and subsurface water. As the less effect of the uplifting, the vadose zone is not very thick and the groundwater is 2-10 m below the ground surface, for example, in the dry season, the water table in Mojiao spring well and wells are 2-2.7m below the groundsurface. The karst water appears on the sides of depressions, some of them runs whole year.

Development of Karst Geomorphology

The surface karst features are as follows:

1 Solutional fissures They are formed by the rainwater or the soil water to dissolve the fractures in the carbonates and generally with less 50 mm wide and several m deep.

2. Karrens They are resulted by the rainwater to corrode the limestone and with the wedth of 10-30 mm and depth of 1-15 mm.

3. Solution cup It is like the semiball or the bowl with the diameter 30-100 mm on the limestone surface. It is resulted from the rainwater and the logging water containing abundant organic matters to strongly dissolve the limestone.

4 Stone teeth It is formed by the soil water with high content of CO_2 and organic acid to strongly corrode the limestone, later the soil was rushed away and the rock is exposed. Its height is generally less than 5 m. If it is taller than 3 m , it is called as the high stone teeth, but it is called as the lower teeth if less than 1 m high. According to the form and the ratioes among the height, wedth and length, they may be classified into canine, molar teeth and ridge stone teeth.

5 Fengcong The conical peaks are commonly sitting on the same baserock in the studied area, it may be divided into the band fengcong with pass between two peaks and the cluster fengcong with the depression developed among the conic peaks. The band fengcong distributes in the belt of Yingdai-Xisa-Banggu and the cluster fengcong in the east researched area.

6 Isolated cone The isolated hills generally develop in the karst basins, valleys and the relic karst high land. There are two isolated hills stand in the Banggu Haizi basin and the hills with 50 -100 m high distribute on the high land near Xiaozhai.

7 Funnel It commonly develops in the depressions, basins, poljes and some on the relic peneplain. The funnels are the places where the fluvial erosion and karstification play very active performance.

8 Sinkhole Most of the sinkholes develop in the lower end of the depressions, basins and valleys. There are 3 types of sinkholes in the area: vertical, delinal and horizontal ones, the vertical sinkhole like Bawei sink, the Luodong sinkhole being horizontal one and Xiabawei sink belonging to the declinal.

9 Depression The depression is defined as the closed and polyboundary negative features which developed among fengcong.

10 Polje It is bounded with the very steep slope hills and with the charater of seasonal or perennial stream and flat floor. such as Liaoxisa polje and Xiafangzhai-Shangbawei-Luodong polje.

11 Karst "Haizi" Term HAIZI means the natural lake in the limestone terrains laying on the marlite in the studied area. Based on the permanent time of the haizi in the year, it may classified into perennial and temporary haizi. The perennial haizi is fed by the permanent springs and can accumulate water whole year such as Zhaojia haizi and Er haizi though there are some solutional fissures and small sinks that cause the haizi leakages. The temporary haizi only can keep water in the rainy season and it will dry out in Winter and Spring as the serious leakage and just few water or nothing to feed it. For example, the Banggu haizi with capacity of 488,000 m³ water, 160,000 m² of water area and 3.0 m of mean depth in the rainy seasons can keep water for 7-9 monthes a year and the Zhaojiagan haizi only works for 3-6 monthes each year. 20 years ago, the local people attempted to drain water out of the haizis for expanding the farm lands by the means of enlarging the sinkholes and solutional fissures. They have gotten negative results decreased the accumulating time and capacity, but in the rainy season they are seriously flooded as the low capacity of draining flood water. The Shuitou haizi is a good example. Later the people planed to block the sinkholes and the fissures to build the depression reservoir, but the water could not be kept longer. Therefore the haizi can not keep the water for long time neither dry out as to become the farmlands.

The compound of karst geomorphology is fengcong depression, fengcong polje and isolated relic plateau (Fig.2).

The fluvial geomorphology developed very well especially in the basin of Chouyang river. The light downcutting caused by the neotectonic movement has created 3 levels of terraces in the Xinjie valley, the first terrace about 2-5 m above the river water level, second 15-30 m higher than

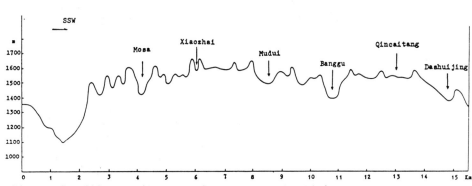

Fig.2 Profile of Karst geomorphology in Xichou

the water level and third 60 m above the water table of the river.

The conduits, caverns and underground rivers developed perfectly in the studied area. The caverns including the Bats cave in Xinzhai and Shengxian dong in Niuchang broadly distribute in limestone terrains. Regionally, there are 3 levels of the caverns developed: first level of caves are about 50 m above the depression bottom, second near the ground surface and third several-tens m below the ground surface. Most of the caves extend along the striking of the formations and faults.

Control of Geologic Structures on karst Development
The framework and compound geometry of karst geomorphology are seriously controlled by the geological structures including the folds, faults and regional joints.

1. Fractures
The distribution of the individual or compound features no matter the depressions or karst hills is closely related to the striking of faults and joints (Fig.3, 4). Based on the statistics of the relations of 563 depressions with the fractures, the distribution of the depressions is coincide with the fault and joints triking (Fig. 5), it suggests the striking of the faults and joints strongly determines the development of depressions and poljes. The depressions distribute along the faults and regional joints. For example, there are more than 20 depressions large or small developed in the zone from Yezhutang to Banggu in the distance of 16 km along Piaopiaochong-Wachang fault. All the depressions take the long shape such as the Yubuna depression 3 km long in NEE and 100-150 m wide and Baonao depression 2.1 km long and 200-250 m wide. The master conduit of the underground drainage system, and the resurgences develop along the faults (Fig.6). The Xisa, Xiaxinming, Niukuo and Bawei long depressions are also developed along the regional joints. Some depressions develop at the cross of the faults or regional joints, for example, the Bawei polje has

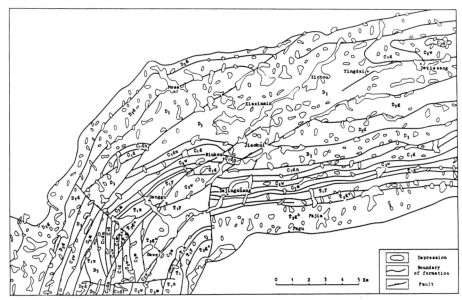

Fig.3 Relationship between the depression and the structure

been produced at the cross of NW regional joint and Banggu NE fault, hence, it is 3.3 km long along the fault and 3.1 km along the joint, the width varies from 100 m to 650 m.

The faults are strongly influenced the framework of the landforms. It is very clear that the karst features may be divided into two different regions taking the Banggu NE fault as the limits. In the eastern area, the formations, folds and faults have not been distroyed by the lateral 3 faults the long depressions and poljes are well developed along the faults. The fengcong hills distribute too along the striking of the faults and the folds. If we separate 3 zones in the eastern region, the depressions in the northern zone are influenced by the NEE and Nw joints, they are strongly controlled by the NEE and NWW joints in the central zone and by the NNE and NNW fractures in the southern, The longitudinal directions of depressions are accordant with striking of faults and joints. Thus regional longitudinal tending of the depressions and karst hills also may be divided into 3 zones as same as 3 zones of the depressions discussed above. In the central zone, the depressions and hills spread in NEE direction, they have arange in NNW direction both in the southern and northern zones or say they are vertical to that in the central zone (Fig.2).

In the west region, the faults and joints are developed very well, especially, the faults distubed the former geological structures and stratigraphy, so that caused the for-mations, folds and faults do not expend very long, generally only 3-5 km or less. The spreading of the formations high dipping greatly changes in the area. Though, the development of depressions are seriously controlled by fractural structures the lithology of the carbonate formations strongly impacts on the karst evolution. The characteristics of the depressions are small, deep and closed compared with that in the eastern region. The positive landforms take the cluster peaks as its predominant feature. The framework of the geomorphology in the

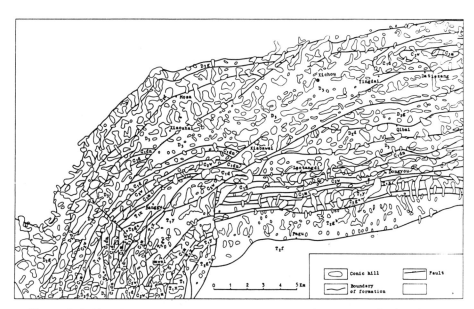

Fig.4 Relationship between the distribution of fengcong and structure

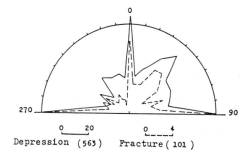

Fig.5 Relations of faults,
joints and depressions

Depression (563)   Fracture (101)

central zone is controlled by NE and NW fractures, while they are strongly influenced by the NS and EW fractures in the eastern and western zones.

2 Effect of folds

As discussed before, the study area is settled on the southern side of Daqingshan-Dongguajing anticlinorium with several subfolds, which has been distroyed by a lot of regional faults. The effects of folds on the karst development are not very significant, but it is still clear in some cases. For example, the Xichou-Yezhutang syncline collects the surface and subsurface water from the 5 southern and northern regions and becomes the water-bearing zone. Near Xisa, the goundwater table is only 2-10m below the ground surface and the springs appearing on the sides of Xisa depression are as the heads of the seasonal and perennial surface streams. The Xisa polje develops in the axial part and fengcong landscape forms on the limbs.

3 Contacts between the different rocks

The evolution of karst geomorphology is more or less effected by the contacts between the different lithological rocks. There are lot of springs and small depressions in the zone from Qincaitang to Fadou developed along the contact between the upper Carboniferous Maping pure limestone with the lower Triassic Ximatang marlite. This contact constitutes the boundary between the NNE or NNW fengcong depression landscape and NEE or NWW fengcong depression landscape. The karst develops very strongly on the contact of carbonate rocks with noncarbonate rocks. For example, the lower Devonian siltypelite contacts with muddy limestone and dolomite near Mosa, the siltpelite makes the karst geoundwater flowing out on the contact. The net-fissures aquifer with water table 1-2 m below the groundsurface has uniformly developed in the Mosa polje with 2.5 km long and 200-600 m wide.

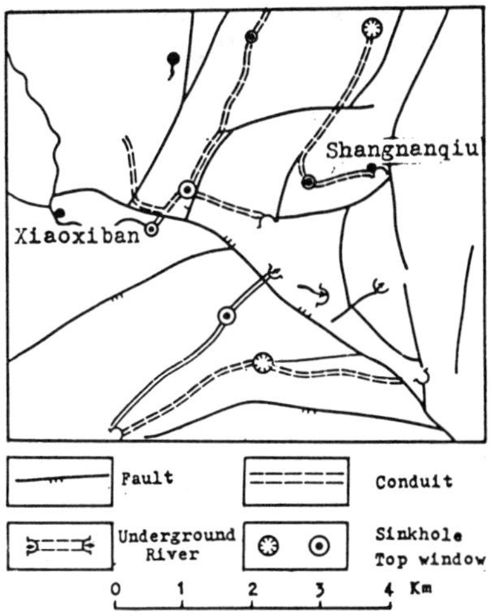

4. Neotectonic movement

Since Quaterny, the crust of earth strongly and intermittent uplifts that caused the local draining base level greatly descended and karstification reacted vertical particularly in the eastern part. A lot of new funnels and sinkholes develop in the older karst basins or depressions. At present, the depressions are been cutting in the eastern region, thus the stone teeth develop in the depressions. In the west region, the vertical karstification is not very strong, the surface

Fig.6 The conduits develop
along the faults

river water table is closed to the ground surface. In the watershed regions from Xichou to Banggu, the effect of karst rejuvenation is small. The large scale of depressions develop very well and the karst residual hills stand on the peneplain near Xiaozhai-Niupengzhai area and Daojiashan.

There are 3 levels of geomorphological surfaces (Fig.4). The first surface with the elevation of about 1700 m above the sea level presents the erosion surface formed in the plateau stage; second level at the altitude of 1600 m a.s.l. formed inthe Stone Forest Stage and third with the elevation from 1400-1450 m a.s.l contitutes the present floor of the depressions. The elevation of the depressions decreases from the watershed to the local drain base level, taking for instance, the Xichou depression with 1472 m a.s.l., Baiwei polje 1317 m and Nanqiu spring at the altitude of 1130 m a.s.l.. In the Xinjie valley, 3 levels of terraces have been developed by Chouyang river, to accord that, 3 levels of caverns are developed in the studied area.

CONCLUSION

The geological structures strongly control the development and distribution of karst geomorphology, some conclusion may be drawn as follows:

(1) The karst geomorphology may be divided into regions with the boundary of Banguo fault, the band karst fengcong and long depressions in the eastern region and the cluster congcong-closed depressings in the western region.

(2) There are 3 geomorphological zones in the carbonate terrains in the studied area, the longitidunal axises of the depressions and conic hills parallel to the striking of formations, folds and regional faults in the central zone, while the long axises of depressions and hills are vertical to the tending of the formations in the southern (or eastern) and northern (or western) zones.

(3) The depressions develop along the faults, regional joints and contacts between the fifferent lithological rocks.

(4) Xisa-Yingdai region is the watershed area of the surface and ground water, and the elevations of the depressions decrease from the watershed area to the local draining base level.

(5) Under the influence of the neotectonic intermittent uplifting, 3 levels of erosion surface have been developed and there are 3 grads of terraces by the Chouyang river valley. The karst rejuvenation simulated the vertical karstification to create the now funnels and sinkholes in the former depressions and poljes and conyons of the surface reiver and underground streams. In the watershed area, the fossil residual karst hills are still kept on the peneplain formed in the Tertiary.

Acknowledgement
The authors should express heartful thanks to Professor Zhang Jiazheng and Miss Du Lihua to work together in the field, and Miss Hu Baohong to make the figgures for this paper.